Advances in Spatial Science

Springer
Berlin
Heidelberg
New York
Barcelona
Hong Kong
London
Milan
Paris
Singapore
Tokyo

Titles in the Series

Manfred M. Fischer · Josef Fröhlich
Editors

Knowledge, Complexity and Innovation Systems

With 44 Figures
and 68 Tables

Springer

Professor Dr. Manfred M. Fischer
Vienna University of Economics and Business Administration
Department of Economic Geography & Geoinformatics
Rossauer Lände 23/1
1090 Vienna
Austria

Dr. Josef Fröhlich
Austria Research Centre Seibersdorf
Business Division Systems Research
Technology-Economy-Environment
2444 Seibersdorf
Austria

Library
University of Texas
at San Antonio

ISBN 3-540-41969-1 Springer-Verlag Berlin Heidelberg New York

Library of Congress Cataloging-in-Publication Data applied for
Die Deutsche Bibliothek – CIP-Einheitsaufnahme
Knowledge, Complexity and Innovation Systems: with 68 Tables / Manfred M. Fischer; Josef
Fröhlich (ed.). – Berlin; Heidelberg; New York; Barcelona; Hong Kong; London; Milan; Paris;
Singapore; Tokyo: Springer, 2001
  (Advances in Spatial Science)
  ISBN 3-540-41969-1

This work is subject to copyright. All rights are reserved, whether the whole or part of the
material is concerned, specifically the rights of translation, reprinting, reuse of illustrations, re-
citation, broadcasting, reproduction on microfilm or in any other way, and storage in data
banks. Duplication of this publication or parts thereof is permitted only under the provisions of
the German Copyright Law of September 9, 1965, in its current version, and permission for use
must always be obtained from Springer-Verlag. Violations are liable for prosecution under the
German Copyright Law.

Springer-Verlag Berlin Heidelberg New York
a member of BertelsmannSpringer Science+Business Media GmbH

http://www.springer.de

© Springer-Verlag Berlin · Heidelberg 2001
Printed in Germany

The use of general descriptive names, registered names, trademarks, etc. in this publication does
not imply, even in the absence of a specific statement, that such names are exempt from the
relevant protective laws and regulations and therefore free for general use.

Hardcover-Design: Erich Kirchner, Heidelberg

SPIN 10835122   42/2202-5 4 3 2 1 0 – Printed on acid-free paper

Preface

In recent years there has been growing scientific interest in the triangular relationship between *knowledge, complexity and innovation systems*. The concept of 'innovation systems' carries the idea that innovations do not originate as isolated discrete phenomena, but are generated through the interaction of a number of actors or agents. This set of actors and interactions possess certain specific characteristics that tend to remain over time. Such characteristics are also shared by national, regional, sectoral and technological interaction systems. They can all be represented as sets of [institutional] actors and interactions, whose ultimate goal is the production and diffusion of knowledge. The major theoretical and policy problem posed by these systems is that knowledge is generated not only by individuals and organisations, but also by the often complex pattern of interaction between them.

To understand how organisations create new products, new production techniques and new organisational forms is important. An even more fundamental need is to understand how organisations create new knowledge if this knowledge creation lies in the mobilisation and conversion of tacit knowledge. Although much has been written about the importance of knowledge in management, little attention has been paid to how knowledge is created and how the knowledge-creation process is managed.

The third component of the research triangle concerns complexity. Although we have no exact definition of 'Complex Systems', there is now an understanding that when a set of evolving autonomous agents interact, the resulting global system displays emergent collective properties, evolutionary and critical behaviour that has universal characteristics. Such systems are found in physics, biology, economics, sociology and geography. There are good reasons to view innovation systems as complex systems. The actors of the system interact with each other, learn, adapt and re-organise, expand their diversity, and explore their various options.

The time seems ripe to bring together contributions from a number of scholars working in different, but related disciplines with the aim of investigating the relationship between knowledge, complexity and innovation systems. The present volume attempts to offer a compact review of current theoretical developments and insights deriving from recent studies. The project originated from an International Workshop held in Vienna, 1-3 July 2000, sponsored by the Austrian Federal Ministry for Transport, Innovation and Technology.

In producing the book, as friends and colleagues, we have benefited from the possibility of exchange of ideas and experience. We have also received useful assistance from the referees who have offered observations and advice in their written reports. The soundness of their comments has contributed immensely to the quality of this volume. We should, in addition, like to acknowledge the timely manner in which contributing authors have responded to our requests, and their willingness to follow the stringent editorial guidelines.

We are extremely grateful for the generous support provided by the Department of Economic Geography & Geoinformatics at the Vienna University of Economics and Business Administration, the Austrian Research Centre Seibersdorf and the Centre for Interdisciplinary Research at the University of Bielefeld. We should also like to thank Ingrid Divis and Thomas Seyffertitz for their capable assistance in co-ordinating the various stages of preparation of the book. Finally, we wish to acknowledge the expert editorial assistance provided by Angela Spence. Her care and attention to the linguistic editing and indexing have considerably enhanced the quality of the work presented here.

Vienna, March 2001

Manfred M. Fischer
*Vienna University of Economics
and Business Administration*

Josef Fröhlich
Austrian Research Centre Seibersdorf

Contents

PART B: Knowledge Creation and Spillovers

PART D: Modelling Complexities

PART E: Policy Issues

19 Does R&D-Infrastructure Attract High-Tech Start-Ups?
Dirk Engel and Andreas Fier

1 Knowledge, Complexity and Innovation Systems: Prologue

Manfred M. Fischer[*] and Josef Fröhlich[**]
[*] Vienna University of Economivcs and Business Administration
  Department of Economic Geography & Geoinformatics
[**] Austrian Research Centre Seibersdorf

The 'systems of innovation' approach has emerged during the last decade as a way of studying of innovation processes as an endogenous part of the economy. The approach is not a formal theory, but a conceptual framework – a framework still in its early stages of development. The idea that underlies this framework is that the economic performance of localities such as regions or nation states depends not only on how business corporations perform, but also on how they interact with each other and with the public sector in creating knowledge and promoting its dissemination. Innovatory firms operate within a common institutional set-up; they also depend on, contribute to and use a common knowledge infrastructure.

Consequently, this approach centres on innovation, knowledge creation and diffusion. Innovation and knowledge creation are viewed as interactive and cumulative processes contingent on the institutional set-up (Fischer 2001a). It departs from the network school of research (Håkansson 1987) in emphasising the role that institutions play in the innovation process (see Edquist and Johnson 1997). The 'systems of innovation' approach concentrates on the concept of the institution at an abstract level, referring to recurrent patterns of behaviour, socially inherited habits and conventions, including regulation, values and routines (Morgan 1997) that help regulate the relations between people and groups of people within, as well as between and outside, organisations.

A *system of innovation* can be thought of as consisting of a set of actors or entities such as firms, organisations and institutions that interact in the generation, use and diffusion of new – and economically useful – knowledge. Within this approach to innovation analysis, different types of system have been defined: firstly, *sectoral or technological systems* that are based on the concept of technological regimes, and have a specific sector or a specific technology as their point of departure, and secondly, *localised systems* which are built on spatial proximity at differing geographical scales – these may be regional, national, European or global systems of innovation. Whether a system of innovation should be sectorally/technologically or spatially defined depends on the context of the study at hand (Fischer 2001a).

Innovation systems are very knowledge intensive. However, the *knowledge* that is used is not necessarily only scientific knowledge. It is often knowledge that may be termed applied, technological or organisational. These different types of knowledge are frequently created and transmitted by different types of institutions, but have to be combined into the production of final outputs.

Defining knowledge is an essential first step when analysing knowledge creation and diffusion within innovation systems. The problem is that definitions of knowledge can vary considerably. To some, knowledge concerns wisdom, the result of learning and experience, to others it is only learning or only the experience; and to others again it involves information or data. Is knowledge something which can be captured in a tangible form such as writing, or is it a process that transpires in a person's mind when he/she is fed information? The answer is not as simple or as obvious as it might appear at a first glance. The question becomes even more difficult in relation to corporate knowledge. Company trademarks and copyrights are certainly the company's intellectual property from a legal point of view, but are they corporate knowledge? Is the data stored in various databases and production systems corporate knowledge? Or is that data the raw material out of which corporate knowledge somehow emerges? Is the information stored in various decision support systems corporate knowledge or simply information? How is information different from data, and do either or both constitute corporate knowledge? Are the insight, wisdom, skill and experience represented by the corporate workforce knowledge or human resources? (Radding 1998).

In principle, scholars trying to define knowledge belong to two different camps. The first view *knowledge as an object* and define knowledge in terms of information theory. The second view *knowledge as process* and see knowledge embodied in the organisation's employees and business processes. The two camps do not overlap. They approach the concept of knowledge from different starting points and use different terminology to define and discuss knowledge creation and diffusion. The contributions to this volume are in line with the second view.

Any discussion of knowledge invariably leads to the question of the relationship between information and knowledge. Our understanding is that information does not become knowledge unless its value is somehow enhanced through interpretation, organisation, filtration, selection or engineering. Information usually takes the form of aggregated numbers, words and full statements. It often combines the numbers and statements in a summarised form which conveys a meaning that is greater than the value revealed by the raw data. Although the exact line between knowledge and information is unclear, several characteristics differentiate the two. Information typically refers to specific situations, conditions, processes or objects. As such, it contains a level of detail and specifity that makes it appropriate to the given task. Beyond the specific situation, the information alone becomes less valuable unless it is transformed into knowledge. Information is generally limited to the context in which it was created. Information is also based on time and, thus, changes continually. As new data is acquired, new information can and should be produced. Knowledge gleaned from yesterday's and today's information can be used to understand tomorrow's information.

In contrast to *information* that may be interpreted as *factual* (Saviotti 1988), knowledge establishes generalisations and correlations between variables. But the number and type of variables and the range of values at which a correlation can be established is limited and differs for different types of knowledge. The extent of correlation can be measured by the span of the piece of knowledge. The smaller the span, the more local or specific knowledge is. The concept of local knowledge has been introduced by Nelson and Winter (1982). More general pieces of knowledge will have a greater span.

The *correlation/retrieval/interpretative* nature of knowledge implies that it is also *cumulative*. The better known a particular field, the easier it is to retrieve and assimilate a piece of knowledge within it. But this also implies an inducement to remain within it and may be a barrier towards moving to different fields (see the first Chapter of PART A for more details).

In most cases, a piece of knowledge can be located somewhere in a range between the completely *tacit* and the completely *codified*. Knowledge is always at least partly tacit in the minds of those who create it. The process of codification is necessary because knowledge production is a collective undertaking that requires communication. The transmitter and the receiver have to know the code to be able to communicate. The codification process for a given subject amounts to the gradual convergence of the scientific community and of other users on common standardised definitions and concepts, on common contexts and theories. The degree of codification differs for different types of knowledge at a given time. Knowledge closer to the frontier, and thus more recent, is likely to be more tacit than knowledge that is already established (Saviotti 1998). The partly tacit character of knowledge is likely to be responsible for the importance that localised networks of personal contacts have for innovation activities of firms in some metropolitan regions.

Codified knowledge is that form of knowledge which is in some way tangible – usually found in print form, such as scientific papers and patent applications. Much knowledge is codified and publicly accessible. But much of the essential knowledge – especially the newer parts that constitute the frontiers – resides in tacit form in the minds of experienced individual researchers or engineers. This person-embodied knowledge is generally difficult to transfer, and is often only shared by colleagues if they know the code through common practice. On the one hand, a given type of knowledge may become more codified as it matures, on the other, the act of embodying it into specific goods and services may reintroduce some tacitness.

The features of knowledge described above imply that systems of innovation tend to have *evolutionary characteristics*, that is, innovation processes are often path dependent over time. It is not clear – even to the actors involved – what the end result will be, in other words which path will be taken. Since innovations occur everywhere in the system and because of the evolutionary character of innovation processes, systems of innovation never achieve an equilibrium. One does not even know whether the potentially best or optimal trajectory is being exploited at all. This means that the notion of optimality becomes irrelevant in the context of innovation systems.

The changes we observe in systems of innovation are quite often discontinuities, radical technological inovations, new types of organisations or activities. That is, *qualitative change* is an essential feature of such systems, especially in the long term. Theories of complexity and evolutionary theories provide on an a priori basis a more appropriate framework for gaining a deeper insight into such systems of innovation. In contrast to neoclassical theory these theories do not constitute a unified body of knowledge, but they are the results of research carried out in different disciplines. The common element amongst these theoretical developments is the attempt to go beyond the Laplacian dream, that in economics is represented by the equilibrium approach, and to develop the theoretical treatment of qualitative change and growing complexity of the reality we observe (for a more detailed discussion see Chapter 2 in this volume). An equilibrium approach attempts to explain why reality is stable, while complexity and evolutionary approaches attempt to explain why reality changes. This is done by emphasising uncertainty, path dependency and multistability, features arising from the out-of-equilibrium nature of innovation systems and innovation processes.

In order to improve our understanding of innovation systems, this volume shifts attention to the world view provided by complexity theory. Complexity is a multidisciplinary concept derived from mathematics and physics that provide convenient tools for modelling complexity in innovation systems. There is however no agreed-upon definition of the term 'complexity'. Rosser (1999) has pointed to the fact that MIT's Seth Lloyd gathered over 45 definitions, most of which are listed in Horgan (1997, p. 303). Many of these emphasise computational or informational measures. The plethora of definitions has led Horgan (1995) to complain that 'we have gone from complexity to perplexity'. This is a serious problem. A broad definition – following Day (1994) – is that a dynamic system is complex if it endogenously does not tend asymptotically to a fixed point, a limit cycle, or an explosion. Such systems can exhibit discontinuous behaviour and can be described by sets of non-linear differential or difference equations, possibly with stochastic elements. But not all such equation systems will generate complexity (Rosser 1999). Nevertheless, Day's definition remains attractive because it includes not only most of what is now generally labelled complexity, but also its non-linear dynamics predecessors: *cybernetics*, virtually developed by Norbert Wiener (1961) and pushed forward by Jay Forrester (1961), *catastrophe theory* developed by René Thom (1975) out of the theory of dynamical systems, and *chaos theory*, the most specular development in modern systems theory.

Many see complexity as a newer and higher stage of analysis for the modelling of discontinuities, as distinct from cybernetics, catastrophe and chaos theories. This begs for a narrower definition of complexity, but no tight definition exists. In the context of the current volume we follow Arthur, Durlauf and Lane (1997) in viewing complexity as characterised by the following features: (i) dispersed interaction among hetereogenous agents acting locally on each other in some space; (ii) no global controller that can exploit all opportunities or interactions in the economy, even though there might be some weak global interactions; (iii) cross-cutting hierarchical organisations with many tangled interactions; (iv) continual adaptation by learning and evolving agents; (v) perpetual novelty as new markets, technologies, behaviours and institutions create new niches in the 'ecology' of the system; and (vi) out-of-equilibrium dynamics with either none or many equilibria and a system which is unlikely to be close to a global optimum. This is a world of bounded rationality, without rational expectations (Rosser 1999).

Although this view has been associated with research carried out at the Santa Fe Institute since the mid 1980s, it was earlier developed in Brussels, Moscow and Stuttgart by chemists and physicists dealing with questions of emergent structures and disequilibrium dynamics. The key figure in Brussels has been the Nobel Prize winning physical chemist Ilya Prigogine. In Moscow it has been the theoretical physicist, Sergej P. Kurdyumov, from the Keldysh Institute of Applied Mathematics, and in Stuttgart the theoretical physicist Herman Haken. Applications have also been made to the social sciences, in particular by Peter Allen and associates in Brussels, and Wolfgang Weidling, Günter Haag and associates in Stuttgart.

Though mainstream economists still prefer to believe that economic systems in general, and innovation systems in particular, are so rational that we can describe them in terms of uniqueness, optimality and equilibrium, a growing number of scholars appear to think that present day instabilities and turbulence in the economy and in society in general require dynamic studies expressely focusing on non-linearities.

It is evident that general features of systems of innovation, such as their persistent asymmetry, historical specificity and the multiplicity of institutional configurations, find a justification in properties such as path dependence and multistability predicted by evolutionary and complexity theories for out-of-equilibrium open systems. Such properties are not easily described within the equilibrium, path independent, reversible world view of neo-classical economics.

The objective of this volume is to provide a better understanding of systems of innovation in general and to investigate in particular the crucial issues of knowledge creation and spillovers. The complex relations between innovation, knowledge and regional development are examined, as well as the problems involved in the modelling of discontinuous processes typical of innovation. The book offers a multidisciplinary view of these issues and has been divided into five parts, which group the chapters according to their main focus. PART A therefore looks essentially at localised systems of innovation, aiming to enhance our understanding of the functioning and dynamics of innovation systems. Innovation and learning are presented as ways in which organisations and localities adapt to new conditions. PART B concentrates on knowledge creation and spillovers, which are fundamental elements of the systems of innovation approach. Special attention is paid to the question of whether or not knowledge spillovers are geographically bounded. PART C attempts to shed further light on the relationship between innovation, knowledge, and regional development, presenting some interesting investigations undertaken in Europe and the United States. As the typical characteristics of innovation systems, such as persistent asymmetries, path-dependence and multistability, are not easily explained in terms of the world view of neo-classical economics, the contributions in PART D shift our attention to an alternative modelling paradigm. This is provided by complexity theory, which enables the modelling of the discontinuities found in economic systems in general, and innovation systems in particular. Finally, in PART E, some of the policy issues relevant to regional development and knowledge infrastructure are investigated.

PART A: Innovation Systems

The concept of innovation systems carries the idea that innovations do not originate as isolated discrete phenomena, but are generated by means of the interaction of a number of actors or agents. These sets of actors and interactions have certain specific characteristics that tend to persist over time. The same characteristics are shared by national, regional, sectoral and technological innovation systems – they can all be represented as sets of actors and interactions whose ultimate goal is the production and diffusion of knowledge. A major theoretical and policy problem posed by innovation systems is that knowledge is generated not only by individuals and organisations [firms, educational organisations, R&D institutions, departments and agencies of the national/regional state, intermediate organisations], but also by their often complex pattern of interaction.

The first chapter in PART A, by *Pier Paolo Saviotti,* shows that the main features of innovation systems can be better explained by evolutionary theory than neo-classical theory. Evolutionary theories are used to derive the main features of the behaviour of actors and agents in an innovation system and, thus, the implications for technology and industrial policies. The general characteristics of localised systems of innovation, such as their persistent asymmetries, their historical specificities and the multistability of the institutional set-up of a territory [e.g. country or region], produce outcomes which can be explained through properties such as path dependence and multistability. Such characteristics do not easily fit into the picture offered by neo-classical economics, based on equilibrium, reversibility and a world not influenced by institutions.

The author argues that neo-classical economics, on the one hand, and complexity and evolutionary theories, on the other, belong to two completely different world views. In the former the world is conceived as deterministic, reversible and, in principle, predictable. The latter emphasises among other things, uncertainty, qualitative changes, irreversibility and path dependence. This alternative view has not yet generated a unified, coherent and elegantly formalised body of knowledge, but provides some important new insights. In Saviotti's chapter the contributions of a range of disciplines and research traditions are analysed, with particular attention to their implications for the social sciences in general and for the innovation system concept in particular.

It is being increasingly recognised that important elements of the innovation process are now transnational or regional rather than national. Behind this, two driving forces are simultaneously at work: the globalisation of factor and commodity markets, and the regionalisation of knowledge creation and learning (Fischer 2001a). This concurs with the view expressed in Ohmae's work on the 'hollowing-out' of the nation-state in an increasingly borderless economic world. The regional rather than the national scale has been identified as the geographical level at which leading-edge business competition is being organised in practical terms (Ohmae 1995).

The second chapter, by Charles Edquist, addresses some basis principles related to localised systems of innovation and the policy implications emerging from these. The author identifies two kinds of policy implication: first, general policy implications that can be derived from the characteristics of this approach, and second, specific policy issues. The general policy implications of the systems of innovation approach are different from those suggested by standard economic theory. This has to do with the fact that characteristics of the two frameworks are very different. The systems of innovation approach shifts the focus away from actions at the level of individual, isolated economic agents [firms and consumers] towards the collective underpinnings of innovation. It addresses the overall system that creates and distributes knowledge, rather than its individual components. Within those systems, innovations are seen as evolutionary processes. The systems of innovation approach can also be used as a framework for designing specific innovation policies. Concrete empirical and comparative analyses are essential for the design of specific innovation policies. The approach is appropriate for this purpose because it places innovation at the very centre of focus and is able to capture the differences between systems.

The globalisation of technology is an ongoing trend likely to have very important implications for national innovation systems, to the point of making them irrelevant. In the last chapter of PART A, *Daniele Archibugi* examines the impact of the globalisation of technology on the European innovation system, looking in particular at the generation, transmission and dissemination of new knowledge. The author distinguishes three major types of globalisation: *first,* international exploitation of nationally-produced technology; *second,* global generation of innovations by multinational enterprises; and *third,* global technological co-operation. The purpose of this taxonomy is to classify individual innovations according to the main methods used to generate and exploit them. The first type includes the attempts of innovators to obtain economic advantages by exploiting their technological competencies in non domestic markets. Both large and small firms take part in this form of internalisation, although larger firms tend to be better equipped to commercialise innovative products in foreign markets. The second type consists of innovations generated on a global scale by a single proprietor. These are innovations produced by multinationals, which receive inputs from different research and technical centres belonging to the same company. The third type of globalisation falls midway between the previous two types. These derive from the growing number of agreements between enterprises, often situated in two or more countries, to develop given technological inventions jointly. It is concluded that globalisation may change substantially the nature of national innovation systems by adding new international links, but is unlikely to eliminate completely national or regional innovation systems.

The author illustrates that the above taxonomy can help to understand shortcomings in the learning economy in Europe, and hence identify policy options. He suggests that one problem is that Europe is less integrated in the worldwide network of knowledge production, transmission and dissemination than the United States. Although economic integration has involved new technologies, until now the European system of innovation has remained somewhat isolated.

PART B: Knowledge Creation and Spillovers

Innovation, knowledge creation and diffusion are central to the systems of innovation approach. Innovation and knowledge creation are viewed as interactive and cumulative processes contingent on the institutional set-up. In the emergent mode of production referred to as the 'knowledge-based' or 'learning economy', knowledge is clearly a crucial element, although knowledge *in itself* does not contribute to economic growth. It has to be incorporated into the production of goods and services. Advances in technologial and organisational knowledge have to be absorbed by firms and applied to the production process and organisation of work. This is true irrespective of whether the new knowledge is created externally [such as in universities or research institutions] or internally [such as in the R&D lab of a firm], and irrespective of whether the knowledge advances embody wholly new knowledge or new combinations of existing knowledge. Clearly, the absorptive capacity of organisations varies substantially and this affects their ability to produce innovations. Viewed from this perspective, knowledge in the form of innovations may be considered an output of learning and economic activities.

Recent debates on the processes of knowledge creation underpinning regional competitiveness stress the importance of proximity and territorial specificity. This is essentially based on the observation that knowledge creation is context-dependent or − more specifically − embedded in socio-economic networks that rely on close interaction and exchange of tacit knowledge. The first chapter of PART B, by *Arnoud Lagendijk*, attempts to throw some light on the debate by examining the significance of the subnational region as a site of economic development and knowledge production. Enriched with empirical evidence, his discussion provides a series of interesting insights, from the role of interaction and competencies to the value of associational processes and governance structures. The contribution points in particular to the need for a social perspective of regional learning that includes concepts of collective goal setting and scaling.

Specific forms of technological learning and of knowledge creation, especially the tacit forms, are both localised and territorially specific. Firms that master non codifiable knowledge are tied into various kinds of networks and organisations through localised input-output relations, especially knowledge spillovers and their untraded interdependencies (Storper 1997). Knowledge spillovers occur when knowledge created by a firm or organisation is not contained solely within that organisation, thereby creating value for other firms or organisations. The spillover beneficiary may use the new knowledge to copy or imitate the commercial products of the innovator, or may use the knowledge as an input to R&D leading to other new products or processes.

The starting point of the next chapter, by *Charlie Karlsson* and *Agostino Manduchi*, is the observation that the current literature on knowledge spillovers in a spatial context tends to be unsatisfactory and, in many respects, even contradictory. Even though there seems to be general agreement that new economic knowledge does in fact 'spill over', there is still substantial disagreement as whether such knowledge spillovers are geographically bounded or not. Hence, the purpose of the contribution is to critically examine the current literature and to make some suggestions on how knowledge spillovers should be analysed in a spatial context. The authors suggest that the starting point for future theoretical and empirical studies of knowledge spillovers in a spatial context should be at the level of properly defined functional regions. Furthermore, it is recommended that these studies should be based on the knowledge network paradigm, which uses the concepts of knowledge production nodes and knowledge transfer links. It is evident that there is a need to reformulate regional technology policy to account not only for public and private R&D, but also the question of how intra- and interregional knowledge spillovers can be stimulated through infrastructure investments and the creation of meeting places for face-to-face contacts.

Indeed, the relationships between knowledge spillovers and space are extremely complex and, at the current state of research, only partially understood. This is partly due to the fact that knowledge spillovers are difficult to measure because they do not leave a paper trail. The chapter, by *Manfred M. Fischer, Josef Fröhlich, Helmut Gassler* and *Attila Varga,* makes a modest attempt to shed some light on the role of space in the creation of technological knowledge in Austria. The study is exploratory rather than explanatory in nature and based on descriptive techniques such as Moran's *I* test for spatial autocorrelation and the Moran scatterplot. Clusters of the knowledge 'output' [measured in terms of patent counts] are compared with spatial concentration patterns of two input measures of knowledge production: private R&D and academic research. In addition, employment in manufacturing is utilised to capture agglomeration economies. The analysis is based on data aggregated for two digit SIC industries and at the level of Austrian political districts. It explores the extent to which knowledge spillovers are mediated by spatial proximity in Austria. A time-space comparison makes it possible to study whether divergence or convergence processes in knowledge creation have occurred in the past two decades. As in the case of any exploratory data analysis, the findings need to be treated with caution and should be viewed only as an initial pre-modelling stage in the enterprise. Future research activities will be devoted to further exploring the issue of local university knowledge spillovers within a refined knowledge production framework (see Griliches 1979).

The final chapter of PART B, by *Elsie Echeverri-Carroll*, shifts attention to knowledge spillovers in high-technology agglomerations. In the 1980s, many empirical studies of U.S. high technology agglomerations were conducted. They stressed that the high level of inter-firm worker mobility and informal communication among engineers contribute to speeding up the movement of ideas. In contrast to this abundance of empirical studies, theoretical models of the relationship between innovative core and non-innovative periphery are still inchoate. The Krugman-type core-periphery models introduced in the New Economic Geography in the 1990s offer little insight on this issue because, although they include pecuniary externalities, they exclude technological externalities, which are essential to high tech regions.

Some useful indications of the economic forces that lead to the spatial concentration of high technology firms can be gained from models developed within the New Economic Development literature of the 1980s. These models must however be used with caution, since they model high tech competitive advantages among countries, not among regions. Both the empirical and theoretical literatures tend to present a dichotomy, suggesting that cities are either specialised in high tech or non-high tech industry. In reality, even in highly innovative places like Austin and San Jose, both high tech and non-high tech firms coexist. In this regard, Krugman warned that there is no reason to assume that the motives for high technology firms to agglomerate should be any different from the motives of concentration of non high technology firms. However, contrary to Krugman's observation, the results obtained in this study suggest that high technology and non-high technology firms locate for different reasons, even when they locate in the same city. They give little support to pecuniary or transport cost-related externalities in explaining the differentiation in location patterns of high tech and non-high tech firms, and strongly support the importance of knowledge externalities.

PART C: Innovation, Knowledge and Regional Development

Today, it is widely recognised that technological change is the primary engine for economic development. Innovation – the heart of technological change – is a process that depends upon the accumulation and development of a wide variety of knowledge. There is general agreement that technology is central to regional change, both positive and negative, and to economic change in terms of job-creation and job-destruction. During the 20th century, the United States experienced very radical changes in the geographical sourcing of invention and new technology. In less than five decades, regions that were previously peripheral or undeveloped turned into the most important domestic sources of new technologies. Such rapid change can be seen as a process of regional inversion, in which the predominance of some regions is overturned, and new areas take their place as the major centres of knowledge production. The rapid rise of the Sunbelt states has no parallels elsewhere, due to the intensity and speed with which the established order was overturned. As recently as the 1960s, few people understood the extent of the changes that had started to occur, or imagined that these would propel some of the Sunbelt states to become the most vibrant areas of the American economy.

The chapter written by *Luis Suarez-Villa* traces and explains the rise of the Sunbelt states, considering the accumulation of inventions over space and time. The concept of innovative capacity is presented and explored, showing how the geography of invention in the United States was transformed over a relatively short period of time. To understand how this process of change occurred, regional and national data on invention patenting spanning over 100 years are presented and analysed. The interaction of three macro-level forces affecting this radical process of change is then considered. One of them is the demographic redistribution of highly-skilled [or knowledge-rich] individuals through internal migration. A second force is the rise of American venture capitalism and its supportive role for new technologies. A third and vital supportive force was the rapid development of the educational and scientific infrastructure, through massive construction and equipment expenditures, which allowed new institutions to rise quickly on the national ladder of recognition for technological excellence. The interaction of these forces and their spatial, temporal and sectoral dynamics was at the root of America's rising innovative capacity, its system of invention and innovation, and its global projection as leader in new technologies.

Three major schools of thought have participated in the debate on innovation, knowledge and regional development: those which concentrate on institutions, those focusing on industrial organisation, and those concerned with technological change and learning. The contribution by *Roberta Capello* belongs to this last school of thought and lies in the tradition of the approach developed by the GREMI group [Groupement de Recherche Européen sur les Milieux Innovateurs]. The central theoretical notion of this approach is that of the *milieu*. In essence, the milieu may be viewed as a context which provides the conditions for agents to be able to innovate and to co-operate with other innovating agents. The milieu is, thus, something like a territorial version of what the American economic sociologist Mark Granovetter has labelled the 'embeddedness' of social and economic processes (Storper 1997). The innovative milieu is described as a system of regional institutions, rules and practices that lead to innovations, with stress on the channels through which collective-learning takes place. In the present study, the author uses a database on innovation in five metropolitan regions in Europe [Amsterdam, London, Milan, Paris and Stuttgart] to analyse similarities and differences between milieu economies and urban agglomeration economies. She tries to discover whether the channels of socialised knowledge creation differ in the two cases. Empirical evidence is presented on the impact of dynamic agglomeration economics and collective learning on the innovation capacity of cities.

Using illustrations from a number of detailed empirical case studies, *Neil Alderman* considers the way in which the knowledge necessary for the successful delivery of complex engineering projects in manufacturing or construction is distributed through a range of different network actors. It is suggested that changes in the nature of the market place for complex engineering projects are creating conditions whereby knowledge sources are becoming geographically more dispersed. This has implications not only for the organisation of projects, but also for regional systems of innovation.

In recent years we have witnessed significant changes in the way major customers procure large engineering products or systems. Increasingly, complete turnkey systems are requested, often associated with maintenance contracts for the lifetime of the equipment. The author examines several such projects in the fields of power generation, railway rolling stock and materials handling for ports. It emerges that traditional manufacturing organisations are being faced with the need to acquire new capabilities and knowledge to enable it to meet such requirements. In order to do this the manufacturer is compelled to enter into new network relationships with a range of supply chain organisations and other partners.

The empirical research underpinning this contribution uses a conceptual framework to analyse the value added system that surrounds a particular project. This recognises that value, as perceived by the customer, is made up of many different components, some of which are services or knowledge rather than actual products. These are provided by a range of different types of actor and have to be assembled through the project network. By implication, in complex engineering projects innovation also arises from the assembly of a diverse set of capabilities: some technological, some managerial, some organisational. The skills associated with such projects are unlikely to be found within the regional innovation system, therefore major projects of this type are often contracted on an international basis. Project partners have to be sought on the basis of capability, rather than familiarity or locational proximity. This means that these projects have to be undertaken within widely differing legal and regulatory regimes. One consequence is that stipulations regarding local sourcing are often an integral component of the contract. Engineering companies find themselves forced to engage with non-local partners in order to build a supply chain, rather source them near the location of their core functions, such as design and project management.

These changes are stimulating many engineering companies to re-think what they perceive to be their core competencies, with the result that we increasingly find such organisations focusing on design and integration tasks at the expense of conventional manufacturing and fabrication activities. This is leading to a divorce with the sources of knowledge required to deliver complex projects and a risk that the loss of the manufacturing function will ultimately erode the knowledge base of the firm. A critical issue for the firm is therefore how it manages the knowledge that is distributed within a project network. Finally, it is stressed that the regional innovation systems within which these companies traditionally operated are also potentially weakened, since key components of the knowledge required for complex projects are lost to the region. Through this process, engineering firms become disengaged from their local environment, as they enter into or create project networks that are geographically dispersed.

If we are to understand why some regions grow and others stagnate, we need to investigate the role played by knowledge, innovation and technological change in regional economic growth. In order to examine the role of knowledge in economic growth, the chapter by *Zoltán J. Ács* places a strong emphasis on knowledge spillovers from the accumulated stock of knowledge. In line with modern Austrian economics, the author argues that existing knowledge needs to be discovered by entrepreneurs and turned into future goods and services in order for growth to take place. The chapter makes a connection between knowledge and entrepreneurship, and provides a novel measure of regional entrepreneurial growth as evidence of the role of entrepreneurial discovery.

PART D: Modelling Complexities

Typical features of innovation systems, such as their persistent asymmetries, their historical specificity and the multiplicity of institutional configurations can be accounted for by properties such as path-dependence and multistability. Such properties are not easily justified in the terms of the world view presented by neoclassical economics. They can however be predicted by evolutionary theories for out-of-equilibrium open systems. Thus, the contributions in PART D shift our attention towards the alternative world view provided by complexity theory.

In the first chapter of PART D, *Günter Haag* and *Philipp Liedl* utilise the Master Equation approach to model the complex and interwoven processes of production and innovation, labour force and capital formation, knowledge production and diffusion. They propose a Master Equation which describes the evolution of the probability function, representing the transition probabilities for well defined states of a dynamic micro-based system of actors. By using, for example, a mean value approach, an elegant link can be established between micro-levels and macro-levels of a system, so that structural changes in dynamic systems can be analysed in a statistically satisfactory way. There are several cogent reasons for using the Master Equation approach in analysing dynamic choice processes in economic systems. The first is its flexibility and generality. The ranges of possible behaviours embodied in Master Equations is almost unlimited. In the second place, this approach makes it possible to take account of synergetic effects in the behaviour of different individuals [such as adaptation processes and learning effects]. The socio-configuration includes the individual transition probabilities based on joint interaction effects. A third major advantage is that it links the micro-level decisions of individuals with the macro-level behaviour of collective variables. Feedback elements, heterogeneity [variation between individuals] and non-stationarity [variation over time] can be taken into account (Fischer, Nijkamp and Papageorgiou 1990). The authors use scenario-based simulations to demonstrate the functioning of the macro-model and illustrate that the development of interlinked firms belonging to different sectors and the existence of stable equilibria depend mainly on the signs of spillover effects between different firms and the impact of the scientific system.

Complex systems usually consist of a large number of interacting entities or agents. The complex behaviour of the system as a whole depends on the complexity of the agents [i.e. the range of possible actions], and on the complexity of the interactions. The next chapter, by *Frank Schweitzer* and *Jörg Zimmermann*, describes - within the framework of a basic model - the interaction between agents as a generalised form of 'communication', which is mediated via a spatio-temporal and multicomponent 'communication field'. As a particular example, the authors investigate decision processes of agents dependent on the information received via the communication field. All agents contribute consistently to this field with their decisions and at their particular location. The information generated this way has a certain life time, representing memory effects, and spreads through the community of agents in a diffusion process which reflects the finite velocity of information exchange. Because the communication field has eigen-dynamics and, thus, differs in time as well as in space, the assumption that every agent is subject to the same information, is no longer valid. The consequences of such heterogeneous information distribution are investigated by means of computer simulations. It is shown that the dynamics of the model result in a process of self-organisation based on communication, which is dependent on different forms of information distribution. In this connection, the authors find the emergence of interesting structures among the agents, such as the emergence of subpopulations sharing the same view on a given subject.

A canonical predecessor to the complexity approach is the model of emergent racial segregation in cities developed by Schelling (1971). In contrast to many of the current complexity models, this can be demonstrated without the help of computer simulations. Schelling (1971) based his model on a rectangle divided into numerous sub-rectangles [i.e. a regular 2-dimensional tessellation], each occupied by a representative of either a white or a black population. He posited movement between these sub-rectangles based on local interactions, and discovered that even a very slight preference by members of either group for living with their own kind would lead to a largely segregated city, even if agents were only acting in relation to their immediate neighbours. This cellular automata model displays an emergent global structure from strictly local effects, one of the fundamental ideas of modern complexity theory. Schelling's approach has been mimicked more broadly by simulating systems of multiple heterogeneous agents that evolve strategies over time in response to the behaviour of their neighbours.

The chapter by *David F. Batten* attempts to describe some basic ideas of agent-based modelling. Today's agent-based simulations have three ingredients: agents with bounded-rationality, an environment in which they 'live', and a set of rules. Although these artificial agents are intelligent and adaptive, the process of interference, learning and discovery that they can embrace is often simplistic. Learning agents can be arbitrarily intelligent, but unless they know other agents' learning methods, they cannot know if their own learning processes are efficient. Agents can only discover the efficacy of their own learning methods by testing them against others. In this chapter, the author reviews some examples of locally interacting agents who reason *inductively* to produce novel, large-scale effects. We cannot 'solve' this kind of agent-based model, instead we must 'evolve' it. Individual beliefs become endogenous, competing within a larger ecology of all agents' beliefs. Furthermore, this ecology of beliefs co-evolves over time, as agents adapt to the choices of other agents and the state of their environment. The contribution concludes with a few suggestions concerning ways we might build more realistic co-evolutionary learning mechanisms into agent-based experiments with artificial societies. Technically, agent-based models are limited by the way they are processed, and parallel machines are required for problems that contain the detail necessary for acceptable applications. Nevertheless, this is still rather more than the many simple pedagogial demonstrations that currently exist.

The final chapter of PART D, by *Michael Sonis*, describes innovation diffusion in socio-ecological and socio-economic systems. In these systems innovation diffusion is generated by the choice of innovation alternatives. A *duality* exists however between the collective choice behaviour of adopters, presenting the *demand pull* component of innovation diffusion, and the behaviour of the innovating system. The latter adopts, generates, supports and introduces the alternative innovation options, representing the *supply push* component of innovation diffusion. Moreover, the choice, generation and spread of innovations occur within a *socially active and economically structured territorial environment* which influences the behaviour of supply-demand components by filtering the information about innovations and through social, physical, cultural, administrative, political and economic stimuli or restrictions.

In Sonis' analysis, four major types of actors participate in the dynamic process of innovation diffusion, generating the deepening complexity of social interactions through a process of complication. These are: a set of *alternative competitive innovations,* generating the new emerging properties of new alternatives, and spreading them within a given territorial unit; different groups of *adopters* of innovations, different groups of choice makers [*innovators and innovating elites,* i.e., different systems supporting, producing and spreading the innovations]; and an *active territorial environment* adjusting the innovations to the structure of socio-economic hierarchical territorial organization. Each type of the major actors is connected with different methodological bases of the behaviour of 'homo socialis'. The mathematical description of the complex behaviour of 'homo socialis' in the choice process can be based on three different approaches, giving the same mathematical form of the innovation diffusion process in real space-time. The author shows that these can be based on the empirical regularities of choice process - the S-shaped change in the portion of adopters of different innovation alternatives; on the first principles of parsimonious human behaviour as collective beings, or on the competitive behaviour of social elites. He considers the implications of this socio-ecological approach for the behaviour of social, economic, cultural, and military elites at the outset of the information era, in relation to their coexistence and collisions, mode locking and ways of achieving stabilisation.

PART E: Policy Issues

Some general policy implications can be derived from the systems of innovation approach. These are essentially related to organisations, the relations between them, institutions, lock-in situations and demand side instruments. One policy implication deriving from the interactive character of innovation processes is that innovation policy should not only focus on the elements of the innovation system [manufacturers, producer service providers, universities etc.], but also on the relations between these elements. This includes among others the relations between firms in science-based industries and universities as well as research institutions. Thus, these interactions should be facilitated by means of policy. In periods of structural change, nation-states and regions may have to redesign the laws and rules governing the relations between universities and firms.

The first chapter of PART E, written by *Dimitrios S. Dendrinos*, directs our attention to options, innovation and metropolitan development in general and, thus, to novel metropolitan development schemes. The central argument of this contribution is that metropolitan development occurs in effect through the use of some highly complex investment instruments, including sophisticated options traded in very informal derivatives markets. The author shows convincingly that the field of development options is an area for innovative public policies. Development options markets are highly innovative, in that they allow new and promising metropolitan development schemes to be effectively and efficiently evaluated within a market context. It is advocated that these markets ought to be far more formal than they are currently, but that they must follow regulatory controls. A number of points are addressed: (i) how development option prices are determined, and how these prices are likely to behave under highly non-linear dynamics. Reference is made to the classical Fisher-Scholes-Merton options prices literature, and the differences and similarities between the two approaches stressed; (ii) the conditions under which the primary metropolitan development markets are likely to operate [especially urban land markets, although markets for environmental pollution and traffic congestion are also discussed], as well as their derivative markets [options and rights]; (iii) the governmental regulations needed to correct the market failures that these sophisticated derivative markets inherently possess; (iv) and finally, the fact that metropolitan plans and urban policies themselves act as if they too are complex options, operating in highly imperfect policy markets.

For many decades there has been continuing, though until now inconclusive, debate on the real impact of government policy [e.g., expenditure, taxation, knowledge infrastructure investments] on economic growth and on the competitive position of a nation-state. Parallel to this debate has been discussion about the impact of growth stimuli on the economic disparities between regions or nations. The question is raised whether overactive innovation policies may in fact have a negative impact on socio-economic convergence.

In the chapter written *by Peter Nijkamp, Jacques Poot* and *Gabriella Vindigni,* a meta-analytic perspective is adopted. The authors analyse the findings of over a hundred studies empirical modelling studies undertaken in different countries and in different time periods. These are systematically investigated using the research synthesis approach advocated in meta-analysis. A recently developed analytical tool from artificial intelligence, rough set analysis, is employed to draw general inferences from the sample of studies. Various sensitivity analyses are carried out to test the robustness of results. The overall finding of the study is that governments tend to have a positive impact on the economic growth of regions or nations, but that the specific mechanisms are not easy to understand, and in some cases ambiguous results tend to emerge.

The final chapter of PART E, by *Dirk Engel* and *Andreas Fier*, discusses whether R&D infrastructure attracts high-tech start-ups. The authors analyse the effects of the R&D infrastructure on the probability of firm foundation in high-technology industries in Eastern Germany. The estimation results confirm that education is crucial for setting up new firms. The proximity to universities with engineering and computer science faculties leads to an increase in the number of high-tech start-ups. Moreover, technology-foundation centres are very important for the concentration of high-tech start-ups.

Like most good research, the contributions in this volume raise as many questions as they answer. The interaction between a strong theoretical foundation and careful empirical analysis seems to us to be the key to continued progress in innovation system analysis. It is hoped that the insights provided will inspire other scholars and practising professionals to further explore the role of knowledge and complexities in innovation systems.

PART A: Innovation Systems

2   Networks, National Innovation Systems and Self-Organisation

Pier Paolo Saviotti
Institute National de la Recherche Agronomique, Université Pierre Mendès

2.1   Introduction

The concept of National Innovation System [NSI] has acquired considerable importance due to the work of Lundvall (1988), Freeman (1987) and Nelson (1993a). The NSI can be defined as the set of institutions and organisations responsible for the creation and adoption of innovations in a country. The origin of the concept of NSI can be related to the observation that the higher intensity of innovation in particular countries, which may have socio-economic systems that are otherwise identical to other countries, is not explained simply by greater R&D expenditure by firms or by public research institutions. We can observe that individual countries show persistent asymmetries at two levels:

- *The structure of output.* Considerable differences exist in the composition of output, and the sectors or sub-sectors in which countries are competitive, with these differences tending to be stable for long periods of time.
- *The institutions.* The institutions and organisations used in different countries to achieve comparable aims, such as the promotion of innovation, can be very different, and again these differences can persist for very long periods of time.

Examples of the first type of asymmetry are Germany's strength in chemicals and luxury cars, Japan's strength in cars and electronics, Italian strength in textiles, footwear, etc. (for these and other examples see Porter 1990). Examples of the second type are the organisation of research institutions in different countries, the involvement of government departments and ministries in the formulation of industrial and technology policies, etc.

The persistence of these asymmetries seems to be independent of the existence of a particular pattern of R&D expenditure or given institutional configuration. This suggests that there is no 'best' solution that can predispose a country to becoming more efficient or innovative than others. If such a solution existed, and if it could be unambiguously correlated with the enhanced performance of a country, then presumably it would be imitated, leading to the convergence of numerous countries on the same policies and institutional configurations.

Therefore persistence of differences implies that even if an optimal solution exists, it is not necessarily perceived as such by all countries, or possibly the obstacles to its achievement are so great that most countries could never succeed. Furthermore, the fact that different countries can achieve comparable rates of growth with very different policies and institutional configurations raises considerable doubts about the existence of a single optimal solution.

On the basis of the above discussion, it seems that there are two aspects of a 'national productive system' which can be used as a basis for international comparison and contrast. On the one hand, there is the set of firms producing outputs, which are characterised by different levels of production efficiency. On the other hand, there are groups of individuals organised into institutions and organisations, that affect the overall national performance as far as efficiency and creativity are concerned. Such institutions show a very high specificity with respect to the circumstances of the country and have a strong resistance to change. They are thus profoundly affected by the history of a country. Innovations are not introduced by individual isolated organisations, be they firms or research institutions, but by the complex interactions of these and other organisations that together form a system. It seems as though the particular patterns of institutional interaction constituting the NSI display features of self-organisation. Thus an NSI is capable of adapting to some changes in its external environment, while preserving its identity and fundamental properties. However, an innovation system can also learn and, under certain conditions, undergo changes in structure. Such changes do not occur only by means of adaptive learning. In adapting to changes in its external environment, the innovation system performs 'search' activities that can lead to partially endogenous structural changes.

These considerations have focused exclusively on national innovation systems. However, some would be equally applicable to other types of innovation system, such as regional or sectoral ones. All these systems display asymmetries equivalent to those previously described for NSIs. Regional innovation systems may possess similar differences in output and institutional set-up, although we would expect the national system to impose a common framework on regional systems. Sectoral systems of innovation can be expected to arise when the constraints internal to the sector outweigh those inherent in the nation state. This does not mean that a sectoral system of innovation is completely independent of its location, but that the variability of its mode of utilisation amongst different countries is less than the differences between sectoral innovation systems. Of course, while these systems are equivalent, in the sense that they are all analysable as systems, they are related to one another in ways that are not yet known and have been very little investigated. This represents a very interesting area of research, but one that will not be pursued further in this chapter. The main problem to be analysed here is the nature of the theoretical explanations appropriate to the systems investigated.

2.2   Complexity and Evolutionary Theories

During the past century our world view has changed considerably. While in this contribution we do not intend to go into these changes in depth, some aspects need to be examined, especially our expectations about the knowledge that we can gain from the external world, whether physical or social. By the mid 19th century, people had become convinced that it was in principle possible to 'know' the physical world. It was thought that both the basic components of the universe and the interaction between them could be understood and predicted. It is important to stress that such knowledge, while considered possible in principle, would be difficult to achieve in practice, because of the difficulties of computation. This conception of a deterministic, reversible and in principle calculable world, was later referred to as the Laplacian dream, after Pierre Simon Laplace who, at the beginning of the 19th century, attempted to give a systematic mathematical formulation to Newton's theories (Mirowski 1989; Prigogine and Stengers 1984, p.28).

This world view has long since been abandoned in physics, though not without some traumas in the profession, but still holds a considerable place in economics. An alternative is provided today by what are generally called the theories of complexity, even though they are not yet as complete as the Laplacian dream purported to be during the 19th century. It has to be admitted at the outset that the complexity theories are not a unified body of knowledge, except for some extremely broad generalisations. What is clear is that our certainty about being able in principle to know the world has been severely dented. Firstly, it was shown by quantum mechanics that our knowledge of the physical world is intrinsically limited, since observations of an unperturbed physical world are in principle impossible. The knowledge we can gain of the physical world is at best in the form of statistical probability. Later theories of non-linear systems and chaos demonstrated that even a system represented by a deterministic equation can behave in a chaotic and non-reproducible way.

The description given so far of the changes in our view of the limits of human knowledge would seem to imply that we were previously highly conceited in believing complete knowledge of the universe could be achieved. The conclusion is that we have since had to scale down our ambitions and recognise that less can be known than we hoped. While there is some element of truth in this, it is not strictly correct to say that we know less or can know less than we hoped. In fact our knowledge of the physical world increased immensely during the 20th century. In fact, during the last century, not only did we accumulate an enormous amount of knowledge, but our world view changed to a very considerable extent. The emerging world view is not yet complete, but it amounts to saying that the world is far more complex than we previously thought and that our possibility to know it is limited. Uncertainty and complexity are recurring words in today's literature.

To summarise this change, we could say that whereas previously we defined the universe as a set of events contained in a given range, and expected to be able to know everything within this range, today we see the universe as being a much wider range than before, but do not expect to know the whole range, even though what we know may well cover a greater range than before. Thus, although we have a clearer perception of our limits, we in fact know a lot more.

As far as the social sciences, and particularly economics, are concerned, the model of knowledge creation and validation was represented by physics. This was not peculiar to the social sciences. Physics was generally considered the model of scientific knowledge, i.e. knowledge that could be rigorously tested. Most epistemology was essentially the epistemology of physics. As pointed out by Georgescu-Roegen (1971) and Mirowski (1989), physics played a very important role in the development of neoclassical economics. Of course, the physics from which neoclassical economists took their inspiration was 19th century physics, and in particular the theory of electromagnetic fields. But although within physics the view of a deterministic universe gradually gave way to that of a complex, uncertain and irreversible universe of which only limited knowledge was possible, the same transition did not occur in economics or, at least, not in neoclassical economics. The research tradition has only recently started to take these changes in world view into account in the form of evolutionary economics (see Nelson and Winter 1982; Dosi and Nelson 1994; Nelson 1995; Saviotti 1996).

Before entering into details of complexity or evolutionary theories, let us point out that while the scope for knowledge may now seem more limited, there is a substantial advantage for economics and the social sciences. 19th century physics studied a world in which the only possible changes were represented by motion, or by the displacement of particles with respect to one another. Change in composition, represented by the emergence of new qualitatively different types of entity, could not easily be taken into account. In economic systems we often observe changes in the form of discontinuities, radical technological innovations, new types of organisations, and new activities. That is, qualitative change is an essential feature of economic development, especially in the long term. Theories of complexity and evolutionary theories therefore provide *a priori* a more appropriate framework for understanding long term economic development.

These theories will be briefly described in the following sections. They are the result of research carried out in different disciplines, but do not constitute a unified body of knowledge. We cannot expect this new world view to give rise to a general theory capable of explaining all observable events in any discipline. In a sense, this would be in contradiction with the newly emerging awareness of the limits of our knowledge. However, the existence of a shared basic framework, consisting of some general concepts, metaphors and tools, can provide inspiration for new questions and promote co-ordination between disciplines. It should however be stated that although the transfer of concepts, models and tools between different disciplines can lead to interesting and novel questions, it cannot provide cross-disciplinary solutions, e.g. answers from discipline A [e.g. biological answers] to problems arising in discipline B [e.g. economic problems].

The attempt to go beyond the Laplacian dream, represented in economics by the equilibrium approach, has involved the development of a theoretical treatment of a world that is complex, changeable and continuously affected by novelty. In the various theoretical developments described below, the only common components are the presence of qualitative change and growing complexity in the economic reality that we observe and study.

Systems Theory and Out of Equilibrium Thermodynamics

A very fundamental distinction can be made between closed and open systems (Nicolis and Prigogine 1989a; Prigogne and Stengers 1984; von Bertalanffy, 1950). The former cannot exchange anything with the surrounding environment while the latter can exchange matter, energy and information. These two types of system have some strikingly different properties, some of which are particularly interesting for the purposes of this analysis.

Closed systems can achieve a state of equilibrium that corresponds to the maximum possible degree of disorder, randomness and homogeneity. Examples of equilibrium in such systems is given by the mixing of two gases or liquids, e.g. when a second gas is introduced into the space containing another gas, or a drop of ink is deposited on the surface of a glass of water. In both cases we expect the heterogeneity constituted by the addition of the second component to the system to gradually disappear as the second component diffuses within the first. The state of equilibrium is constituted by a homogeneous gas or liquid. As long as these systems are kept isolated from their environment, we expect the equilibrium state to be stable. In other words, we do not expect the second gas to separate out spontaneously from the first, or the drop of ink to separate from the water.

We can imagine obtaining an open system from a closed one by gradually increasing the exchange the system has with its environment. As we start doing this, the equilibrium of the system will be disturbed, leading to a number of possible outcomes, of which instability or turbulence is one. Another is the achievement of one or more stationary states, i.e. states characterised by the invariance of a number of their properties. We can consider the rate of exchange with the environment as a measure of the distance from equilibrium. While these steady states could be confused with the equilibrium states attained by closed systems, they are in reality very different. Both equilibria and steady states show a temporal invariance of their properties, but while an equilibrium is characterised by the maximum possible disorder or randomness, that is not necessarily the case for a steady state.

As we move away from equilibrium, the system may undergo a number of transitions, leading to qualitative changes in its structure. The concept of structure itself involves the existence of separable and distinguishable components within the system, i.e. system heterogeneity. If we create an open system by gradually increasing the interactions of a closed system with its environment, we start from a homogeneous state, in which individual components cannot easily be separated or distinguished. Conversely, when we add a second gas or a drop of ink to the first gas or to water respectively, we move from a heterogeneous to a homogeneous state. That is, in moving *towards* equilibrium, a closed system loses structure, while an open system may acquire a structure as it moves *away* from equilibrium.

This difference is very important for the social sciences. The systems we observe are usually highly structured, and the changes they undergo over time often lead to modifications in this structure, represented by the emergence of new components and the disappearance of pre-existing ones. This is especially true if we observe long term developments. However, we should bear in mind that the rate at which innovations are being introduced into socio-economic reality is now increasing sharply, thus shortening the time scale at which these qualitative changes can be observed. If it is true that only open systems are subject to changes in structure, then most of the systems involved in socio-economic reality are likely to be open systems. These may undergo transitions in which their structure and properties change radically or qualitatively. When interaction with the environment increases and the distance from equilibrium becomes large enough, bifurcations may occur, i.e. transitions to totally new states. In the vicinity of this kind of transition, the system displays fluctuations or becomes highly indeterminate. The final state of the system cannot be determined a priori, as the transition can have multiple outcomes. We can at best predict the probabilities of the different states after the transition. There is thus an element of indeterminacy in the evolution of open systems. Furthermore, the transitions of open systems are both irreversible and path dependent.

So far we have compared the approach to the modelling of equilibrium in closed systems with that in open systems. As pointed out by Allen (2000), a model needs some simplifying assumptions in order to reduce the complexity of reality to a manageable simplicity. The more restrictive the assumptions, the simpler and apparently more predictive the models. The five most common assumptions made in models are:

- the possibility of defining the boundary separating the observed system from its environment,
- the possibility and stability of a taxonomy describing and classifying the components of the system,
- the assumption of homogeneity, i.e. that all individual sub-components are identical or have a diversity that is distributed around the average,
- the overall behaviour of the variables can be described by the smooth average rates of individual interaction events,
- the stability of equilibrium.

If all five assumptions are made, we obtain a model which is incapable of accounting for feedback processes, non-linear effects etc. If we eliminate the assumption of equilibrium and then gradually remove the others, leaving only the first hypothesis, we move gradually from mechanical to non-linear dynamic systems, then self-organising systems and evolutionary complex systems. On the way we [apparently] lose predictive power, but acquire the ability to study systems that are more complex and hence more realistic.

Models of evolutionary complex systems rather than being detailed descriptions of existing systems are more concerned with exploring possible futures. What is their relevance for the social sciences? The first answer is that biological and social life cannot be explained by means of laws which apply to closed systems only. The increasingly complex structures that we observe in modern societies are incompatible with the homogeneity and randomness of the equilibria in closed systems. In fact, it is easy to find examples of open systems in socio-economic reality. Hardly any organisation can be conceived of as a closed system. A firm, or any other organisation, continuously exchanges matter, energy and information with its environment. Interestingly Chandler (1977, Chapter 8) considers the rate of throughput as the most important variable determining the organisation of a firm. Theories of complex systems and of phenomena which are far from equilibrium are required to explain the features of biological and social life. Of course, this does not imply that all the systems involved are open. Even though completely closed systems are an abstraction difficult to reproduce in reality, some types of 'real' behaviour can resemble that of systems with very limited degrees of openness. Thus, it is quite likely that biological and social life consists of a mixture of systems of different degrees of openness.

A number of implications follow from the above. Social systems show evidence of self-organisation, in the sense that they can adapt to fluctuations in their environment. 'Adaptation' implies that the system is not destroyed but remains recognisably itself. This form of adaptation, called self-regulation or homeostasis, involves maintaining certain critical variables within pre-determined ranges (von Bertalanffy 1950). Examples are the maintenance of body temperature or the sugar level in the blood of biological organisms. Thus self-organisation refers to the maintenance of the structure of a system despite some degree of change. Of course, no system needs to be eternal, and social systems are in this sense more labile than biological ones. Political systems, the organisational structures of firms and other organisations, are fairly stable with respect to the average life span of their members, but can in the longer term undergo discontinuous changes, leading to the collapse of the system's structure and its replacement by another system. Such transitions correspond to bifurcations. Radical innovations, new paradigms (Dosi 1982), dominant designs (Abernathy and Utterback 1975), and technological regimes (Nelson and Winter 1977) represent discontinuities in technological and social evolution.

Other implications of the general considerations above are that the evolution of social systems may be irreversible and path dependent. These properties are also found in social systems (David 1985; Arthur 1989). When a technology produces increasing returns, it can end up by dominating other technologies even if the latter are more efficient. On the other hand, the presence of increasing returns in the creation of knowledge may be responsible for the continuation of economic growth in the long term (see for example Romer 1987, 1990).

To summarise, in this subsection we have contrasted the equilibrium approach, normally used in economics, with an approach derived from out of equilibrium thermodynamics and from systems theory. We have argued that the latter is required in order to explain the evolution of socio-economic systems. The differences in the two approaches are more than a question of their techniques or equations. In economics, an equilibrium approach attempts to explain why reality is stable, while an evolutionary approach attempts to explain why reality changes (Metcalfe 1997).

Ideas from Biology

The general properties predicted by systems theory and by non-equilibrium thermodynamics are displayed by biological, economic and social systems. Systems theory and non-equilibrium thermodynamics provide a theoretical justification for all these disciplines/research traditions. However, this does not imply that we can deductively infer the properties of biological or economic systems from systems theory and non-equilibrium thermodynamics. Historically the theoretical legitimation comes ex-post, and each discipline/research tradition has developed concepts appropriate to its observation space.

Biology started much earlier than economics to tackle the problem of qualitative change: the origin of species, the emergence of new species and the extinction of existing ones, are examples of such problems. If economics intends to deal with qualitative change, it seems legitimate to borrow ideas and metaphors from biology. Some of the most important biological ideas that can be applied to the study of economic evolution are variation, selection and inheritance. *Variation* is the process which gives rise to new species, some of which will survive the process of *selection*. The number of different species surviving determines the diversity of the system. In order to survive, species have to adapt to their environment. While it was previously believed that only the fittest would survive, modern opinions imply that also the tolerably fit can survive (Hodgson 1993). In economic systems, R&D or research activities in general, contribute to variation, while regulation and competition are the main forces responsible for selection. *Inheritance* refers to the conservation of some traits in subsequent generations of a species. Such conservation depends in turn on the transmission of genetic material from one generation to the next. Of course, firms and research laboratories do not have DNA, but organisational routines have been interpreted as the equivalent of the genetic heritage of biological organisms. The use of a *population approach*, as opposed to the typological approach commonly used in economics, is common in biology (Saviotti and Metcalfe 1991).

All these concepts and processes currently used in biology constitute a good basis for the analysis of qualitative change and of the heterogeneity of agents, problems which are central to an evolutionary approach in economics. In other words, economics and biology have a considerable degree of similarity, both structurally and in terms of overall goals of knowledge. However, it is not possible to mechanically transfer concepts and models between different disciplines or research traditions. Adaptation of the general concepts is required to suit the specific context. Thus, whereas in biological systems variation is considered to be blind or random, corresponding to Darwinian evolution, it acquires a Lamarckian character in economic systems, due to their intentionality and purposeful character (Saviotti and Metcalfe 1991; Hodgson 1993; Nelson 1995; Dosi and Nelson 1994). Biology can be a very powerful source of inspiration for evolutionary economics, in the sense of allowing us to formulate new questions and problems, but does not provide answers to economic problems.

Organisation Theories

The term organisation theory is used here to refer to a number of theories of the firm, as well as theories and concepts which have emerged in management science and business history. These theories have two aspects in common: first, they differ from neoclassical theories because they do not assume optimising behaviour, second, they open up the black box of the firm, or other organisations, by explicitly introducing the influence of organisational structure. Behavioural theories of the firm deal with satisficing behaviour and internal conflicts (Simon 1947, 1957; Cyert and March 1962). The distinction between strategy and structure, and the emergence of different forms of organisational structure [the U and M form] have been studied by Chandler (1962, 1977). Penrose (1959), McKelvey (1982), Teece (1986), Tushman and Anderson (1986), for example, have stressed the importance of competencies. Satisficing behaviour, routines and selection rules have been introduced into their evolutionary scheme by Nelson and Winter (1982). The growing role played by knowledge creation and its impact on the performance of firms has become an important aspect developed mainly within the research tradition of evolutionary theories.

Economic Antecedents of Evolutionary Theories

In the past a number of economists had intuitions which represent antecedents of modern evolutionary theories. For example, Marshall is often quoted as having said that 'the mecca of the economist lies in economic biology rather than in economic dynamics' (Marshall 1949). Marshall recognised that 'economics, like biology, deals with a matter, of which the inner nature and constitution, as well as the outer form, are constantly changing' (1949, p.637), a clear reference to the qualitative and structural changes which occur in economics. However, in spite of recognising the value of a biological metaphor, Marshall did not use it and relied more on economic statics than on economic dynamics.

Herbert Spencer was amongst the first to develop an evolutionary approach to social development. Whilst some of his ideas can be interpreted in a pro-aristocratic, racist and sexist light, others remain relevant for modern evolutionary developments. Spencer (1892, p.10) defined evolution as 'a change from an indefinite, incoherent, homogeneity, to a definite, coherent heterogeneity through differentiation'. Spencer thought that evolution necessarily involves progress, and that complexity is generally associated with fitter and more adaptable forms. These considerations anticipate the formation of structure and growth in variety.

Veblen made a very explicit use of a biological metaphor (Hodgson 1993). For him 'idle curiosity' was the source of diversity or mutation the evolutionary process. The institution became the unit of selection, but also in the replicator. He suggested that institutions were characterised by relative stability and continuity through time, and could thus transmit diversity from one period to the next, ensuring that selection had relatively stable units on which to operate. Variation, selection and inheritance were thus present in Veblen's analysis.

Schumpeter defined economic development as the creation of new combinations of productive means by entrepreneurs (Schumpeter 1934, p.66). For him these new combinations are new products, new processes, new markets, new sources of raw materials and new organisational forms. All these new combinations give rise to products, processes, etc. which are qualitatively different from those that preceded them. In more modern terms, one would say that Schumpeter attached great importance to radical innovations as ingredients of economic development. Thus in his view, qualitative change and the generation of economic diversity are central to long term economic development. Furthermore,

Schumpeter stressed the non-equilibrium aspects of capitalist development. The creative destruction which 'incessantly revolutionises the economic structure from within, incessantly destroying the old one, incessantly creating a new one' is one of the fundamental mechanisms of capitalist economic development (Schumpeter 1942). Curiously Schumpeter rejected the use of a biological metaphor in economics.

Great importance was attached by Hayek to the role of rules. He spoke of the 'genetic primacy of rules of conduct' (1982, Vol. 3, p.199, cited in Hodgson 1993, p.164). A rule is defined by Hayek as a regularity of conduct of individuals. The durability of rules is due to replication through imitation. This mechanism accounts for the much faster rate of cultural evolution compared to the sluggish biotic process of genetic change and selection (Hodgson 1993, p.165). The selection procedure for rules, however, is quite interesting. Rules are selected on the basis of their human survival value, that is, they are indirectly selected through the association with a particular group. Also, for Hayek the idea of spontaneous order, which he compared to the concepts of autopoiesis, cybernetics, homeostasis, self-organisation and synergetics (Hayek 1988, p.9, in Hodgson 1993) was central. In support of spontaneous order, he quotes Prigogine and his school (Hayek 1982, Vol. 3, p.200, cited in Hodgson 1993).

The ideas of all these economists were compatible with and sometimes anticipated further evolutionary developments. As previously pointed out, evolutionary theories are still at an initial stage and have not yet achieved the degree of definition of neoclassical economics. Their comparative advantage, which justifies efforts towards further development, lies in the explanation they offer of situations characterised by qualitative change, radical uncertainty and by the heterogeneity of agents and techniques. Qualitative change will be one of the main topics in the rest of this chapter. Thus variety will be analysed as a concept that allows us to treat analytically qualitative change. As qualitative change can only occur by means of the creation and utilisation of new forms of knowledge, such phenomena are fundamental for the performance of NSIs.

While evolutionary theories have some advantages in the analysis of qualitative change, radical uncertainty and heterogeneous agents, they do not necessarily escape all of the tensions inherent in economics and in the social sciences in general. With respect to determinism, in neoclassical theory it is assumed that the economic system moves towards equilibrium and remains there, except for temporary displacements. This leaves agents the freedom only to optimise (Hodgson 1988). But even in evolutionary theories, path dependence may be seen as compelling agents to stay on a path which they have not chosen. However, such determinism in evolutionary theories is never complete. First, even after having chosen a given path or trajectory, agents still have a considerable amount of residual freedom which influences their performance. Of course, such freedom does not allow them to redesign radically the technological or conceptual system on which they base their competitive capabilities, but it can manifest itself in terms of incremental innovation. Second, in the vicinity of transitions leading to qualitative change fluctuations lead to a very high uncertainty, destroy previously accumulated competencies (Tushman and Anderson 1986) and temporarily disrupt path dependence. In these conditions agents' freedom is very considerable. Of course, agents are not necessarily aware of being in a transition phase. In these conditions uncertainty usually means greater risk and greater opportunities than in a mature, stable market. Transition phases represent conditions more favourable to entrepreneurial than to routine behaviour (Winter 1984). Thus, evolutionary theories leave greater room for uncertainty, intentionality and individual freedom than neoclassical ones.

We now summarise the main differences between neoclassical and evolutionary theories. *First,* evolutionary theories deal with qualitative change, or change in the composition of the system, resulting from the balance of variation, creating new 'species', and selection, based on differential adaptation. Inheritance too affects the rate and type of qualitative change. *Second,* they accept the existence of uncertainty, path dependency and multistability, all features arising from the out of equilibrium nature of systems and processes. *Third,* they assume that there exists heterogeneity of agents, requiring a population approach, emphasising not only representative agents and mean values of properties, but also their distribution within a population. These features provide a more realistic analysis of innovation systems and also help to explain some of their main properties, such as historical specificity and the multiplicity of institutional configurations, both of which are impossible to account for in terms of neoclassical theory. In the following sections, the implications of these general concepts for innovation systems will be discussed in greater detail.

2.3   Implications of Evolutionary Theories for the NSI

Economic development not only allows us to produce existing goods and services more efficiently and in greater quantities, but also to create new goods and services which are qualitatively different from existing ones. In the course of economic development, the composition of the economic system thus changes. In fact most of the objects of consumption that we use in everyday life, and the activities required to produce them, have been created within the last hundred years. The importance of this changing composition of the system depends on whether the composition is only an effect of previous economic development or also a determinant of future economic development. There is considerable evidence that the composition of an economic system is an important determinant of its future economic development. For example, studies by Salter (1960) and by Cornwall (1977) showed that structural change could make a very important contribution to economic growth. The efforts made and resources allocated in many countries to create new industrial sectors, either by supporting the relevant search activities or by providing the resources and environmental conditions required for their creation, are proof of an implicit admission of a relationship between strcuture and growth. Yet our theoretical understanding of how composition affects economic development is very limited. More recent studies by

Fagerberg (2000) and by Fagerberg and Verspagen (1999) show that the role played by structural change in economic growth in the period 1970-1990 was radically different from that of the periods studied by Salter and by Cornwall (1925-1950 and 1950s to 1960s respectively). A limit of these analyses, pointed out by the authors, is the use of statistical information that was not designed to measure changes in the composition of the economic system. While there is not the space here to go into the question in exhaustive detail, it is important for the purpose of this chapter to observe that there is evidence that the composition of the economic system can be an important determinant of its growth and development, but that our theoretical understanding of the mechanisms by which this occurs is very limited.

In order to improve our understanding of the role played by the composition of the economic system, we need first to define one or more variables capable of measuring this composition and its changes over the course of time. The variable used in this chapter is variety. Qualitative change modifies the composition of the system. New entities, qualitatively different from the old ones, appear and the old ones disappear. Variety measures the number of these entities. If we adopt a simplified representation of the economic process as a set of *actors* [individuals and institutions] performing *transformation activities* [transforming inputs into outputs] thus generating new types of *output*, we can define variety as the number of actors, activities and objects necessary to describe the system (Saviotti 1991, 1994, 1996).

Several aspects of variety need to be distinguished. Here we consider three types of variety, defining *output variety* (V_q) as the number of distinguishable types of output, *process variety* (V_p) as the number of distinguishable types of processes used to produce V_q, and *institutional variety* (V_i) as the number of distinguishable types of institutions. The overall variety of the economic system, as defined above, is the combination of all these types of variety. Another relevant distinction is that between national (V_N) and world (V_w) variety. These distinctions must be introduced because the types of variety can behave differently in the course of economic development. For example, standardisation or modularisation may cause process variety to decline, while output variety may be increasing. Here it should be stressed that an important component of process variety comes from the competencies used to produce the required output. This type of variety is discussed far more explicitly by Cohendet and Llerena (1997).

Variety and Economic Development

Most theories of growth, while not denying the presence of qualitative change, do not take it into account. Most of these theories are macroeconomic, and hence the composition of the system is not one of the variables that they consider. Schumpeter's theories, stressing the fundamental role played by radical innovations, imply a very close relationship between qualitative change and economic development. Of course, economic development can be conceived as comprising a large number of aspects, such as demand, competition, trade, etc.

Variety is in principle related to all of these aspects of economic behaviour. However, in this chapter the discussion will concentrate on a limited number of aspects. For a more extended discussion, the reader is referred to Saviotti (1994, 1996). We start the discussion with two hypotheses about the role of variety in long term economic development.

HYP (1): Growth in variety is a necessary requirement for long-term economic development.

HYP (2): Growing variety [new sectors] and growing productivity [in older sectors] are complementary, and not independent aspects of economic development.

It must be pointed out that these two hypotheses are valid only in the long run and at a sufficiently high level of aggregation. At a lower level of aggregation, variety in some cases does not grow but decline. For example, the emergence of dominant designs (Abernathy and Utterback 1975, 1978), of new technological regimes (Nelson and Winter 1977), technological paradigms (Dosi 1982) and of standardisation tend to reduce the variety within a given technology. But while variety may fall within a given technology, the greater efficiency afforded by this trend contributes to the accumulation of resources required for the generation of 'new species', leading to growth in variety at the highest level of aggregation in the system. The situation here bears a strong similarity to the role of productivity growth in agriculture, which allowed the accumulation of resources to be invested in the nascent process of industrialisation (Kuznets 1965).

The second hypothesis refers to complementarity between the search for greater efficiency in existing routines and the undertaking of research activities in order to generate new types of output and processes. This can be reformulated as complementarity between the circular flow and innovations in Schumpeter's theory of economic development. The former hypothesis can be likened to the Schumpeterian process of creative destruction, in which old species disappear and new ones emerge. Creative destruction leads to qualitative change, but not necessarily to growth in variety. At the same time, growth in variety can take place without creative destruction, simply through the addition of new species to old ones. Here we should observe that hypotheses *HYP (1)* and *HYP (2)* are concerned with *net variety* surviving the process of selection. These hypotheses imply that the rate of creation of new species is greater than the rate of extinction of existing ones, not that such rate of extinction is zero.

Growth in variety, and the nature of creative destruction can be more accurately represented by means of elementary processes in technological evolution. These are processes such as substitution [of one technology for another], specialisation and the emergence of completely new products. In principle, we can reconstruct any complex process of technological evolution as a combination of elementary processes. Each elementary process has different implications for variety: for example, substitution does not change variety at all [an old product disappears and a new one takes its place], while the emergence of completely new products gives the maximum possible contribution to variety. Overall variety can therefore only increase if variety-creating processes [e.g. new products] predominate over variety-conserving processes [e.g. substitution] (Saviotti 1988, 1991, 1994, 1996). Hence, creative destruction exists, but it creates more than it destroys. In other words, it has to be pointed out again that the variety discussed in the hypotheses (*HYP (1)* and *HYP (2)*) is *net* variety, the outcome of the combined processes of variation and selection. In economic systems, variation is essentially created by research activities, and all those activities which scan the environment searching for alternatives to existing routines or products. Variation creates a large number of potential species/technologies, accompanied by new routines, only some of which are sufficiently adapted to the environment. Those adapted least well are eliminated by selection. Selection is the result of a series of processes, such as competition and regulation.

There are several types of empirical evidence which indicate that variety has been increasing in the course of economic development [family trees of technologies, classes of patents and of international trade, etc.]. For a more detailed discussion, the reader is referred to Saviotti (1996). The two hypotheses can be justified, for example, on the basis of Pasinetti's model of economic development (1981, 1993). If we take an economy constituted by a constant set of economic activities in which technical change leads to constant productivity growth, and in which demand for existing goods and services tends to saturate. In such an economy an imbalance would arise in the long run because it would be possible to produce all the demanded output using only a part of the existing resources, including labour. This imbalance, which would constitute a barrier to long term economic development, could be imputed to technological change. However, technological change also plays a second role, that of creating completely new types of outputs and activities. If this happens, that is if net variety increased, the new activities could utilise the resources made redundant by the imbalance productivity growth/demand growth in old activities. On the other hand, variety can only grow if productivity growth in older activities allows the accumulation of the resources to be invested in research activities which will lead to new varieties.

This treatment has focused exclusively on the structural conditions leading to growth in variety. But even if the new goods, services and sectors created could be correctly identified ex-ante, which is unlikely because uncertainty is at its maximum near transitions, the adaptation of the economic system to the new structure/composition cannot be taken for granted. New human resources need to be created to produce the new goods or services. The creation of these resources requires in addition investment, and institutional and organisational adaptation. While the required investment can come from the increased efficiency of pre-existing sectors and from the supra-normal profits of the temporary monopolists in the new goods/services, the required patterns of institutional and organisational adaptation are more difficult to predict. Here, we can simply state that a set of development paths leading to greater variety has a higher economic development potential than another set in which the level of variety is conserved. The study of the institutional and organisational conditions required for the growth of variety would be an interesting subject for further research!

Variety and the International Competitiveness of Countries

If we accept that growing variety is a necessary requirement for long term economic development, it follows that the income share of existing sectors can be expected to fall gradually in the course of time. We can also expect, as stated above, that however limited the extent of specialisation in a country, its national output variety V_i will be lower than the world output variety V_w at a given time:

$$V_j \leq V_w \tag{2.1}$$

If the variety of world output keeps increasing we can expect that, although individual countries tend to specialise, this specialisation cannot remain constant and must reflect the new goods and services emerging in the world economy. In general we expect national variety to increase when world variety increases. If countries aim to keep an almost constant share of world income, or, in the case of developing or industrialising countries to catch up, then the ratio of national to world output Q_i/Q_w must, at the least, remain constant or increase gradually.

$$\frac{d(\frac{Q_j}{Q_w})}{dt} \geq 0 \tag{2.2}$$

Considering that:

$$\frac{Q_j}{Q_w} = \frac{\sum_{i=1}^{V_j} p_{ij}q_{ij}}{\sum_{i=1}^{V_w} p_{iw}q_{iw}} = \frac{V_j \overline{p_j q_j}}{V_w \overline{p_w q_w}} \tag{2.3}$$

where $p_j q_j$ and $p_w q_w$ are the prices and quantities produced in country j and in the world economy respectively, and the bars above $p_j q_j$ and $p_w q_w$ indicate average quantities. We can see that there are several possible catch up strategies:

(i) V_j/V_w increases and $\dfrac{\overline{p_j q_j}}{p_w q_w}$ increases,

(ii) V_j/V_w falls but $\dfrac{\overline{p_j q_j}}{p_w q_w}$ increases enough to compensate,

(iii) V_j/V_w increases enough to compensate for the fall in $\dfrac{\overline{p_j q_j}}{p_w q_w}$

Strategies (i) and (iii) are equivalent to a form of de-specialisation. In the former case this is accompanied by an increase of unit prices and/or of quantities produced, while the latter case corresponds to lower prices and/or quantities produced. Case (i) corresponds to the case of a country previously specialised in low technology goods and services, now entering high technology sectors at a rate higher than the world economy. Case (iii) corresponds to the case of a country that undergoes the reverse transition, adding lower technology goods and services to high technology ones, in such a way that the increase in variety can more than compensate the fall in prices or quantities produced. Strategy (ii) corresponds to the case of a country that becomes increasingly specialised and concentrates on high price goods and services and/or produces large quantities of them.

The catch-up strategies outlined above can be further distinguished into more cases, if we consider that the same change of the ratio $q_j p_j / q_w p_w$ can be achieved in different ways. Thus the ratio can increase either if p_j increases and q_j remains constant or falls, or if p_j remains constant or falls and q_j increases. The two cases would correspond to entering a new sector in the very early phases of its lifecycle [high prices, small q_j] or to exploiting scale economies [high q_j, low p_j]. The complete set of possible catch-up strategies is shown in Appendix A-2.1. Here, however, we can remark that although catch-up can be based on a specialisation, such specialisation cannot be static. As the world economy evolves and its variety grows, no country can afford to completely ignore new sectors. Thus, national variety will have to increase together with world variety, though not necessarily at the same rate. If we bear in mind that small countries tend to specialise more than large ones, we can conclude that in the long run, especially for large countries, to keep an almost constant ratio of national to world variety is a wise policy.

Interestingly, we can interpret in terms of variety some of the policy recommendations formulated in the past. If we look at the case of a country whose economy has always been based on the exploitation of natural resources, the introduction of sectors that transform these natural resources into finished products, i.e. manufacturing, constitutes an increase in national variety. Considering the two extreme development strategies that can be envisaged here, i.e. concentration on sectors exploiting natural resources or addition of new manufacturing sectors, the former involves a declining percentage of world output variety, while the latter can lead to a growing national output variety. Seen in this light, strategies of industrialisation aimed at import substitution lead to growing national output variety and thus, possibly, to a constant share of world output variety. From Equation (2.3) we can also estimate the likelihood of a strategy based on a growing specialisation and thus on a falling share of world variety. Strategy (ii) can be implemented only if $p_j q_j / p_w q_w$ increases enough to compensate the fall in V_j. However, if V_j/V_w falls because country j concentrates on traditional sectors and does not incorporate new ones, it is quite likely that $p_w q_w$ will increase more rapidly than $p_j q_j$, since the prices of the new goods and services are likely to be higher at the beginning of their life cycle. We can now understand why countries concentrating predominantly on natural resource based sectors encounter increasingly unfavourable terms of trade. Thus, strategy (ii) will be viable only if the process of increasing specialisation takes place by rapidly incorporating emerging sectors and by becoming less reliant on traditional ones.

Following the above reasoning based on variety, we can understand how for Cornwall (1977) manufacturing should be considered the engine of economic growth. In countries whose economies are still mainly concentrated on the primary sectors, the development and diffusion of manufacturing leads to an increase in variety. We can also understand why for the more recent study period [1973-1990] (Fagerberg and Verspagen 1999) this is only true for industrialising countries. The highly industrialised countries that already had a well developed manufacturing sector could not increase their output variety in this way. The general rule seems to be that a country needs to keep a constant or growing share of world output variety in order to keep its per capita income at least constant with respect to other countries, rather than simply seek to expand manufacturing activities. In fact the new information and communication technologies exert as strong an influence on services as on manufacturing. Following Fagerberg and Verspagen (1999) we can stress that: (a) structural change can occur within manufacturing and (b) the distinction between manufacturing and services has become irrelevant. Our analysis of variety is based on activities, which can encompass manufacturing as well as services.

Knowledge

Innovation systems are very knowledge intensive. The knowledge used is not necessarily only scientific knowledge, quite often it is what could be called applied, technological or organisational knowledge. These different types of knowledge are often created and transmitted by different types of institutions, and are combined in the production of final outputs. In studying these flows of knowledge, a number of generalisations have emerged, applicable to all the various types of knowledge, whether scientific or more empirical. For example, knowledge and theories are *correlational structures* (Saviotti 1994b, 1996) in that they establish correlations between variables. However, the number and type of variables and the range of values over which correlation can be established is limited, and differs for different types of knowledge. The extent of correlation can be measured by the *span* of the particular theory/piece of knowledge. This amounts to saying that knowledge has a *local* character: the smaller the span, the more local or specific the knowledge. On the contrary, more general pieces of knowledge will have a greater span.

In a more limited sense, the concept of local knowledge was introduced by Nelson and Winter (1982). They claimed that firms are better able and more inclined to do 'similar' things to those done in the past. This is because correlations within a given discipline or research tradition are based on observations of given phenomena, and will have a greater chance of being applicable to similar phenomena. This implies some form of path dependence in the construction of disciplines or research traditions and the presence of co-ordination difficulties when using the results of different disciplines or research traditions.

Knowledge is also a *retrieval* or *interpretative* structure (Saviotti 1994b, 1996). As opposed to knowledge, information is of a factual nature. However, information can only be retrieved or interpreted by those actors possessing the right conceptual categories. Information corresponding to a new paradigm cannot be interpreted by actors who know only the old paradigm. Given the local character of knowledge, the retrieval ability of a particular institution is proportional to the similarity between the external knowledge to be acquired and the present internal knowledge possessed by the institution. The correlational/ retrieval/interpretive nature of knowledge implies that it is also cumulative. The better known a particular field, the easier it is to retrieve and assimilate a piece of knowledge from within it and to progress. However, this also implies an inducement to remain in the field and a barrier to moving to different fields. The idea of Cohen and Levinthal (1989, 1990) that R&D contributes to the absorptive capacity of a firm, corresponds to this function of knowledge, as does the idea that competence destroys technological change (Tushman and Anderson 1986).

Knowledge can also be *tacit* or *codified*. Knowledge generation is a collective enterprise requiring communication between individuals and groups. For codified knowledge, communication is easier and less costly. However, there are costs involved in codifying knowledge and in acquiring the code once it has been established. The actors who establish the code are in a position of temporary monopoly. Furthermore, codification occurs gradually during the development of a discipline. The degree of tacitness [or codification] is an important component of the appropriability of knowledge. We can distinguish an intrinsic component of appropriability, resulting from the internal features of the knowledge itself, and an institutional component, resulting from the institutions that attempt to increase the appropriability of specific knowledge. We can expect the degree of codification of a given piece of knowledge to increase with its age, and thus its intrinsic appropriability will decrease systematically. If this limited appropriability were to occur in a core technology, overall appropriability could still be increased by combining the core technology with complementary assets. Furthermore, a piece of knowledge that is already highly codified at the beginning of its industrial utilisation can remain very appropriable if few actors know the code and to acquire it is very costly (Saviotti 1998).

New knowledge is generated by means of *search activities*. These are the activities that firms and other institutions use to scan the environment to search for alternatives to their present routines and for new activities. These are analogous to R&D and can be considered a form of learning-by-not-doing! From another point of view, all activities can be classified as either search activities or routines. It is possible to represent all types of search activities [basic, applied, and development] by means of the two variables: 'range of search' and 'probability of finding the required outcome' (Saviotti 1998). Basic search would have a wide range and a low probability, while development would be much more narrowly focused and have a much higher probability of success. We can expect knowledge fields to have a lifecycle. They begin by scanning a wide range, then as they achieve some success, they tend to focus on this neighbourhood and become gradually more focused, local and specific.

The features described above can lead to barriers to entry. Let us assume that only the last unit of knowledge is useful for production purposes. Given the interconnectedness of knowledge, in order to learn the last unit, all the previously created pieces of knowledge have to be learned. The more similar the external knowledge is to the existing knowledge base of a firm, the higher the rates of knowledge accumulation and the higher the barrier. In other words, these features of knowledge imply *cumulativity, irreversibility* and *path dependence.* The same features that can be predicted for socio-economic systems in general on the basis of evolutionary theories are obtained by means of considerations on knowledge.

Networks

According to the previous considerations, the composition of a country's economic system has to adapt to developments in the world economy if the country is to remain reasonably competitive. Furthermore, to change the composition of an economic system is becoming more and more dependent on the creation of new knowledge. Clearly this has important implications for the nature of the institutions. Institutional rigidities which hinder the required adaptation can slow down the process of economic development. The following considerations refer mainly to those institutions which are more closely related to innovation.

Great emphasis was placed above on the possibility that the structure of an economic system might change discontinuously at times. The structure emerges as the individual components of the system which are initially combined in a random way, then begin to assemble in particular organisations whose interactions are determined by institutions. This would of course greatly limit the actual number of interactions with respect to the possible maximum. Thus we can in principle imagine that within a given system more than one structure may emerge. Of course, a structure may at times become unstable due to changing environmental conditions. Indeed some of these changes, such as important innovations, may be endogenously created and yet capable of destabilising the same structure that created them. We can therefore imagine a situation in which innovations created within a given structure act as fluctuations and lead to the creation of a new, qualitatively different structure. Once created, the new structure will still be loosely organised, but we can expect its organisation to gradually become more interconnected and rigid. We can also expect endogenously created innovations to arise 'outside' existing institutions and organisations, in the sense of requiring new firms to produce them and new institutions to define the rules of their use. Such rules will not be available ex-ante, but will be created simultaneously with the diffusion of an innovation.

Institutions themselves should not be interpreted only as constraints limiting the creation and diffusion of innovations, but also as facilitators of innovations (Edquist and Johnson 1997). Only in presence of new institutions will the innovation acquire its full scope or economic weight. Yet it is possible that once the new institutional infrastructure is in place it could hinder the transition to a different structure when the set of innovations has exhausted its potential. As hypothesised by Perez (1983), innovations and institutions may be expected to have intrinsically different dynamics. Innovations develop first and the relevant institutions follow after the delay during the formation of techno-economic paradigms. In this view, the new structure needs new or modified institutions. Its emergence as well as its subsequent development are due to the co-evolution of technology and institutions (Nelson 1994).

The concept of network allows us to give a more specific meaning to the NSI. This was defined as a set of institutional actors and of their interactions, having as their ultimate goal the generation and adoption of innovations. Institutions and organisations shape the patterns of inter-individual interactions. Networks, consisting of a set of actors and the links connecting them, are an analytical category which allows us to study such patterns of inter-individual interactions more systematically. In this sense, the particular networks within a social system at a given time represent its structure. Their existence and the qualitative transitions between different networks can be explained on the basis of evolutionary theories, although they were developed outside these theories (see Powell 1990; Barley and Freeman 1991; DeBresson and Amesse 1991; Callon 1992, 1993). Networks can also be a very useful tool for the development of these theories and for their application to the study of NSIs. In spite of their appeal, networks have been given different meanings by the authors cited and a unified definition and treatment of this subject is still lacking. Here we use the term networks to indicate a combination of actors and links.

In this general sense networks have a very widespread existence in socio-economic systems. Any reasonably stable and repeated pattern of interaction can be represented as a network. If actors can be either individuals or institutions, networks constitute the structure of a socio-economic system. Although the concept of networks was developed independently of evolutionary theories, their existence can be justified on the basis of the previous considerations on open out of equilibrium systems. A network corresponds to a given structure and a discontinuous/qualitative change will lead to the formation of a completely different network. Networks are related to the discussion above because learning and knowledge creation are collective phenomena and involve the creation of some kind of network. Relationships between actors in these processes are not impersonal, and display both persistence and structure formation.

We can then re-interpret all the previous considerations in terms of networks. First, an NSI can be conceived as a set of nested networks. We can imagine important innovations arising within a given NSI but 'independently', or against the rules of the NSI. Innovations represent fluctuations within the given system. Fluctuations might then induce a transition to a different innovation system, which initially are only imperfectly formed and would need the creation and development of a new set of institutional infrastructures. The new system would be characterised by new organisational actors and new links. We can expect that as the new system develops, its connectivity [the density of existing links] and the centrality of different actors will undergo systematic variations. For example, we could expect the connectivity of the new system to increase as the new organisations responsible for the production and regulation of innovations establish links with the rest of the economic system. The new institutions would allow the full scope of the innovation to be achieved. Yet as the new innovation system develops in this way, it may in turn become unstable and through new fluctuations give rise to further transitions.

Two important aspects of networks are their stability and their efficacy. Efficacy here implies the achievement of the goal(s) of networks, and must be clearly distinguished from the concept of efficiency. Stability may result from continued efficacy or from phenomena of histeresis. A network which is stabilised and does not perform its function any more is a policy problem. In this sense network dynamics, for example related to the barriers stabilising existing networks or hindering the formation of new ones, are very important.

If networks are an important ingredient of the good economic performance of a country, then their creation and maintenance are a concern of industrial policy. But, if networks are created out of equilibrium, this implies a radical change in the basic philosophy of industrial policy. In this case, the overall policy goal would not be to eliminate market imperfections and lead the system back to equilibrium, but to move the system deliberately away from equilibrium (Carlsson 1992, 1994). The problem would become how to create the right networks, a problem amplified by the existence of multistability. This property implies that more than one institutional configuration can lead to an equivalent outcome. The properties of a good network will be some form of synergism or autocatalyis. Synergism could result, for example, if each actor produced a generalised output which was a more or less appropriate input for another actor. For example, financial institutions can produce the appropriate type of financial input, knowledge institutions the right kind of knowledge input etc. Obviously, these are only preliminary considerations. The role, nature and dynamics of networks need to be studied in much greater depth. The purpose of these observations is to place networks in the context of evolutionary theories, and in the meantime to point towards their policy implications in situations of high knowledge intensity.

Convergence and Divergence

An important question about the NSI is whether ongoing globalisation will lead to a complete homogenisation of the world economic system, thus making the concept of NSI redundant. The problem here can be reformulated as the balance convergence/divergence or homogeneity/heterogeneity. Based on the previous discussion of evolutionary theories, the problem can be approached in a number of ways.

First, according to Prigogine and Stengers (1984), the phenomena taking place in a system can be classified as either *forces* or *fluxes*. Forces would be, for example, chemical reactions, giving rise to local heterogeneity constituted by a gradient in the concentration of the new product. This creates a potential for diffusion, a type of flux, which tends to homogenise the concentration of the new product. In terms of socio-economic systems innovations are forces, which lead to an uneven distribution of the capacity to create and use these innovations. International trade, technology transfer and diffusion are fluxes, which tend to distribute innovations uniformly over the world economic system. Complete convergence would be obtained only if forces or innovations ceased completely and only fluxes remained. What is implied by the term globalisation is only an acceleration of the fluxes present in the system, and not the disappearance of innovation. Certainly the degree of convergence would increase if fluxes increased more than forces.

Second, if we represent NSIs as networks of institutional actors and their links, complete convergence implies that all existing links are destroyed and new links formed. In view of the considerable historical continuity of institutions, the probability of this happening is not very high. We can expect that new institutional actors will emerge, new links will be formed and that some of the old ones will disappear, but a complete disappearance seems unlikely.

Third, if world variety, V_w, increases, the number of possible specialisation strategies/output asymmetries possible for each NSI increases. If we take into account the previous discussion on institutional diversity and the possibility of multistability, we have to conclude that a further divergence of NSIs is possible.

Fourth, to the extent that the diffusion of knowledge and creation of further innovations are linked, we can expect a faster diffusion of knowledge to lead to an increasing rate of creation of innovations. Yet, such innovations are unlikely to be evenly distributed in the world economic system. Increasing returns to adoption lead to localised patterns of settlement of the firms producing the innovations (Arthur 1994). In particular, as far as knowledge exchange is concerned, the presence of tacit elements tends to raise communication costs and usually requires personal contact in learning processes, thus favouring geographic proximity of firms producing similar goods or services. In fact, there is increasing recognition that the region [a subset of a nation state] is becoming a more appropriate geographical unit for the creation of leading edge business competitiveness than the nation state (Fischer 2000; Ohmae 1995). We can therefore expect globalisation to be accompanied by faster diffusion of knowledge and by a more localised pattern of establishment of the firms producing the innovations.

To sum up, due to the dynamic balance of forces and fluxes, as well as the network character of NSIs in presence of growing diversity and multistability, and increasing returns to adoption in processes of knowledge exchange, we cannot necessarily expect the complete convergence of NSIs and other innovation systems. Some of the boundaries that hinder the movements of goods, services and individuals [e.g. national boundaries] may become more permeable, while more localised patterns of industrial development will occur at the regional level. It is thus very unlikely that the world economic system will move towards greater homogeneity, eliminating the differences in per capita income, and in scientific, technological and industrial capabilities that have emerged in the course of four centuries.

2.4   Conclusions

The main aims of this chapter have been, *first*, to show that the existence and the major features of innovation systems can be better explained by evolutionary than by neoclassical theory, and, *second*, to derive from evolutionary theories the main implications for the behaviour of actors and agents in an innovation system, and thus for technology and industrial policies. The general features of NSIs, such as their persistent asymmetry, historical specificity and the multiplicity of different institutional configurations which may be present even in countries with comparable economic systems, can be explained through properties such as path-dependence and multistability, as predicted by evolutionary theories for out of equilibrium open systems. Such properties are not easily accounted for by the world view of neoclassical economics.

It is argued that neoclassical economics, on the one hand, and complexity and evolutionary theories on the other, belong to two different world views. In the former, the Laplacian dream, the world is conceived as deterministic, reversible and in principle calculable. An alternative world view, to which complexity and evolutionary theories belong, stresses uncertainty, qualitative change, irreversibility and path dependence amongst it most important characteristics. This alternative world view has not yet produced a unified, highly coherent and elegantly formalised body of knowledge. In this chapter the contributions of a number of disciplines and research traditions have been examined, paying particular attention to their implications for the social sciences. The disciplines and research traditions analysed have a relationship of hierarchical complementarity, in the sense that some [systems theory and non-equilibrium thermodynamics] predict the fundamental properties and concepts of the others. Biology provides an important inspiration for evolutionary theories of economic and technological change due to their similarity, both structurally and in terms of knowledge goals. Organisation theory has supplied evolutionary economics with a number of very important concepts.

This importance follows from the Schumpeterian emphasis on the role of innovation. Innovations, leading to qualitative change, are not only an epiphenomenon of economic development, but are one of its most significant determinants. In other words, there is a relationship between the composition of the economic system and its capacity to develop. In this contribution, it is argued that the composition of the economic system, which can be represented analytically by diversity, is an important policy concern for the NSI. A number of implications for the balance between old and new activities are discussed, including those for the relationship between routine and research activities

The institutional and organisational dimensions of the NSI are highly relevant in this context. Evolutionary theories can make a very important contribution to the analysis of this problem, although they have not yet bridged the gap between empirical and theoretical analysis of institutions. Networks can be a very useful tool for establishing this connection. Although introduced independently, networks can be justified in terms of evolutionary theories. Networks represent the interactions of individual and institutional economic agents, and constitute the structure of socio-economic systems. Developments in network dynamics should allow us to make progress in the theoretical analysis of organisations and institutions.

Globalisation is an ongoing trend which could have very important implications for the NSI, to the point of making the concept itself irrelevant. In this chapter such implications are discussed from the viewpoint of evolutionary theories. It is argued that globalisation implies a faster rate of diffusion of new practices, but that this is not likely to eliminate the local heterogeneity created by innovation. It is the balance between the rate of innovation and the rate of diffusion which determines the degree of convergence of economic systems, and thus the relevance of the NSI concept. It is concluded that, although globalisation can change the nature of the NSI substantially, for example by adding new international links and by making the systems more interactive, it is unlikely to completely remove national or local specificities. In fact, while a higher rate of knowledge diffusion can weaken some boundaries that hindered flows of goods, services and knowledge, it is likely to be accompanied by a more localised pattern of production and innovation at the regional level.

Appendix A-2.1:
Different Catch-Up Strategies for Countries at Different Levels of Economic Development

V_j/V_w	$p_{ij}q_{ij}/p_{iw}q_{iw}$	Type of Strategy	Further Subdivisions	Conditions for Economic Progress	Comments/ Problems
↑	↑	Creative despecialisation, increasing variety		Always	No problems if required resources, human capital and NSI available
↓	↑	Virtuous specialisation	$P_{ij}{\uparrow}\,q_{ij}{\downarrow}$, high-tech, up-market goods $P_{ij}{\downarrow}\,q_{ij}{\uparrow}$, efficiency, scale economies	Increasing value and efficiency need to more than compensate for fall in variety	Can it be maintained indefinitely?
↑	↓	Vicious despecialisation	$P_{ij}{\uparrow}\,q_{ij}{\downarrow}$, high-tech, up-market goods $P_{ij}{\downarrow}\,q_{ij}{\uparrow}$, efficiency, scale economies	Increasing variety needs to more than compensate for fall in value and efficiency	Problems with Accumulation of human capital

In columns 1 and 2 a larger arrow means a greater change. Thus in the second row the price-quantity ratio increases more than the variety ratio falls. The third column indicates the ways in which the corresponding strategy described in the second column can be achieved. For example, the virtuous specialisation strategy can be achieved either by means of high prices and small quantities or by low prices and large quantities.

3 Innovation Policy in the Systems of Innovation Approach: Some Basic Principles

Charles Edquist
Department of Technology and Social Change, Linköping University

3.1 Introduction

This contribution aims to deal with innovation policy from the perspective of the systems of innovation approach. The intention is to address basic principles related to national, regional or local as well as sectoral systems of innovation and the policy implications emerging from these. Hence, there will be a very strong emphasis on policy in what follows. However, it is essential, of course, that any observations should be adapted to the specific conditions in particular sectors, regions or countries to be fully relevant[1].

3.2 Innovation Policy

Innovation Policy is public action that influences technical change and other kinds of innovation. It includes elements of research and development [R&D] policy, technology policy, infrastructure policy, regional policy and education policy. This means that innovation policy goes beyond science and technology [S&T] policy, which mainly focuses on stimulating basic science as a public good. Innovation policy also includes making use of the knowledge produced – for socio-economic purposes.

It is necessary to go beyond S&T as a determinant of innovation when designing innovation policies, since many innovations emerge outside the formal S&T system – for example, in everyday economic routine activities. At the same time, innovation policy is a part of what is often called industrial policy. But industrial policy is a term that is burdened with a lot of deadwood in many countries because of vain efforts to provide support from public finances for old and dying industries. The term innovation policy is naturally associated with change, flexibility, dynamism and the future. Innovation policy should therefore serve as a midwife, not provide support towards the end of a life.

3.3   Objectives of Innovation Policy

Let us now turn to the objectives of innovation policy. These objectives have normally been economic ones – like economic growth, productivity growth, increased employment and competitiveness. However, innovation policy can also have non-economic objectives, such as social, military or environmental ones. There may be conflicts between the objectives when several are being pursued at the same time. For example, an exclusive emphasis of economic growth might lead to environmental consequences that preclude 'sustainable' development. Or, the objective of productivity growth might conflict with the creation of employment, an issue to which I shall return later.

These objectives are, of course, determined in a political process. However, it is absolutely essential that they are specific and explicitly formulated. Otherwise, the policy may lack a sense of direction and become a victim of pressure groups and nepotism. Afterwards, if the objectives were not specified clearly in the beginning, it will also be impossible to judge whether the policy was a success or a failure. Policy learning, i.e. the possibility that politicians and policy-makers learn from their successes and failures – which is so important – then becomes impossible.

3.4   Reasons for Innovation Policy Intervention

What, then, are the reasons for public intervention in the field of innovation in a market economy? Most innovations are carried out by manufacturing firms. This is quite natural, since firms are most familiar with their own production processes, as they are with the markets for their products. It is often argued that national, regional and local governments should rely on firms to the largest possible extent in the field of innovation.

However, markets are not independent from governments. It is important that governments help to make markets function well. This is because the appropriate balance between the activities of governments and markets in the field of innovation depends on how well the markets function. Governments can make markets operate better by creating laws and rules about competition, contracts, property, patents and so on. However, even if markets function well and companies are innovation-oriented, there are certain things in the field of innovation that firms cannot achieve efficiently, if at all. Therefore, there is a role for government innovation policies in all countries.

One fundamental question this raises is what should be performed by the state or public sector and what should not[2]. In other words, what should be the division of labour between the state, on one hand, and markets and companies, on the other? As I see it, two conditions must be fulfilled for public intervention to be justified in a market economy.

Firstly, the market mechanism and firms must be failing to achieve the objectives formulated, i.e. a *'problem'* must exist. A problem exists when firms and markets do not automatically realise the objectives that have been politically determined. There is no reason for public intervention if the firms and the markets are fulfilling the objectives, i.e. if there are no problems. This is in line with the principle that innovation policy should complement firms and markets, not replace or duplicate them. I will come back to these 'problems' several times later on.

Secondly, the state [national, regional, local] and its public agencies must also have the *ability* to solve or mitigate the problem. If the public sector does not have this ability, there should of course be no intervention, since the result would be a failure. In other words, this condition is an attempt to make sure that political failures are avoided to the greatest possible extent[3].

There may be two reasons why public intervention cannot solve or mitigate a problem. One is that it is simply not possible to solve the problem by political means. In this case, any type of intervention would, of course, be in vain – and the problem would remain. The other reason is that the state might first need to *develop* its ability to solve the problem. For example, a detailed analysis of the problems and their causes may be necessary means of acquiring this ability. Or, new policy instruments might need to be created. For example, the creation of new organisations and institutions to carry out the intervention might be necessary. A patent office is an example of such an organisation and a patent law is such an institution.

3.5   The Role of Different Kinds of Innovation

In mainstream economic theory, the notion of innovation is often assumed to be limited to process innovations. However, the category of innovation is extremely complex and heterogeneous. It is therefore useful to make analytical distinctions between different categories of innovation. Technological innovations are not the only ones that are important for economic growth and employment.

A useful taxonomy is to divide innovations into new products and new processes, as follows:

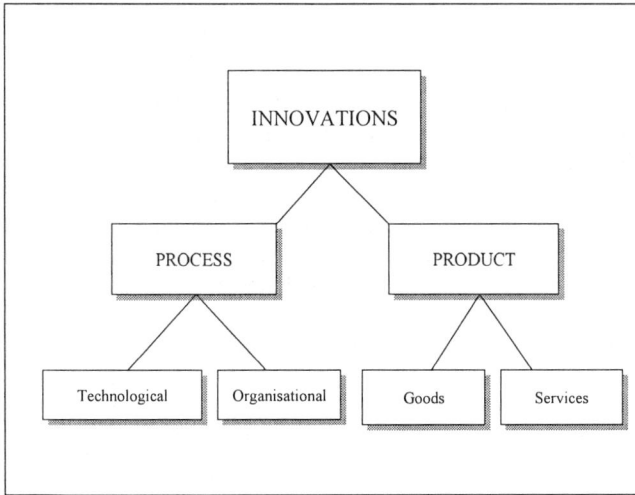

Fig. 3.1 A taxonomy of innovations

Product innovations may concern either goods or services, depending on what is being produced. Process innovations, on the other hand, may be technological or organisational. This depends on how goods and services are produced. Some product innovations are transformed into process innovations in a 'second incarnation' ['second appearance'], but this concerns only 'investment products' – not products intended for immediate consumption. For example, an industrial robot is a product when it is produced and a process when it is used in the production process. Product and process innovations are also closely related to each other in many other ways. In spite of this, it is important to make distinctions between different kinds of innovation, i.e. to disaggregate and pursue the analysis at a micro and meso level. In this taxonomy, only goods and technological process innovations can be considered 'technological innovations' of a 'material' kind. Organisational process innovations and services are 'intangibles'. It is crucial to take these intangible innovations into account as well, since they are increasingly important for economic growth and employment!

The distinction between process and product innovations is particularly important to a policy for increasing employment. The main reason is that – on the whole – process innovations are labour saving whereas product innovations are employment creating. This is true even when counter-acting compensation and substitution effects have been taken into account[4]. Product innovations that neither substitute an existing product nor are later used as process innovations have the greatest positive effect on employment creation. Product innovations are also the main mechanism behind changes in the production structure – another issue that I shall come back to later. It is therefore dangerous to ignore product innovation, as is often done by mainstream economics.

3.6   Systems of Innovation

The 'systems of innovation' [SI] approach is a fairly new approach for the study of innovations. A system of innovation can be defined as including all the important factors that influence the development, diffusion and use of innovations as well as the relations between these factors (Edquist 1997). These factors can be studied in a national, regional or sectoral context. In other words national, regional and sectoral systems of innovation coexist and complement each other. Initially, the SI approach was dominated by a national perspective Later regional studies and sectoral ones became more important.

The SI approach has diffused surprisingly fast in the academic world as well as in the field of public innovation policy-making. The OECD, the European Union, UNCTAD, UNIDO and many individual countries have used it for policy purposes. In Sweden, a Public Agency for Innovation Systems [VINNOVA] was established in January 2001. 'Systems of innovation' are simply at the centre of modern thinking about innovation and its relation to economic growth, competitiveness and employment[5].

3.7   General Policy Implications of the SI Approach

In the SI approach, a long-term perspective is natural and important. This is because innovation processes take time, sometimes decades. They also have evolutionary characteristics, i.e. the processes are often path dependent. It is often not clear – even to the actors involved – what the end-result will be, i.e. which path will be taken over time. The SI approach has adopted this major contribution from evolutionary theories of innovation.

An evolutionary theory of innovation generally contains the following components:
- The point of departure is the existence and reproduction of entities such as genotypes in biology, or a certain set-up of processes and products in innovation contexts.
- Certain mechanisms introduce novelties into the system, i.e. mechanisms that create diversity: mutations and innovations.
- There are also mechanisms that select among the entities present in the system: diversity is reduced. Together, the selection mechanisms constitute a 'filtering' mechanism that functions in several stages and leads to a new set-up consisting, for example, of processes and products. Then the process starts all over again.

Since innovations occur – to a greater or a lesser extent – everywhere in a system, and because of the evolutionary character of innovation processes, an innovation system never achieves equilibrium. We do not even know whether the potentially 'best' or 'optimal' trajectory is being exploited at all, since we do not know which one it would be. This means that the notion of optimality is irrelevant in a system of innovation context. We cannot specify an optimal or ideal system of innovation. This means also that we cannot compare an existing SI with an ideal or optimal one – which is done in the market failure approach in traditional economics.

We can identify two main kinds of policy implications:

- The systems of innovation approach contains *general* policy implications, which can be derived from the characteristics of the approach. They are general in the sense that they are of a 'sign-post' character. They can serve as rules of thumb and point out relevant issues.
- The systems of innovation approach also provides a framework of analysis for identifying *specific* policy issues. It is helpful in identifying the 'problems' that need to be the object of policy and in specifying how innovation policies to solve or mitigate these problems could be designed.

We first address the general policy implications and issues. The SI approach emphasises that innovation is a process that involves more than individual firms and other organisations. Interaction and interdependence is one of the most fundamental characteristics of the SI approach. Let us give an empirical example. In a survey carried out in the region of East Gothia in Sweden, about 50 percent of all firms made a product innovation during the period 1995 to 1997. Of these firms, 76 percent developed the new product in *collaboration* with other organisations, e.g. firms and public organisations (Edquist, Ericsson and Sjögren 2000). This provides a strong support to one of the central characteristics of the systems of innovation approach.

The interactions in a system of innovation take place between the most important components of the system. These are organisations and institutions. Organisations in this context are 'actors' or players with an explicit objective, like firms, universities and government agencies, which interact in the creation of innovations. The framework created by institutions, consisting of laws, norms, routines, technical standards and so on, helps to shape this interaction between organisations. Institutions are not organisations, but the *rules of the game*. Two general policy issues, which emanate from the SI approach, are therefore:

(1) *Organisational actors might need to be created, redesigned or abolished.*
(2) *Institutional rules might need to be created, redesigned or abolished.*

In any system of innovation, it is important from a policy point of view to study whether the existing organisations and institutions are appropriate for promoting innovation. How should institutions and organisations be changed or 'engineered' to induce innovation? This dynamic perspective on institutions and organisations is crucial in the SI approach, both in theory and in practice. Not only organisational change, but also the evolution and design of new institutions was very important in the development strategies of the successful Asian economies as well as in the ongoing transformation of Eastern Europe. Hence organisational and institutional changes are particularly important in situations of rapid structural change.

A general policy implication which derives from the fact that much learning and innovation is interactive, is that this interaction should be targeted much more directly than is normally the case in innovation policy today.

(3)   *Innovation policy should not only focus on the elements of the systems, but also – and perhaps primarily – on the relations between them.*

This includes the relations between various kinds of organisation, but also those between organisations and institutions. For example, the long-term innovative performance of firms in science-based industries is strongly dependent upon the interactions of these firms with universities and research institutes. These interactions should be facilitated by means of policy – if they are not spontaneously functioning smoothly enough. This can partly be done by changing the laws and rules which govern the relations between universities and firms. Incubators, technology parks, and public venture capital organisations may also be important. This means that the public sector may create organisations to facilitate innovation. At the same time, however, it needs to establish the rules and laws that govern these organisations and also its relations with private ones. In other words, the state has a double role.

We argued earlier that the SI approach considers innovation processes to be evolutionary and path dependent. From this follows the danger of negative 'lock-in' situations, that is patterns or trajectories of innovation, which lead to low growth and decreasing employment. This may apply to patterns of learning and production specialisation of firms, industries, regions, and countries. The next general policy issue is therefore:

(4)   *That innovation policy should ensure that negative lock-in situations are avoided.*

What can governments do then to support transitions from dead-end trajectories? The answer is that:

(5)   *Governments should facilitate changes in the production structure.*

In manufacturing industries across the world, there has been a growing divergence between those with high levels of product innovation and those concentrating on process innovation. The former are clearly the 'growth industries', which have experienced net gains in employment. The process innovation oriented sectors of production tend to be 'declining' industries with net employment losses (Edquist, Hommen and McKelvey 2001).

There are three mechanisms through which the production structure can change through the addition of new products:
• Existing firms may diversify into new products [examples are Japan and South Korea].
• New firms in new product areas may grow rapidly [this mechanism is more common in the USA].

• Foreign firms may invest in new product areas [Ireland might be a good example here].

If a time period, country, firm or region is dominated by process innovations, this constitutes a tendency to decrease employment. If product innovations dominate, there is an opposite tendency to generate new employment. So, if the objective of innovation policy is to secure job creation:

(6)   *Governments should support structural changes in the direction of production sectors dominated by product innovations rather that process innovations.*

Such a policy must not, however, be directed towards preventing process innovations. Process innovations are necessary for increases in productivity and they might also provide the basis for product innovations. To prevent them, in the present era of globalisation and competition, would be self-defeating. Firms, regions or countries doing so would be overtaken and go bankrupt.

Those sectors that generate most product innovations are generally the R&D intensive sectors – or high-technology sectors – in manufacturing and the knowledge intensive sectors in service production. These are often the 'new' sectors emerging in our economies. Which means they are increasingly becoming 'learning economies' or 'knowledge-based economies'. The knowledge intensive sectors are of far greater strategic importance for job creation than other sectors, since they are engaged in the creation of new products and new markets. In general, the demand for new products grows more rapidly than for old ones. The implication is that firms, regions and countries producing new products tend to do so for markets that are growing rapidly. Growing markets mean an increase in demand and output – which reinforces the intrinsic employment creation effect of product innovations.

Governments should therefore create opportunities and incentives for changes in the production structure. They should promote sectors characterised by high knowledge intensity and a high proportion of product innovations. Policy issues in this context concern how policy-makers can help develop alternative patterns of learning and innovation and nurture emerging sectoral systems of innovation. A key issue here is therefore the choice between supporting existing systems – with their historically accumulated knowledge bases – and supporting the development of radically new products and sectoral systems. Radical innovations and the emergence of new sectoral systems of innovation seem to be more of a problem for markets and private firms than reproduction and incremental innovation in established sectors[6]. We also know that large-scale and radical technological shifts, i.e. shifts to new trajectories, have rarely taken place in the OECD countries without public intervention. This is true for most of the electronics sector as well as for aircraft and biotechnology, also in the USA. Therefore:

(7)   *Governments should primarily be proactive, supporting the emergence of new product areas and new sectoral systems of innovation.*

This is particularly important in a world characterised by rapid technological change and economic globalisation. Whether it is in a national, regional or local context, the support of new sectoral systems is the key. This means that geographically delimited systems of innovation should be combined with sectoral perspectives. National or regional policies should include sectoral specialisation strategies. It is important to develop geographically defined systems that promote development in new or immature sectors.

These arguments are closely related to the final general policy issue we wish to mention:

(8)   *Governments need to intervene early in the development of product innovations and new sectoral systems of innovation.*

Such intervention at an early stage in the product cycle may have a tremendous impact. In the case of the creation of the NMT 450 mobile telecommunications technical standard in the Nordic countries about twenty years ago, this proved to be extremely important. It was crucial for the emergence of the mobile telephone industry and for the fact that both Ericsson and Nokia became global leaders in this field[7]. On the other hand, there are many examples showing that massive government support to old and dying industries have had limited effect. Often it has only marginally delayed the death of these industries. One example is the Swedish shipyard industry in the late 1970s and early 1980s. The cost of the support to the shipyard industry was several hundred times greater than the cost of developing NMT 450.

The general policy implications of the SI approach are different from those suggested by standard economic theory. This has to do with the fact that the characteristics of the two frameworks are very different. The SI approach shifts the focus away from actions at the level of individual, isolated units within the economy [firms and consumers] towards the collective underpinnings of innovation. It addresses the overall system that creates and distributes knowledge, rather than its individual components. Within these systems, innovations are seen as evolutionary processes.

The general policy issues listed above can serve as sign-posts, suggesting where to look for problems and possible solutions in innovation policy-making. However, they are not a sufficient basis for designing specific innovation policies. These policy issues do not tell a policy-maker exactly what to do in order to improve the functioning of the system. The SI approach as such cannot provide this, but neither can any other approach or theory. Let us take standard economic theory as an example. Market failure analysis in standard economic theory argues that a completely competitive, decentralised market economy would provide sub-optimal investment in knowledge creation and innovation. Firms underinvest in R&D because of uncertainty and appropriation problems. This leads, for example, to a case for public subsidies for knowledge creation, or for the creation of intellectual property rights. This links up nicely with the 'linear model' approaches and economists, and hence policy-makers often consider this to be a justification – or theoretical foundation – for governments to subsidise R&D.

However, the policy implications that emerge from the market failure theory are actually not very helpful for policy-makers from a practical and specific point of view. They are too blunt to provide much guidance. They do not indicate how large the subsidies should be or within which specific area one should intervene. Also, they say almost nothing about *how* to intervene, i.e. which policy instruments should be used. Standard economic theory is therefore not much help when it comes to formulating and implementing specific R&D and innovation policies. It only provides general policy implications, e.g. that basic research should sometimes be subsidised.

3.8   The SI Approach as a Framework for Designing Specific Innovation Policy

The SI approach can be used as a framework for designing specific innovation policies. we shall now, very briefly, outline how this can be done. Previously, we concluded that a necessary condition for public intervention in processes of innovation is that a 'problem', which is not automatically solved by markets and firms, must exist[8]. Substantial analytical and methodological capabilities are needed to identify these problems.[9]

Systems of innovation can be quite *different* from each other, e.g. with regard to specialisation of production, resources spent on R&D, etc. In addition, organisations and institutions constituting elements of the systems may be different. For example, research institutes and company-based research departments may be important organisations in one country [e.g. Japan], while research universities may perform a similar function in another [e.g. the United States]. Institutions such as laws, norms, and values also differ considerably between national systems.

In the SI approach, the considerable differences that exist between various systems of innovation are stressed, rather than ignored. These differences may be between national, regional as well as sectoral systems of innovation. This makes it not only natural but vital to compare different systems. Without such comparisons, it is impossible to argue that one system is specialised in one way or the other, or that a system performs well – or badly. Comparisons are therefore the most important means for understanding what is good or bad, or what is a high or a low value for a variable in a system of innovation. However, since we cannot specify an optimal or ideal system of innovation, comparisons between an existing system and an ideal or optimal one are not possible. The existence of a 'problem' cannot be identified in this way.

The only possible comparisons are between existing systems. The comparisons must be genuinely empirical and very detailed. They are similar to what is often called benchmarking at the level of the firm. Such comparisons are crucial for policy purposes. They can identify the problems that should be subject to policy intervention. However, to know that there is reason to consider public intervention is not enough. It is only a first step. It only indicates where and when intervention is called for. It says nothing about how this should be pursued. In order to be able to design appropriate innovation policy instruments, it is also necessary to know the causes behind the identified problems – at least the most important ones[10].

Within the systems of innovation framework, the identification of the underlying causes of the problems is the same as identifying deficiencies in the functioning of the system. It is a matter of identifying those functions that are missing or inappropriate and which have led to the problem in terms of comparative performance. Let us call these deficient functions 'system failures'. When we know the causes behind a certain problem – for example low economic performance - we have identified a system failure.

There are at least three main categories of system failure:
• Organisations in the system of innovation might be inappropriate or missing.
• Institutions may be inappropriate or missing.
• Interactions or links between these elements in the system of innovation might be inappropriate or missing.

Not until policy-makers know the character of the system failure, do they know whether to influence or change organisations, institutions, interactions between them – or do something else. Therefore, an identification of a problem should be supplemented with an analysis of its causes as part of the analytical basis for the design of an innovation policy. Benchmarking is not enough.

To sum up, concrete empirical and comparative analyses are absolutely essential for the design of specific innovation policies. National, regional, local and sectoral systems of innovation must be systematically compared with each other in a very detailed way. Only in this way can specific innovation policies be designed. In the effort to design innovation policy, there is no substitute for concrete analyses of concrete conditions! The SI approach is an analytical framework suited for such analyses. It is appropriate for this purpose because it places innovation at the very centre of focus and because it is able to capture the differences between systems.

Endnotes

1   This chapter is largely based on two other studies: Edquist (2001) and Edquist, Hommen and McKelvey (2001). Detailed references to the work of others are given in these two publications. Some background references can also be found in Edquist (1997), Edquist and Riddell (2000) and Edquist and McKelvey (2000).
2   The 'state' may be national, regional or local.
3   However, political mistakes and failures cannot be completely avoided. Public actors may fail, just as markets and private actors do.

4 This has been shown in Edquist, Hommen and McKelvey (2001) where the relations between [different kinds of] innovations, [different kinds of] growth and [different kinds of] employment are dealt with.

5 Edquist and McKelvey (2000) is a collection of 43 articles which deals with:
- national, regional and sectoral systems of innovation, including case studies,
- theoretical origins of the systems of innovation approach: interactive learning, evolutionary theories, institutional theories,
- innovations, growth and employment, and
- public policies and firm strategies.

6 If incremental innovation is not a problem for most traditional sectors, then there should be no public intervention there – see Section 3.4.

7 Incubators, technology parks and the financing of new technology-based firms, as well as the creation of standards, are examples of policy instruments relevant for the early stages of innovation.

8 This means that neutral or general policies are normally irrelevant; selectivity is necessary if specific problems are to be solved or mitigated.

9 Such capabilities are also needed to design policies that can mitigate the problems.

10 A causal analysis might also reveal that public intervention is unlikely to solve the problem identified, due to lack of ability on the part of the government.

4 The Globalisation of Technology and the European Innovation System

Daniele Archibugi with Assistance of Alberto Coco
Italian National Research Council

4.1 Introduction

The process of European integration is based not only on a customs union, a common agricultural policy and a single currency, but also on the circulation of knowledge among individual member countries. This concerns not merely the interchange of cultural and social values, but also the belief that economic growth and welfare in Europe is closely linked to the ability to generate and diffuse new technologies. It is therefore not surprising that a major policy concern of governments, business and trade unions is the search for ways to promote scientific and technological activities and to foster innovation in firms.

There are growing signs that Europe in this respect is falling behind both Japan and the USA (see Gambardella and Malerba 1999; Fagerberg, Guerrieri and Verspagen 1999; Vivarelli and Pianta 2000), especially in the most dynamic sectors of the economy. It is increasingly evident that Europe is failing to keep pace with developments in the new economy based on the intensive use of ICTs [Information and Communication Technologies]. For example, while ICT expenditure in the USA and Japan is respectively 8.0 and 7.5 percent of GDP (OECD 1999b, p. 21), in Europe [with some exceptions, such as Sweden, which is over 8 per cent] the figure is below 6 percent. Looking at the share of ICT in registered patents (OECD 1999b, p. 25), this contrast is confirmed. Patenting in these fields in the USA is growing rapidly [average annual rate of growth of 20 percent from 1992 to 1998].

These are just a few of the indications that the United States and Japan have proved far more dynamic in the newly emerging fields and industries. In the 1980s we saw the dramatic rise of Japan and other East Asian economies in hardware technologies linked to ICT (for an overview, see Freeman 1987), and in the 1990s the United States managed to recover its traditional economic leadership in knowledge-intensive industries by exploiting and disseminating ICTs in the service sector. So while Japan and the other East Asian economies continue to have a prominent position in the generation of the 'hardware' component of ICTs, and the USA now has a dominant position in the 'software' component, Europe has a leading role in neither. It is therefore not surprising that this is causing concern and that a substantial part of the European Union budget is being devoted to promoting scientific and technological advance.

The fact that Europe is composed of many member states which retain substantial autonomy clearly poses a problem for the development of an effective innovation strategy. In fact, Europe is characterised by a lack of cohesion in its innovation system. While some regions of the Union are strongly integrated in terms of knowledge transmission, others continue to be peripheral or excluded from the major technology transfer flows. One of the core issues which needs to be addressed both at the national and at the European policy levels is therefore how to integrate the different local and national components into a single innovation system comparable to the American or Japanese system.

In the knowledge generation process, as in any aspect of economic and social life, Europe is not separate from other parts of the world. The dissemination of new ideas, know-how and technical expertise does not respect national frontiers nor the European borders. Although Europe appears to be adjusting with difficulty to the technological landscape characterised by the so-called 'new economy' (see Soete 2000), the forces of globalisation are affecting science and technology along with numerous other facets of life. So how will the dynamics of globalisation affect the European economy with respect to the generation, transmission and dissemination of new knowledge? Will this process make it possible to bridge the gaps between Europe and the USA or Japan? In an attempt to provide some answers to these questions, this chapter[1] discusses the following issues:

- to identify the various meanings of the 'globalisation of technology';
- to report some empirical evidence on various forms of the globalisation of technology;
- to explore to what extent and in which direction the globalisation of technology is affecting the European Union;
- to assess the role of science, technology and innovation policies carried out at the European level for the benefit of European welfare, competitiveness and growth.

4.2   The Globalisation of Technology

Globalisation is one of the most widely used recent neologisms. It has been applied virtually to every aspect of human life (for an overview, see Held et al. 1999). Unsurprisingly, a very large literature has been devoted to the globalisation of technology (for overviews, see Lundvall and Borrás 1998; Archibugi, Howells and Michie 1999). On the one hand, the new technologies are a fundamental vehicle for the transmission of information and knowledge between regions. The existence of the Internet, satellites and new forms of telecommunications make it possible to transfer information at low or negligible costs from one part of the world to another. This is the material condition that allows globalisation in areas as different as finance, production, fashion, the media and culture. At the same time, the production and dissemination of inventions and innovations has become much more global in scope than in the past.

We need to clarify two aspects of the globalisation of technology: the first relates to its time dimension, the second its spatial relevance. Regarding the former, it should be noted that while flows of knowledge among human societies have always existed, they have become much more intense over the last thirty years due to the massive diffusion of new information technologies. Concerning the spatial dimension, it is significant that some regions of the world, notably North America, Europe and Japan, have been far more affected than others (the majority of innovations produced worldwide have been concentrated in these three major regions: 85 percent of patents operative in 1998).

In recent studies, we have attempted to distinguish different aspects of the globalisation of technology with a view to measuring them quantitatively and providing an appropriate policy analysis on each dimension (see Archibugi and Michie 1995; Archibugi and Iammarino 1999; 2000). The taxonomy identifies three main categories:

• the international exploitation of nationally-produced technology;
• the global generation of innovations by multinational enterprises [MNEs]; and
• global technological collaboration.

The aim of this taxonomy is to classify individual innovations according to the main methods used to generate and exploit them. The categories are not mutually exclusive at the firm level[2] since enterprises, especially larger ones, often generate innovations following all the three procedures described. Moreover, although each category is meant to classify the generation of an individual innovation, it also relates to information on how knowledge is transmitted geographically. The three categories are described in Table 4.1.

The first category includes attempts by innovators to obtain economic advantage by exploiting their technological competencies in markets other than the domestic one. We have labelled this category 'international' as opposed to 'global' since the innovation preserves its own national identity, even when diffused and marketed in more than one country. Both large and small firms take part in this form of internationalisation, although large firms are generally better equipped to commercialise their innovative products in foreign markets.

The second category represents the global generation of innovations and includes innovations generated by single proprietors on a global scale. Innovations in this category are not generated in one single country; on the contrary, they receive inputs from different research and technical centres belonging to the same company. Most innovations fitting into this category are produced by MNEs which are ususally large or even giant-sized corporations. For small firms it is rather problematic to generate innovations globally.

Recently, another form of globalisation of innovative activities, falling midway between the two categories described above, has asserted itself. These are the growing number of international agreements between enterprises, often situated in two or more countries, to jointly develop given technological inventions (Mytelka 1991; Dodgson 1993). The need to cut the costs of innovation has created new forms of industrial organisation and proprietary arrangements which are now expanding beyond the technological sphere as such. Both large and small firms are active in this form of transmission of knowledge. Small firms, in particular, can use it as an alternative way of innovating while preserving their separate ownership.

Table 4.1   A taxonomy of the globalisation of innovation

Categories	Actors	Forms
International Exploitation of Nationally Produced Innovations	Profit-seeking firms and individuals	Exports of innovative goods Cession of licences and patents Foreign production of innovative goods internally designed and developed
Global Generation of Innovations	Multinational Firms	R&D and innovative activities both in the home and the host countries
		Acquisition of existing R&D laboratories or green-field R&D investment in host countries
Global Techno-Scientific Collaborations	Universities and Public Research Centres	Joint scientific projects Scientific exchanges, sabbatical years International flows of students
	National and Multinational Firms	Joint-ventures for specific innovative projects Production agreements with exchange of technical information and/or equipment

Source: Elaboration on Archibugi and Michie [1995]

In reality, in adopting this approach, enterprises have imitated a method of generating and transmitting knowledge typical of the academic community. The academic world has always had a transnational range of action, with knowledge being transmitted from one scholar to another, then disseminated, without economic compensation being invariably necessary.

4.3   The Quantitative Significance of the Globalisation of Technology and the Position of Europe

Having asserted that a number of different phenomena come under the umbrella of globalisation of technology, it is important to measure the quantitative importance of each of its main components. This will also allow us to assess

Europe's contribution to the overall process, the impact of each of these forms and the policy implications. This is particularly important as a basis for the design of appropriate policies at the local, national or European levels. It should however be remembered that in the majority of cases the available indicators do not provide full information on the categories of globalisation singled out here. Nevertheless, the evidence discussed is able to give a good idea of the main significance and trends in the globalisation of technology.

International Exploitation of Technology

For a quantitative assessment, it is useful to distinguish the embodied and disembodied components, since both play a crucial role in firms' strategies (Evangelista 1999). The former is captured by traditional international trade indicators, the latter by indicators of the transmission of know-how such as patents, trade of licences, technical assistance and so on.

Table 4.2 recapitulates the evidence on this form of globalisation. Although all commodities include a technological component, some are more technology-intensive than others. From the available classifications, it emerges that high-tech industries absorb more than one fifth of the world trade in manufacturing (Guerrieri 1999; World Bank 1999). This share has more than doubled in the last twenty-five years. The position of Europe in this respect is hardly satisfactory. Its market share [that is, the ratio of national exports to world exports] in science-based products declined from 48.6 percent in 1970 to 33.8 percent in 1995 (Fagerberg, Guerrieri and Verspagen 1999, p. 12).

One indicator of disembodied knowledge is represented by patent statistics and, in particular, international patent flows. Patents are extended to foreign markets to sell products that embody the innovation and also to sell the innovation itself. Each invention is, on average, patented in more than four countries, as shown by the rate of diffusion across countries, that is the ratio between external patent applications [e.g. applications presented by inventors in foreign patent institutions] and resident applications [e.g. those presented by inventors in their own country] (OECD 1999a). There was a dramatic growth in external patent applications from 1990 to 1996, equal to 19.8 percent a year (OECD 1999a). This growth rate was substantially higher than in industrial R&D expenditure, which was 5.4 per cent a year from 1992 to 1997 (OECD 1999a). The trend in external patent applications is therefore not related only to a faster pace of technological change, but also to the increasing propensity to exploit the results of innovation in overseas markets. The markets where pay-off for technological investment is sought are becoming more and more global. From a geographical point of view, US inventors have increased their propensity to extend their patents in foreign countries more than European inventions [9.5 per cent versus 8.0 per cent].

The other side of the coin is represented by the inflow of technology through patents registered in national patent offices. From 1990 to 1996 the annual growth rate of patents applications by non-residents increased constantly: the annual average growth rate for OECD countries is equal to 11.7 per cent (OECD 1999a, table 73). In 1996 Europe as a whole had a dependence ratio [non-resident/resident patent applications] of 1.1 against 1.0 in the United States. In 1998 the ratios became 1.2 for Europe and 0.8 for the USA (WIPO 1998; NSF 2000) showing a substantial increase of foreign dependency for Europe with respect to the USA.

Table 4.2   Empirical evidence on the international exploitation of nationally produced innovations

Indicator	Source	Stock	Trends
International Trade	World Bank [1999]	High-tech exports in 1997 absorb 19.6% of manufacturing exports: 32% in USA, 26% in Japan, under 20% in Europe	Growth of 10% in the world from 1991 to 1997
	Fagerberg et al. [1999] Guerrieri [1999]	Science-based products absorb 21.5% of world trade in manufacturing	European share in science-based products declined from 48.6% in 1970 to 33.8% in 1995
Patents Extended to Foreign Countries	OECD [1999a]	On average 4.3 extensions for each patent in 1996	Annual average growth of 19.8% in the period 1990-96
Inward Flows of Patents	OECD [1999a] WIPO [1998]		Annual average growth of non-resident patents by 11.7% from 1990 to 1996
Patents in the Triad (the US, Japan and Europe)	EPO [1999]		International flows of patent applications between the three blocs increased by 43% from 1997 to 1998
Patents in High-Tech Fields	EPO [1999]	Japan and the USA have a larger and increasing share of patents at the EPO in high-tech fields than Europe	

Finally, it is interesting to look at patents in the fields with the greatest technological component, since patents in high technology fields registered today are likely to become the exports of tomorrow. If we consider the geographical origin of patents in high-tech fields registered at the European Patent Office, the penetration of United States and Japan is more substantial than for patents in all fields. In 1998 US inventors accounted for 36 percent of high-tech patents [against 29 per cent in all fields] and Japan 22 percent [against 17 percent in all fields]. Europe accounts for 50 percent of applications presented at the European Patent Office in all fields, but its share declines to 36 percent in high-tech fields (EPO 1998).

Global Generation of Innovations by MNEs

Thanks to a number of research studies carried out on indicators based on R&D expenditure and on patents, fresh evidence is available on the global generation of technology. As shown in Table 4.3, the foreign R&D in OECD countries is on average equal to 14 percent of the total business R&D (OECD 1999a, p. 39), indicating that R&D is overall less internationalised than production. In terms of inward flows of R&D by MNEs, there are large variations over countries. At one extreme is the Japanese economy, which is the most typical case of an innovation system in which domestic firms concentrate their investment in their home country and the presence of foreign firms is sporadic. At the other extreme is Australia, where foreign firms account for nearly half of total business R&D (OECD 1997).

Table 4.3   Empirical evidence on the generation of innovations by MNEs

Indicator	Source	Stock	Trends
A. R&D Flows			
Inward Flows of R&D by Foreign MNEs	OECD [1997, 1999a, 1999b]	In OECD, average inward flow is 14 % of total business R&D. Foreign affiliates account for 1 % [Japan] to 46 % [Australia] of R&D in manufacturing. Greater R&D intensity in national firms than foreign ones	Significant increase in Europe Increase in USA Moderate increase in Japan
Outward Flows of R&D in Host Countries by MNEs	USA Survey on R&D, National Science Foundation [2000]	7 – 10 % of R&D of US firms is undertaken abroad [1980-96]	US overseas R&D has increased more than domestic R&D
B. Patents Generated in Foreign Subsidiaries by Large Firms			
Outwards Flows	Patents granted in USA by a sample of large firms, Patel and Vega [1999]	12.6 % of patents generated in foreign subsidiaries of large firms [1992-96]; 22.7 % in EU vs 8 % in USA and 2 % in Japan Residents invent 8% of their total owned patents abroad [US 9%, EU 4% and Japan 2%]	Small but constant increase, European firms are increasing the number of overseas inventions both in Europe and the US Growth of 33% from '80s to '90s, approx. 2.9% a year
Inward Flows	Patent applications at the EPO OECD [1999b]	Foreign residents own 8% of total patents invented domestically [EU 7%, USA 5 % and Japan 3 %]	Growth of 33 % from '80s to '90s

In the three main European countries, Germany, France and the United Kingdom, foreign firms account respectively for 16.5, 14.9 and 18.5 percent of the total national R&D expenditure in manufacturing industry [while the bulk is still national]. Greater importance is played by foreign affiliates in Spain (see Molero 1995), where they account for nearly one third of total R&D intensity of the manufacturing industries. Overall, individual European countries are slightly more open than the United States to foreign R&D, while the penetration of foreign firms in Japan is still negligible [1.4 percent of the national R&D of manufacturing industries only].

Is there any R&D-based reason why a national government should prefer foreign firms to domestic ones? Are host MNEs generally keener than domestic enterprises to invest in R&D? One way to check this is to compare the R&D intensity of national and foreign affiliated firms. In the United States, home and host firms have the same R&D intensity. In all other countries, with the exception of Australia, the ratio of R&D expenditure to turnover of foreign affiliated firms is lower than for national firms (OECD 1997).

In the European countries examined, the R&D intensity of national firms [strongly dominated by the so-called national champions] is substantially higher than that of foreign firms. In Germany foreign affiliated firms report a ratio which is almost half that of national firms [it is however significant that in Germany also foreign affiliates have a very high R&D intensity] and in all other countries it is always lower. In other words, there is robust evidence that domestic firms are more R&D intensive than foreign ones.

This does not necessarily imply that R&D location by foreign companies is to the detriment of the R&D investment of national companies. Although foreign firms might be less keen to invest in R&D in the country, there is no evidence that they crowd out the investment of national firms. On the contrary, there is abundant evidence that R&D centres tend to agglomerate. A high level of R&D activity by national firms might therefore induce the location of foreign firms and viceversa (Cantwell 1995). In fact, the R&D intensity of foreign firms is higher in those countries where the R&D intensity of national firms is also high.

Patel and Vega (1999) have produced some evidence, based on US patent statistics, on the share of innovative activities carried out in foreign countries. This shows that European firms as a whole have a far larger share of patents granted by foreign subsidiaries than either the United States or Japan [22.7 percent versus 8.0 and 2.6 percent respectively of the total in a sample of large firms], indicating that large European firms are much more international in the scope of their innovative activities than their American and Japanese competitors.

It is equally interesting to identify the geographical origin of this R&D investment, and as far as Europe is concerned, to discover how much comes from other European firms and how much from outside Europe. In fact European firms tend to distribute their activities between the United States and Europe. It appears that firms with a presence in all major European countries (i.e. excluding the smaller countries such as Belgium, Finland, Austria and Norway) tend to have a higher level of technological activity in the United States than in other European countries. The preference for location in the United States is particularly significant for German firms [14.1 percent versus 6.5 percent of the total patents of German-based large firms] and British firms [38.1 percent versus 12.0 percent of the total patents of British-based large firms].

Global Technological Collaboration

Some evidence based on the available statistics concerning global technological collaboration in the business sector as well as the educational or academic area is given in Table 4.4. For the business sector, we have relied on the classic database developed at Merit by John Hagedoorn and his colleagues (see Hagedoorn 1996). This shows that as much as 60 percent of the total strategic technology alliances recorded are international in scope. This form of generating technological knowledge has considerably increased its significance and the number of recorded agreements nearly doubled between 1981-86 to 1993-98.

The share of alliances taking place within the United States is high and increasing rapidly: 48.3 percent of all strategic technological alliances recorded in the period 1993-98 occurred between American firms, against 24.0 percent from 1981-86 (NSF 2000). Moreover, the US has strong ties on both the Atlantic and the Pacific shores. On the contrary, the share of strategic technological alliances within Europe has substantially declined: they accounted for 19.2 percent from 1981-86 and less than 10 percent from 1993-98 (see Table 4.5). It seems that policies at the European level to foster co-operation in R&D and innovation have not been able to reverse the propensity of European firms to search for American partners. This is a fact that requires further policy analysis [see the next section].

Table 4.4   Empirical evidence on global techno-scientific collaboration

Indicator	Source	Stock	Flows
International Inter-Firm Technical Agreements	Hagedoorn [1996] National Science Foundation [2000]	60% of inter-firm technical agreements are international: 9.9% are intra-Europe, while 28.1% involve US-European alliances	Nearly doubled from 1981-86 to 1993-98 Growth of US-European and intra-USA agreements Decrease of intra-European agreements
Number of Foreign Students Enrolled in Higher Education in Developed Countries	Unesco [1995]	Share of foreign students from 1% to 17% according to country	Increase in absolute terms constant as share of total students
Number of Foreign Post-Graduate Students in the USA	National Science Foundation [1996]	24% students enrolled in post-graduate courses are foreign [1994]	Increase of 4% in a decade
Internationally Co-authored Scientific Papers	European Commission [1997] National Science Foundation [2000]	14.9 of papers at world level are internationally co-authored. Higher in Europe than the US or Japan. Strong intra-European co-authorship	Number nearly doubled from 1986 to 1996 [rise of 6.6% a year] Growth of intra-European collaboration; decrease of US-European collaboration

Partnerships and collaboration promoted by public research institutions and universities play an equally crucial role in the international dissemination of knowledge. Indicators based on the number of undergraduate and postgraduate students and internationally co-authored scientific papers show a substantial increase in the last decade. A dramatic increase in the internationally co-authored papers – also facilitated by the diffusion of Internet and e-mail – is evident in all countries. In just one decade, the percentage of internationally co-authored papers in the world has nearly doubled (NSF 2000).

Table 4.5 Distribution of strategic technology alliances between and within major economic blocs [1981–98]

Year	1981-1986		1987-1992		1993-1998	
	[no.]	[%]	[no.]	[%]	[no.]	[%]
Interregional Alliances						
Eur-Jap	146	7.9	148	5.7	127	3.7
Eur-US	415	22.5	665	25.8	971	28.1
Jap-US	329	17.8	334	13.0	292	8.4
Subtotal	890		1,147		1,390	
Intraregional Alliances						
Europe	354	19.2	449	17.4	344	9.9
Japan	160	8.7	90	3.5	53	1.5
USA	443	24.0	892	34.6	1,672	48.3
Subtotal	957		1,431		2,069	
Total	**1,847**		**2,578**		**3,459**	

Source: National Science Foundation

As shown clearly in Table 4.6, European countries have a much higher rate of international collaboration than the United States or Japan. This is perhaps not surprising given the smaller size of the scientific communities in each individual European country. From a dynamic viewpoint, however, it should be noted that the rate of increase has been higher in the United States and Japan than in Europe. The academic community in Europe is an asset which can and should be exploited to increase the international circulation of knowledge and know-how.

Does the academic community also share the same preference of European firms for American rather than European partners? Table 4.7 gives some data on the distribution of internationally co-authored collaborations across collaborating countries. Europe is by far the greatest collaborator for the American academic community. In 1995-97 as many as 60.3 percent of the US internationally co-authored papers involved a European partner. But European individual countries have stronger ties with other European countries than with the United States. Moreover, comparing the first period [1986-88] to the second one [1995-97], it emerges that, proportionally, intra-European collaboration is increasing, while European-American collaboration is decreasing for all EU member countries.

Table 4.6   Percentage of internationally co-authored papers published in selected countries in all papers

Country	1986-88	1995-97	Growth
United States	9.8	18.0	84%
Japan	8.1	15.2	88%
European Union			
United Kingdom	16.7	29.3	75%
Germany	20.7	33.7	63%
France	22.2	35.6	60%
Italy	24.0	35.3	47%
Netherlands	21.3	36.0	69%
Sweden	24.0	39.4	64%
Denmark	25.9	44.3	71%
Finland	20.9	36.1	73%
Belgium	31.2	46.6	49%
Austria	27.1	43.6	61%
Ireland	28.9	41.9	45%
Spain	18.8	32.2	71%
Greece	27.6	38.3	39%
Portugal	37.6	50.8	35%
World	7.8	14.8	90%

Source: National Science Foundation [2000]

Note: The world totals are lower than those for individual countries because in this case each internationally co-authored paper is counted only once. In 1997 each internationally co-authored paper involved an average of 2.22 countries.

A contrasting tendency therefore emerges: while the European business community has an increasing propensity for technological alliances with US firms, the European academic community has a growing propensity for intra-European partnerships.

4.4   Policy Analysis

Before examining specific policy instruments, it is important to identify the objectives of public policies at the national or European level in relation to each of the three dimensions of technology globalisation singled out above. First of all, there is a crucial dimension common to each of the three components. This can be formulated as follows: 'It is in the interest of a given territorial authority to promote the exchange of both embodied and disembodied knowledge when this offers new learning opportunities'.

The basic assumption of this statement is that the key to a nations' long run economic growth and welfare is increased learning. Globalisation in technological activities provides advantages to individual nations if it allows them to learn (this is the main message emerging from Archibugi and Lundvall 2000).

Table 4.7   Distribution of internationally co-authored papers [1986-88 and 1995-97]

Country	Year	USA	Jap	EU:	UK	Ger	Fra	Ita	Neth	Swe	Den	Finl	Belg	Aus	Irel	Spa	Gre	Por
United States	1986-88		8.2	54.9	12.7	11.8	8.3	5.7	3.4	4.1	1.7	1.2	1.9	1.0	0.3	1.7	0.9	0.2
United States	1995-97		9.6	60.3	12.4	12.8	8.9	6.7	4.2	3.5	2.0	1.6	2.0	1.4	0.4	3.1	0.9	0.4
Japan	1986-88	54.0		33.3	7.0	10.2	5.1	2.1	2.0	2.0	0.8	0.6	1.5	0.8	0.2	0.8	0.2	0.0
Japan	1995-97	45.6		39.4	9.1	9.9	5.7	3.5	2.7	2.6	1.1	1.1	1.2	0.7	0.2	1.1	0.3	0.2
European Union																		
United Kingdom	1986-88	33.9	2.9	46.3		10.2	8.2	6.1	4.8	3.7	2.4	1.1	2.4	0.9	1.6	2.8	1.3	0.8
United Kingdom	1995-97	30.6	4.7	60.2		12.6	10.7	7.8	6.5	3.8	3.0	1.5	3.1	1.4	1.8	5.0	1.7	1.3
Germany	1986-88	31.1	4.1	47.9	10.2		9.5	5.5	5.1	3.7	2.3	1.4	2.8	4.1	0.3	1.9	0.8	0.3
Germany	1995-97	30.0	4.9	58.6	11.9		11.5	7.1	5.9	3.8	2.7	1.7	3.0	4.5	0.5	3.7	1.6	0.7
France	1986-88	28.9	2.7	54.7	10.7	12.5		8.6	4.0	2.9	1.5	0.7	6.0	1.0	0.4	4.6	1.1	0.7
France	1995-97	26.1	3.5	66.7	12.7	14.4		10.1	5.0	3.1	1.9	1.4	6.2	1.4	0.5	6.9	1.8	1.3
Italy	1986-88	35.7	2.0	64.3	14.5	13.1	15.5		4.3	3.9	2.0	0.9	3.6	1.7	0.4	3.6	0.6	0.2
Italy	1995-97	32.6	3.5	76.3	15.4	14.8	16.7		5.7	3.6	2.7	1.7	3.5	2.1	0.6	6.5	1.9	1.1
Netherlands	1986-88	31.1	2.7	73.7	16.4	17.5	10.4	6.2		3.9	2.3	2.0	9.4	1.9	0.7	2.2	0.4	0.4
Netherlands	1995-97	29.2	3.9	85.4	18.4	17.6	11.8	8.1		4.6	3.5	2.3	9.5	1.8	1.0	4.6	1.0	1.2
Sweden	1986-88	36.1	2.7	61.2	12.0	12.1	7.4	5.4	3.7		8.8	5.8	2.5	1.2	0.2	1.2	0.7	0.2
Sweden	1995-97	28.8	4.5	73.2	12.6	13.5	8.8	6	5.4		9.1	7.3	3.7	1.4	0.5	3.1	0.9	0.9
Denmark	1986-88	29.6	2.0	73.2	15.4	14.8	7.3	5.5	4.4	17.2		3.4	1.6	1.0	0.2	1.6	0.4	0.4
Denmark	1995-97	29.0	3.4	94.1	17.9	16.4	9.6	8.1	7.3	15.9		4.5	3.0	2.1	0.7	5.5	2.3	0.8
Finland	1986-88	33.1	2.4	71.3	11.2	14.7	5.6	4.1	6.1	18	5.4		2.6	1.8	0.4	0.9	0.3	0.2
Finland	1995-97	32.1	4.8	89.0	12.4	14.9	9.3	7.1	6.7	17.7	6.3		3.8	2.1	0.8	3.7	2.4	1.8
Belgium	1986-88	25.9	3.0	83.7	11.8	13.8	22.8	7.5	13.7	3.8	1.3	1.3		2.0	0.6	3.3	1.1	0.7
Belgium	1995-97	22.9	2.9	97.6	14.1	14.8	23.8	8.2	15.5	5.0	2.4	2.2		1.4	1.0	5.3	2.1	1.8
Austria	1986-88	25.8	3.0	78.5	8.4	38.6	7.0	6.5	5.1	3.4	1.4	1.6	3.7		0.3	2.3	0.2	0.0
Austria	1995-97	25.1	2.8	82.1	10.1	34.5	8.5	7.6	4.8	3.0	2.6	1.9	2.2		0.5	4.3	1.7	0.4
Ireland	1986-88	22.3	1.8	78.2	42.6	7.9	8.8	4.1	5.2	1.5	0.9	1	2.9	0.9		1.4	0.4	0.6
Ireland	1995-97	21.8	2.5	101.3	40.6	12.3	10.2	7.4	8.1	3.2	2.9	2.3	5	1.5		4.8	1.6	1.4
Spain	1986-88	28.9	1.9	79.6	18.2	12.6	22.5	9.9	4.2	2.4	1.6	0.5	4.3	1.6	0.3		0.3	1.2
Spain	1995-97	25.4	1.9	84.5	16.9	13.1	19.5	11.1	5.5	3.1	3.2	1.6	3.9	2.0	0.7		1.7	2.2
Greece	1986-88	42.0	1.1	69.2	22.5	14.5	14.9	4.4	2.3	3.6	1.0	0.5	3.7	0.4	0.3	0.9		0.2
Greece	1995-97	31.2	2.5	122.1	23.5	23.3	21.0	13.8	5.1	3.9	5.6	4.1	6.6	3.3	1.0	7.0		3.9
Portugal	1986-88	24.2	0.4	88.4	29.3	11.6	20.8	3.4	4.8	1.8	2.5	0.7	5.4	0.2	0.9	6.6	0.4	
Portugal	1995-97	21.0	2.1	126.5	25.9	15.8	22.2	11.5	8.9	5.6	2.9	4.4	7.9	1.1	1.3	13.4	5.6	

Source: National Science Foundation [2000]

Note: Row percentages may add to more than 100 because articles are counted in each contributing country and some may have authors in three or more countries. [Rows show percentage of total coauthorships in the country.]

Although the benefits associated with each knowledge-intensive 'transaction' will not be equally distributed among the participating nations, the aim of public policies should be to involve national economic agents in knowledge exchange. To put it another way, a bad deal is better than no deal at all. In the globalising economy, it is very easy for a firm, an academic circle or an entire industry to become marginalised and excluded from the main knowledge flows. Since the pace of change is so rapid, such exclusion can easily jeopardised the competitive position of the economy.

Nevertheless, a basic distinction needs to be made between cross-border transmission of knowledge which does or does not allow endogenous learning. In the long run, it is not in the interest of a community to systematically acquire knowledge from abroad if the conditions to replicate it autonomously are lacking. This does not necessarily mean that each country should become self-sufficient in the generation of knowledge. No country, not even the United States, is able today to produce all the knowledge it uses. All countries are more or less specialised in selected science, technology and production niches, but none is self-sufficient (for a quantitative assessment, see Archibugi and Pianta 1992). It is in the interest of each country to develop some recognised strengths in certain technology-intensive sectors to compensate fields where the country is dependent on knowledge and technology generated abroad. The main advantages and disadvantages associated with each of the three suggested categories are shown in Table 4.8.

Table 4.8   Impact of the globalisation of innovation on national economies

Categories	Impact	
	Inward Flows	**Outward Flows**
International Exploitation of Nationally Produced Innovations	Low profile of national institutions	Expansion of market and areas of influence
	Low learning in consumption goods	Maintenance of national technological advantages
	Medium learning in capital goods and equipment	
Global Generation of Innovations by MNEs	Acquisition of technological and managerial capabilities	Missed technological opportunities for internal markets
	Increased dependence on strategic choices of foreign firms	Strengthening of competitive position of national firms
Global Techno-Scientific Collaboration	Increase in techno-scientific flows and sources of innovation	
	For developed countries, diffusion of their knowledge	
	For developing countries, acquisition of knowledge and learning opportunities	

Source: Based on Archibugi and Iammarino [1999]

There is also a specific European dimension that should be considered. The European Commission has devoted an increasing part of its budget to research and technological development: from 2.5 percent in the First Framework Programme

[1984-87] to 4.6 percent of the Fifth Framework Programme [1998-2002] (Sharp 2000). However, the resources made available by the European Commission still account for less than 6 percent of the overall European Union's R&D budget. This indicates that a European policy for science and technology cannot rely on the European Commission's budget only, but should also directly involve national governments and authorities.

Policies Concerning the International Exploitation of Technology

The international exploitation of national technological competence has traditionally produced conflicts between governments and firms, as Friedrich List knew very well. Concerning the inward flows of technology-intensive products there is generally low learning in consumption goods, while there is a more significant learning in the import of capital goods and equipment.

It is obviously an advantage for a country to exploit its technological innovations in foreign markets, since this leads to the expansion of internal production and the areas of influence. A large market share, moreover, allows economies of scale and scope and therefore preserves and develops expertise in fields of excellence. There is a long and controversial practice of export incentives and today trade rivalry is gaining importance in technology-intensive sectors at the expense of traditional sectors: agriculture and raw materials are losing importance vis-à-vis electronics and software (see Scherer 1992; Tyson 1992).

International trade rivalry is not only shifting between industrial sectors, it is also changing its nature. There is increasing concern by policy makers for disembodied knowledge. Intergovernmental negotiations and litigation seems ot be more and more related to the violation of intellectual property rights, copyright infringements and similar issues rather than the physical transfer of commodities across borders. In this area, there is a definite need to redefine the rules of the game (see David 1999; David and Foray 1996).

Within Europe, government policies are somewhat limited by the integration acts adopted. The single market should favour intra-European trade and make it more difficult for individual countries to protect their own internal market from other European countries. But the available data show that, inasfar as technology-intensive products are concerned, there is also a strong propensity to trade with the United States and Japan. From this point of view, European policies aimed at creating a European 'technological identity' have, so far, not been successful. This, however, does not necessarily mean that a strategy of European protectionism should be implemented. It is more important to increase European production and expertise in the knowledge-based industries than to limit high-technology imports from other countries.

There is also a strong need to redefine the rules for the trade of disembodied knowledge. One important step would be to agree on a common European patent law and practice. As crucial components of contemporary knowledge, and most notably software, are currently outside the scope of patent protection, the legal framework for intellectual property rights protection in Europe should be substantially wider than provided for by present patent legislation.

While it is relatively easy for large firms to set up international networks to sell their know-how and to acquire know-how from other firms, small firms need support for both the marketing of their innovations and monitoring of international technological developments which might be relevant for their business.

Policies for the Global Generation of Innovations by MNEs

What should the attitude of governments be towards:
- national firms locating their R&D and innovation centres abroad,
- home-based MNEs investing in R&D and innovation at home?

There are both advantages and disadvantages associated with each of these policies. On the one hand, it is certainly an advantage if MNEs hosted in a country invest in innovative projects as they can help to up-grade its technological competence. On the other hand, there is a risk that the activities of MNEs will crowd out national firms. However, the danger of crowding out national firms is more associated with FDI [Foreign Direct Investment] in the country than with the *technological component*. A strong presence of, for example, foreign automobile companies can be an obstacle to the development of a national automobile industry. However, once a foreign company is established in a country, there is no disadvantage associated with technology-intensive activities within the company. Governments might have their own reasons to encourage or discourage FDI but, once FDI is hosted, it is certainly an advantage to adopt policies to foster a strong technological component. The only problem is when a substantial amount of national R&D is carried out by MNEs, as this increases the dependency on the strategic choices of foreign firms, which may have preferential ties with the governments of their home country. Once R&D investment by host MNEs is accepted, there is a wide range of public policies which should be carried out in order to secure the benefits to the nation and the loyalty of foreign firms.

Equally controversial is R&D investment in foreign markets by so-called 'national champions'. On the one hand, this can be seen as a lost technological opportunity for the home country, but on the other hand, it might be an open window into technologically dynamic countries that strengthen the competitive position of national firms.

From the public policy perspective, it is worth exploring the reasons that induce firms to locate part of their R&D and innovative activities overseas. Sometimes this is related to the lack of adequate infrastructure or human resources in the home country [and therefore has direct policy implications]. On the other hand, national firms may wish to keep a window open on the technological opportunities of other countries. In this case, it is important that public policies encourage the dissemination of know-how acquired abroad.

As stated above, it is only large MNEs which generate innovations globally. Small and medium sized firms do not generally use this channel since they do not have the organisation or the financial resources to invest in overseas R&D labs. But this does not mean that small and medium sized firms do not need to acquire technical information from other countries. They might sometimes manage to bridge the gap by using other forms, most notably cross-border collaboration.

It is rather difficult at present to envisage a common European policy for the generation of innovations. Since the European Union is based on the free circulation of capital, each nation competes with the others in order to attract investment. Likewise, nations compete to attract the part of FDI that has the largest knowledge component. So far, government policies vis-à-vis MNEs have been national in scope.

Policies for Global Technological Collaboration

In the case of global technological collaboration, the distinction between inward and outward flows disappears, since each country simultaneously receives and provides some expertise. Of the three forms of the globalisation of technology discussed, this is the most typical example of a positive sum game, as all the members involved can increase their expertise and the externalities associated with it. It is therefore comprehensible that an inter-governmental organisation, such as the European Union, has focussed policy action in this area. Global technology alliances do not provoke direct conflicts among the participating countries, since all of them can potentially benefit.

However, this does not mean that the advantages and disadvantages will necessarily be equally distributed among the participants. As in many marriages of convenience, it may well be that one of the partners obtains greater benefit. Since the learning potential of each partner will be different, the partner with greater knowledge will have more to teach, but will also be quicker to learn from others. Public authorities are not in a position to detect the learning potential involved in each collaboration. Therefore it is important for a country to become a focus of exchange of knowledge and technical expertise rather than attempt to secure positive returns from each exchange. In fact, available evidence has shown that the countries with the highest share of scientific and technical alliances are those with the highest technological potential. This is hardly surprising, since prospective partners are sought among those who have already an accumulated knowledge.

Probably, the most significant policy for fostering cross-border collaboration has been implemented by the European Union. The bulk of the European Union's financial resources for science and technology have been devoted to schemes of a collaborative nature. This has however been combined with the competitive selection of projects. The European Commission selects the best projects through competitive bids from those applicants willing to collaborate with teams in other countries. There is a strong economic rationale for applying a combination of competitive and co-operative incentives of this kind. Firstly, the competitive nature of the selection process allows the funding of the most promising projects. Secondly, the requirement of cross-border collaboration helps to disseminate and diffuse knowledge across regions with a view to achieving greater cohesion.

Despite this, it has been shown that the propensity of European firms for American partnerships has not substantially altered. On the contrary, the share of strategic technology partnerships of European firms in the United States increased in the 1990s, often at the expense of intra-European collaboration. But the case of the academic community is equally significant: it seems that policy efforts have managed to increase the number of intra-European partnerships.

The propensities of the business world described above should raise serious policy concern. Are R&D funds managed at the European level too low in comparison to the R&D funds managed at the national level to provide visible benefits? Should the general philosophy of promoting intra-European business collaboration be revisited? There is a significant opportunity to use the European academic community as a vehicle for greater access to global knowledge networks. This suggests that policies that increase public/business co-operation in Europe might also lead to increase international co-operation among firms.

It has often been discussed whether European schemes to foster collaboration and partnerships should be limited to member countries or should also be open to prospective collaborators from other regions, most notably the United States. The view suggested in this chapter is that the key discriminating factor should be associated with learning. It might be in the interest of Europe to involve and fund the participation of selected non-European partners if this provides additional learning potential.

4.5   Conclusions

The taxonomy of the globalisation of technology proposed in this chapter can help understand the reasons for the gap between Europe and the United States and Japan in the learning economy, as well as inform policy actions. Although the evidence reviewed is fragmentary, a few clear signals do emerge.

- *First*, Europe is not at the core of the globalising learning economy. It is less integrated within the world markets than the United States in key dimensions of knowledge production, transmission and dissemination. Moreover, Europe is losing ground in almost all the dimensions involved.
- *Second*, the analysis of the three categories relating to the globalisation of technology provide some indications about the most effective focus for policy making, especially when this concerns a supra-national level, as in the European case. In spite of a good mixture of competitive and co-operative incentives, the EC policies have not managed to increase the likelihood of European countries to opt for partners within Europe. If we recall that the European Commission's budget for Research and Technological Development is less than 5% of total European expenditure, this is hardly surprising.
- *Third*, it has been shown that the centripetal pull of the American economy has been strongly felt by European business over the last decade. Many European firms have shown a preference to locate R&D and innovative activities in the United States rather than in other European countries. Likewise, they have become keener to sign strategic technological alliances with US counterparts than with European ones. However, it has also been shown that the European

academic community has a stronger and increasing propensity towards intra-European collaboration.

If the European Union wishes to be more fully integrated also in terms of knowledge creation, it would be wise to make a greater effort towards the generation and transmission of knowledge, and also to achieve stronger co-ordination between the various political actors. European countries have joined a monetary union, but are leaving the management of knowledge to the national level. If it is true that knowledge is becoming the driving force of the globalising learning economy, a stronger policy action in this field is needed. Lundvall (2000) has suggested the setting up of a European High Level Council for Innovation and Competence Building with authority and powers comparable to the European Central Bank. This proposal is certainly utopian, but the political actors involved should bear in mind the widespread consensus that we are moving into an age dominated by globalisation and knowledge. Today the management of knowledge deserves the same attention as the management of money.

Endnotes

1   A first version of this chapter was prepared for the European Commission's Directorate-General Enterprises, Expert Group on 'Innovation Policy in a Knowledge-Based Economy' (March 2000). I wish to thank the members of the group and the co-ordinators Robin Cowan and Gert van de Paal for their helpful comments. This work has also been discussed within the STATA Network 'Mesias', co-ordinated by José Molero, and at the International Workshop on *Knowledge, Complexity and Innovation Systems* [Vienna, July 1-3, 2000]. Special thanks go to Marcela Bulcu and Ingrid Divis for their help in the preparation of this chapter.
2   There is an additional category that should be added to this taxonomy, namely the innovations that are generated and used within the boundaries of a state. However, since the purpose of the taxonomy is to describe and interpret the globalisation of technology, innovations that do not cross borders are not considered.

PART B: Knowledge Creation and Spillovers

5 Scaling Knowledge Production: How Significant is the Region?

Arnoud Lagendijk
Faculty of Policy Studies, University of Nijmegen

5.1 Introduction

Territoriality is an important dimension of knowledge production and application. Despite the ethereal nature of the product itself, knowledge is thought to originate from, and be anchored to, particular places. Over recent decades various studies have focused on and explored the spatial dimension of knowledge production. This includes work on national and regional innovation systems, on 'innovative milieu' and spatial clusters, or what Moulaert and Sekia (1999) call 'Territorial Innovation Models'. These studies have shaped a perspective that stands in marked contrast to the notion of 'footloose' knowledge that travels at the speed of light through the world's electronic highways. So, for knowledge production like for other forms of production, space still seems to make a difference.

A territorial unit that has gained much prominence in the debate is the region [as a subunit of a nation] (Fischer 2001). The notion that the region is an appropriate level at which to examine knowledge production is gaining increasing credence. The aim of this chapter is to discuss the relevance of the regional level for knowledge production and to look at the various ways in its appropriateness can be determined. This will be done as follows. The chapter starts by exploring approaches in which knowledge production is functionally related to the role the region plays in the wider ['global'] economic system. In a functional approach, it is the logic of the wider economic system [e.g. 'global market logic'] that primarily determines the role and significance of system parts such as regions. Two variants of this approach will be discussed here, under the labels of 'regional competencies' and 'regional innovation systems'.

In contrast to the functional approach, the appropriability of the region can also be defined in more political terms. Instead of the wider system, regional scaling of knowledge production is then associated with the interests, intentions and positions of particular actors at the regional as well as other spatial levels. Finally, partly building upon, as well as responding to these perspectives, the appropriateness of the region can be interpreted as *socially and discursively constructed*. In this view, it is the story-telling about the position of the region itself that has contributed to the role of territory and the region in knowledge production.

In the following sections we shall explore each of these perspectives, culminating in a comparative overview (Table 5.1). The purpose of this discussion is to provide greater clarity in the debate on knowledge production and territory, as well as to contribute to a non-functionalist approach to regional development, and, more specifically, to the relationship between the region and knowledge development. Such clarification is needed because, as will be argued below, much of the literature in the field pays scant attention to the premises and perspectives underlying the regional scaling of knowledge production. What is also apparent is a tendency to conflate different perspectives. In particular, the more strategic [political] and functional [economic] dimensions of knowledge production are often inappropriately combined and intermingled. There is a need, therefore, to juxtapose different perspective and acknowledge their distinctive impact upon the literature about regional development and knowledge production (MacLeod and Goodwin 1999).

Table 5.1   Characterisation of regions as sites of knowledge production

Site Label	Core Approach	Unit of Reference	Core Theoretical Concepts	Authors
Natural	Competence, Repository of Knowledge	Firm	Resource-Based, Learning Organisation	Maskell (1999) Storper (1997) Malmberg (1999) Lawson (1999)
System of Innovation	'Systemness' of Innovation	National System of Innovation	Technological Paradigms; Evolutionary-Institutional Approach	Howells (1996) Edquist (1997) Cooke and Morgan (1998) Lundvall (1992a)
Political	Governance	Political Agent	Trust-Building, Associational Approaches	Keating (1998a) Lovering (1999)
Social Construction	Discursive and Social Constitution	Nexus of Interaction	Scaling, Networks, Loose Coupling, Anchoring	Keating (1998a) Macleod (1999) Ritaine (1998)

5.2   The Region as a 'Natural' and Unique Site of Knowledge Production

The perspective that has been most prominent in shaping and promoting the literature on 'Territorial Innovation Models' pictures the region as a prime site for knowledge production on the grounds of what is essentially a functional logic, based on socio-economic arguments. Due to changes in industrial organisation, notably the rising importance of networking and learning, the region has come to be seen as a highly appropriate level for knowledge production. More broadly, a shift can be observed in the capitalist system tending towards a 'reflexive' and 'learning' economy. In this context, the region presents the fundamental level for constructing an effective 'supply' architecture for innovation and learning-based competitiveness (Storper 1997). Departing from traditional approaches in which growth was primarily considered as an exogenous factor, recent approaches to regional development are inspired by new economic models that *internalise* processes of cumulative growth on the basis of self-induced productivity increases and innovation (Grossman and Helpman 1991; Echeverri-Carroll, see Chapter 8). More specifically, the resurgence of the region can be attributed to two prime factors.

First, regions naturally present the economic bases for firms that are disposed towards *agglomeration* [clustering], externalisation, and dynamic specialisation based on inter-firm networking (Cooke and Morgan 1998; Piore and Sabel 1984). Such firms derive their learning capabilities and competitiveness from using specific local pools of knowledge and labour, and from close interaction with related firms. The quality of the social environment, and especially the creation of relationships based on trust, are core conditions for enduring economic success. From a spatial standpoint, the region presents a natural site for sustaining the associated political and social processes. The region, in the words of Cooke and Morgan (1998, p. 29), embodies the 'most effective scale at which to nurture the high-trust relations that are essential for learning and innovations'. According to Maskell (1999, p.36) 'being embedded in a mesh of local connections helps firms to survive and thrive', also by reducing complexity and uncertainty, and facilitating the exchange of tacit knowledge.

The second factor is the observed shift in the nature of production and, more fundamentally, of capitalism. Inter-firm linkages have become more important, as part of learning and innovation processes. This is due to increased complexity of production [including services] and the tendency of firms to externalise non-core activities through various types of networking. For similar reasons firms have also become more reliant on the quality and dynamics of environmental factors, i.e. developments in the labour market, institutional factors and technology support (Oinas 1998). The shift in capitalism involves the growth of *meta-capacities*, of complex reflexive organisational capabilities underlying the 'learning economy'. These capacities have not only grown at the level of businesses and business networks, but also at the level of societies placed within a competitive context. The learning economy is therefore, 'an ensemble of competitive possibilities, reflexive in nature, engendered by capitalism's new meta-capacities, as well as the risks or constraints manufactured by reflexive learning of others' (Storper 1997, p.31). These meta-capacities are nurtured through what Storper calls 'relational assets'. The region, in turn, has emerged 'as a site of important stocks of relational assets' (Storper 1997, p.44).

Thus, to summarise, it is in two main respects that the region emerges as an essential site of economic development, and, more specifically, of knowledge production. Firstly, because processes of externalisation and networking have fostered the region as an essential economic base, or 'milieu', for business development. Secondly, the region itself has become the locus of society-based *learning capacities* that help it to improve and sustain its competitive position. Asheim and Cooke (1999, p.152) summarise this line of argument as follows: 'the combination of territorially embedded Marshallian agglomeration economies, the interplay of tacit and codified disembodied knowledge and untraded interdependencies could constitute the bases for a new form of socially created comparative advantage for regions'.

Inspired by the new insights into processes of learning at the levels of businesses, networks and societies, combined with concepts from institutional approaches, various authors have set out to define new concepts of regional development. Storper depicts the region as a complex social world in which *conventions* play a fundamental and distinctive role. Conventions are forms of pragmatic reason which underpin specific forms of collective action, including the shaping of a local 'learning economy'. If they become more manifest, conventions evolve into institutions. Other authors have resorted to concepts from business management literature, notably [core] *competencies* and *resources*. In line with the seminal work of Prahalad and Hamel (1990), the concept of *core competencies* is associated with deeper organisational capacities geared to collective learning, and to aligning streams of production, work and technology. Lawson (1999) recognises most explicitly that this resonance of business concepts in the regional literature amounts to a *competence theory of the region*. In Lawson's approach, both regions and firms can be understood as ensembles of competencies. These competencies, in turn, are emergent features of social interaction. Similarly, Maskell (1999) advocates a resource-based approach to regional development in which regional competitiveness is based on heterogeneous assets rooted in localised learning processes (see also Maskell and Malmberg 1999).

While the terminology and ideas on social interaction, communication and causality differ, there is one aspect in which the literature on regional development resonates with that of business development. In the resource-based approach to business development, resources underpinning competitiveness are characterised by (1) *heterogeneity* [or uniqueness] and (2) *immobility* [due to the difficulty of copying] (Barney and Hesterly 1999). Immobility is attributed, among other factors, to the fact that resources emerge from socially complex settings, characterised by a specific culture and organisational behaviour. The literature on endogenous regional development adopts the same stance towards competitiveness (Amin 1999). In Storper's view (1997, p.170): 'economic viability is rooted in assets [including practices and relations] that are not available in many other places and cannot easily or rapidly be created or imitated in places that lack them'. Being competitive means being able to outrun forces of imitation on the basis of learning-oriented conventions and relational assets. Equally, in Maskell's (1999) view, competitiveness is based on non-imitation, that is, on continuous dynamic improvements outperforming competitors. In other words, regions should prevent specific, localised assets from becoming ubiquitous (Maskell and Malmberg 1999). The tacit dimension of knowledge production is one core element serving this purpose.

Cooke and Morgan (1998) expand the concept of regional assets by focusing, more explicitly, on the level of regional governance. To keep abreast of innovation, regions need to develop *associational capacity*, that is, the capacity to collaborate within ensembles of firms, state organisations and other institutions, and to use knowledge as a strategic resource. Learning-based agglomeration economies, together with associational activities underpin regional 'asset stock accumulation' (p. 210) that is somehow unique and prevents imitation. Such assets thus 'emerge' through social interaction with the help, or to use Cooke and Morgan's word 'animation' of the state. In line with the resource-based approach, it is not so much the pattern but the *process* of competition that counts. Cooke and Morgan (1998, p. 82) thus stress 'the need for vigilance in a world where information is communicated instantly and yesterday's new trick becomes tomorrow's old news' as part of 'intelligent governance' at the regional level.

How should we assess this notion of the region as a 'natural' site of economic development and 'knowledge production'? The various new approaches in regional development theory deserve recognition for the way they have related regional competitiveness to rich, socially-oriented concepts of regional development, and the way they highlight the importance of the relational dimension of production. Recent explorations into the notion of cultural competencies that support the social reproduction of successful place-based communities confirm this idea (Scott 1999). However, the way these ideas relate to the business literature is a cause for concern. The resource-based business approach is essentially geared to explaining the success of the firm in the 'outer' market place, accepting that this underpins the long-term survival of the business firm. The approach can thus be regarded as *structural-functionalist*, that is, focusing on system-given roles and ends, namely the success of a firm in the 'market place' as the wider system [the external environment]. While firms represent social entities and should be approached as such, a functional approach may help to understand their location and performance.

The problem with applying the same idea to regions is that regions, like nations, cannot be compared with firms. Their existence and survival is not intrinsically or wholly linked to how they perform in a wider environment. Regions and nations present complex social and political arenas, in which not only means but also ends of social behaviour are co-determined. Moreover, from an economic perspective, regions and nations do not only embody *production*, like firms, but also consumption. Indeed, a functionalist account that only underpins the production, or more precisely, the entrepreneurial side of the economy [how to compete] may represent, to use Krugman's well-known phrase, a 'dangerous obsession' (Krugman 1994). Even if entrepreneurship at the regional level is considered as more diffused in comparison with single business organisations, as institutional accounts of regional development tend to do, the approach does not escape its structural-functionalist overtone.

The essential problem with the competence-based development approach is thus that, besides its sole focus on production, it does not pose the questions 'for what' and 'for whom'. It takes regional competitiveness as system-given, and hence as an inevitable and unambiguous end. In the new regional approaches much attention is given to social interaction, but this is confined to 'how to compete' and 'how to learn' in order to adapt to what are, in essence, uniform external pressures. Obviously, regions in the contemporary world are confronted with competitive pressures that they cannot simply ignore or evade. But the analysis should not forego the discussion of 'ends'. In their analysis of 'Territorial Innovation Models', Moulaert and Sekia (1999) also detect an uneasy co-habitation of a functional logic and local community aspirations, where the latter receives limited attention. In their view, even if some attention is paid to wider social goals, the analyses tend to be quickly refunctionalised. Learning and 'knowledge production' is framed within the context of creating competitive assets, not of social assets which may for instance be geared to balancing economic and social development and environmental needs. This reasoning may even be prone to a kind of circular logic. The success of regional development is defined and explained in terms of what allegedly successful regions present as their key economic aims and approaches.

5.3   The Region as a System of Innovation

Another approach that has supported the notion of the region as a site of knowledge production stems from the innovation systems perspective. Originally the innovation system approach was developed to support the idea that, rather than towards total convergence, technological progress would still work out differently in different countries (Nelson 1993b; Zysman 1996). Such differences reside primarily in the way essential parts of the national economy work together in a system-like way, from the broader education/training sector and financial system, to elements with a direct impact on innovation, such as research and development, user-producer interactions, and the role of supporting institutions (Lundvall 1992a). In contrast to conventional linear approaches to innovation, this approach stresses the *interactive* dimension of innovation, and therefore its dependence on particular social and institutional set-ups (Fischer 2000). Combining these modern insights into innovation with ideas from economic geography, notably the revival of agglomerative forces, inspired the notion of regional innovation systems, as well as, in its wake, 'learning regions' (Morgan 1997). Note that 'system' now refers to the *internal* make-up of the unit under study, i.e. the region, rather than the *external* environment, as in the previous section.

Besides the emphasis on institutions, the innovation systems approach has also found inspiration in the evolutionary perspective in economics (Cooke 1998; Edquist 1997). Although initial formulation of the innovation systems approach started from a neoclassical framework, the adoption of an evolutionary approach, using Schumpeterian ideas on technological development, provided more insight into the process of learning and the position of business firms. In an *evolutionary approach*, firms are seen as differentiated organisations that develop and learn [as well as unlearn] on the basis of their capabilities and routines. In an innovation systems approach, firms are creative – and sometimes destructive – agents that learn not only from their own activities but also from interaction and confrontation with other firms and agencies. Therefore, the innovation system approach can be considered as complementary to the learning-oriented resource-based conception of the firm discussed above. The complement consists of setting out the relational and social environment in which firms operate, and in focusing on the collective, i.e. system level, processes and outcomes rather than the development trajectories of specific firms.

At the regional level, two conceptual types of innovation systems may be distinguished (see Howells 1996). The first type is a *scaled-down* version of national innovation systems. In the view of Cooke (1998), the region presents a highly manageable level at which to identify the core relationships that make up spatial innovation systems. The region is thus considered as an appropriate level for nurturing the kind of institutions and interactions that support 'collective' or 'systematic learning' and innovation. In the words of Asheim and Cooke (Asheim and Cooke 1999, p.145): 'Based on modern innovation theory it could be argued that firms in territorial agglomerations can develop their own competitive advantage based on innovative activity, which is a result of socially and territorially embedded, interactive learning processes'. The focus on the social embedding of interactive innovation processes has also brought in the notion of the regional 'milieu' – the wider social-cultural setting in which firms and agents operate – as an essential factor in shaping innovative behaviour.

It is in this scaled-down version, inspired by evolutionary approaches, that the combination between the system level and resource-based theory of the business firm has been most influential. Much of the relevant regional literature shows attempts to weave the macro and micro-level together, with the help of intermediary concepts to articulate the meso level, such as innovation networks, innovative milieu, social capital and institutional thickness (Oinas and Malecki 1999). For instance, Cooke and Morgan (1998) argue that, at the regional level, the user-producer system, intermediate institutions and social capital foster associational action within firms and inter-firm networks as well as between firms and their institutional milieu. More fundamentally, the authors seek to develop the 'system' dimension linked to innovation through reflexivity and organisational-institutional intelligence. In a similar fashion, Oinas and Malecki (1999) emphasise the network characteristics of spatial innovation systems, in which interaction is fostered by proximity and socialisation effects.

The second approach attaches a different role to the region. Rather than a scaled-down model of national systems, regions feature as *subsystems*. In Howells' (1996) view, the essential issue for the regional level is how [national] educational, enterprise support and regulatory environments are delivered 'on the ground', which depends primarily on local institutional capacity. Howells' idea of a regional innovation system is as an arena for localised learning, which derives specific advantages from locally rooted forms of [mostly] tacit knowledge, set within a larger - national - context. Regions thus become 'innovation chambers', 'R&D laboratories' or 'business support areas' of a larger system. In policy terms, the region presents an instrumental level to create a more effective enterprise support system through contextualisation and customisation. Not only does the regional level allow closer interaction between firms and support organisation, it can also serve spatial development goals. As part of national policies targeting imbalances in regional development, support programmes can be differentiated and customised between regions, in quantity as well as contents. In this way, spatial policy can be focused on more balanced economic development by selectively strengthening the development potential of regions through nurturing local knowledge production and use, instead of relying on simple spatial fiscal redistribution.

A good example of the second approach can be found in the European approach to regional innovation as manifested in the RIS/RITTS programme (Lagendijk and Rutten 2001). Originating from the RTP [Regional Technology Plan] pilot project developed in 1994, the RIS [Regional Innovation Strategy] and RITTS [Regional Innovation and Technology Transfer Strategies] programmes are a joint initiative by two Directorates General of the European Commission, respectively DG XVI [Regional Development] and DG XIII [Technology]. Although the conceptual background for the RIS/RITTS programmes is largely based on the endogenous development models and innovation system perspective already presented, it is essential to understand that the regional initiatives are intended to serve wider European goals. These goals differ between the two DGs. For DG XVI the prime goal is spatial cohesion across Europe. A core condition for its RIS programme is thus the fact that the support is spatially selective, assisting only those regions where at least 50 percent of the population is eligible for the European Regional Funds [ERDF]. The main aim of DG XIII is the improvement of Europe's innovative potential. The relevant RITTS programme is therefore, not spatially selective, but takes on board regions as 'laboratories' for interactive innovation. Much attention is paid to capturing and exchanging the lessons learnt across Europe, for instance through the obligatory use of consultants.

In conclusion, compared with the competence perspective of the region presented in the previous section, the innovation systems approach tends to focus more on the interaction within regional networks, and how these interactions bring about systemic effects in terms of innovative and competitive performance (see Fischer 2000). Rather than considering the region as a business unit, applying the business concepts of resources and competencies directly to the region, the innovation system approach makes a clearer distinction between concepts pertaining to the business level [like resources] and the regional level [varying from institutions, network characteristics, regulatory elements to milieu and social capital]. Yet, the difference between the two perspectives should not be overstated. While the competence approach is based on the metaphor of the firm, and the second approach adopts a system perspective, both approaches view the region as a unit that fits a certain market-economic logic (Moulaert and Sekia 1999). Like the competence approach, the innovation systems approach focuses solely on the production side of the regional economy, set within the context of global competition. There is still a hint that the region will derive its competitive edge by acquiring some specific, unique position in the global market-place based on interactive learning and innovation.

Moreover, the 'scaled-down' version of the regional innovation system appears to adhere to a functional logic similar to the competence approach. The approach takes the goal of regional competitiveness derived from innovative capacity, as given and unquestioned: the region is subject to structural demands of [global] market logics. Admittedly, this point does not apply to the second approach above. In the innovation subsystems approach, the regional level is mobilised for specific purposes, notably national and international aspirations towards improved business support and spatial cohesion [e.g. by means of RIS/RITTS]. It is only through understanding these goals that the use and implementation of the innovation systems approach can be understood. Yet, it is also apparent in this second case that the region is still approached in an instrumental way. The region is serving goals that are set and defined at higher spatial levels. This introduces the question of how the ends of knowledge production and use at the regional level are actually determined, by whom, and for whom. This political question is the subject of the next section.

5.4   The Region as a Political Site of Knowledge Production

A political perspective on the region as a site of knowledge production and use starts from a very different perspective from the region as a 'natural site' of knowledge production. In the latter case, there is little discussion about the meaning of the region. In the 'natural site' approach, the significance of the region stems from organisational-economic trends and 'global' imperatives. The region has been elevated to a core site of knowledge production in the global economy as a result of the shift to either flexible dynamic modes of production or to interactive modes of innovation, in both cases linked to the notion of social embedding and the development of competence. Through a politically oriented approach, this structural-functionalist view may be avoided. The significance of the region is no longer seen as reflecting any kind of structural imperative or necessity derived from its position in the global economic system, but as the outcome of political processes. In a political perspective, certain groups and actors have *interests* associated with promoting the region as a site of knowledge production. In addition, certain historical and contextual reasons have allowed these interests, at least in recent times, to be effectively pursued in a strategically selective manner, particularly by the shaping of coalitions oriented towards the region. The analysis therefore focuses on examining the actors, interests and factors involved in the shaping of such coalitions. This does not mean that the idea of regions playing certain roles and having certain functions in a wider context should be totally rejected. It only means that such roles and functions are politically shaped, rather than system-based.

We start with the factors that have supported and shaped the growth of regional coalitions. These can be captured by the common denominator of 'regionalisation'. Regionalisation can be understood as a set of pressures and motives that have raised and promoted the position of the region in the politico-administrative environment. According to Wright (1998), the factors and processes underscoring regionalisation vary enormously across time and space. Nevertheless, we can divide these into broad categories, which come down to three broad transformations. The first transformation is a change in the role and position of the state due to ideological, budgetary and organisational pressures. This has resulted in a different approach to the relationship between the state and civil society, as well as to the significance of other spatial administrative levels – international to regional and local. The second transformation is that of the market, notably as a result of processes of privatisation, international integration [such as Europeanisation] and liberalisation. The third shift is related to civil society, which is in search of new structures of governance now that old pillars such as the church, political parties and class are crumbling. Another consequence of societal shifts is the rise of new demands to policy output. This includes economic performance suiting employment as well as consumer needs, environmental issues, and quality and safety of public space. A final issue is the growing disenchantment with politics, in many countries particularly at the central level.

Together, these transformations have created incentives for decentralising political power and, even more so, operational responsibility. They have also induced the growth of new governance structures, notably partnerships between the state, civil society and market actors at the regional level (Healey et al. 1997). Such partnerships play a role, for instance, in the regionalisation and customisation of enterprise support and innovation policies. In recent times, such partnerships have tended to broaden towards regional 'development coalitions', which play a more strategic role in the context of regional development. Such coalitions are comparable with the concept of 'regimes' in urban development theory (Stone 1989). Apart from general processes of devolution, a major factor in the more strategic role of such regional governance structures is the position of more proactive regional development agencies and their capacity to act (Keating 1998b). While many of these agencies were traditionally focused on attracting foreign investors, they have increasingly shifted attention to competence-based development agendas, with emphasis on innovation, SME development and training of the local workforce. The region has thus emanated as a more strategic domain that, through local interaction, networking, and consensus building, is shaping its own capabilities, or 'system of action', to influence regional social and economic development according to its own perceptions and ambitions (Gualini 1999). In the context of larger structures, such as the EU, regions not only present significant political *arenas*, but also, through the proactive role of local governments, development agencies, partnerships, and political agents, influential political *actors* (Keating 1998a).

Understanding regionalisation against the background of broad transformation in the role of the state, society and market should not only be read as a top-down story of decentralisation, devolution and the facilitation of regional institutional change. There is also an important bottom-up dimension to regionalisation (Keating 1998a) which concerns the level of local actors, interests and culture. In many regions, local ambitions towards building stronger regional identities and enhancing territorial solidarity have encouraged the shaping of regional governance structures and regional development agendas. In this context, the political aim to be distinctive and unique, also in economic terms, could be considered as the collective expression of a territorially rooted culture, not just a requirement for competitiveness. Regionalisation is thus linked to the phenomenon of *regionalism*, in which the regional agenda is driven by socio-cultural factors, such as the shaping of a local identity.

In particular, a regionalist perspective may help to throw light on the notion and meaning of regional uniqueness in economic development. In a structural-functionalist reasoning that follows a market-economic logic, regional uniqueness reflects a strategic concept or means of achieving the goal of competitiveness. Oinas and Malecki (1999, p.9) express this logic as follows: 'as regional actors try to create competitive advantage, they actually try to distinguish themselves from others'. So, uniqueness is born out of the economic necessity to be competitive in the wider system. In a regionalist perspective, this logic may actually be reversed. The ambition of regional actors to distinguish themselves from others – a socio-cultural phenomenon – may well be expressed by pursuing competitive advantage based on unique regional assets. Rather than seeing territorially rooted assets being mobilised as part of a competitiveness strategy conditioned by a market-economic logic, the reverse logic stresses the shaping of a regional competitiveness strategy as part of a process of local political mobilisation and identity shaping. So, the regional competitiveness agenda originates from the intention to become a social unit different from others. The issue of 'competitiveness' should thus be reinterpreted as a *means* serving politically shaped local goals rather than an *end* conditioned by the 'global market system'!

The region, however, is not the only level at which interests and goals regarding regional development are defined, articulated and pursued. Admittedly, one can point to cases where the ends [regional distinctiveness] and means [creating competitive advantage] are both regional. Yet, the remarks made above also indicate that regionalisation is a process involving various spatial scales. In particular, the role and interests of the nation state in devolving operational responsibility to the regional level should not be overlooked. In this context, Wright (1998) regards as one of the paradoxes of regionalisation the fact that it is not automatically associated with hollowing out the central state. On the contrary, regionalisation has also been employed by the central state for 'problem dumping', as a convenient transfer of intractable problems and costs, for instance in the area of environmental clean-up and social polarisation. Although no tendencies have appeared so far to decentralise basic social welfare programmes, regionalisation has for instance involved the shifting of responsibilities for the labour market (Fagan 1996). In line with Wright's paradox, regionalisation can thus go hand-in-hand with strengthening of the central state. The main question is this: whose interests are effectively represented and served? More specifically, what are the relationships between the regional, national and international levels?

Taking into account the intentions and actions of actors at different spatial levels, rather than deducing the position and role of the region from the wider system, hints at a non-functionalist approach to regional development. The emphasis on 'regional competitiveness' in such an approach is not seen as the logical response to the recent evolution of the global market system. Rather, 'regional competitiveness', and the associated interest in local knowledge development, has emerged as a dominant concept that coincides with the interests, intentions and actions of actors engaged in regional development. The attention thus shifts from system characteristics to the context in which such actors work, and develop and exchange their ideas. One interesting question, for instance, concerns the kind of perspectives and principles from which new ideas about regional development are developed. To start with 'competitiveness', this term can be associated with the rise of the neo-liberal agenda, with its emphasis on the supply-side of the economy at the expense of issues like distribution and consumption. To give another example, recent interest in the concept of 'development coalitions' to support competitiveness strategies has been inspired by the socio-economic perspectives focusing on the interaction between various types of social actors [state, private sector, civil society]. Another crucial concept, 'devolution', has been promoted as part of the tendency to reduce the role of the central state in supply-side policy, particularly in relation to economic support. Finally, concepts like 'competitiveness', 'development coalitions' and 'devolution' may be linked to endeavours to shape regional identities.

When we take the interests and actions of political actors as a starting point, rather than the global market, a crucial question remains – how have issues like competitiveness and knowledge strategies become so prevalent? Indeed, does the emergence of such dominant strategies not prove that, in the end, the wider system presents a very limited number of solutions from which actors can choose? And will these solutions simply be those that help units such as regions to accommodate to the demands of the wider system, such as the global market? There may, however, be an alternative explanation for the prevalence of certain concepts and agendas. Rather than presenting a uniform response to structural demands, it may stem from the way perspectives and ideas are exchanged between actors at a global level. Here again the issue of scaling is paramount. On the one hand, practical notions and concepts of regional development are shaped and applied in the context of particular regions. For instance, regional actors will develop specific innovation strategies based on, and geared towards, the specifities of the own region. On the other hand, region-specific innovation concepts will be inspired by more generic ideas and concepts that are exchanged between regions, or filter down from national or international bodies (Lagendijk and Cornford 2000). Such generic notions can generally be associated with prevalent development perspectives, such as supply-side or Keynesian orientation. Accordingly, in seeking to understand the role of actors' interests, intentions and positions in shaping regional development concepts, we must take into account both a local and a more 'cosmopolitan' level. We shall explore this issue further in the last section on 'social construction'.

5.5   Intermezzo: The Region as a 'Spatially Reified' Site of Knowledge Production

The first two approaches discussed here, based on the idea of uniqueness and innovation systems, both attach great importance to the region as a spatial unit. This revolves around notions such as the 'regional milieu', 'agglomeration economies', 'institutional thickness', social embedding and the exchange of tacit knowledge facilitated by proximity. Hence space, or more specifically, regional space, plays a substantial role (Fischer 2000). Moreover, even a more politically oriented approach which emphasises how political actors conceive and act on the role of the region, requires the regional level to have some substance. For ideas about regional development to evolve and be influential, the region must somehow facilitate the creation of arenas in and around which political actors interact and build coalitions. But before turning to the last perspective, we present in this section an intermezzo, querying the significance of the region as a spatial unit. The claim put forward here is that the literature tends to overplay the relevance of the region. The critical point is whether substantial support can be found for the notion that the region represents the most suitable level for issues such as the support of relational structures, socio-economic embedding, innovative milieus, associational structures, innovation laboratories, and identity shaping. If such support is found to be lacking, the literature can be said to have fallen into the trap of *spatial reification* – erroneously linking social processes to a specific spatial level or, even worse, of anthropomorphism, i.e. regarding the region in its entirety as a core agent in social processes (Keating 1998b).

In seeking support for the region as a significant spatial unit, we look first at how the literature on the region as a site of economic development (Section 5.2) and innovation (Section 5.3) actually endorses the relevance of this spatial level. In these functional approaches, two concepts stand out: *proximity* and *homogeneity*. The significance of proximity is based on the notion that distance has a strong influence on the shaping of relational structures supporting knowledge production and application. This is illustrated in Italian industrial districts, where proximity has helped to secure and lubricate continuous communication between firms (Sengenberger and Pyke 1992). More generally, according to Maskell (1999, p. 48): 'innovation processes - requiring a high level of interaction, dialogue and exchange of information - may be conducted long distance, but is often less expensive, more reliable and easier to conduct locally'. The main reason for the latter is the need to exchange tacit knowledge. Similarly, Amin (1999), following an institutional view on relational assets, stresses the fact that competitiveness, which is increasingly based on learning and the use of tacit knowledge, is rooted in relations of proximity.

The second argument is that the region presents an entity which is socially and culturally *homogeneous*, often backed by a dedicated institutional-political structure. Such an entity nurtures processes of collective learning through supporting common understanding and the creation of trust. Again this idea has emerged through the work on industrial districts, especially through the notion of spatial embedding of social relations (Harrison 1992). In their key contribution on the role of networks, co-operation, and trust in economic activities, Powell and Smith-Doerr (1994) observe that industrial districts display a form of trust which radically reduces the cognitive complexity and uncertainty associated with most business dealings. However, they also warn that the simple fact of proximity between business actors reveals little about the mode of organisation. So, it is the *specific social fabric* that should be the starting point, not just the mapping of distances.

In more recent work, authors tend to combine the proximity and homogeneity view. Such a synthesis can be found, in particular, in literature inspired by socio-economic ideas. Cooke and Morgan (1998), for instance, put forward three arguments to support the significance of the regional scale. First, because of the importance of interaction and the exchange of tacit knowledge, the backbone of interactive learning process consists of externalised and specialised agglomerations. Second, regionalised social and political capital represent vital components of associational activities. And third, supported by the first two factors, the region has become an important site of 'asset stock accumulation', the shaping of unique resources that help local production networks to keep ahead. The regional scale thus plays a specific role within a multi-level perspective: 'Firms are increasingly forced towards sub-national interaction amongst suppliers and innovation support organisations, especially where tacit knowledge is being exchanged, and pulled towards global, or at least transnational, interaction for learning of a more codified nature, acquisition of more standardised inputs, and, of course, for sales' (Cooke and Morgan 1998, p. 202).

In expounding his 'Regional World', Storper (1997, p.181) also articulates a multi-level perspective in which territorialised 'economies of interdependencies and specificities' are juxtaposed against a global 'flow economy'. He claims that while globalisation is driving the flow economy, territorialisation results from the necessary relations of proximity in learning-based production systems. In particular, regional spaces are suitable for the development and rooting of 'non-cosmopolitan' knowledge, since this kind of knowledge can only be interpreted and applied in a concrete, localised context, and is largely tacit. Because of the difficulty of imitation, non-cosmopolitan knowledge underpins sustainable competitiveness. 'Cosmopolitan' knowledge, on the other hand, can be easily used by agents across the globe, virtually turning it from an asset into a liability. 'Cosmopolitan' knowledge supports quick substitution between firms and territories in the global flow economy, hence eroding competitiveness. In essence, territorialisation is based on an intermingled process of creating rooted resources and agglomerative forces: 'Territorialisation is thus not equivalent to geographical proximity or agglomeration, although such agglomeration may be at some times be the *cause* and at others the *effect* of territorialisation: it is the *effect* where scarcities of key resources such as labour and technology draw producers to a place, and when non-substitutabilities keep them there; it is *cause* when the transactional structure of production draws producers into an agglomeration, and the key dimensions of the production become relation-specific and key to its ongoing efficiencies' (Storper 1997, p. 180, my italics).

What emerges from the above is that the significance of the region is based on a clear distinction between the regional scale and higher spatial scales. The shaping and exchange of tacit or 'non-cosmopolitan' knowledge, as well as the nurturing of associational processes takes place primarily at the regional level. The global scale, on the other hand, presents the system level at which the competitive game is played out. Admittedly, Storper (1997, p.71) somewhat qualifies the association of 'non-cosmopolitan' knowledge and the regional level: 'Non-cosmopolitan knowledge[...] can be 'localised' in a restricted technological, organisational, or professional space that is, in certain interpretative networks that transcend local geographical space', but, in Storper's view, this is confined merely to a small number of incidences. Fundamentally, differentiation in resources and conventions arises most markedly at the regional level, between the 'localised conventional-relational' worlds of regions (Storper 1997, p.48). While Storper acknowledges that not all regions manifest successful cases of 'localised conventional-relational' worlds, he sees the region as the prime level at which such worlds emerge.

The critical point arising from this debate is the connection between [tacit] knowledge production and the regional level. Is Storper right in signalling that most networks producing and using 'non-cosmopolitan' knowledge are rooted in regional worlds? Is the exchange of tacit knowledge effectuated most effectively and cheaply at the local level? Obviously, many examples can be found to support the significance of regions, from industrial districts, high-tech areas to specific cluster developments. Indeed, the literature referred to above is replete with such examples. However, it is equally easy to find networks that span countries, continents or even the globe, that meet similar criteria. Research by Cornford, Naylor and Robins (2000) on the new computer and video games, for instance, shows how important national and international meeting points and exchanges are in disseminating core knowledge between producers in different territories. Also in many other sectors, such as pharmaceuticals, aerospace, and computer programming, one can observe the exchange of tacit knowledge across the globe, through dedicated exchange channels and meeting points. Another example even closer to home, is that of academia. Academic conferences and other information and exchange channels play a vital role in exchanging both codified and tacit forms of knowledge vital for the development of individual academics and research centres. Even Asheim and Cooke (1999, p. 156) thus warn that 'the claim for the superiority of tacit knowledge could lead to a fetishisation of the potentials of local production systems.'

Obviously, in all these cases one could point at the fact that many innovative activities are concentrated in regions. The question is to whether the conventions and institutions that underpin these innovative activities should really be regarded as part of regional worlds, or whether they should be linked to the relevant professional networks and market areas. To what extent do the international markets and networks of computer programming, for instance, embody sector-specific conventions, and to what extent are they territorially differentiated? Various authors have indeed questioned the strong emphasis on the *territorial* differentiation of learning-supporting conventions stemming from the 'necessary relations of proximity'. In his extensive study of UK engineering companies,

Alderman (Chapter 11 in this volume) shows that the business environment is more than just regional, especially where it concerns more crucial forms of knowledge exchange. This leads to a rejection of a competence approach at the regional level. In Alderman's view, the innovation culture within an engineering establishment results not so much from the local environment as from the internal [corporate] culture of the enterprise. This evolves by means of constant interaction between the company and its environment in different forms [through the market, suppliers, etc] at various spatial levels. The work of Alderman, among others, has led Malecki, Oinas and Park (1999, p. 262) to conclude that: 'strong local networking may be less common - and necessary - than had been thought'. Hence, space is not necessarily a determinant factor for business, but a variable of a more endogenous nature that can be managed and manipulated in many ways. This clearly invokes the more politically approaches presented in the last section.

Another challenging contribution to the debate, notably on the theme of [non-] cosmopolitan knowledge, comes from the work of Kanter. Rather than nurturing non-cosmopolitan knowledge, Kanter (1995, p. 22) advocates the development of a class of 'cosmopolitans' defined by their 'ability to command resources and operate beyond borders and across wide territories'. The task of cosmopolitans is to tap knowledge and scan market developments across the globe. Competitiveness, in Kanter's view, is based on local learning triggered by *world concepts*, i.e. the standards, practices and innovations that define success in global markets. Non-cosmopolitans are those who reject world concepts because they are thought to damage local identity and options. Global connections are thus essential, not only for learning, but also for supply chain development. This is illustrated as follows: 'Proximity has not disappeared as one of the criteria for suppliers; but it has been joined by so many other criteria that it no longer confers an automatic advantage' (Kanter 1995, p. 97). Space, accordingly, is not only an endogenous variable, it is also just one variable among many.

Focusing on the role of cities, Crevoisier (1999, p. 67) also finds that proximity and agglomeration do not necessarily coincide: 'explanations of technological change in metropoles highlight the advantages of proximity but fail to specify exactly why and how proximity would serve innovating actors in the city specifically'. In Crevoisier's view, urban agglomerations represent sites of social-economic interaction and learning. Such sites are institutionalised and materialised in trade fairs and other meeting places. It is essential, however, to consider how these sites also form part of wider spatial networks. The significance of the local as exchange point is also stressed in the work of Echeverri-Carroll and Brennan (1999, p. 30). These authors explicitly set out to test the proximity thesis, which assumes that intellectual breakthroughs will generally only travel over short distances. The results show a mixed picture. Regions and cities holding core positions in knowledge accumulation, such as Silicon Valley, confirm the strong relationship between knowledge production and proximity. However, in other places it is accessibility that counts most, endorsing the significance of 'cosmopolitan' relations.

Pointing out that knowledge production is not significantly bounded by proximity is an important step, but represents only half of the story. A serious shortcoming of the studies referred to above is their failure to provide details on the process of knowledge accumulation over distance [for more details, see Chapter 6 in this volume]. This, indeed, seems to present a more general gap in innovation research. While much research has focused on the local level, less attention has been devoted to learning and innovation over distance (Oinas 1998). Such research should be combined with an exploration of the social and cultural dimensions of long-distance networking between different types of professional and knowledge communities, paying much attention to how space is endogenised and manipulated through social and political action.

From a regional perspective, a relaxation of the 'proximity imperative' has important consequences for the perception of the region in processes of knowledge production. The emphasis shifts from internal interaction and 'asset stock accumulation' to the capacity to create external linkages, and to trace and absorb external knowledge (Lagendijk 1999b). In this view, the region presents just one scale at which knowledge production occurs. It is not only at the regional level that 'worlds of production' emerge. Similar processes take place at other spatial scales, including global networks. Moreover, such a shift in emphasis also has implications for what the regional scale stands for. The region as a site of knowledge production does not refer to the region in its entirety. Indeed, while there may be cases, such as the emblematic industrial district, which come close to the idea of an organic whole where the social fabric is closely interwoven with the core industrial activities (Lagendijk 2000), in most regions interactive processes of knowledge production are localised in specific networks, which are nested in networks of information exchange at higher spatial levels. Collective learning is thus a process that takes place *within* regions, but is not generally an attribute *of* regions. To what extent other sections of the region benefit from the economic success achieved by collective learning depends on regional socio-economic factors and political intentions (Hudson 1999).

What emerges from this discussion is that in relation to the both questions, of the significance of the regional scale and the 'proximity imperative', the jury is still out. There is no doubt that certain 'conventional-relational' worlds supporting collective learning can be localised at the regional level. This includes the well-known stories of successful regions in both high-tech as traditional economic activities. The question remains, however, to what extent the emphasis on the regional scale is due to a neglect of the development of 'conventional-relational' worlds at higher spatial levels at which collective learning takes place. Another question concerns the kind of factors to which the significance of the regional level should be attributed. Is it the social nature of innovation that demands close interaction, and thus requires proximity? Or is it a process of deliberate scaling of innovation processes at the regional level, induced by political processes of regionalisation and downscaling of innovation policy, by keen management? In other words, is the emphasis on the regional level the result of a social and political construction in which *proximity* is one of the components, and therefore part of a process of management and manipulation of space? This perspective is the subject of the last section.

5.6   The Region as a 'Socially Constructed' Site of Knowledge Production

In many ways, the term 'social construction' has become an concept which is overused and somewhat slippery. Yet it may still serve to draw a distinction between the more objectivity-oriented, essentialist approaches on the one hand and the approaches that stress the social underpinning and historical specificity of social phenomena on the other. In the context of regional development, social constructivist approaches have been employed in a variety of ways, supporting some of the perspectives already mentioned as well as providing new insights into the meaning of space and scale. It is especially the latter that is of interest here.

A social constructivist perspective of the region is not a denial of the constraining and enabling effects of spatial co-location and proximity. While the physical aspects of space, and the kind of dependencies they produce, should not be disregarded, the social constructivist perspective views the concept and contents of the region as produced essentially by the social interaction between actors based in and around the region.

Regions do not represent natural units, except perhaps in the case of some islands. According to Keating (1998b), there is no universal territorial 'oneness'. A region represents a nexus of interaction and is, to some extent, internally coherent because of the nested position it holds in national and global levels of territory. The significance of a region is not the structural outcome of a given internal coherence and uniform set of external pressures, but of social and political processes. From an institutional angle, regions can also be understood as institutional ensembles. Such ensembles present the framework of political action, and temporarily stable outcome, of a plethora of regional actors and organisations. According to the urban regime approach, it is the interests, perspectives and actions of the actors making up the dominant regional coalitions, as well as their confrontations with the interpretations of other regional actors, that determine what is understood as a region and how a region is 'enacted'. This leads to the notion that the shaping of regional perspectives and development strategies is not dependent only on the mobility of capital and other phenomena of globalisation. Rather, the significance of the region and direction of regional development strategies are determined by the way these and other factors are represented and mobilised by strategic actors in the region (MacLeod and Goodwin 1999).

Obviously, many of these issues can be recognised in the approaches discussed above. For instance, Maskell (1999), in his work on interactive innovation, states that clustering and co-location of innovative agents lead to a socially constructed framework, including a sharing of culture, routines, and conventions. The result is a local 'club' of rivals that, through collective tacit learning, create unique collective assets, like 'club goods', which are difficult to imitate. The work of Storper referred to earlier follows a similar line of reasoning, through its emphasis on reflexivity and the territorial embedding of relational assets and learning-oriented institutions. An essential feature of his approach is the emphasis on relationships in proximity and how these are mediated by, and contribute to the development of conventions, i.e. to a shared local culture and circulation of knowledge.

Proximity is not just a physical attribute, but a conceptual creation drawing on a collective social basis. In other words, regions represent arenas for defining social constructs of a relational nature (Gualini 1999). The resulting organisational form is a 'social construction marked by relations of trust between individuals and/or tensions between social groups or divergent interests' (Dupuy and Gilly 1994, p.5). Local innovation systems are thus socially constructed entities created on the basis of physical, organisational and institutional proximity.

Yet, while great efforts are made to explain the basis for regional economic success in terms of social interaction and institutional development, the meaning and significance of the regional level itself is frequently not questioned. Interactions within the region seem to receive primary attention at the expense of relationships and processes that unfold at other spatial levels. The richness of the analysis of regional processes is thus qualified by the way the regional scale and role are treated as ontologically given as a core setting for social relations. Indeed, a similar problem can be observed in the case of urban development theories for urban regime theory, in which, according to Macleod and Goodwin (1999, pp. 702-3), 'the accent placed on alliances *internal* to regime often leads to a neglect of the determinate social forces operating beyond a particular urban scale.' The danger thus lies in being trapped in a reductionist picture, in which the region is spatially reified and even endowed with the characteristics of a single agent, a phenomenon described by Keating (1998b) as *anthropomorphism*.

The upshot is that, to understand the position of the region as a site of economic development and knowledge production, it is not only the processes within the region that should be addressed, but also the shaping of the region itself as a relevant level for action. Like spaces, geographical scales are actively constructed and produced through social processes including socio-political struggle and the enactment of particular 'representational practices' (MacLeod and Goodwin 1999, p. 711). In particular, processes of regionalisation as well as territorial integration at international levels [e.g. EU] should be seen as part of 'scale politics', that is, the deliberate scaling up and down of socio-economic and political processes to pursue specific interests and mobilise particular resources. One example of such scaling strategy is the devolution of capital-labour regulation from the national to lower spatial and sectoral levels, to restrict the impact of national bargaining (Swyngedouw 1997). Another example is 'problem dumping' as part of the regionalisation processes mentioned earlier.

Bottom-up initiatives to 'close ranks' and establish dominant coalitions, and development agencies at the regional levels can also be viewed as manifestations of 'scale politics'. Forging alliances at the local level presents an effective way for economic and political actors to mobilise resources and pursue common objectives. According to a social constructivist perspective, regions represent the embodiment of potential resources and potential political work (Ritaine 1998), relevant to regional ['bottom up'] as well as supra-regional ['top-down'] scales. In many cases, a scalar perspective will be most appropriate for understanding the position of the region. This does not only include a combination of processes at different spatial levels, such as the creation of development coalitions to achieve local ambitions, in line with the EU objective to strengthen and mobilise Europe's regions as against the nation state. It also involves a complex process of up- and down-scaling resulting from the continuous drives for [dis]empowering and [de]mobilising resources at different spatial levels. In research terms, this means that the views of regional actors about regional development always need to be assessed in the context of their relationships, practical as well as mental, to other spatial levels.

This confluence of the internal socio-economic and political conditions with the external position of the region is clearly illustrated by the current shaping of regional development coalitions and strategies (Keating 1998b). As a result of a shift from broader social to specific pro-business attitudes, at least in the public discourse, development strategies and coalitions are being increasingly framed in the context of the regional desire to be 'competitive' in the global market place. The discursive constitution of this 'local coalescing against external threats' plays an important role in promoting the business interests in regional strategies and institution-building. More essentially, it also serves in a self-reinforcing manner to advocate a business logic in regional development perspectives, pitting the specific area-bound rationality of locally rooted and networked business against the placeless rationality of global flows of business, trade and finance. This is at the expense of other, more socially oriented strategies. The shift to pro-business attitudes, according to Keating (1998b, p. 150), 'in turn restructures the arena in a way favourable to capital rather than labour. In the context of interregional competition, business is also able to make universal claims to embody the general interest in growth, while labour is left making particularist claims to a share of the social product'. In particular, the universal claims made about growth and competitiveness, which have resulted in the dominance of pro-business concepts of regional development, have played an essential role in the social construction of regions as well as regional development perspectives and practices.

The emphasis on the shaping of regional development concepts has drawn attention to the role of another community – the research community. As illustrated above, regions have become an important object of inquiry and also linked to issues of economic development and innovation. Indeed, much of the knowledge supporting the idea of the region as an important site of knowledge production has been shaped in the context of continuous interaction between academics, consultants, development practitioners, civil servants and politicians at different spatial levels [regional, national, international, notably EU]. This represents an extensive social world of knowledge development and exchange, through research and policy networks, professional communities and associations [such as the European Association for Development Agencies and global organisations disseminating 'best practice' on regional learning etc.] as well as innumerable seminars and conferences (Lagendijk 1999a; Lagendijk and Cornford 2000). In other words, the knowledge of regional development, and the perspective it produces about the role of the region, is also a social construction (MacLeod 1999). It is perhaps ironic that, whereas the shaping of new academic perspectives on regional development has undoubtedly been part of a multi-scalar process – involving local, regional, national, continental, and global levels – the ideas produced tend to be uni-scalar. What is also telling is how the academic discourse appears to have contributed to the shift towards pro-business and 'competitiveness' approaches, partly as result of its close affiliation with policy-making and the ability to translate local development phenomena into universal observations and claims (Lovering 1999).

A social constructivist view of the region also has consequences for thinking about regionalisation. Rather than representing the outcome of economic and political processes, regionalisation drives the reasoning, interests and coalitions that favour regionalisation (Wright 1998), in practice as well as conceptually. Consequently, as part of a social construction process that has made regionalisation a dominant concept, regionalisation has become a self-reinforcing phenomenon: the concept feeds the practice of regionalisation, and vice versa. Regionalisation is thus not a process to grasp and exploit given opportunities offered by 'associational economies' germane to the regional level. On the contrary, Wright (1998) point outs that, rather than *responding to* the disposition of competence, regionalisation *exposes* regional competence problems. This involves, in particular, competencies directly related to the shaping of associational structures, such as durable and legitimate leadership, financial resources, budgetary and technical expertise, knowledge of key networks, local entrepreneurial expertise, adequate public infrastructure, and a regulatory and social environment suitable for public-private alliances.

The significance of the constant interaction between conceptual development and the practice of regional development shows once more how wary one should be of strong logics - whether economic or political - that support the pertinence of the regional scale. The logic of regional development, and the way it bears on innovation processes, is in no way the determinate or natural outcome of wider social, economic and political processes. Rather, the region should be seen as a context which, by nesting itself in a complex array of scalar processes, provides an anchor for specific development and innovations processes. Such anchoring should not be understood in terms of universal processes of embedding, but as a specific process of local network creation in the pursuit of specific goals of regional development [economic growth or restructuring, employment, technology absorption etc]. It is the discursive and social construction of the region as a platform for social exchange that shapes the region as a nexus of interaction. Such a construction is based not on any substantial advantages inherent in the regional scale but on the strategic intent, of regional actors for instance, to create a regional nexus serving local socio-economic ambitions.

So, the question is not how regions can *apply* the knowledge captured in 'Territorial Innovation Models', it is how regions can create a learning environment that, through social interaction and institutionalisation, will shape local ideas and actions favouring economic development. The most universal claim that can be made on the basis of this discussion is that such an approach requires a multi-scalar perspective, based on ties and interaction at various spatial levels. Using network terminology, Keating (1998b, p. 75) thus advocates the importance of 'loose coupling' and the interweaving of local and global networks, instead of 'persistence of an over-enclosed system of collective social order in which micro-constitutional regulation is too exclusive and localised learning is overemphasised'.

5.7   Outlook: The Region as a Site of Policy Innovation

The last section can be brief, since the line of argumentation was presented in the introduction and the conclusion of the argument has already given in the previous section. This chapter has presented various perspectives on the significance of the region as a site of economic development, and, more specifically, knowledge production. The results of the investigation are summarised in Table 5.1. The discussion has tried to indicate that all approaches contain valuable insights and concepts, from the role of interaction and competencies in innovation to the value of associational processes and governance structures. However, it has also made a case for an approach that is multi-scalar and non-functionalist, rejecting the idea that the significance of the region reflects a natural outcome of universal economic and political processes.

Regional learning, consequently, should not be approached in an instrumental way, that is, as a means to achieve system-given goals of innovative capacity and competitiveness. Nor should regions be used as innovation 'laboratories' building on assumed 'associational powers'. This also bears on the role of policy. Policy must not be seen as something facilitating and supporting learning for predefined goals at an ontologically pre-given scale, but as an integral component of the learning processes. There is a need, accordingly, for a more social and policy-oriented approach to regional learning and for 'territorial innovation models' that include concepts of collective goal setting and scaling. A competence perspective, of the kind presented in the first sections, may help in developing such an approach. However, rather than seeing these competencies – and hence also *conventions* – as directly linked to regional competitiveness, they are seen as part of regional processes of social interaction and institutionalisation supporting regional learning and policy innovation. It is time now to elaborate a more social and scalar perspective of regional learning, in line with the work of Keating (1998a, 1998b) and Moulaert (Moulaert and Sekia 1999).

Acknowledgement. The author would like to thank Elsie Echeverri-Carroll and Olivier Kramsch for their critical and thoughtful comments on earlier drafts of this chapter.

6 Knowledge Spillovers in a Spatial Context – A Critical Review and Assessment

Charlie Karlsson and Agostino Manduchi
Jönköping International Business School, Jönköping University

6.1 Introduction

Since the pioneering works of Schumpeter (1934, 1942), a considerable amount of research has investigated the relationship between the pace of innovative activity and the pace of economic development. Although increases in the stock of knowledge – as well as the constantly improving possibility of exploiting any *given* stock of knowledge - have been widely regarded as the driving force of economic growth in developed countries (Kuznets 1966; Teece 1981), economists have only recently made serious efforts to model the process of knowledge creation. Many recent contributions to economic growth theory have however remarked that although we generally model the level of technological knowledge as a public good, with no restrictions imposed on the spatial extension of the spillovers associated with it, the stocks of technological knowledge do differ significantly across countries and regions (see for example Romer 1994).

Why is this so? What follows from the fact that a desirable good can at the same time be non-rival and, to a large extent, inaccessible to certain countries and regions? Conventional growth theory gives no good answers to this question. A growing body of literature suggests however that phenomena like these may be explained by economic geography. In his presentation of economic geography, Krugman (1991b, 1995) states: 'What is the most striking feature of the geography of economic activity? The short answer is surely concentration ... production is remarkably concentrated in space'. The implications of the concentration of economic activity for economic growth have recently been studied by, among others, Krugman (1991a, 1991b) and Lucas (1993). The costs associated with transactions over space tend to increase with distance and the other factors that can limit the possibility of effectively communicating for agents located in different regions (Krugman 1991b). 'Spatial frictions' of this type can in turn influence the accessibility of knowledge and the rate of economic growth in each region (Andersson and Mantsinen 1980). These models thus link increasing returns to scale resulting from externalities within a geographically bounded region to higher rates of growth. One important source of such externalities is represented by the knowledge spillovers resulting from R&D and other economic activities.

If we assume that knowledge is not freely accessible at every point in space, the location of knowledge production and the character of knowledge spillovers become key explanatory factors for the different pace of economic growth across countries and regions. Over the last 15 years, considerable attention has been given to the potential role of knowledge spillovers in determining both the endogenous character of growth and the patterns of trade. Thus, for example, Krugman (1987) and Young (1991) examine the consequences of the spillovers associated with learning-by-doing, whereas Grossman and Helpman (1991) analyse the consequences of the spillovers resulting from R&D activities. It has been shown that assuming knowledge spillovers of this type can have dramatic effects on the equilibrium pattern of trade and production. The standard Heckscher-Ohlin framework has a unique equilibrium. On the other hand, the new models typically feature multiple equilibria, with different consequences for the relative welfare of the trading countries. In order to obtain such multiple equilibria, one must generally assume that knowledge spillovers are intra-national in scope. It has also been shown that the importance of knowledge diffusion tends to increase as regions become more integrated, due to interregional trade in goods. The intensity if the interregional spillovers can have very important welfare implications. Thus, for example, free trade in goods may harm the development in a region if the interregional knowledge diffusion is not intensive enough (Bretschger 1999).

The latest developments notwithstanding, the literature on knowledge spillovers is still, in many respects, unsatisfactory. The knowledge concept used is seldom defined. Knowledge spillovers are often dealt with by imposing highly restrictive and arbitrary assumptions; furthermore, the exact mechanisms through which the spillovers take place are not discussed. While the fact that new economic knowledge spills over is uncontroversial (Wheeler and Mitchelson 1991), substantial disagreement exists regarding whether such knowledge spillovers are geographically bounded or not (Audretsch and Feldman 2000). Furthermore, in those studies that deal with knowledge spillovers in a spatial context, the spatial element is seldom modelled in a satisfactory way and often suppressed. This in turn makes it difficult to draw precise conclusions as regards regional technology policy from the existing theoretical and empirical literature. This state of the art is especially unsatisfactory in an era of rapidly growing knowledge production, in which the tendencies towards decentralisation and specialisation in knowledge production inevitably bring about an increased need to understand how newly produced knowledge can be transmitted and exchanged in efficient ways, without destroying the incentives for further knowledge production.

The purpose of this chapter is to critically analyse the current literature on knowledge spillovers and to suggest some possible new directions for the spatial analysis of knowledge spillovers. The chapter is organised as follows. Section 6.2 discusses aspects of the knowledge concept in economic theory. The concept of knowledge spillover is presented in Section 6.3. In Section 6.4 knowledge spillovers are discussed within an industry life cycle framework. Empirical evidence on knowledge spillovers is presented in Section 6.5. Section 6.6 gives examples of how knowledge spillovers might be treated formally within a spatial context by using the accessibility concept. Knowledge spillovers and spatial interactions are discussed in Section 6.7. The issue of how to model the influence of knowledge spillovers in knowledge production functions is tackled in Section 6.8, while Section 6.9 provides a discussion of regional technology policy in the presence of knowledge spillovers. The main conclusions of the chapter are summarised in Section 6.10, which also contains some suggestions for future research.

6.2   Knowledge in Economic Theory

A review of the literature on knowledge spillovers suggests that some authors use the concept of knowledge in their analysis, while others use the concept of information. Only a few authors however define the specific concept that they use. Even worse, some authors use the two concepts interchangeably. This state of affairs makes it very difficult to evaluate the different contributions - whether they are theoretical or empirical in scope.

In this contribution we maintain the view that it is vital, when discussing knowledge spillovers, to make a clear-cut distinction between information and knowledge. We suggest that information should be defined as such data that can be easily codified and therefore transferred and accessed, for example via the Internet. Thus, information consists of those uncomplicated messages and routinised data which are easy to manipulate and to store, for either a longer or a shorter time (Kobayashi, Sunao and Yoshikawa 1993).

Knowledge, on the other hand, consists of information that is difficult to codify, generally due to its intrinsic indivisibility. It is therefore difficult to transfer without direct face-to-face interaction. So face-to-face contacts are a necessary, but of course not sufficient condition for knowledge transfer. Von Hipple (1995) persuasively demonstrates that highly contextual and uncertain knowledge, i.e. what he refers to as 'sticky knowledge', is best transmitted via [preferably frequent] face-to-face interactions[1]. This is in line with the claim by Teece (1998) that knowledge assets are often inherently difficult to copy. Proximity matters in connection with sticky knowledge because, as Arrow (1962) points out, such knowledge is generally non-rival, and the knowledge developed for any particular application can easily spill over and find additional applications if the obstacles to communication are not too great.

While the costs of transmitting information may be invariant to the distance, the cost of transmitting knowledge increases together with the distance. Since the transmission of knowledge requires face-to-face contacts, knowledge exchange requires an extensive amount of somewhat diffused movements throughout various transportation networks[2]. Knowledge exchange is an essentially interpersonal activity that is defined here as any action that can contribute to the process of the disclosure, dissemination, transmission, and communication of knowledge. This presupposes that knowledge is neither shared ubiquitously nor passed around at zero cost (Teece 1981).

In terms of economic theory, knowledge is not restricted to the technical aspects of know-how for firms, but also includes components such as institutional and organisational know-how. A comparison of the knowledge concepts used in endogenous growth theory, and in regional economics and economic geography, reveals that such concepts are quite different. In endogenous growth theory, knowledge is normally defined as an output generated by R&D and/or by learning-by-doing. In regional economics and economic geography, on the other hand, authors often work with a knowledge concept that is in line with that of Marshall's. Such a concept is much broader than the former one and includes, for example, market knowledge. In later sections, we shall explore how this broader concept of knowledge might be dealt with so as to obtain a starting point for the discussion of knowledge spillovers.

In our view, it is useful to distinguish three concepts of knowledge: (a) scientific knowledge, namely basic scientific principles; (b) engineering knowledge – or blueprints – namely inventions that can be directly used in the production of goods and services; (c) entrepreneurial knowledge, in other words, business relevant knowledge about products, business concepts, markets, customers, and so on.

In dealing with the different concepts of knowledge it is essential to characterise them according to the degree to which they are *rivalrous* and *excludable* (see Cornes and Sandler 1986). A purely rivalrous good has the property that its use by one firm or person precludes its use by another, whereas a purely non-rivalrous good has the property that its use by one agent in no way limits its use by another. Excludability relates to both technology and legal systems (Kobayashi and Andersson 1994). A good is excludable if the owner can prevent others from using it. While conventional goods are both rivalrous and excludable, pure public goods are both non-rivalrous and non-excludable.

Scientific knowledge has the character of a pure public good, although it is generally only available to those with the relevant scientific training. Hence, access to scientific knowledge can differ between regions, due to an unequal supply of scientifically trained labour. However, legal restrictions can also be imposed, at least for commercial applications. This is illustrated by the increasing propensity in recent years to try to patent basic scientific discoveries in e.g. biotechnology (*The Economist*, April 8th, 2000).

Engineering knowledge may be perceived and even deliberately created as a non-rivalrous, partially excludable good (Romer 1990). Its non-rivalrous character stems from the fact that engineering knowledge is inherently different from other economic goods. Once the costs of creating new 'engineering knowledge' have been incurred, this knowledge may be used over and over again at no additional cost. It is in this sense that engineering knowledge is non-rivalrous. The partially excludable character of engineering knowledge stems from the fact that firms generally protect new inventions by having patents issued on them. However, patent applications – and therefore patents – must be quite detailed. This opens up opportunities for the competitors to imitate or to 'invent around' patents, so that as a matter of fact engineering knowledge *may* be accessible for intellectual purposes.

Entrepreneurial knowledge is most often the result of learning-by-doing, and can generally be viewed as a non-rivalrous, partially excludable good. Access to entrepreneurial knowledge can be limited in various ways, for example by trying to preserve 'business secrets'. There appears to be a growing tendency to protect business ideas by means of patents (*The Economist*, April 8th, 2000).

The processes by which the different types of knowledge are made available by their creators to other individuals or firms take place in spatial networks, i.e. 'knowledge networks' (Batten, Kobayashi and Andersson 1989; Kobayashi 1995). These consist of a set of nodes and a set of links connecting them. The nodes are represented by human settlements such as towns, cities and metropolitan regions, providing different instances of functional regions[3]. The nodes can be characterised by their endowment of knowledge production capacities and related activities, including knowledge infrastructure such as universities, meeting infrastructure, stocks of knowledge and human capital, local knowledge networks, and so on. The links include transportation as well as communication channels. The spatial perspective adds a further dimension to knowledge transfers. Partial excludability of the new knowledge is not only a result of patents, business secrets, and so on but also a consequence of limited physical accessibility.

One further important issue concerning knowledge spillovers is the relative importance of the existing stock of knowledge and the flow of new knowledge. A substantial part of the literature emphasises the role of knowledge stocks[4]. However, in our view, the critical role in the promotion of innovative activities in a market environment is that played by the output of *new* knowledge. Good access to knowledge stocks may be a prerequisite for innovative activities, but what really matters is knowledge about new techniques, new products, new production processes, new competitors, new customers, new business concepts, and so on. Without access to the new knowledge created in the market place it seems extremely difficult for firms to stay competitive in dynamic markets.

6.3   Knowledge Spillovers – A General Introduction

Griliches (1994) defines knowledge spillovers as 'working on similar things and hence benefiting much from each other's research'. Knowledge spillovers have been traditionally identified with those aspects of research and development [R&D] externalities in which ideas discovered within any given research project affect the productivity of other projects. The basic idea here is that the creation of new knowledge by one firm has positive external effects on the knowledge production activities of other firms, either because knowledge cannot be kept secret, or because patents do not guarantee full protection from imitation[5]. However, a second type of knowledge spillover originates from production activities (Udayagiri and Schuler 1999). Knowledge spillovers are not tied to direct compensation. The character of the externality stems from the fact that the protection of proprietary knowledge is incomplete. Spillovers lead to increasing returns to scale at the economy-wide level, thereby causing the competitive equilibrium of a decentralised economy to be socially inefficient (Arrow 1962; Romer 1986; Smolny 1999).

Knowledge diffusion can be described as a special type of communication related to the diffusion of messages that contain new ideas, concepts, blueprints, and so on (Rogers 1983). As the creation of knowledge is spatially concentrated, it is obvious that the diffusion of knowledge plays a decisive role in regional development. Marshall (1920) identifies the flow of information and ideas between firms of a certain region as one of the main reasons for the spatial concentration of economic activities. Henderson (1996, p.31) observes that 'information is subject to spatial decay' and assumes that agglomeration benefits, including knowledge spillovers, arise from localised external economies of scale. Glaeser et al. (1992, p.1127) argue that geographic proximity facilitates knowledge spillovers, because 'intellectual breakthroughs must cross hallways and streets more easily than oceans and continents'. Feldman and Audretsch (1999, p.410) maintain that 'knowledge spillovers not only generate externalities, but the evidence suggests that such knowledge spillovers tend to be geographically bounded within the region where the new economic knowledge was created'. Lucas (1993) emphasises that the most natural example of the role of geographically bounded externalities, from the point of view of the mechanics of economic growth, is provided by metropolitan areas, where communication is facilitated by the compact nature of the geographical unit. Indeed, Lucas asserts that the only compelling reason for the existence of cities is the presence of increasing returns to agglomerations of resources, which make these locations more productive due to - among other things – the intensity of the knowledge spillovers.

The existence of increasing returns at the regional level, driven by intra-regional knowledge spillovers in the production of knowledge, may have dramatic long-term effects. Increasing returns create a tendency for any given ranking of the competitive positions of firms and regions to persist over time. Positive feedback mechanisms reinforce the winners and challenge the losers.

What can be said then about interregional knowledge spillovers? Palander (1935) observed long ago that one of the most remarkable features of modern urban structures is the frequency and extension of the interactions between activities carried out in different cities. These interactions presuppose of course the possibility of communicating across cities. The question then arises why are interregional knowledge spillovers often so limited in scope in an age in which the telecommunications revolution has lowered the marginal cost of information exchange between different locations to levels very close to zero, and the evolution of the national and international air travel networks has significantly reduced the travel costs and times? If our attention to scientific knowledge, we limited could indeed say that interregional knowledge spillovers are both substantial and rapid. The reason for this is that the international scientific community is organised as one big knowledge network, relying for example on scientific conferences and journals. Are there reasons why engineering knowledge should be different from entrepreneurial knowledge from this point of view? Some authors claim that the differences have actually been overstated, and that there is abundant evidence that information and knowledge networks that enhance business efficiency can be and often have been widely diffused geographically (Hansen 2000).

But there exist considerable differences between regions, viewed in their entirety, in the extent to which they produce knowledge, as well as in the way in which they communicate knowledge internally. Differences also exist from the point of view of the possibility of access to the knowledge generated in other regions. These differences partly reflect the differences in intra- and interregional communication and transportation infrastructure between regions and are reflected in the empirical observation that different regions – even the more developed ones – tend to specialise in different economic activities (Dollar and Wolff 1993). It also reflects differences in the supply of knowledge handlers across regions.

At the micro-level we can distinguish between spillovers across *products* and spillovers [or diffusion] across *firms*. Spillovers across products can occur either between firms, or within a single firm – if the latter produces the full range of products affected by a given spillover. Indeed, a limited appropriability of the value of the knowledge spillovers [which can be obtained either via patents, or by trying to prevent the employment of 'experts' by competitors] may suggest the opportunity to internalise the spillover by producing multiple products within the firm, thereby taking advantage of economies of scope.

Spillover across firms occurs when the knowledge generated by one firm is 'borrowed' by other firms. Such spillovers are not necessarily associated with the exchange of inputs between firms, as they may be technologically related. The greater the spillovers, the closer the relationship which can be expected (Karlsson 1997). Thus 'intellectual-scientific-technological' regions (Griliches 1991, p.15) become as important as geographic regions, from the point of view of the identification of the spatial extension of spillovers (see Olsson and Frey 2000).

In the case of spillovers across firms, we must distinguish between intra- and inter-industry spillovers (Feldman and Audretsch 1999). Here the relevant question is: 'Does the specific mix of economic activities undertaken within any particular region matter?' This question is important because recent debate has focused precisely on the mix of economic activities carried out within agglomerations, and on how externalities generated by knowledge spillovers are shaped by this mix. Despite the general consensus that knowledge spillovers within a given region stimulate dynamic externalities, there is no agreement as to the precise way in which this occurs.

One point of view is that portrayed by Glaeser *et al.* (1992). Their study considers the factors that influence innovative activities in urban regions and identifies two relevant models in the economics literature. The so-called *Marshall-Arrow-Romer* model formalises the insight that the concentration of a particular *industry* within a specific urban region (Lösch 1954) promotes intra-regional knowledge spillovers across firms and therefore stimulates innovation in that particular industry. The basic assumption here is that knowledge spillovers mainly take place across firms within the same industry.

An alternative view regards inter-industry spillovers as the most important source of new economic knowledge. Specifically, Jacobs (1969) argues that the agglomeration of firms in urban regions fosters innovations due to the diversity of knowledge sources located in such regions. Thus, the variety of industries within an urban region can be a powerful engine of growth for that region, and the exchange of complementary knowledge across diverse firms and economic agents leads to increasing returns to new economic knowledge.

A related controversy has to do with the influence of market structure on R&D behaviour. The Marshall-Arrow-Romer model predicts that local monopoly is superior to local competition because it enhances the ability of firms to appropriate the economic value accruing from their innovative activity. By contrast, Jacobs (1969) and Porter (1990) assume that, due to knowledge externalities[6], competition is more conducive to innovative activity. Since the early 1980s, several authors have provided an analytical framework relating market structure under oligopolistic competition to the nature of inventive activities (Dasgupta and Stiglitz 1980; Kamien and Schwartz 1982). These authors argue that both the market structure and the nature of innovative activities are endogenous, as they both depend on factors such as R&D technology, demand conditions and the nature of capital markets. However, they generally assume that knowledge is monopolised by the firm that produces it, i.e. that the results of R&D activities are fully appropriable, thus disregarding knowledge spillovers. Although these models paved the way for investigations of the relationship between market structure and R&D intensity, the lack of an explicit spatial dimension makes it impossible to use them as a basis for the analysis of the geographic features of spillovers and their consequences. However, Kobayashi (1995) provides an analytical framework relating market structure to knowledge accessibility in knowledge networks.

The position taken in the debate summarised above clearly has important implications from the point of view of technology policy. If the specialisation hypothesis is correct, then policy should focus on developing a narrow set of economic activities within a functional region in order to generate large dynamic externalities. On the other hand, if the diversity hypothesis is correct, then a functional region composed of a diverse set of economic activities will generally yield larger dynamic externalities. The key policy issue would then become how to identify commonalties and how to enhance diversity.

Research on R&D spillovers focuses on diffusion of R&D knowledge across firms. For example, growth models have been developed whereby the number of products [and/or product quality] increases over time due to the innovative activity of profit-seeking firms. In these models, decreasing returns to innovation never set in because the innovative activities of firms not only lead to new products, but also contribute to a general stock of knowledge upon which subsequent innovators can build. Over time, the foundation of general knowledge grows, allowing more differentiated products to be introduced without a continual increase in the research resources[7]. This is a typical knowledge spillover, since the benefits of innovation accrues not only to the innovator, but 'spills over' to other firms by raising the level of knowledge upon which new innovations can be based[8]. Thus, knowledge spillovers represent an endogenous engine of growth.

Production-related knowledge is created in the design, manufacture and sales of a product in a process known as 'learning-by-doing' (Arrow 1962). Even if the firm where learning-by-doing takes place can fully enjoy the contribution of this knowledge to the productivity of its own processes, a part of it spills over to other firms and sectors (Griliches 1991). In general, production-related knowledge spillovers have two important characteristics. Firstly, they are 'joint products' (Rosen 1972) in that a firm undertaking a production process creates two outputs: the physical product and knowledge about the production process. Secondly, the knowledge created cannot be easily codified, and is therefore difficult to patent.

6.4   Knowledge Spillovers over the Industry Life-Cycle

Many studies of knowledge spillovers seem to have overlooked the existence of a coherent and compelling theoretical framework, as well as the increasing amount of empirical evidence supporting the idea of a prototypical industry life cycle (Klepper 1992). This framework, by considering the stage of the life cycle in which an industry is operating, helps to provide answers to questions such as *who* innovates, *how much* innovative activity is undertaken and *where* the innovative activity takes place. In this context, the location of the firms carrying out innovative activities is important as it determines the access to the available knowledge inputs – including the intra- and interregional knowledge spillovers which contribute to the generation of innovative activity.

The industry life-cycle has been described by Williamson (1975, pp.215-216) in the following fashion: 'Three stages are commonly recognised in an industry's development: an early explorative stage, an intermediate development stage, and a mature stage. The first or early formative stage involves the supply of a new product of relatively primitive design, manufactured on comparatively unspecialised machinery, and marketed through a variety of exploratory techniques. Volume is typically low. A high degree of uncertainty characterises the business experience at this stage. The second stage is the intermediate development stage in which manufacturing techniques are more refined and market definition is sharpened; output grows rapidly in response to newly recognised applications and unsatisfied market demands. A high but somewhat lesser degree of uncertainty characterises market outcomes at this stage. The third stage is that of a mature industry. Management, manufacturing, and marketing techniques all reach a relatively advanced degree of refinement. Markets may continue to grow, but do so at a more regular or predictable rate... Established connections with customers and suppliers [including capital market access] all operate to buffer changes and thereby to limit large shifts in market shares.

Significant innovations tend to be fewer and are mainly of an improvement variety'.

Klepper's (1992) study of the evolution of innovative activities over the industry life cycle distinguishes three general tendencies. Firstly, the innovative activity tends to be more intense during the earliest phases of the life cycle. Secondly, in the earlier stages, the most recent entrants – possibly the smaller firms in the industry – account for a disproportionate share of the major product innovations that are introduced. Thirdly, as the industry evolves towards maturity, a distinct shift emerges in the distribution of the innovative activities across firms, with a larger and larger part of the innovations introduced by the more established and larger firms. The overall pace of the innovative activity also tends to decrease, as the industry grows older.

Given Klepper's results, it is natural to return to the questions regarding the location of the innovative activity. In the absence of spatially bounded knowledge spillovers, the location of innovative activity would not matter. However, if knowledge spillovers are spatially bounded, it might matter a great deal. In the latter case, the propensity for innovative activities to cluster spatially will be influenced by the stage of the industry life cycle. The theory of knowledge spillovers suggests that the propensity for innovative activity to cluster spatially should be stronger in those industries in which tacit knowledge plays a relatively important role. The reason is that tacit knowledge can be transmitted only informally and typically demands direct and repeated contact. The role of tacit knowledge for innovative activity is presumably more important during the early stages of the industry life cycle, before product standards have been established and a dominant design has emerged.

6.5   Knowledge Spillovers – The Empirical Evidence

Going through the existing empirical literature on knowledge spillovers, it appears that most authors are willing to consider the evidence related to either one or the other thesis, but that they are reluctant to perform comparative tests of alternative hypotheses. Thus, while many studies evaluate the importance of intra- and interregional knowledge spillovers, the question of whether intra- or interregional knowledge spillovers are more important as an engine of growth is generally not addressed. One exception is Branstetter (1996), who compares the two types of spillovers and finds robust evidence that intra-national knowledge spillovers are stronger than international spillovers. In any case, the possibility that the relative importance of these two types of spillovers may differ between different types of knowledge and between different types of economic activities is seldom tested. Also, few authors consider the possibility of interactions between intra- and interregional knowledge spillovers.

One reason for the lack of comprehensive studies of intra- and interregional knowledge spillovers might be Krugman's contention (1991a, p.53), that economists should abandon any attempts to measure knowledge spillovers because '…knowledge flows are invisible, they leave no paper trail by which they may be measured and tracked'. However, the difficulties associated with the direct measurement of spillovers are not necessarily insurmountable. In fact, Jaffe, Traitenberg and Henderson (1993, p.578) point out that 'knowledge flows do sometimes leave a paper trail', for example in the form of patented innovations and the introduction of new products.

It is nevertheless true that some data-related problems tend to get in the way of a detailed analysis of the spatial dimension of the spillovers. For example, in the United States, the National Science Foundation only reports data on R&D expenditure at the level of the state - and not for finer geographical units, such as individual cities. The situation is no better, from this point of view, in the European Union. Literature-based innovation indicators too – generally thought to be more closely associated with the occurrence of spillovers than the more conventional proxies based on patents – have only been constructed for relatively few geographic areas and-or time-periods[9].

The empirical evidence does support the existence of significant dynamic externalities associated with knowledge spillovers from the research efforts of both private and public institutions. In earlier works (Griliches 1958), social returns to investment in R&D were computed to estimate the spillovers characterising R&D effort. Several investigations followed, using the same general methodology (e.g. Mansfield et al. 1977). A geographical dimension was added to the research on spillovers across firms by Jaffe (1989) in his study showing that corporate patent activity increase as a result of R&D expenditures undertaken by universities within the same state.

A fundamental question addressed by the research on knowledge spillovers is whether these spillovers are spatially bounded or not. Several studies, using the knowledge production function approach, suggest that knowledge spillovers tend to be geographically bounded within the region where the new economic knowledge was created (Jaffe 1989; Feldman 1994a). Jaffe, Trajtenberg and Henderson (1993) find that patent citations tend to occur more frequently within the state in which the patents were issued. An implicit assumption of the knowledge production function approach is that innovative activity tends to take place in those regions where the direct knowledge-generating inputs are the greatest and where knowledge spillovers are more intense. One obvious explanation – stressed by Jaffe, Trajtenberg and Henderson (1993) – for the observed differences across industries in the propensity for innovative activity to cluster geographically is that the location of production is inherently more concentrated in some sectors, for reasons that have nothing to do with spillovers. Thus, in explaining why the propensity for innovative activity to cluster geographically varies across industries, one must control for the presence of other factors affecting the geographical concentration of the location of production.

Audretsch and Stephan (1996) and Audretsch and Feldman (1996) find that the propensity for innovative activity to cluster geographically tends to be greater in industries where new knowledge plays a more important role. Using innovation data on the introduction of new products, Feldman and Audretsch (1999) find evidence of a tendency towards clustering both in the firms' location choice and in the geographical distribution of the innovations. Although there is fairly widespread agreement that knowledge externalities do exist in urban regions, the studies disagree about whether such externalities primarily reflect interactions among different industries or interactions among local firms in the same industry. Also, in the latter case, it is not clear whether knowledge externalities reflect benefits from competition or explicit or implicit co-operation among firms (see also Beardsell and Henderson 1999; Black and Henderson 1999; Drennan 1999; Glaeser et al. 1992; Henderson 1997; Henderson, Kuncoro and Turner 1995; Porter 1996; Rauch 1993; Santiago and Lobo 1999).

A number of studies (Anselin, Varga and Ács 1997, 1999, 2000; Varga 1998, 2000, 2001) suggest that the propensity for innovative activity to cluster geographically may display special characteristics if universities are present in the area. Using different analytical methodologies, these studies generally find evidence of positive agglomeration effects in the form of local academic technology transfers in high-tech sectors such as Electronics and Instruments. Such effects appear however to become relevant only if the size of the agglomeration is larger than a certain 'critical mass'.

On a related note, a study by Zucker, Darby and Brewer (1998) on a sample of California biotech firms shows that the timing and the location of innovations are mainly explained by the presence of scientists actively involved in basic research who publish in academic journals. These results point to the role played by universities in encouraging local economic development. Another study by Zucker, Darby and Armstrong (1998) shows however that such a role, at least in the specific case considered, is mainly associated with exchanges [in the form of employment contracts] between firms and star academic scientists, rather than with spillovers.

Audretsch and Feldman (1995 and 2000) find empirical support for the hypothesis that the propensity for innovative activities to cluster spatially is influenced by the stage of the industry life cycle. The presence of some types of knowledge sources, such as universities, often tends to characterise the innovative clusters in the initial phase of an industry's life cycle, but not in later stages. Other knowledge sources, such as skilled labour, appear to promote innovative clustering throughout the whole life cycle. Audretsch's and Feldman's most striking result is perhaps the finding that during the mature and declining stages of the life-cycle, the increases in the geographical concentration of production tend to increase – rather than decrease – the degree of dispersion of innovative activity. An explanation suggested by the authors is that new ideas might need 'new space,' at least during the mature and declining stages of the industry life cycle. Overall, the positive agglomeration effects typical of the early stages of the industry life cycle appear to become less important during the later stages of the cycle.

Irwin and Klenow (1994) examine the relative strength of intra- and international spillovers associated with learning-by-doing in the semiconductor industry. They find that learning-by-doing spillovers do exist, but that such spillovers are much less important than the effects of the production experience accumulated within the firm[10]. While learning-by-doing is certainly important, the economic gains from the refinement of production techniques are probably to a great extent product-specific. Over time, the contribution of the R&D-intensive sectors of the economy to the creation of producer and consumer surplus tend to increasingly involve the introduction of new and better goods and services, rather than the more efficient production of existing goods and services. It is difficult to think of this kind of fundamental innovation as if they were driven principally by learning-by-doing.

The relevant empirical literature provides abundant evidence that long-distance co-operation can generate dynamic externalities. One example is the study by Appold (1995) showing that the agglomeration of manufacturing establishments is often erroneously interpreted as evidence in favour of the existence of locally bounded externalities. Using a random sample of almost 1000 U.S. metalworking plants, Appold found that the establishments' performance was enhanced by successful inter-firm co-operation, but that neither the performance nor the extent of co-operation were associated with the location of the units within the same agglomeration. Angel (1995) provides another example. He found that about one-third of a random sample of 495 manufacturers in the chemical, instruments and electronics industries was engaged in some form of collaborative technology development activity with customers, suppliers or other firms. However, only a small proportion of the partnerships were established with local firms; most of such partnerships were either national or international in scope. Further evidence is given by Hansen and Echeverri-Carroll (1997), who studied the importance of various types of formal and informal inter-firm relations for the business performance of high-tech establishments. They have found that the most important relationships are with non-local firms engaged in the same activity, followed by relationships with local firms in the same activity and local firms in other activities. Feser (1998) reports on five other studies, which indicate that the role played by linkages internal to metropolitan areas is generally not as important as that of external linkages.

Other work supporting similar conclusions are by Fischer (1999), Cooke (2000) and Fischer and Varga (2000). In particular, Fischer (1999) and Cooke (2000) further qualifies the conclusions on the limited relevance of co-operation with local firms. They observe that in relation to the information flows which are relevant for innovation, the most important links are very often those between the firms and their customers or suppliers – regardless of their locations – rather than between local firms in general.

The geographical division of labour in the Swedish electronics industry also contradicts the conventional wisdom that agglomeration is especially advantageous in the case of high technology industries (Suarez-Villa and Karlsson 1996). The creation of new enterprises in hinterland areas has led to an increased geographical dispersion of the firms. Many indicators show that, on average, firms involved in research-intensive production located in such areas tend to perform better than those located in clusters. A study of the early use of microelectronics in products by plants in the Swedish engineering industry shows somewhat similar results (Karlsson and Olsson 1998). They showed that location in a large, densely developed region does have a positive significant effect on product innovation in large firms, but a negative [and non-significant] effect on product innovation in small firms. Thus, small firms seem effectively to enjoy early access to new technologies, even if they are located outside the large urban areas.

A closer, critical look at the existing empirical studies of intra- and interregional knowledge spillovers reveals that the results of many studies can easily be misinterpreted. One fundamental criticism which can be made is that the regional scale used in many of the American studies is unsuitable. The basic spatial unit of observation used is the State. But a State rarely qualifies as a functional region, so we must be careful not to make unwarranted inferences from the findings of such studies. If we make the reasonable assumption that the borders between the regional units adopted should coincide with barriers to economic interactions, we are left with two relevant spatial units, namely the country and the functional urban region.

Hansen (2000) illustrates some serious weaknesses in the empirical studies on intra-regional knowledge spillovers. He argues that the extensive recent literature on the importance of dynamic knowledge externalities in urban regions, especially large regions, has not directly measured these externalities, nor has it taken sufficient account of long-distance co-operation as a source of dynamic externalities. Many researchers, reporting on allegedly successful agglomerations, may have ignored the possibility that firms not located in such areas, but with characteristics similar to those located in agglomerations, may be just as productive and innovative. Hansen maintains that the lack of consistent findings in these studies is probably a consequence of their reliance on location quotients and similar measures of sectoral concentration or diversity to explain employment, income or other indicators of growth that do not directly capture knowledge-related information flows or dynamic externalities. Hence, more precise and reliable studies would require disaggregated empirical studies of how knowledge is transmitted from person to person.

Another critical remark concerns cause-and-effect relationships. Even if it were possible to establish the existence of knowledge spillovers within functional urban regions, we could not conclude that the firms' location choices are determined only by the possibility of taking advantage of spillovers. The location decisions may in fact be governed by reasons totally unrelated to spillovers, such as access to a pool of qualified labour.

6.6   Knowledge Spillovers – Towards a Formalisation

The intensity of knowledge diffusion in general, and of knowledge spillovers in particular, depends upon different factors and mechanisms. The extent to which knowledge spillovers actually occur can be assumed to depend upon the knowledge accessibility of individual firms. Accessibility measures can be regarded as the spatial counterparts of discounting. Such measures thus represent the distribution of activities in a simple way that imposes a very clear structure upon the relationship between activities and their environment. Various frictional effects arising from geographical, social, political, educational and/or psychological distance between knowledge-handling workers or knowledge centres determine the accessibility of knowledge.

According to Marshall (1920), the time required by the transmission of information flows grows with distance. Hence, a first distinction that we make is that between intra- and interregional knowledge accessibility. Relative to intra-regional knowledge accessibility, we can further distinguish between access to three different knowledge sources, namely intra-industry knowledge accessibility, inter-industry knowledge accessibility and accessibility of knowledge created in universities and in other specialised R&D institutions.

Regarding intra-industry knowledge accessibility, let us consider the whole set of firms ($i=1, \dots m$) belonging to a given industry j within an urban region r. The knowledge production within an individual firm i is denoted by A_{ij}. A_{ij} denotes knowledge created both by R&D and by learning-by-doing. For a given infrastructure capacity, $\bar{I}^r$, the intra-regional knowledge accessibility within a given industry rises as the A_{ij}-values are increased as a result of either a higher level of R&D or increased production. If congestion applies on the infrastructure, knowledge accessibility is of course reduced. For a given firm i in industry j the knowledge accessibility within the industry within the urban region r can be described as follows:

$$a_{ij}^r = \sum_{k=1}^{m_j} \exp\left\{-\lambda_j^1 \, d_{ijk}\right\} A_{kj} \qquad\qquad i \neq k \qquad\qquad (6.1)$$

where d_{ijk} represents some relevant distance measure and λ_j^i is a measure of distance sensitivity. From (6.1) it is possible to calculate an average knowledge accessibility value for industry j in region r

$$\bar{a}_j^r = \left\{a_{1j}^r + a_{2j}^r + \dots + a_{m_j j}^r \right\}/ m_j \qquad\qquad (6.2)$$

where m_j is equal to the number of firms in industry j.

In a similar way the total knowledge accessibility within an industry j within a region r can be calculated as follows:

$$\hat{a}_j^r = \sum_{i=1}^{m_j} a_{ij}^r \qquad\qquad (6.3)$$

Let us now turn to inter-industry knowledge accessibility within a certain region. Consider the whole set of industries ($j=1,...,n$) within an urban region r with its pertinent set of firms ($i=1,..., m, m+1,..., M$). The knowledge production within a firm i in industry j in region r is denoted by A^r_{ij}. For a given firm i in industry j the inter-industry knowledge accessibility within the urban region r can be described as follows:

$$b^r_{ij} = \sum_{l=m+1}^{M} \exp\left\{-\lambda^{II}_j d_{il}\right\} A^r_{lk} \qquad k \neq j \qquad (6.4)$$

where d_{ij} as above represents some relevant distance measure and where λ^{II}_j is a measure of distance sensitivity. The total inter-industry knowledge accessibility for industry j within a region r can be calculated as:

$$\hat{b}^r_j = \sum_{i=1}^{m_j} b^r_{ij} \qquad (6.5)$$

The third type of intra-regional knowledge accessibility that we shall consider here is accessibility for a given industry to knowledge created at universities and other specialised R&D institutions. The knowledge produced within each R&D institution may be denoted by R_k ($k=1,...,N$). For a given firm i in industry j the knowledge accessibility as regards universities and other specialised R&D institutes can be modelled as follows:

$$c^r_{ij} = \sum_{k=1}^{N} \exp\left\{-\lambda^{IR}_j d_{ijk}\right\} R_k \qquad (6.6)$$

where d_{ijk} represents some relevant distance measure and λ^{IR}_j is a measure of distance sensitivity for contacts with universities and other R&D institutions. The total intra-regional accessibility to knowledge created within universities and other R&D institutions to a particular industry j can be modelled as:

$$\hat{c}^r_j = \sum_{i=1}^{m} c^r_{ij} \qquad (6.7)$$

Interregional knowledge accessibility between two regions r and s can be modelled in a similar way to the intra-regional knowledge accessibility. For a given firm i in industry j in an urban region r the interregional intra-industry knowledge accessibility can be described as follows:

$$a_{ij}^{rs} = \sum_{s=1}^{m} \exp\{- \lambda_j^{III} d_{ij}^{rs}\} A_j^s \qquad\qquad i \neq k, \, r \neq s \qquad\qquad (6.8)$$

where A_j^s represents knowledge created within industry j in region s, d_{ij}^{rs} represents some relevant distance measure, and λ_j^{III} is a measure of distance sensitivity. Using A^s as an overall indicator of knowledge generation in region s the accessibility in region r to knowledge created in region s can be described by:

$$a^{rs} = \exp\{- \lambda^{II} d^{rs}\} A^s \qquad\qquad (6.9)$$

Assuming that there exist S regions the total interregional knowledge accessibility for a region r is calculated as:

$$\hat{a}^r = \sum_{s=1}^{S} a^{rs} \qquad\qquad (6.10)$$

If we now consider, as an illustration, the situation of an individual firm i belonging to industry j located in a functional urban region r, using equations (6.1) and (6.8) it is obvious that a_{ij}^r often can be significantly larger than a_{ij}^{rs} since in most cases ΣA_j^s is significantly larger than ΣA_{kj}. For the vast majority of industries the knowledge produced in each functional urban region is only a small share of the total knowledge created within the industry. This means that even if the influence of distance decay is taken into account, the empirical results showing that firms consider interregional knowledge spillovers to be more important seems quite plausible. Another factor not considered in most empirical studies is that intra- and interregional knowledge spillovers, respectively, may be important for different types of creative activities within a firm as well as having different importance for different types of knowledge. The situation may also differ between regions as a result of region-specific factors.

6.7   Knowledge Spillovers and Spatial Interaction

Knowledge spillovers are the result of social interactions between businessmen, researchers and so on, i.e. between people we can characterise as 'knowledge-handlers'. We assumed above that [possibly frequent] face-to-face contacts are a prerequisite for knowledge transfer and thus for knowledge spillovers. To be able to better understand the conditions for this kind of interaction, it might be useful to see what insights spatial interaction models might provide. Interestingly enough, few studies of knowledge spillovers use such models to analyse the foundations for spatial interaction.

One popular way to model spatial interactions is to use gravity type models. In these models, interregional relations are perceived as interaction between discrete spatial entities. The frequency and intensity of such interactions are assumed to be governed by general principles that also influence the behaviour of individual units within each mass[11].

Given the basic assumption that knowledge transfer – and therefore knowledge spillovers – require frequent face-to-face contacts, it is natural to assume that the spatial entities in this case are measured in terms of the number of knowledge-handlers. Considering the case of interregional interaction, one might assume that the knowledge flow from region s to region r, A_{sr}, increases with the number of knowledge-handlers in the origin region, O_s, and the number of knowledge-handlers in the destination region, D_s, and as the spatial separation between s and r, t_{sr}, decreases. Hence, knowledge interaction opportunities are assumed to be directly related to the population of knowledge-handlers in each region, and inversely related to the interregional separation. These considerations can be summarised by the following gravity formulation:

$$A_{sr} = \alpha \ O_s^{\beta_1} \ D_r^{\beta_2} \ e^{\lambda t_{sr} + \varepsilon} \qquad\qquad (6.11)$$

with $\beta_1, \beta_2 > 0$, $\lambda < 0$ parameters, e a stochastic term. This general formulation can be extended in various ways. It is easy, for example, to introduce the effects of barriers to knowledge flows. Two alternative formulations are presented in equations (6.12) and (6.13), where B is a barrier dummy. The barrier dummy can be interpreted as a measure of the affinity between regions in various respects [culturally, technologically, and so on].

$$A_{sr} = \alpha \ O_s^{\beta_1} \ D_r^{\beta_2} \ e^{\lambda t_{sr} + \gamma B + \varepsilon} \qquad \beta_1, \beta_2 > 0, \lambda < 0 \qquad (6.12)$$

$$A_{sr} = \alpha \ O_s^{\beta_1} \ D_r^{\beta_2} \cdots e^{\lambda t_{sr}(1 + \gamma B) + \varepsilon} \qquad \beta_1, \beta_2 > 0, \lambda < 0 \qquad (6.13)$$

Given the assumption that human interaction is needed for knowledge transfer – and hence for knowledge spillovers to take place – it seems natural also to consider the conditions for intra- and interregional interaction as a part of the analysis of knowledge spillovers.

6.8   Modelling the Influence of Knowledge Spillovers

To model the influence of knowledge spillovers on knowledge production Griliches (1979) introduces the concept of 'knowledge production function'. The knowledge production function links the inputs in the innovation process to innovative outputs. According to Griliches, the critical innovative input is new economic knowledge, and R&D should be viewed as the greatest source of new economic knowledge. The knowledge production function provides a formal economic framework for the study of the effects of both private R&D and R&D conducted within universities. It can be modelled using a Cobb-Douglas production function with two inputs:

$$AR = a_0 (RD)^{a_1} (UR)^{a_2} \varepsilon \qquad\qquad (6.14)$$

where AR stands for economically useful new knowledge, RD is industrial R&D, UR stands for research performed by universities and ε is a stochastic error term.

Jaffe (1989), Audretsch and Feldman (1994, 1996) and Feldman (1994a, 1994b) modified the knowledge production function approach to a model specified for spatial and product dimensions:

$$AR_{sj} = \beta (RD_{sj})^{\beta_1} (UR_{sj})^{\beta_2} (UR_{sj} \times GC_{sj})^{\beta_3} \varepsilon_{sj} \qquad\qquad (6.15)$$

where GC measures the geographic coincidence of university R&D and industrial R&D. The unit of observation for estimation was at the spatial level, s, a state, and industry level, j.

The traditional knowledge production function approach tends to be used at an aggregated level, and does not consider the knowledge spillovers made possible by knowledge accessibility as defined in this chapter. Machlup (1980) defined knowledge production as any activity through which someone in a firm or an organisation learns of something he or she had not known before, even if others knew about it. Knowledge production can involve both the creation of new knowledge and the search for a new understanding of old knowledge. Knowledge production implicitly presumes the exchange of knowledge among persons. The formation of something new demands the amalgamation of different concepts and different pieces of knowledge. This creative feature of the process of knowledge exchange can be viewed as a dynamic synergy. Hence, knowledge production activities demand a high degree of accessibility to other knowledgeable persons. We argue here that the effects of knowledge spillovers on the output from R&D carried through within an industry should also be considered. The R&D output from industry j in region r could be modelled as follows:

$$R_j^r = R_j^r\left(K_{Rj}^r, H_{Rj}^r, \hat{a}_j^r, \hat{b}_j^r, \hat{c}_j^r, \hat{a}^r\right) \tag{6.16}$$

where K_{Rj}^r is capital, H_{Rj}^r the input of human capital in R&D activities in industry j, and $\hat{a}_j^r$, $\hat{b}_j^r$, $\hat{c}_j^r$ and $\hat{a}^r$ are respectively defined in Equations (6.3), (6.5) and (6.10). Likewise, the total knowledge output from all specialised R&D institutions, i.e. universities and similar institutions, in region r, $\hat{R}^r$, can be modelled as follows:

$$\hat{R}^r = \hat{R}^r\left(K_{RR}^r, H_{RR}^r, \hat{c}_j^r, \hat{a}^r\right) \tag{6.17}$$

where K_{RR}^r is the total capital used and H_{RR}^r the total volume of human capital engaged in specialised R&D institutions in region r. For the specialised R&D sector we assume that the important knowledge spillovers come both from within the sector and from other regions. However, knowledge flows not only influence knowledge production, but also have a direct effect on the output of an individual industry j in region r:

$$y_j^r = y_j^r\left(K_j^r, H_j^r, R_j^r, \hat{a}_j^r, \hat{b}_j^r, \hat{c}_j^r, \hat{a}^r\right) \tag{6.18}$$

where y_j^r is the output of industry j, K_j^r is the capital input in industry j, H_j^r is the input of human capital in industry j and R_j^r is the output from R&D activities within industry j. The basic idea behind this formulation is that knowledge spillovers also may have a direct effect on the output of the industry.

6.9 Technology Policy in a World with Knowledge Spillovers

The analysis of knowledge spillovers has important implications for public policy. Knowledge spillovers represent positive externalities, in the sense that the economic and social return from a firm's R&D activities and/or the knowledge generated by learning-by-doing do not benefit only the firm itself. Hence there is a justification for governments to encourage R&D and/or knowledge production by using subsidies and other measures.

The framework presented in this chapter offers a new perspective for the discussion of technology policy. We maintain that technology policy should be examined within this broader framework – rather than being limited to issues regarding R&D and higher education. Also, infrastructure policies involving intra- as well as interregional communication and transportation networks must be put on the agenda.

It is also obvious that simple solutions, such as providing broadband Internet access for everyone, will not be adequate. Rather, attention must be paid to possible complementarities between the communication and transportation networks, and between infrastructure investments and investments in R&D and higher education. In this connection, it must also be acknowledged that policies aiming to stimulate knowledge spillovers, for example via cluster formation, may reduce the private incentives for R&D. Extended legal protection of inventions, larger public investments in R&D or subsidies for privately undertaken R&D may thus become necessary in order to counter the effects of such distortions.

6.10 Conclusions

The starting point for this contribution was the observation that the current literature dealing with knowledge spillovers in a spatial context appears to be unsatisfactory in many respects, and even contradictory. Hence, the purpose of the chapter was to critically examine the literature on knowledge spillovers and, if possible, make some suggestions as regards the analysis of knowledge spillovers in a spatial context.

To summarise our findings, we can make the following observations concerning the treatment of knowledge spillovers in the current literature. The concept of knowledge as such is often not dealt with in a satisfactory way. Definitions are lacking, and some authors do not distinguish clearly between information and knowledge. Turning to the theoretical discussion of knowledge spillovers, we find that these are frequently treated superficially, and that the channels and mechanisms through which the spillovers take place are not made explicit. The role of market structure for knowledge spillovers is rarely considered in detail, and the same is true for the dynamic relations between knowledge spillovers and market structure. It would seem to be very important to work from an industry life cycle perspective, since the structure of the knowledge generated by each industry in any given stage of its life cycle may not match the knowledge required by the same industry at the same point in time.

Despite the general agreement that new economic knowledge does in fact spill over, there appears to be substantial disagreement about whether such knowledge spillovers are geographically bounded or not. Such diverging views can be partly traced back to the different definitions of 'geographical boundaries' used in the various studies. Furthermore, the definitions actually used often fail to provide a useful criterion for the identification of 'functional regions', which we claim to be the most relevant units for the spatial analysis of knowledge spillovers. Furthermore, studies dealing with knowledge spillovers in a spatial context often fail to model the spatial element in a satisfactory manner. We suggest that knowledge spillovers should be studied by employing traditional concepts and methods from regional science, such as knowledge accessibility and spatial interaction models. Hence, studies of spatial knowledge spillovers should be based on the knowledge network paradigm. These suggestions also have implications for the modelling of the knowledge production function in a spatial context. The network approach stresses the need to broaden the scope of regional technology policy by including new topics, not limiting it to issues concerning public and private R&D. In particular, regional technology policy must be able to deal with the question of how intra- and interregional knowledge spillovers can be stimulated by means of infrastructure investments and the creation of opportunities for face-to-face contacts. Any intervention must take into account the need not to distort the incentives for further private knowledge production. These issues are of course very important in an era characterised by rapidly growing knowledge production and by strong tendencies towards decentralisation and specialisation.

As regards further research, we suggest that future theoretical and empirical studies of knowledge spillovers in a spatial context should take as their starting point properly defined functional regions rather than administrative regions. Furthermore, these studies should incorporate the knowledge network paradigm, with its knowledge production nodes and knowledge transfer links. Knowledge networks provide a basis for studying the various types of intra- and interregional interactions that are a necessary condition for knowledge spillovers to occur. This would give the opportunity to understand the mechanisms behind knowledge spillovers, but also to understand the relevance and the spatial extension of knowledge spillovers under different scenarios.

Endnotes

1   This type of knowledge is often referred to as tacit knowledge.

2   Historically, the transfer and/or communication of rich information has required proximity and specialised channels to customers, suppliers, and distributors. However, we must acknowledge the possibility that new developments are undermining the traditional chains and business models, and that new structures – generally less dependent on physical communication channels – might increasingly become an economically viable option (cf. Teece 1998).

3   The definition of a functional region is based upon the spatial interaction patterns of economic agents in a country. A functional region is fundamentally characterised by its size, by the density of economic activities, social opportunities and interaction options, and by the frequency of spatial interaction between the actors within the region (Johansson 1997).

4   A serious problem with the knowledge stock concept is the common assumption that past knowledge never becomes obsolete. This assumption has the unrealistic implication that that every increase in the resources to R&D leads to an increase to the growth rate of the economy. In reality, the stock of knowledge does depreciate over time due to technological obsolescence (Caballero and Jaffe 1994; Jones 1995).

5   In the literature there has been a focus on the effects of knowledge spillovers on knowledge production. One can, however, imagine that new knowledge produced by one firm spills over directly to the production functions of other firms, thereby generating positive external effects.

6   It should be emphasised that the term 'local competition' is used in Jacobs (1969) to denote competition for new ideas carried by the economic agents. This interpretation is at variance with that prevailing in the industrial organisation literature, whereby 'local competition' is generally used to denote competition within product markets. An increasing number of firms provides greater competition for new ideas, and greater competition across firms also facilitates the entry of new firms specialising in new, specific product niches. This is because the necessary complementary inputs and services are likely to be available from small specialist niche firms, and not necessarily from large, vertically integrated producers.

7   With two countries, the total stock of knowledge can be expressed as $K_{it} = \lambda_h \int exp \ (\lambda_h (\tau - t) \ n_i (\tau)) \ d\tau + \lambda_f \int exp \ (\lambda_f (\tau - t) \ n_j (\tau)) \ d\tau,$ where λ_h is the rate of knowledge spillover within country i, λ_f is the rate of spillover between countries, n_i is the number of innovations at home by time τ, and n_j is the number of innovations abroad. If $\lambda_h = \lambda_f$, then firms in both countries will have access to the same level of knowledge even if knowledge generation occurs more rapidly in one of the countries. If they are different, then firms in the country with a higher initial level of innovation will have a greater stock of knowledge on which to draw and will, according to the previous equation, become more productive in subsequent innovation. Expressing knowledge in this way assumes that home innovations always differ from innovations from abroad.

8   There are, however, also differences related to patenting, as far as the consequences are concerned.

9   We are grateful to an anonymous referee for this observation.

10 They estimate $c_i = vE^{\beta t} e^{ui}$, where c is marginal cost and E is the cumulative production experience of the firm. Here experience also includes production external to the firm, such that $E_i = Q_i + \alpha (Q_c - Q_i) + \gamma (Q_w - Q_c)$, where Q_c is the production experience of all domestic producers and Q_w the production experience of all world producers. Thus α captures the effect of intra-national spillovers and γ the effect of international spillovers.

11 For an interesting attempt to model human interaction in the knowledge society using a micro-economic approach, see Kobayashi and Fukuyama (1998).

7 The Role of Space in the Creation of Knowledge in Austria An Exploratory Spatial Data Analysis

Manfred M. Fischer*, Josef Fröhlich**, Helmut Gassler**
and Attila Varga*
* Vienna University of Economics and Business Administration
   Department of Economic Geography & Geoinformatics
** Austrian Research Centre Seibersdorf

7.1 Introduction

The systems of innovation approach has recently received considerable attention as a promising conceptual framework for advancing our understanding of the innovation process in the economy. A system of innovation may be thought of as a set of actors such as firms, other organisations, and institutions that interact in the generation, diffusion and use of new – and economically useful – knowledge in the production process. Territorially based systems of innovation build on spatial proximity in terms of both spatial distance and contiguity – as either regional [subnational], national or global systems of innovation. The central idea underlying territorially based systems is that the economic performance of territories depends not only on how business corporations perform, but also on how they interact with each other and with the public sector in knowledge creation and dissemination. Knowledge creation is viewed as interactive and cumulative process contingent on the institutional set-up (Fischer 2001a).

The concept of territorially based systems of innovation evolved first in a national context (Freeman 1987), and then in a regional context (see, for example, Cooke, Uranga and Etxebarri 1997, Brazcyk, Cooke and Heidenreich 1998). It is being increasingly recognised that important elements of the process of innovation have become transnational and global, or regional rather than national. The driving forces behind this are two processes that are simultaneously at work: the process of globalisation of factor and commodity markets, and the regionalisation of knowledge creation and learning. Specific forms of technological learning and knowledge creation, especially the tacit forms, are both localised and territorially specific. The firms that master knowledge which is not fully codifiable are tied into various kinds of networks with other firms and organisations through localised input-output relations, as well as knowledge spillovers and their untraded interdependencies (Storper 1997).

Knowledge spillovers occur because knowledge created by a firm or other organisation is not normally contained solely within that organisation, but is also exploited by other firms. The spillover beneficiary may use the new knowledge to copy or imitate the commercial products of the innovator, or may use it as an input to R&D leading to the development of other new products or processes. Three vehicles of such spillovers may be distinguished: first, the scientific sector with its general scientific and technological knowledge pool; second, the firm specific knowledge pool; and, third, the business-business and industry-university relations that make them possible (Fischer 2001a).

In this chapter we make a modest attempt to shed some light on the role of space in the creation of technological knowledge in Austria. The study is exploratory rather than explanatory in nature and is based on descriptive and exploratory techniques such as *Moran's I* test for spatial autocorrelation and the Moran scatterplot. Clusters of the output of the knowledge creation process [measured by patent counts] are compared with spatial concentration patterns of two input measures of regional knowledge production: private R&D and academic research. In addition, we consider employment in manufacturing to capture agglomeration economies. The analyses are based on data aggregated by two digit SIC industries and at the level of Austrian political districts to explore the extent to which knowledge spillovers are mediated by spatial proximity in Austria. A time-space comparison will make it possible to study whether divergence or convergence processes in knowledge creation have occurred between the years of 1982 and 1998.

The remainder of the chapter is structured as follows. Section 7.2 introduces the exploratory tools used to analyse the spatial data and describes the data on which the study is based. Section 7.3 focuses on the identification of spatial clustering patterns of knowledge production in the last two decades, while Section 7.4 relates spatial distribution of knowledge inputs to spatial patterns of knowledge output. The final section summarises the research findings and points to directions for future research.

7.2   Methodology and Data

This contribution builds on the proposition that spatial clustering of knowledge production is induced by geographically bounded knowledge externalities: the larger the intensity of knowledge spillovers among the actors of a spatial [national, regional or local] innovation system, the higher the degree of spatial clustering of knowledge production. In order to shed some light into this issue, we have used the normalised Herfindahl index first to measure the degree of spatial concentration of both some input and output measures of knowledge production utilizing political districts as the basic spatial units of analysis.

To assess the extent to which the variable of interest is concentrated at the level of spatial units, the Herfindahl index in its normalised version is used in this contribution. This index is defined as $HI = 1 + \ln \Sigma_i S_i^2 / \ln n$, where S_i stands for the share of the measurement of the variable of interest in basic spatial unit i of the national total and n denotes the number of basic spatial units. A major advantage of this index is that it can provide a basis for straightforward comparisons as it ranges between 0 and 1. The index takes the value of 0 if the variable of interest is evenly distributed across regions and the value of 1 if it is completely concentrated in one basic spatial unit.

Spatial autocorrelation [also referred to as spatial dependence or spatial association] in the data can be a serious problem, rendering conventional statistical analysis tools invalid and hence requiring specialised spatial analytical tools. This problem occurs in situations where the observations are non-independent over space, that is where nearby basic spatial units are associated in some way. Sometimes, this association is due to a poor match between the spatial extent of the phenomenon of interest such as knowledge production in the current context and the administrative units for which data are available. Sometimes, it is due to a spatial spillover effect. The complications are similar to those found in time series analysis, but are exacerbated by the multi-directional, two-dimensional nature of dependence in space rather than the uni-directional nature of time dependence. Avoiding the pitfalls arising from spatially correlated data is crucial to good spatial data analysis (Fischer 1998, 2001b).

Exploratory analysis of area data involves identifying and describing different forms of spatial variation in the data. In the context of this contribution, special attention has been given to measuring the spatial association between observations for one variable. The presence of spatial association can be identified in a number of ways: a rigorous method is to use an appropriate spatial autocorrelation statistic, a more informal one to use, for example, a scatterplot and plot each value against the mean of the neighbouring areas. In the former approach to spatial autocorrelation the overall pattern of dependence in the data is summarised into a single indicator, such as *Moran's I* or *Geary's c*. Both of these require the choice of a spatial weights matrix [also referred to as contiguity matrix] that represents the topology or spatial arrangement of the data and manifests our understanding of spatial association.

In this current study *Moran's I* statistic is used. *Moran's I* is based on cross-products to measure value association:

$$I = (n / S_0) \sum_i \sum_j w_{ij}(x_i - \mu)(x_j - \mu) / \sum_i (x_i - \mu)^2 \qquad (7.1)$$

where n stands for the number of observations, x_i denotes an observation on a variable x at location i, w_{ij} is an element of the spatial weights matrix ($i=1,\ldots, n$; $j=1,\ldots, n$), μ the mean of the x variable, and S_0 the normalising factor equal to the sum of the elements of the weights matrix:

$$S_0 = \sum_i \sum_j w_{ij}$$ (7.2)

For a row-standardised spatial weights matrix – the preferred way to implement this test – the normalising factor S_0 equals n [since each row sums to 1], and the statistic simplifies the ratio of a spatial cross-product to a variance. The neighbourhood or contiguity structure of a data set is formalised in a spatial weights matrix $(w_{ij}) = W$ of dimension equal to the number of observations (n), in which each row and matching column correspond to an observation pair (i, j). The elements w_{ij} of the weights in the matrix W take on a non-zero value [1 for a binary matrix, or any other positive value for general weights based on the distance view of spatial association] when observations i and j are considered to be neighbours, and a zero value otherwise. By convention, the diagonal elements of the weights matrix, (w_{ij}), are set to zero. Note that the row-standardized weights matrix is likely to become asymmetric, even though the original matrix may have been symmetric.

Tests for spatial autocorrelation for a single variable in a cross-sectional data set are based in this study on the magnitude of *Moran's I* that combines the value observed at each basic spatial unit with the values at neighbouring locations. Basically, *Moran's I* is a measure of the similarity between association in value and association in space [contiguity]. Spatial autocorrelation is considered to be present when the statistic for a particular map pattern has an extreme value compared to what would be expected under the null hypothesis of no spatial autocorrelation. We are interested in instances where large values are surrounded by other large values, or where small values are surrounded by other small values. This is referred to as *positive spatial autocorrelation* and implies a spatial clustering of similar values.

The exact interpretation of what is 'extreme' depends on the distribution of the test statistic under the null hypothesis, and on the chosen level of the Type I error, that is on the critical value for a given significance level. Two main approaches are used in the study to determine the distribution of a test for spatial autocorrelation under the null hypothesis. The first, and most widely used assumption, is that the data follow an uncorrelated *normal* distribution. If this is not the case, the so-called permutation approach is adopted. This utilises the data themselves to construct an artificial reference distribution by resampling the data over the basic spatial units [that is by allocating the same set of observations randomly to the different locations]. The degree of 'extremeness' of the *Moran I* statistic for the observed pattern can then be assessed by comparing it to the frequency distribution of the random permutations. A simple rule of thumb can be based on a so-called pseudo significance level. This is computed as $(T+1)/(M+1)$ where T denotes the number of values in the reference distribution that are equal to or more extreme than the observed statistic, and M is the number of permutations carried out [M may be taken to be 99, for example].

Since the *x*-variable is in deviation from its mean, *Moran's I* is formally equivalent to a regression coefficient in a regression of Wx on x. The interpretation of I as a regression coefficient provides a way to visualise the linear association between x and Wx in form of a bivariate scatterplot of Wx against x, termed as a *Moran scatterplot* (Anselin 1997). The *Moran scatterplot* can be augmented with a linear regression [as a linear smoother of the scatterplot] that has *Moran's I* as slope, and can be used to indicate the degree of fit, the presence of outliers etc. in the usual manner. The lower left and upper right quadrants represent clustering of similar values. By contrast, the upper left and lower right quadrants contain non-clustering observations. Points in the scatterplot that are extreme with respect to the central tendency reflected by the regression slope may be outliers in the sense that they do not follow the same process of spatial dependence as the other observations. Leverage points are observations that have a large influence on the regression slope. If the regression has a positive slope [that is, positive global spatial association], points further than two standard deviations from the center (0, 0) in the upper left and lower right quadrants are considered in this study as outliers. Observations that are in a two standard deviations distance from the centre in the lower left and upper right quadrants are leverage points.

The interpretation of *Moran's I* as a regression coefficient clearly illustrates the way in which the statistic summarises the overall pattern of linear association, in the sense that a lack of fit would indicate the presence of local pockets of non-stationarity. It also indicates that the global measure of spatial association may be a poor measure of the actual dependence in the process at hand. *Local* measures of spatial association such as the *local Moran* statistic (Anselin 1995) are suitable for detecting potential non-stationarities in a spatial data set, for example, when the spatial clustering is concentrated in one subregion of the study area only. The local Moran for an observation *i* may be calculated as follows:

$$I_i = (x_i - \mu) \sum_j w_{ij} (x_j - \mu)$$
(7.3)

where w_{ij} denotes the (i,j)th element of a spatial weights matrix in row-standardized form. Significant local *Moran's* I_i detect non-random local spatial clusters where observation *i* is the center of the cluster. Significance tests are based on the permutation approach [see above].

Exploratory spatial data analysis in this study focuses explicitly on the spatial aspects of both input and output measures of the knowledge production process. Given the supposedly micro scale of interactions in knowledge production, the spatial level of data aggregation should be as low as possible. Due to data availability restrictions we were forced to choose political districts as the basic units of analysis in this study. Two input measures of knowledge production are considered: R&D expenditures in manufacturing and university research expenditures. Additionally, manufacturing employment is included to proxy agglomeration effects on knowledge production in an unspecified form. Patent count data are used as indicators of knowledge output despite their widely known drawbacks and problems (Basberg 1987; Pavitt 1988; Griliches 1990; Archibugi 1992; Archibugi and Pianta 1996; Fischer, Fröhlich and Gassler 1994).

Raw data on Austrian patents filed between 1982 and 1998 were provided by the Austrian Patent Office [APO]. The data files contain information on the application date, name of the assignee(s), address of the assignee(s), name of the inventors(s), location of the inventor(s), one or more International Patent Classification (IPC) codes and some information on the technology field of the patent application. Since location information on the inventor(s) was not always provided, the address of the assignee was used for tracing patent activity back to the region of knowledge production. It is common in Austria that the location of both the assignee [usually the firm where the inventor has a job] and the inventor are very near, and often in the same political district. Deviation from this pattern was found only for large multiple-location companies where patent applications were submitted by the companies' headquarters. For these cases, patents were re-distributed to the addresses of the inventors when these were located in different political districts. In the case of multiple assignees located in different political districts, we followed the standard procedure of proportionate assignment. We used the MERIT concordance table (Verspagen, Moergastel and Slabbers 1994) between patent classes (International Patent Classes, IPC) and industrial sectors (ISIC) to match the patent data with the two-digit ISIC codes.

Finally, we needed data on the amount of university research relevant to each two-digit ISIC industry. There are great differences in the scope and commercial applicability of university research undertaken in different scientific fields. Academic research will not necessarily result in useful knowledge for every industry, but scientific knowledge from certain academic institutes [especially those operating in the transfer sciences] is expected to be important for specific industries. To capture the relevant pool of knowledge, scientific fields/academic disciplines have been assigned to relevant industrial fields at the broad level of two-digit ISIC industries using the survey of industrial R&D managers by Levin et al. (1987) to measure the relevance of a discipline to an industry. For example, product innovation activities in drugs (ISIC 24) is linked to research in medicine, biology, chemistry and chemical engineering.

Unfortunately, university research expenditure data disaggregated by scientific disciplines are not available in Austria. But they can be roughly estimated on the basis of two types of data provided by the Austrian Federal Ministry for Science and Research: *first*, national totals of university research expenditures 1991 disaggregated by broad scientific areas [natural sciences, technical sciences, social sciences, humanities, medicine, agricultural sciences], and, *second*, data on the number of professional researchers employed in 1991 [that is, university professors, university assistants and contract research assistants] disaggregated by scientific areas and political districts. The best that can be done is to break down the university research expenditure data to the level of scientific disciplines disaggregated by political districts using the following disaggregation procedure:

$$R_{DP} = \frac{R_{AN}}{P_{AN}} P_{DP} \qquad (7.4)$$

where R_{DP} denotes university expenditure in a specific discipline D and in political districts, R_{AN} national research expenditure in a particular scientific area A, P_{AN} national total of professional researchers in scientific area A, and P_{DP} the number of professional researchers working in university institutes belonging to discipline D and located in political district P.

7.3   Time-Space Patterns of Knowledge Production in Austria

During the last two decades knowledge production in Austria, measured in terms of patent applications, shows an apparent stability both spatially and by industry. Table 7.1 shows the sectoral distribution of patent applications in two time periods: 1982-1989 and 1990-1997.

Table 7.1   Sectoral distribution of Austrian patent applications in the periods 1982-1989 and 1990-1997

	Time Period		Percentage Change from 1982-1989 to 1990-1997
	1982-1989	1990-1997	
Sectoral Share of Patents in Total Patents in Manufacturing			
Machinery	26.02	24.52	-5.75
Metal Products excluding Machines	18.18	19.97	9.87
Instruments	9.48	10.64	12.27
Transportation Vehicles	9.23	8.47	-8.29
Chemistry and Pharmaceuticals	8.33	7.3	-12.39
Electrical Machinery	6.86	6.54	-4.73
Construction	5.53	5.26	-4.88
Stone, Clay and Glass Products	3.73	3.39	-9.1
Paper, Printing and Publishing	2.53	3.29	30.07
Electronics	2.61	2.78	6.46
Basic Metals	2.62	2.52	-3.73
Textiles and Clothes	1.87	1.38	-26.49
Computers and Office Machines	0.77	1.35	75.95
Food, Beverages, Tobacco	0.83	1.12	34.05
Rubber and Plastics	0.94	1.03	9.87
Oil Refining	0.29	0.25	-11.77
Wood and Furniture	0.18	0.19	5.39
Correlation Coefficient	0.99		
Total Number of Patent Applications in Manufacturing	15,019	14,251	-5.11
Normalised Herfindahl Index of Sectoral Concentration	0.30	0.29	
Share of Vienna in Manufacturing Total [as percentage]	32.16	34.05	

Source: Austrian Patent Office

It can be seen from Table 7.1 that knowledge production concentrates in mechanical areas of manufacturing, especially in machinery. High technology fields such as electronics, computers or chemicals and pharmaceuticals are significantly less represented. This corresponds to the sectoral structure of manufacturing production (Gassler 1993). However, no apparent specialisation is present at the sectoral level, as indicated by the Herfindahl index (0.30).

Neither the total number of patents in manufacturing [about 1,800 per year] nor the ranking of manufacturing sectors [as shown by the high correlation of sectoral shares in the two time periods] have changed meaningfully from the 1980s to the 1990s. It is also clear from the table that knowledge production is predominantly concentrated in Vienna, the capital of Austria, as shown by its share of more than 30 percent of the national manufacturing total.

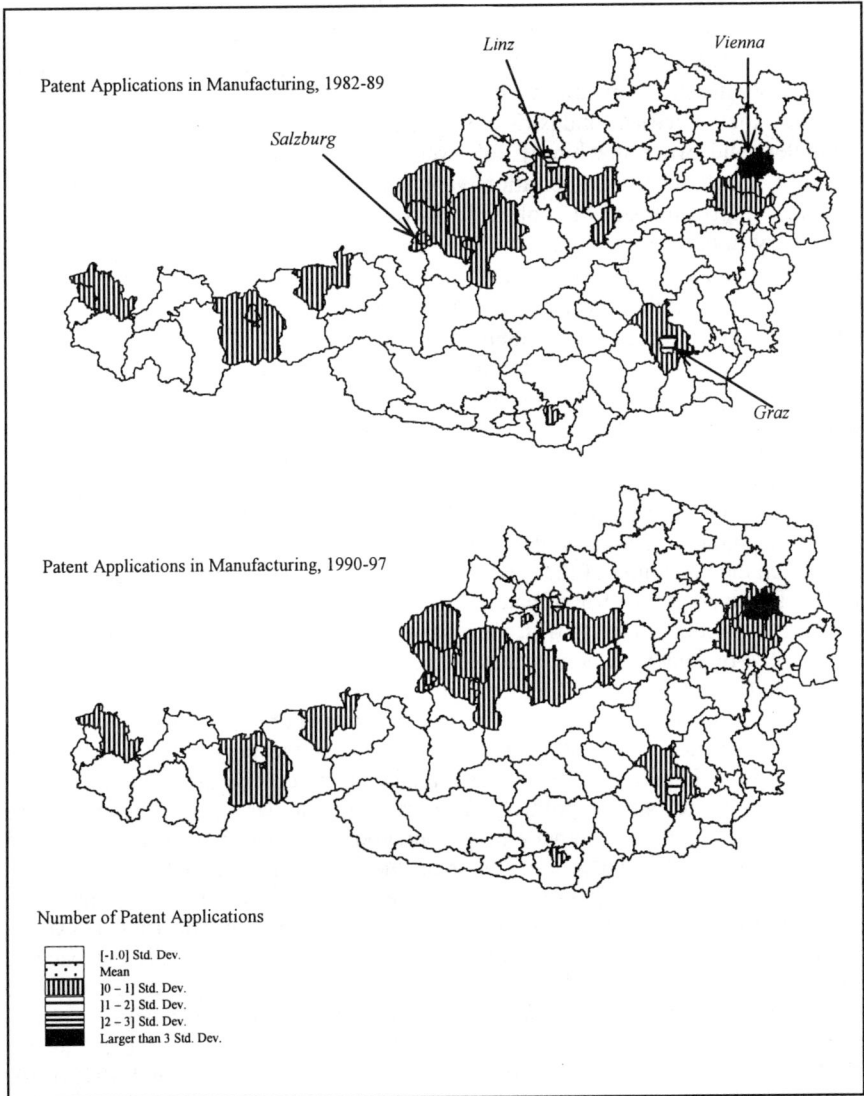

Fig. 7.1   Spatial distribution of Austrian patent applications in manufacturing in the periods 1982-1989 and 1990-1997

Spatial distribution of knowledge production also shows clear stability during the time period of the study. As indicated in Fig. 7.1, there are three larger concentrations of patents and some smaller ones. The three large areas of knowledge production constitute about two-thirds of the total number of Austrian patents. These include the metropolitan area of Vienna [i.e., the city of Vienna and the political districts building the urban fringe] with more than 30 percent of the national knowledge output; the Salzburg and Linz regions with 21 percent, and the Graz region with 8 percent of national knowledge production (see Fig. 7.1).

Fig. 7.2a provides insights into regional concentration tendencies of Austrian knowledge production for four different manufacturing areas over the period of 1982-1998 measured by means of the normalised Herfindahl Index. There is evidence that electrical industries [including electronics, electrical machinery, computers and office machines], followed by mechanical sectors [such as metal products, machinery, transportation vehicles and instruments] concentrate in a relative small number of political districts, whereas chemistry and drugs [chemistry and pharmaceuticals, rubber and plastics, and oil refining] together with traditional sectors [food, beverages and tobacco, construction, stone, clay and glass products, textiles and clothes, paper, printing and publishing, and wood and furniture] tend to spread more widely over the country.

Interestingly, the level of spatial concentration did not change meaningfully during the eighties, whilst the nineties brought a notable decrease in geographical concentration especially in traditional and chemical sectors. This change was induced by a transformation in the spatial structure of Austrian patenting activities. Even though the overall level of knowledge creation was about the same in 1998 [1,637 patents] as it had been in 1982 [1,597 patents], the share of total knowledge output in political districts which had had an above-average level of knowledge creation in the beginning of the period had decreased significantly by the end of the 1990s. The average number of patents diminished from 32 to 24 [a decrease of 25 percent] in political districts with above-average level of knowledge production in 1982, while regions with below-average patent applications at the beginning of the time period expanded their patenting activities from an average of 8 to 14 patents by 1998 [an increase of 88 percent]. As a result, the share of those political districts that had above average levels of knowledge creation at the beginning of the period decreased from 72 percent of the national total to 52 percent by the end of the 1990s.

The extent to which political districts with similar levels of knowledge production locate in each other's neighbourhood was measured by the Moran's I statistic for the four manufacturing areas for the period of 1982-1998, and is shown in Fig. 7.2b. A general trend of increasing spatial dependence among neighbouring political districts emerges with no significant variation across industries. However, values of Moran's I remain rather low during the entire period of study and become significant only at the beginning of the 1990s. Some sectoral differences are highlighted in this respect. While for traditional sectors clustering became significant [at the 10 percent level] between 1991 and 1996, this took place for the electronic and mechanical areas during the period of 1995-1998 and 1996-1998, respectively. There was no period of significant spatial clustering for chemical sectors. Overall, the results in Fig. 7.2b show a low level of spatial dependence among neighbouring political districts in Austrian manufacturing knowledge production.

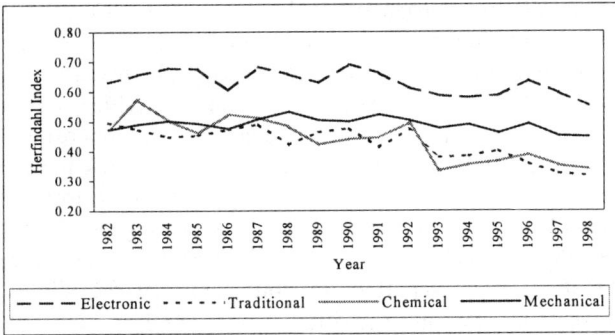

Fig. 7.2a   Geographical concentration of patents for four manufacturing areas, measured by the normalised Herfindahl index [1982-1998]

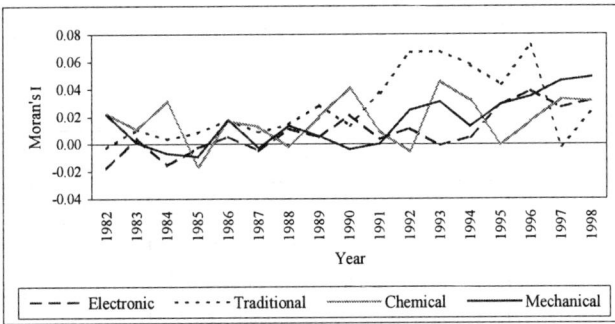

Fig. 7.2b   Spatial association of patents across political districts for four manufacturing areas, measured by Moran's I statistics [1982-1998]

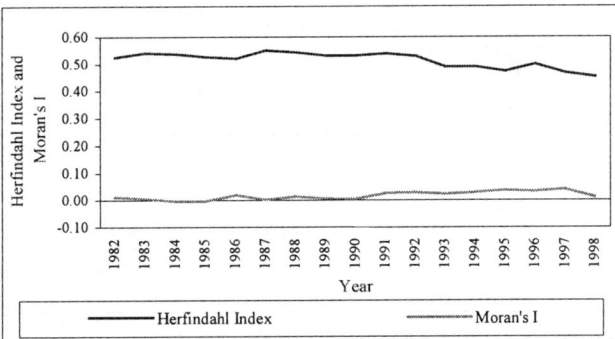

Fig. 7.2.c   Geographic concentration and spatial association across political districts of patents in manufacturing in Austria [1982-1998]

Fig. 7.2c illustrates the values of the two compatible measures of spatial clustering of knowledge production: the normalised Herfindahl index of geographical concentration and *Moran's I* statistic of spatial dependence, both calculated at the level of Austrian political districts for manufacturing over the period of 1982–1998. The fact that patenting activities did not expand significantly during the period of study together with the opposite trends of the two measures in the 1990s suggests that relocation of knowledge production [as indicated by a decrease in the Herfindahl index] took place from core areas of patenting to their neighbouring political districts [as suggested by the positive trend in *Moran's I* statistics] resulting in increased spatial concentration of knowledge creation. It is important to note here that the slight increase of clustering in Austrian patenting activities in the period of 1982–1998 does not seem to be the outcome of a dynamic, self-reinforcing process induced by local environments with many knowledge externalities leading to expanding clusters of knowledge production and an overall growth in knowledge output. Instead, it is characterised by a spatial shift of knowledge production to neighbouring peripheral areas, while the overall level of knowledge output stays largely unchanged.

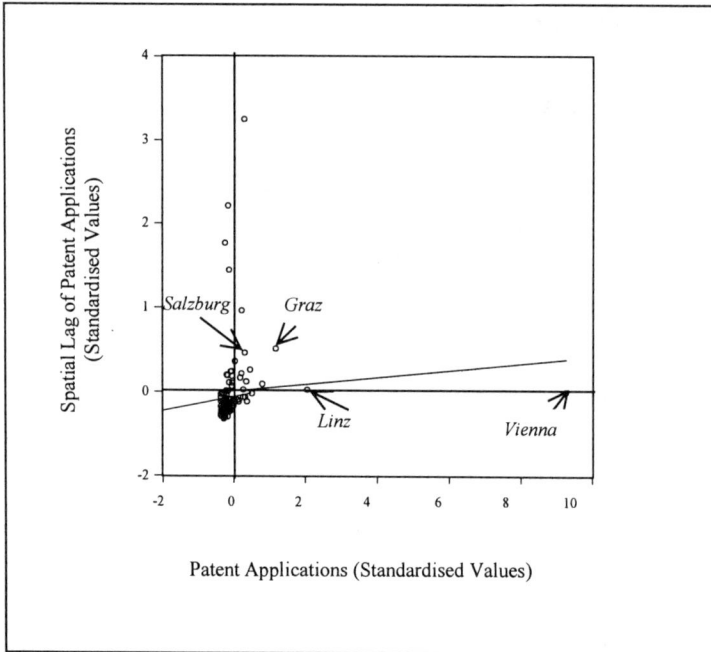

Fig. 7.3   Moran Scatterplot: Austrian patent applications in manufacturing [1998]

The *Moran scatterplot* of Austrian patents in 1998 in Fig. 7.3 shows spatial patterns of Austrian knowledge production at the end of the study period. The units of observation are Austrian political districts. The horizontal axis represents standardised values of patent counts while on the vertical axis average values of the same variable in neighbouring political districts are given [i.e. a row-standardised simple contiguity matrix is used for calculations]. The positive slope of the regression line reflects a positive value of *Moran's I* indicating an overall tendency of positive spatial association among neighbouring political districts. This tendency is predominantly supported by spatial clustering of political districts where lower than average level of knowledge creation takes place [as indicated by the high concentration of observations in the lower left quadrant of the scatterplot]. Leverage points in the upper right quadrant [i.e. political districts with above average patenting activity neighbouring similar regions] include Salzburg, Linz and Graz.

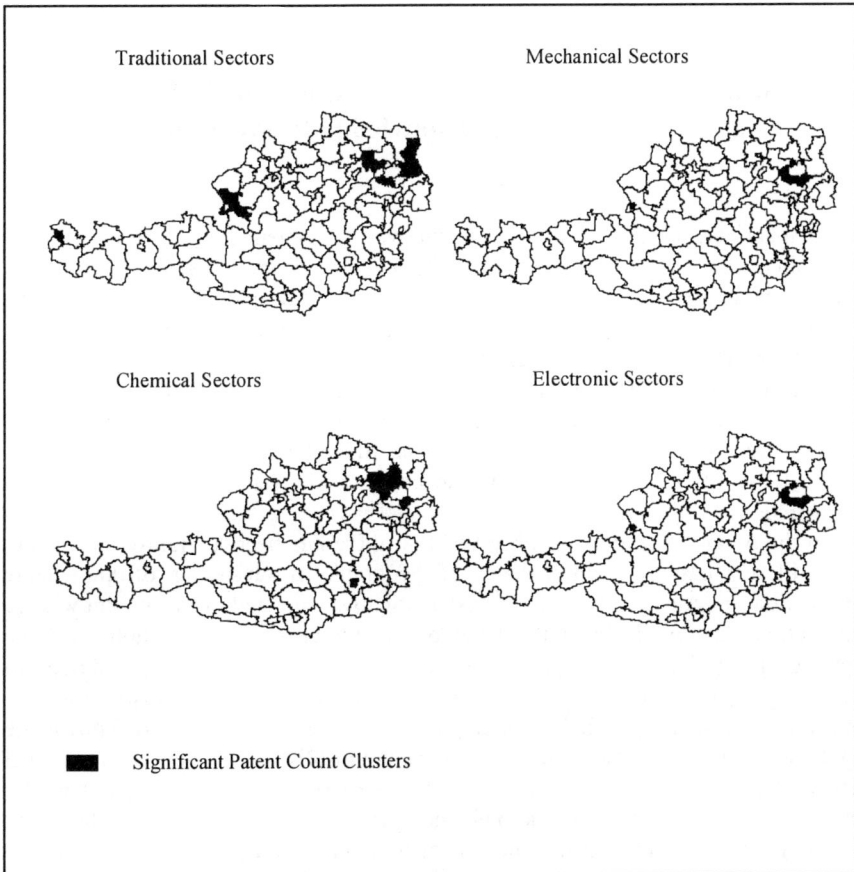

Fig. 7.4 Clusters of high values of patent counts for four manufacturing areas, measured by significant values of the local Moran statistics [1998]

It is very clear from Fig. 7.3 that Vienna is an outlier in Austrian knowledge production. The standardised value of the number of patents in Vienna is 9 times higher than the respective Austrian average. On the other hand, it is also demonstrated that Vienna is surrounded by political districts with levels of patenting activities around the average [i.e. the mean value of patents in its neighbourhood equals the national average].

Significant clusters of patenting activity in four manufacturing areas in 1998 are shown in Fig. 7.4. Significance at $p < 0.05$ is based on 1,000 random permutations. A row-standardised simple contiguity matrix has been used for calculations. Dark areas stand for core political districts of spatial clusters. The largest clusters are formed in traditional sectors whereas mechanical and electronic concentrations are relatively small. The Vienna metropolitan area is a significant cluster in all areas of manufacturing. Other clusters are formed around Salzburg [traditional, mechanical and electronic sectors], Graz [chemical sectors] and Dornbirn at the western border of the country [traditional sectors].

7.4   Local Inputs to Innovation – An Assessment of Their Relative Significance in Knowledge Production

As emphasised in the literature on innovation systems, production of new technological knowledge is not simply the outcome of the independent efforts of firms to innovate, but is also influenced by knowledge interactions with various actors in the system including other firms, and private and public research institutions. However, knowledge flows are very difficult [if not impossible] to trace empirically. Different methods have been proposed in the literature to measure knowledge flows at least partially such as patent citation analysis (Jaffe, Trajtenberg and Henderson 1993), analysis of patterns of co-patenting or co-publications (Hicks and Katz 1996) and counts of industry technology alliances (Haagedoorn 1994).

In the previous section we observed that there has been a slightly increasing, but still a relatively modest level of geographical clustering of knowledge production in Austria. Since no systematically collected data on knowledge interactions are available at the level of the regions, in this section we have applied an indirect approach to assess the significance of local inputs to knowledge production. A positive association in the spatial distribution of patenting and local knowledge inputs is taken to be an indication of knowledge spillovers existing in the production of economically useful new technological knowledge. Industrial R&D and university research are considered as potentially providing direct inputs to knowledge production, whereas manufacturing employment is included in the analysis as a proxy for unspecified agglomeration effects. Analysis is based on data aggregated at the level of Austrian political districts. In order to account for the time necessary to come up with patentable inventions, following the industrial experience reported for example in Edwards and Gordon (1984), a two-year time lag is applied between knowledge inputs (1991) and knowledge output (1993).

Table 7.2 provides a general profile of sectoral distribution of the three proxy variables of inputs to knowledge production: R&D in manufacturing and university research expenditures as well as the auxiliary variable of manufacturing employment. The three variables evidently follow different patterns of sectoral specialisation. Whereas R&D in manufacturing concentrates in electronics, university research focuses mainly on chemistry and pharmaceuticals, and instruments. On the other hand, about fourty percent of manufacturing employment is in the machinery, food and wood sectors. However, the overall sectoral concentration is not very strong, especially in manufacturing employment as indicated by the corresponding Herfindahl Index. Low values of correlation coefficients with patent counts in manufacturing suggest that sectoral distribution of knowledge production at the country level only vaguely follows the respective patterns of R&D, university research and employment.

Table 7.2   Sectoral distribution of R&D in manufacturing, university research and manufacturing employment [1991] [ranking follows patent orders in 1990-1997 in Table 7.1]

	R&D Expenditure in Manufacturing	University Research Expenditure	Manufacturing Employment
Manufacturing Sectors [a]			
Machinery	11.78	11.22	12.64
Metal Products excluding Machines	3.11	9.07	10.09
Instruments	0.73	59.49	3.80
Transportation Vehicles	7.05	21.58	4.62
Chemistry and Pharmaceuticals	15.22	62.41	4.13
Electrical Machinery	7.67	11.81	4.18
Stone, Clay and Glass Products	5.04	4.06	6.06
Paper, Printing and Publishing	2.00	na	7.19
Electronics	29.68	11.81	2.22
Basic Metals	4.28	9.07	5.14
Textiles and Clothes	1.72	na	9.43
Computers and Office Machines	1.98	25.27	0.14
Food, Beverages, Tobacco	2.20	1.55	12.45
Rubber and Plastics	5.86	9.23	4.13
Oil Refining	1.37	9.23	0.37
Wood and Furniture	0.32	0.80	13.40
Manufacturing Total	16.25[b]	6.41[b]	0.72[c]
Normalised Herfindahl Index of Sectoral Concentration	0.31	na	0.15
Correlation with the Sectoral Share of Patents in 1990-1998	0.15	na	0.34

Notes: a   denotes column percentage [for R&D expenditures in manufacturing and employment] and percentages of total university R&D expenditures [for university research expenditures]. Given that certain university institutes are allocated to more than one manufacturing sector, sum of percentages is not 100 in the third column.
  b   is in terms of 10^9 ATS.
  c   is in terms of 10^6 persons.

Fig. 7.5 shows that, though by and large the spatial distribution of patent counts follows the geographical patterns of industrial and university R&D as well as manufacturing employment, there are notable differences in pattern matching. A deeper understanding of the geographical patterns of Austrian knowledge production may be gained by calculating correlation coefficients between patent counts and each of the input measures [including the auxiliary variable of employment] at the level of political districts and for four manufacturing areas[1], in order to account for the supposedly different characteristics of the innovation system of the metropolitan area of Vienna [the definite positive outlier in Austrian knowledge production] and the three major cities supporting the overall positive clustering tendency of patent counts [i.e., Salzburg, Linz and Graz].

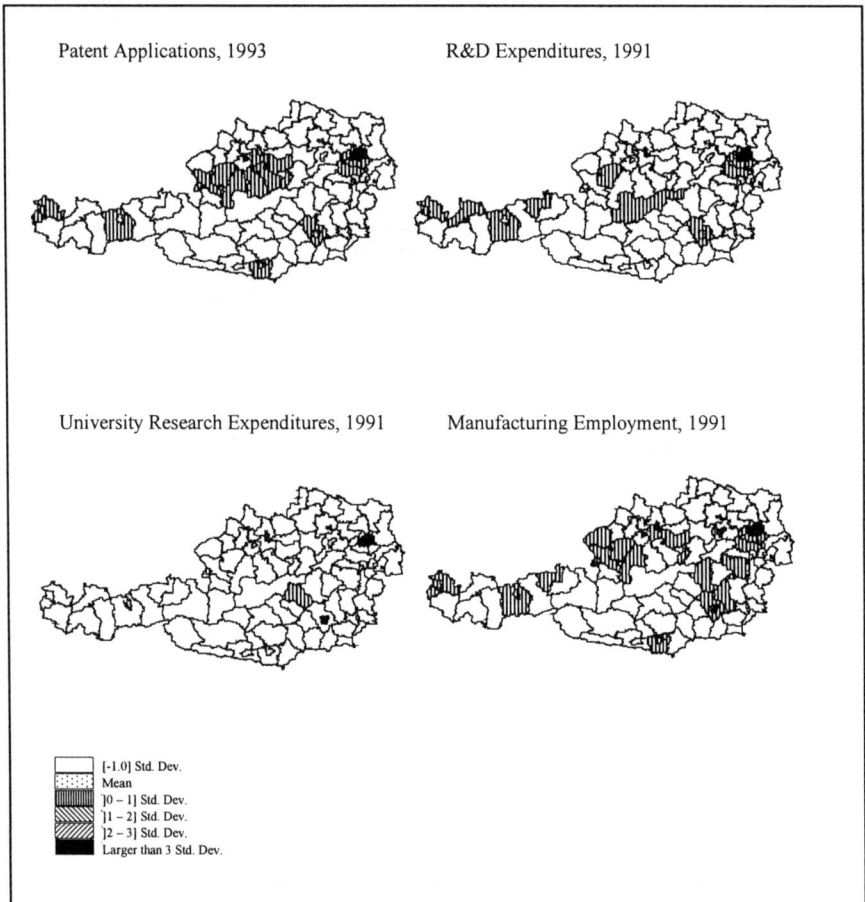

Patent Applications, 1993   R&D Expenditures, 1991

University Research Expenditures, 1991   Manufacturing Employment, 1991

[-1.0] Std. Dev.
Mean
]0 – 1] Std. Dev.
]1 – 2] Std. Dev.
]2 – 3] Std. Dev.
Larger than 3 Std. Dev.

Fig. 7.5   Spatial distribution of patent applications, private R&D expenditures, university research expenditures and manufacturing employment in Austria

Fig. 7.6 shows correlation coefficient values for three different sets of observations: the whole sample, political districts excluding Vienna and political districts excluding Salzburg, Linz, Graz and Vienna. The following three major observations can be derived from this figure. *First*, the four manufacturing areas exhibit dissimilar correlation patterns. Considering only those coefficients calculated for the whole sample, patent counts in electronic sectors are highly correlated with all the three measures of local knowledge inputs, while knowledge production in chemicals is more related to local employment and R&D. On the other hand, in mechanical and traditional sectors the highest correlations are observed with employment and university research. *Second*, the data shown in the figure suggests that the outlier position of Vienna in knowledge production might well be the result of its comparatively strong reliance on local knowledge inputs.

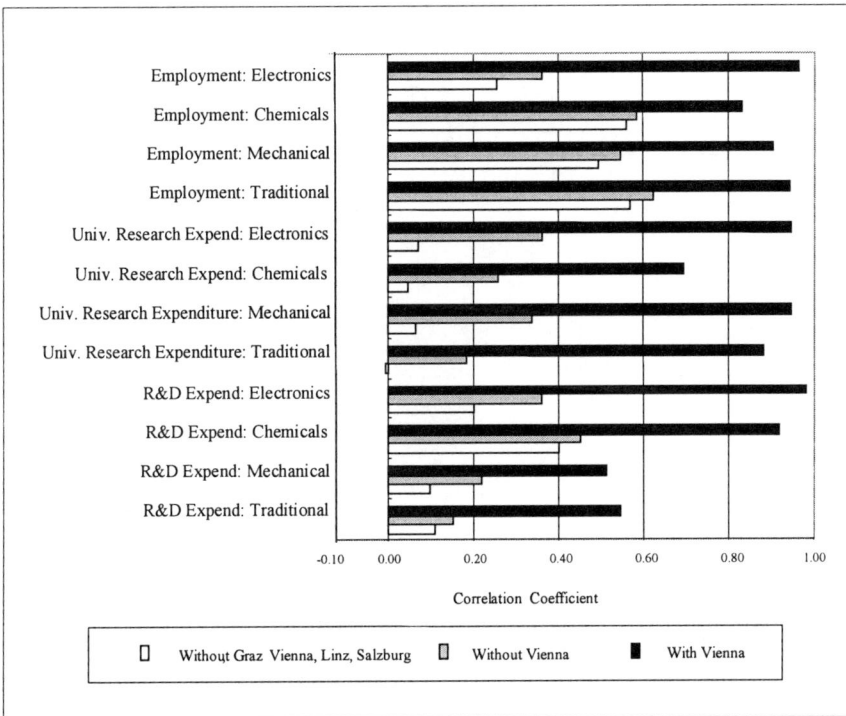

Fig. 7.6 Correlation between patents in 1993 and selected measures of potential local inputs to innovation in 1991 at the level of Austrian political districts

After excluding Vienna from the sample, correlation coefficients decrease significantly, especially the R&D measures. The smallest falls are observed in correlations with local employment, with the exception of the electronic sectors. *Third*, regarding the degree to which knowledge production in the three major Austrian cities exhibit distinct characteristics relative to the rest of the sample [excluding Vienna], dissimilar patterns are observed for the research variables, but not for employment.

It is important to note that regional knowledge output increases faster than any of its local inputs. This might be taken as a sign of the existence of regionally mediated knowledge flows. In Fig. 7.7 we can see scatterplot diagrams of patents and R&D in manufacturing, university research and manufacturing employment. Data are arranged in increasing order of the variables on the horizontal axes.

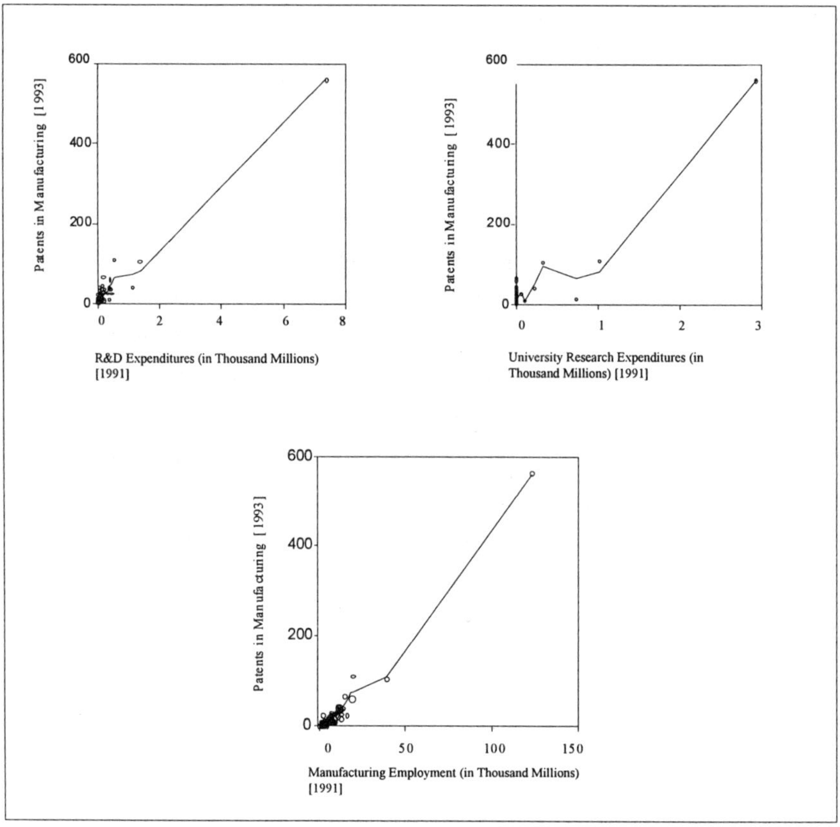

Fig. 7.7   Scatterplots with curves of nearest neighbour fit [Loess fit] for patents in manufacturing related to R&D in manufacturing, university research and manufacturing employment in Austria

Additionally, in order to give an indication of the direction and size of the change in patents in manufacturing, we have also estimated the curves of nearest neighbour fit [Loess fit]. For each data point in the sample, a locally weighted polynomial regression has been estimated. This is a local regression, since we have used only the subset of observations which lie in the neighbourhood of each point to fit the regression model (Cleveland 1994). In case of increasing returns in knowledge production, Loess fit curves show an exponential growth in patents.

The only variable for which increasing returns dominate the entire sample is manufacturing employment. This shows that the higher the concentration of production in an area, the higher the probability of knowledge-related linkages arising among firms, resulting in a higher than proportional increase in knowledge production. However, this relationship cannot be observed for R&D in manufacturing and university research linkages throughout the whole sample. Some degree of potential research spillover effects might be present in larger cities, and they seem to have a definite role in Vienna [the highest point in each scatterplot]. However, Fig. 7.8 indicates no signs of significant interregional linkages.

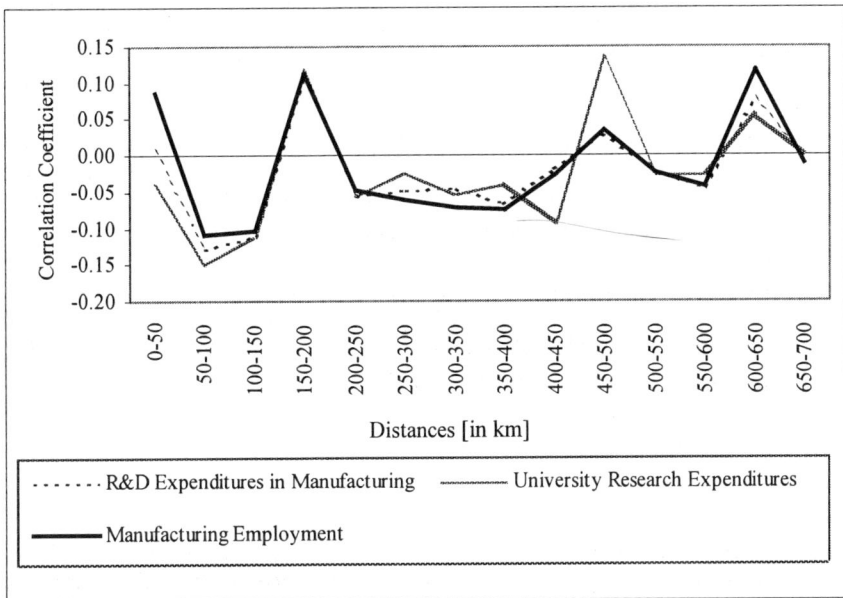

Fig. 7.8 Cross-regional correlation patterns between patent applications [1993] and knowledge inputs [1991] in increasing distances from the patenting political district

7.5 Conclusions

In recent years, the role of space in general and of spatial externalities in particular has gained an increasingly prominent position in mainstream economics, partly stimulated by the visibility of Krugman's work on the 'New Economic Geography' (Krugman 1991b). Of course, the importance of space is not new to geographers and regional scientists. Based on descriptive and exploratory techniques [*Moran's*

I test for spatial autocorrelation and the Moran scatterplot] in this chapter we have made an initial attempt to analyse the effect of space in the creation of knowledge. Clusters of the output of the knowledge creation process [measured in terms of patent counts] are compared with spatial concentration patterns of three input measures of local knowledge production: R&D in manufacturing, university research activities and manufacturing employment.

Empirical evidence shows that knowledge production in Austria tends to focus largely in mechanical areas of manufacturing rather than in high-tech fields such as electronics or computers. It is interesting to note that this pattern has changed little during the past two decades. We have been able to identify only a weakly growing trend of clustering. However, this does not appear to be so much the outcome of a dynamic process generated by intensive knowledge flows at the local level, as the consequence of a spatial shift in knowledge production. There is no doubt that Vienna with its strong presence of high quality research organisations and R&D in manufacturing dominates the knowledge creation process. Some smaller clustering tendencies were discovered around Salzburg, Linz and Graz.

Geographic stability of knowledge generation characterised by weakly expanding clusters may well be the outcome of relatively undeveloped linkages among the major actors of the Austrian innovation system as suggested by the limited role of local knowledge flows in the most parts of the country. Cluster generating increasing returns appear to result largely from between-firm knowledge diffusion rather than from knowledge spillover effects.

As in the case of any exploratory data analysis, the above findings need to be treated with caution and should be viewed only as an initial pre-modelling stage in the endeavour. Future research activities will be devoted to shedding further light on the issue of local university knowledge transfer by refining Griliches-Jaffe knowledge production approach (see, Griliches 1979; Jaffe 1989), modelling knowledge spillovers in form of a spatially discounted external stock of knowledge, and employing spatial econometric tools for model specification and estimation (see Fischer and Varga 2001)

Acknowledgements. The authors gratefully acknowledge the grant no. 7994, provided by the Jubiläumsfonds of the Austrian National Bank, and the support received from the Department of Economic Geography & Geoinformatics at the Vienna University of Economics and Business Administration, and the Austrian Research Centers Seibersdorf. They also wish to express their thanks to Christian Rammer, Doris Schartinger, Norbert Böck [Austrian Research Centre Seibersdorf], Werner Hackl [Austrian Chamber of Commerce, Vienna] and Karl Messman [Austrian Central Statistical Office, Vienna] for assisting in various phases of data collection.

Endnote

1   Traditional sectors include food, beverages and tobacco [ISIC 15-16], construction [ISIC 45], stone, clay and glass [ISIC 25], textiles and clothing [ISIC 17 and 18], paper,

printing and publishing [ISIC 21 - 22], wood and furniture [ISIC 20 and 36]. The mechanical sectors include basic metals [ISIC 27], instruments [ISIC 33], transportation vehicles [ISIC 34 - 35], machinery [ISIC 29], metal products [ISIC 28]. The chemical sectors consist of rubber and plastics [ISIC 25], chemistry and pharmaceuticals [ISIC 24] and oil refining [ISIC 23], whereas the electronic sectors include electronics [ISIC 32], electrical machinery [ISIC 31], and computers and office machines [ISIC 30].

Appendix A-7.1:

Patent Applications, R&D Expenditures, University Research Expenditures and Employment in Manufacturing for the 99 Austrian Political Districts

Political District	Patent Applications (1993)	Industrial R&D Expenditures in 10^3 ATS (1991)	University Research Expenditures in 10^6 ATS (1991)	Employment (1991)
Eisenstadt (Stadt)	6	0	0	596
Rust (Stadt)	0	0	0	41
Eisenstadt-Umgebung	4	32,344	0	1,776
Güssing	0	1,000	0	1,023
Jennersdorf	1	0	0	1,427
Mattersburg	4	10,548	0	3,461
Neusiedl am See	5	14,771	0	1,731
Oberpullendorf	1	4,390	0	2,555
Oberwart	0	3,978	0	5,096
Klagenfurt (Stadt)	27	13,527	36	7,113
Villach (Stadt)	11	25,919	0	5,647
Hermagor	2	160	0	1,045
Klagenfurt Land	22	0	0	2,251
Sankt Veit an der Glan	5	8,160	0	5,162
Spittal an der Drau	5	90,711	0	4,655
Villach Land	8	52,886	0	3,687
Völkermarkt	1	3,200	0	3,236
Wolfsberg	5	18,586	0	4,497
Feldkirchen	2	1,439	0	1,702
Krems (Stadt)	5	52,877	0	4,057
Sankt Pölten (Stadt)	7	33,383	0	8,333
Waidhofen (Stadt)	7	4,595	0	1,606
Wiener Neustadt (Stadt)	5	36,376	0	5,143
Amstetten	36	107,121	0	12,255
Baden	38	348,885	0	13,350
Bruck an der Leitha	2	34,450	0	2,343
Gänserndorf	13	7,225	0	4,711
Gmünd	9	0	0	5,514
Hollabrunn	1	770	0	1,743
Horn	7	1,456	0	2,279
Korneuburg	16	26,586	0	5,579
Krems (Land)	2	0	0	1,823
Lilienfeld	1	3,521	0	3,253
Melk	9	9,790	0	4,714
Mistelbach	8	0	0	3,697
Mödling	32	196,105	0	10,616
Neunkirchen	13	67,802	0	8,637
Sankt Pölten (Land)	14	51,303	0	7,303
Scheibbs	2	3,600	0	2,847
Tulln	4	28,057	0	3,445
Waidhofen an der Thaya	1	11,930	0	3,168
Wiener Neustadt (Land)	7	7,618	0	5,515
Wien-Umgebung	23	305,350	0	10,303
Zwettl	4	0	0	2,233
Linz (Stadt)	101	1,375,777	218	39,068
Steyr (Stadt)	39	1,124,624	0	11,399
Wels (Stadt)	28	35,720	0	9,744
Braunau am Inn	14	158,617	0	12,958
Eferding	5	3,772	0	2,725
Freistadt	1	420	0	2,571

			ctd.
Gmunden	42	133,864	0	11,832
Grieskirchen	14	51,170	0	5,883
Kirchdorf an der Krems	23	17,706	0	7,065
Linz-Land	21	102,877	0	16,499
Perg	13	23,580	0	4,894
Ried im Innkreis	7	50,189	0	6,108
Rohrbach	4	3,650	0	3,817
Schärding	8	33,760	0	4,239
Steyr-Land	12	9,314	0	3,317
Urfahr-Umgebung	10	0	0	2,658
Vöcklabruck	56	386,655	0	19,110
Wels-Land	9	79,982	0	7,511
Salzburg (Stadt)	37	41,309	137	10,594
Hallein	12	123,539	0	6,642
Salzburg-Umgebung	31	22,640	0	10,490
Sankt Johann im Pongau	14	21,155	0	5,200
Tamsweg	1	0	0	1,044
Zell am See	7	32,316	0	4,575
Graz (Stadt)	105	519,747	1,288	19,544
Bruck an der Mur	7	99,697	0	9,246
Deutschlandsberg	9	114,536	0	5,595
Feldbach	3	6,705	0	4,050
Fürstenfeld	2	12,416	0	2,308
Graz-Umgebung	25	461,144	0	9,425
Hartberg	4	10,400	0	4,929
Judenburg	14	79,326	0	6,633
Knittelfeld	3	19,529	0	3,805
Leibnitz	4	3,017	0	5,377
Leoben	9	48,238	176	6,755
Liezen	7	191,806	0	6,040
Mürzzuschlag	6	26,212	0	6,336
Murau	4	0	0	1,837
Radkersburg	0	383	0	1,249
Voitsberg	13	40,615	0	4,010
Weiz	9	142,596	0	7,566
Innsbruck (Stadt)	15	5,692	907	5,637
Imst	5	14,050	0	2,352
Innsbruck (Land)	35	422,458	0	13,247
Kitzbühel	10	22,031	0	3,233
Kufstein	10	356,486	0	9,382
Landeck	0	0	0	1,776
Lienz	5	9,147	0	4,043
Reutte	5	183,676	0	2,722
Schwaz	18	102,295	0	7,303
Bludenz	5	24,674	0	7,075
Bregenz	65	180,774	0	14,763
Dornbirn	23	191,232	0	13,117
Feldkirch	20	134,127	0	10,918
Vienna	541	7,374,721	3,652	122,960

Sources: Patent applications data come from the Austrian Patent Office; industrial R&D data from the Austrian Chamber of Commerce; university research data from the Austrian Federal Ministry for Science and Research; employment data from the Austrian Central Statistical Office.

8 Knowledge Spillovers in High Technology Agglomerations: Measurement and Modelling

Elsie Echeverri-Carroll
The Red McCombs School of Business, The University of Texas at Austin

8.1 Introduction

Malecki (1980) observed that one of the characteristics of innovative or high-tech firms is that they tend to cluster in relatively few places. In the United States, for instance, a large proportion of the high tech industry is concentrated in Silicon Valley. Indeed, if we use patents to measure innovative activity in the United States, we see that about 50 percent of U.S. patenting activity occurs in only six states: California [where Silicon Valley is located], New York, Texas, Illinois, Michigan, and New Jersey.

When and why does high technology manufacturing concentrate in just a few regions? We claim that this key question may be closely related to the more general question that core-periphery models are trying to answer: When and why does manufacturing concentrate in some regions, leaving others relatively underdeveloped? Our hypothesis is particularly relevant now that the core-periphery model, introduced in regional economics in the 1950s (Myrdal 1957; Hirschman 1958), has been reworked in the 'new economic geography' that emerged in the 1990s (Krugman 1991a, 1991b; Fujita, Krugman and Venables 1999).

The new core-periphery models are built on the assumption that externalities which lead to the emergence of a core-periphery pattern are often *pecuniary*, associated with demand or supply linkages, rather than purely *technological* [knowledge-related]. However, most of the empirical literature on the growth of high technology regions stresses the importance of knowledge [or technological], not pecuniary externalities. In this regard, we believe that the growth models developed in the new economic development literature, that consider innovations as endogenous (Grossman and Helpman 1991), could provide much better insights into the relationship between high technology agglomeration, knowledge spillovers, and regional growth.

Researchers doing empirical work on the importance of knowledge externalities for high technology agglomerations and regional growth confront a difficult issue: how to proxy the 'immeasurable' knowledge externalities. Whereas pecuniary externalities are visible and therefore measurable, technological externalities, being invisible, are hard to measure. Krugman (1991a) points out that the difficulty in measuring technological knowledge spillovers is because they do not 'leave a paper trail'. Others have challenged this assumption and attempted to quantify knowledge spillovers through various proxies, including linkages between patent citations (Jaffe 1989; Jaffe, Trajtenberg, and Henderson 1993) or the effect on a firm's development of new products from other firms' R&D (Audretsch and Feldman 1996).

Here, we take a rather different approach in proxying knowledge spillovers. We believe that technological spillovers occur mainly between high technology firms, such as those in Silicon Valley. Thus, we asked a sample of these firms to assess the relative importance of various proxies of Marshallian agglomeration economies, including proxies for knowledge externalities. Before doing this, however, we had to decide how to define high technology firms.

Previous studies have used SIC code classifications to identify high technology firms. However, not all firms within an industry designated as high-tech are actually high-tech. For instance, the criteria used in a high-tech industry classification might include higher-than-average investments in R&D or a high proportion of engineers and scientists in R&D, but some of the firms within that SIC code will not satisfy these criteria. Thus, a bias is introduced into analyses that simply use SIC classifications to define firms as high tech. We found this bias to be quite large, as more than 50 percent of firms in our sample did not comply with the criteria that defined their industry group as high tech. Our results suggest that researchers should be cautious in formulating conclusions based on studies that use a SIC classification for high technology firms.

In our empirical study of the relative importance of knowledge externalities for the location of high technology firms, we take into consideration Krugman's (1991b) argument that there is no reason to assume that the motives for the agglomeration of high technology firms differ from the motives of non-high technology firms. In particular, we divide our sample of firms into two groups – high technology and non high technology – and test whether their reasons to agglomerate spatially are different.

Section 8.2 of this contribution contrasts the 'traditional' and 'new' core-periphery models and discusses their limitations as a way of explaining the growth of high-tech regions. Section 8.3 gives an overview of the data and the definitions of high-tech firms. Section 8.4 presents some statistical results on the relative importance of agglomeration economies for high-tech and non high-tech firms. Section 8.5 gives conclusions and suggestions for future research.

8.2   Core-Periphery Spatial Models

New economic development theory suggests that sustained long-term economic growth depends on the continuing capacity of firms to innovate (Grossman and Helpman 1991). In turn, the capacity to innovate will depend in part on the characteristics of the region in which a firm locates – in particular, on the relative abundance of skilled labour. Firms with little capacity to innovate make use of unskilled labour, and thus these firms tend to concentrate in regions where this kind of labour is abundant. Well-known examples of these two kinds of region are: Silicon Valley, where a high proportion of firms developing new products and processes locate, and the U.S.-Mexico border, where a large proportion of firms producing few, if any, innovations locate. In this regard, the spatial structure can be characterised by a 'core', where most innovations take place, and a 'periphery', where fewer innovations take place. One might think, therefore, that core-periphery models would be useful to explain this unbalanced spatial pattern. We label the two kinds of core-periphery models found in the literature as 'traditional' and 'new'.

Before discussing core-periphery models, we need to answer a key question: Why is modelling the imbalances in spatial structure important? Krugman (1998) notes that one of the obvious benefits of modelling is that it serves the purpose of placing geographical analysis squarely in the economic mainstream. Indeed, modelling has certainly stimulated the emergence of a new wave of theorising and, to a lesser extent, empirical work (see, for instance, Keller 2000).

The Traditional Core-Periphery Model

Researchers studying spatial structure before the 1980s observed that a cumulative growth process tended to benefit some regions [the core] at the expense of others [the periphery]. They associated economic growth with intra-regional sectoral linkages (Perroux 1955; Hirschman 1958; Myrdal 1957) or with interregional sectoral linkages through exports (Richardson 1979). The general consensus was that unbalanced spatial patterns were the result of the degree of input/output linkages between sectors in a particular space.

At that time, it was expected that regions with good inter-sectoral linkages would consistently show higher rates of growth than regions with poor inter-sectoral linkages. Richardson (1979) developed a very simple model of cumulative growth that could be easily linked to a dynamic version of an export-based model, assuming that exports were the leading force [and, in a more rigid position, the sole determinant] of regional growth.

Richardson's model is an example of the kind of partial equilibrium analysis that was conducted before the 1980s. The main characteristic of what we call *traditional* core-periphery models is that the process of cumulative growth is *exogenous* to the models. For example, there exists an exogenous power structure that favours unbalanced spatial patterns. Indeed, one could argue that the exogenous nature of the variables that cause cumulative growth in these models was one of the main reasons that regional economics remained outside mainstream economics until the 1990s. Also, as Krugman (1998) points out, the scarcity of models that incorporate increasing returns to scale and imperfect competition before the 1980s constrained the introduction of space into mainstream economic analysis.

Most economic models developed before the 1980s assume perfect competition and constant returns to scale. Without economies of scale at the level of the plant, producers would have no incentive at all to concentrate their activity. They would simply supply consumers from many local plants, taking into consideration transportation costs. As Krugman (1998) notes, increasing returns are central to the story of spatial concentration.

In Richardson's model, the cumulative growth process can be described by the three simple linear equations shown in Table 8.1. An increase in exports [output linkage] will escalate regional growth rates. Increasing regional growth rates (y) – used as a proxy for agglomeration economies and increasing returns to scale – induce higher productivity rates (r) (equation (1)). At the same time, higher productivity rates reduce the rate of growth of efficiency wages (w), defined as an index of monetary wages over an index of productivity (equation (2)). Finally, higher efficiency wages stimulate faster growth rates (y) (equation (3)). In this simple model, cumulative growth occurs because labour productivity is higher in large cities [due to larger agglomeration economies and returns to scale] than in small ones. Further, higher labour productivity will decrease efficiency wages, lower efficiency wages will generate more growth, and the process will feed itself continuously.

The New Economic Geography: Core-Periphery Models and Pecuniary Externalities

Krugman (1998) states that the key reason for the emergence of the new economic geography is the fact that imperfect competition is no longer regarded as impossible to model. In the 1980s, the 'new' industrial organisation literature developed tractable models of imperfect competition, and since 1990 the 'new' economic geography has used these models to explain endogenously generated spatial patterns. This new literature insists on models which are in general equilibrium and in which spatial structure emerges from invisible-hand processes. In other words, these models emerge as a result of the inter-temporal actions of self-interested economic agents. Fujita, Krugman, and Venables (1999) summarise the toolbox of technical tricks used in the new economic geography [and in the new trade and new growth theories]: Dixit-Stiglitz modelling of monopolistic competition[1], iceberg transportation costs[2], evolutionary game theory, and the use of the computer.

A simple model developed in Krugman (1991a) is widely regarded as having given birth to the new economic geography. Fujita, Krugman, and Venables (1999) [hereafter referred to as FKV] present a multi-location version of this model. Four equations of the FKV model, for the two regions case, are presented in Table 8.1. These four equations do not define a full economic model, but they nevertheless imply some of the most important relationships that drive the results of the model. This model's instantaneous solution can be thought of as determined by the simultaneous solution of the four equations for the number of regions, 4R equations in total. Obviously, not much can be said about the solution of this non-linear system of equations in the general case[3]. But FKV gain considerable insight by examining an obvious special case, that of a two-region economy in which agriculture is evenly divided between regions. This model still does not look very tractable, consisting as it does of eight simultaneous non-linear equations. The final methodology used is numerical analysis, giving specific values to the parameters: $\sigma=5$, $\mu=0.4$, and three different values for transportation costs (T): $T=1.5$ [low]; $T=1.7$ [intermediate]; and $T=2.1$ [high].

Agglomeration economies are introduced in these equations as forces that attract labour to a specific location – in other words, forces that determine spatial differences in real wages. Ottaviano and Puga (1997) call these migration-induced demand linkage models. The key question in these models is therefore: What forces maintain real wages continuously higher in the core region and, hence, continuously attract labour to this region? These forces can be explained through two effects, the *price index effect* and the *home market effect*.

The price index effect is related to the *supply* of manufactured goods. That is, a region with a larger manufacturing sector also has a lower price index for manufactured goods simply because a smaller proportion of this region's manufacturing consumption bears transportation costs. As seen in equation (2) in Table 8.1, a region with a larger manufacturing sector (n) has a lower price index for manufactured goods (G). A region with a lower price index for manufactured goods would also have higher real wages (ω) as represented by equation (4), therefore this region will attract workers. More workers means more manufacturing plants (higher n) since $n = L^{\mu}/\mu$. Thus, a cumulative growth process is generated.

The home market effect is related to the *demand* for manufactured goods. It says that, other things being equal, the nominal wage rate will tend to be higher in the larger market. In equation (3), we see that, other things being equal, a larger home market (y) will have higher nominal wages (w). Higher nominal wages also mean higher real wages (ω), as seen in Table 8.1 (equation (4)), and higher real wages will attract workers to this region. Manufacturing workers demand manufacturing goods (increase in y), thus a cumulative growth process is generated.

Suppose that for some reason a firm decides to move production from one region to another. Ottaviano and Puga (1997) point out that the dynamics of Krugman's (1991b) model can be understood by using this model to answer the following question: How does this affect the firm's profitability? The model uses the price index and the home market effects to estimate the final effect upon profitability. On the one hand, the presence of one more firm will increase competition in the product and labour markets of the region receiving the firm, thus tending to *reduce* local profits and make relocation unprofitable. On the other hand, the increase in the number of local varieties tends to attract workers [price index effect], easing competition in the labour market. Moreover, because manufacturing workers demand manufacturers, there tends to be an increase in the local demand for manufactured goods [home market effect], easing competition in the product market. The final result is that local profits tend to *increase* and therefore to make relocation profitable.

Table 8.1   Classical examples of traditional and new core-periphery models

Richardson's (1979) Core-Periphery Model	Fujita, Krugman, and Venables' (1999) Core-Periphery Model
(1) Rate of Productivity Growth Equation	(1) Regional Income Equations
$$r = a + b\,y$$	$$Y_1 = \mu\lambda\,w_1 + (1-\mu)/2$$ $$Y_2 = \mu\,(1-\lambda)w_2 + (1-\mu)/2$$
(2) Efficiency Wages Equation	(2) Price Index of Manufactured Goods Consumed in each Region
$$w = c - d\,r$$	
(3) Rate of Output Growth Equation	$$G_1 = [\lambda w_1^{1-\sigma} + (1-\lambda)(w_2 T)^{1-\sigma}]^{1/1-\sigma}$$ $$G_2 = [\lambda(w_1 T)^{1-\sigma} + (1-\lambda)w_2^{1-\sigma}]^{1/1-\sigma}$$
$$y = e - f\,w$$	or
Where the variables of the model are:	$$G_1 = [n_1(p_1^{M})^{1-\sigma} + n_2(p_2^{M}T)^{1-\sigma}]^{1/1-\sigma}$$ $$G_2 = [n_1(p_1^{M}T)^{1-\sigma} + n_2(p_2^{M})^{1-\sigma}]^{1/1-\sigma}$$
r = rate of growth of productivity y = regional output growth rate w = rate of growth in efficiency wages	(3) The Nominal Wage Rate of Workers in each Region:
$a, b, c, d, e,$ and f are parameters.	$$w_1 = [Y_1 G_1^{\sigma-1} + Y_2 G_2^{\sigma-1} T^{1-\sigma}]^{1/\sigma}$$ $$w_2 = [Y_1 G_1^{\sigma-1} T^{1-\sigma} + Y_2 G_2^{\sigma-1}]^{1/\sigma}$$
	(4) The real wage in each region is given by the following equations:
	$$\omega_1 = w_1 G_1^{-\mu}$$ $$\omega_2 = w_2 G_2^{-\mu}$$
	Where the variables of the model are:
	Y = regional income W = nominal wage G = price index λ = Region 1's share of manufacturing $(1-\lambda)$ = Region 2's share of manufacturing
	The parameters of the model are
	μ = a constant representing the expenditure share of manufactured goods σ = elasticity of substitution among varieties ($\sigma > 1$) T = transportation costs

Whether the overall effect of entry is to increase profitability of local firms [encouraging further entry] or to lower that profitability [leading to exit] depends on the parameters of the model. These parameters are: (a) consumers' share in manufactured goods (μ); (b) elasticity of substitution among varieties ($\sigma > 1$); and (c) transportation costs (or the fraction of goods that arrived: $\tau < 1^4$). The convergence or divergence of regions will also be determined by where firms end up locating.

Could the New Core-Periphery Models Explain Cumulative Growth in High Technology Regions?

The introduction of agglomeration economies as pecuniary externalities in general equilibrium models has proved an important step in understanding how the behaviour of economic agents could lead to a core-periphery spatial structure with all manufacturing in only one region, or to a diversified spatial pattern in which manufacturing is divided evenly between two regions. Can we use these models, however, to help us discover which characteristics of a region contribute to the development of a long-run comparative advantage in high technology?

The answer is no. We cannot use the new core-periphery models because they are developed using only two sectors, manufacturing and agriculture, both of which produce a good that is rival and excludable. A rival good has certain technical characteristics, so producers can completely appropriate the benefits of the good. An excludable good is one for which property rights can be developed to assure complete appropriability of its benefits. High technology firms use knowledge [i.e. investment in R&D] as an input to create new products and processes. This knowledge cannot be completely appropriated by its owner, so 'technology' is a good with different economic characteristics. In essence, it is a good that generates knowledge externalities.

What kinds of issue need to be accounted for in a spatial model that tries to identify those characteristics contributing to the development of a long-run comparative regional advantage in high technology? Some insight on these issues can be gained by studying models that try to answer the same question at the *country* level – in particular, models that consider economic growth as the effect of ongoing invention of new differentiated products in the wake of Romer (1990) and Grossman and Helpman (1991). These models, developed in the context of the new economic development literature, incorporate research and development as an endogenous sector – an economic activity, like any other, that functions on the basis of maximising profits – and account for the fact that R&D activities produce a rival and excludable good and therefore generate technological externalities.

In using these models, however, we must take into consideration that they are designed to study specialisation among countries and not among regions. We expect to find lower labour mobility between countries than between regions. Moreover, we expect to find that transportation costs play a less significant role in cross-regional than in cross-country interactions.

A key contribution to the development of models of growth and innovations in the new economic development literature was the paper by Dixit and Stiglitz

(1977) because it allowed the possibility of incorporating the diversity of goods in demand and production functions. Romer (1990) used Dixit and Stiglitz's formulation to model the innovation process in the context of microeconomic behavioural assumptions for consumers and producers. Grossman and Helpman (1991) used both results to develop theoretical models that study the effects of innovations and trade on a country's sustained long run economic growth.

There are two important assumptions in Grossman and Helpman's type of model. First, innovation results from intentional research undertaken by a firm in response to economic incentives. Second, resources and knowledge are combined to produce new knowledge, some of which then spills over to the research and development community, thereby facilitating the creation of still more knowledge. In this way, knowledge spillovers create the possibility that the potential for furthering technical understanding does not diminish over time.

Both theoretical studies (Grossman and Helpman 1991) and empirical ones (Suarez-Villa and Walrod 1997; Saxenian 1994; Echeverri-Carroll 1997, Echeverri-Carroll and Brennan 1999; Fischer 1999; Oden 1997) have supported the hypothesis that knowledge spillovers are essential for explaining the spatial concentration of high technology firms, such as those found in California's Silicon Valley and in Austin, Texas, which actively develop new products and processes. This finding is not surprising because knowledge is the main input in the innovation process.

It could also be deduced from this analysis that knowledge spillovers are not important for non-innovative or non high technology firms, which is why they tend to cluster in areas with an abundant supply of unskilled labour, such as the U.S.-Mexico border region. In short, high-tech and non high-tech firms should tend to locate in different cities with different pools of knowledge. As mentioned in the introduction, Krugman (1991b) has however warned that there is no reason to assume that the motives for high technology firms to agglomerate are different from the motives of non-high technology firms.

We observe that in cities like Austin, Texas and San Jose, California, agglomerations of *both* high-tech and non high-tech firms coexist. In this regard, we can ask: Do high-tech firms and non high-tech firms located in the same city enjoy similar benefits? A positive answer to this question would support Krugman's (1991b) hypothesis that high-tech and non high-tech firms locate for similar reasons. If the answer is no, this suggests that the variables determining the location of each type of firm are different, contradicting Krugman's hypothesis.

The importance of Marshallian externalities [including knowledge externalities] for high-tech and non high-tech firms located in the four largest metropolitan areas in Texas have been examined. But before we present the results, however, we must answer an important question: Which firms do we consider high-tech?

8.3   Data and the Definition of High Technology

What *is* high technology? The main issue involved in deciding whether to classify an industry as 'high tech' or not concerns finding a proxy for technology. Once this proxy has been found, classification becomes an easy task, as it usually entails dividing industries into two groups: those with above average values for the selected proxy [the high-tech group] and with below average values [the non high-tech group]. The problem, however, is in defining technology, as it is a broad concept that can be associated with, among other things, new products, new processes, knowledge-intensive services, accumulation of know-how, skilled labour, machinery and equipment, and investment in R&D.

For our study we chose Markusen, Hall, and Glasmeier's (1986) definition of high-tech industries, which is based on the percentage of high 'human capital' jobs in an industry. [High human capital jobs include engineers, technicians, scientists, mathematicians, or some combination thereof]. This selected 100 four-digit high technology SIC codes with a percentage of skilled labour above the average for all industries. This definition offers the advantage that human skills are strongly correlated with other indicators of 'technological' performance, such as R&D investment, the stock of know-how, and new product development.

One problem, already mentioned, associated with using SIC classification to identify high technology *firms*, is that not all manufacturing establishments within a SIC group will fulfill the criteria used to identify high technology *industries.* For example, although SIC 3674 [semiconductors] is clearly a high-tech industry because it has a higher than average number of engineers and scientists, not all establishments within this SIC group in fact have a higher than average proportion of engineers and scientists.

How significant is this bias? In other words, what proportion of manufacturing establishments within a high technology SIC are really high-tech? To our knowledge, no previous study has addressed this issue. So we constructed an original database that allowed us to reclassify manufacturing establishments *within* high-tech SICs according to the percentage of engineers and scientists. We then designated two kinds of firm: 'high-tech' firms were those that had a higher percentage of engineers and scientists than the national average, and 'non high-tech' firms were those that, although listed within a high tech SIC, in fact had a percentage of engineers and scientists equal to or below the national average. In other words, they were essentially not high-tech.

Our original database of 1,772 high technology establishments came from the 1995 edition of the *Directory of Texas Manufacturers* [DTM], published by the Bureau of Business Research[5]. These data include DTM information on the top five four-digit high technology SIC codes both by volume of employment and number of firms in the five metropolitan areas of Austin, Dallas, Fort Worth, Houston, and San Antonio[6].

The empirical part of our study was accomplished in two phases. During the first phase, performed in the summer of 1996, we interviewed CEOs and managers at 23 high technology firms in Austin, Dallas, Fort Worth, Houston, and San

Antonio. The companies were carefully selected to comprise a representative sample of the population of high-tech companies; thus, we chose companies of different sizes and in different sectors. The second phase, carried out in the fall of 1996, involved mailing questionnaires to 1,772 high technology establishments. The target respondent was the CEO [president] or general manager. We employed the total design method [TDM], a technique designed to produce a minimum response rate of 40 percent for mail surveys (Dillman 1978)[7]. This technique has been used successfully in other studies on the high technology industry (Lyons 1995). A total of 374 high technology manufacturing establishments responded to our questionnaire, a response rate of 21 percent. This rate can be considered high for a study of high-tech industries, which are highly competitive as they need to develop new products and processes in very short time periods and operate on the just-in-time principle. Industry executives therefore have little time for non-profitable activities such as responding to a questionnaires! Moreover, they have no pecuniary incentive to do so.

8.4   Statistical Results

This section describes the statistical results on the importance of agglomeration externalities in our sample of high technology firms. Our analysis was performed with the objective of examining the influence of two factors, Marshallian externalities and pecuniary externalities, on high-tech and non high-tech firms.

Because not all firms chose to answer to our questionnaire, we acknowledged the possibility of bias in our sample. The problem was minimised by sending blind questionnaires so that individual firms could not be identified. This was to reduce the likelihood of receiving responses only from firms less sensitive to releasing information. We also tested for sample bias: we expected high-tech and non high-tech firms to be statistically differentiated in the variables associated with agglomeration economies, but similar in relation to variables linked to non agglomeration economies.

As shown in Table 8.2, the number of respondents was very similar for the two groups of firms. For both groups, most firms have fewer than 100 employees, are independent, and are located in the Dallas-Fort Worth and Houston metropolitan areas. Moreover, both types of firm show a similar distribution in their start-up date. In spite of these similarities, the two groups revealed some statistically significant differences in terms of the relative importance of agglomeration economies for their performance and in terms of their innovation performance.

According to Marshall (1920), the benefits offered by a city are associated with three kinds of variable: technological spillovers, a pool of workers with specialised skills, and the availability of specialised inputs and services. Krugman (1991b) mentions that pecuniary externalities, suggested much earlier by Weber (1929) and Isard (1956), could also play a key role in the agglomeration of firms in space. The proxy we employed for these pecuniary externalities is ease of transportation and distribution of products. Proxies for all agglomeration economies are shown in Table 8.3.

Are there any differences between high-tech and non high-tech firms in the relative importance of agglomeration economies? We asked the two groups of firms in our sample to evaluate the relative importance of agglomeration economies to their performance on a scale from one [not important] to five [very important]. The results, presented in Table 8.3, indicate that the groups were statistically significantly different from each other in terms of *all* proxies for Marshallian agglomeration economies, including those for knowledge spillovers.

Table 8.2   General characteristics of firms in the sample

	High Tech (number of firms)	Non High-Tech (number of firms)
Employment Size		
1-20	80	98
21-99	57	64
100-499	29	17
500-999	7	2
1,000-5,000	6	0
>5,000	2	1
Year Manufacturing Started		
Before 1970	40	46
1970s	44	53
1980s	76	54
1990s	20	28
Kind of Manufacturing Facility		
Independent firm	138	161
Branch plant	44	17
Location		
Austin-San Antonio	37	34
Dallas-Fort Worth	59	89
Houston	86	60

The average responses on the relative importance of different kinds of Marshallian economies [including knowledge externalities] in their performance were consistently higher for high technology than for non high technology firms. The hypothesis that there are differences between the mean values of high-tech and non high-tech firms for all the variables measuring technological externalities was accepted by t-tests at the 0.05 and 0.10 level.

Table 8.3 Mean responses on importance of agglomeration economies for performance by high-tech and non high-tech firms

Proxy		High-Tech Firms[a]	Non High-Tech Firms[b]	Pro>ITI[c]
Marshallian Externalities: Specialised Services				
Specialised Business Services	357	2.86 (180)	2.63 (177)	0.0763**
Temporary Help Establishments	360	2.42 (182)	2.08 (178)	0.0085**
Marshallian Externalities: Specialised Labour				
Availability of Technical Personnel	356	3.17 (180)	2.69 (176)	0.0002*
Other Firms to Attract Skilled Labour	354	3.02 (180)	2.73 (174)	0.0245*
Marshallian Externalities: Knowledge Externalities				
Information from Local Universities	355	2.32 (181)	2.06 (174)	0.0404*
Engineers (from local universities)	357	2.36 (181)	1.94 (176)	0.0006*
Engineers (from local firms)	353	2.58 (178)	2.09 (175)	0.0001*
Accessibility to Frequent Flights	344	2.67 (174)	2.42 (170)	0.0735**
Pecuniary Externalities				
Central Location for Product Distribution	361	3.10 (181)	3.31 (180)	0.1357
Other Externalities				
Foci for Firm's Sales and Purchases	353	2.60 (178)	3.19 (175)	0.0001*
Quality of Life	353	3.69 (180)	2.56 (173)	0.2391

Notes: a *High-tech firms* are those who responded that they have a proportion of engineers and scientists above the U.S. average. Numbers in parentheses indicate the number of firms.

 b *Non high-tech firms* are those who responded that they have a proportion of engineers and scientists equal or below the U.S. average. Numbers in parentheses indicate the number of firms.

 c Prob > absolute value of t

 * Significant at the 0.05 percent level.

 ** Significant at the 0.10 percent level.

The two groups also show significant differences in terms of product innovations. We asked both groups of firms: When you compare the number of new products your plant developed and brought to market in the last two years with the number of new products your industry group developed and brought to market in the last two years, is your plant above, in, or below your industry average? We gave values of zero for the answer *above* the industry group average, one for *in* the industry group average, and two for *below* the industry group average. Thus, a small value indicates that most firms in a group will be above their industry group average.

The average response for high technology firms was 0.77, indicating that most plants within this group were developing and commercialising products at a higher rate than their industry group average. In contrast, the average for non-high tech firms was 1.08, indicating that most plants in this group were developing new products at a slower than average pace. These differences were statistically significant at the one percent level. In sum, results from our sample gave some support to the hypothesis that high technology and non high technology firms locate for different reasons, even when they locate in the same city. This contradicts Krugman's (1990b) hypothesis that high-tech and non high-tech firms may locate for similar reasons.

As already mentioned, one of the parameters that Krugman (1991b) used to explain concentration was transportation cost. But, as Glaeser (1994) points out, this did not mean that Krugman claimed that transport costs were driving most locational decisions, it simply suggested that transport costs were one of the forces behind agglomeration that could be most easily be modelled explicitly. The hypothesis that there are differences between mean values of high-tech and non high-tech firms for the variable measuring pecuniary externalities [central location for product distribution] was rejected by t-test at the 1.0 percent level. In sum, central location, a proxy for transportation costs, is one of the most important agglomeration economies for *both* kinds of firms.

8.5   Conclusions

As already mentioned, Marshall (1920) was the first to identify the economic benefits which accrue to firms when they locate in cities with a high concentration of similar firms. One of the most remarkable concentrations of specialised firms in the United States is found in Silicon Valley with its high concentration of high technology firms. Equally remarkable spatial concentrations may be found among carpet producers around Dalton, Georgia, jewellery producers around Providence, Rhode Island, and financial service providers in New York. Successful milieux of specialised firms are also found in Europe, particularly in Italy [i.e. the Third Italy], Denmark, and Spain.

Another well-known spatial agglomeration of specialised firms is located in the U.S.-Mexico border region, where over 3,000 plants called *maquiladoras*[8] employ a total of more than one million Mexicans. The growth dynamics of these firms in the U.S.-Mexico border region are unprecedented. The 540 *maquiladora* plants in Mexico in 1979 had increased to 3,500 by 2000, while the number of people employed at these plants had increased from 111,365 to 1,242,800 during the same period.

All these spatial agglomerations of specialised firms would enjoy *similar* Marshallian benefits: a large supply of specialised services and labour and an intense exchange of ideas. It is probably in this regard that Krugman (1991b) argues that there is no reason to assume that the motives for the agglomeration of high technology firms to differ from the motives of non-high technology firms. Not all of these agglomerations of specialised firms, however, would possess *similar* potential for long-term sustainable growth.

The new economic development theory suggests that sustained long-term economic growth depends on a firm's continuing capacity to innovate (Grossman and Helpman 1991). In this context, only regions like Silicon Valley, where firms innovate at unprecedented levels, can maintain long-term growth. The focus of the new economic development theory is to understand how the core grows.

The new economic geography goes a step further by trying to explain when and why a core and a periphery emerge. Can the new economic geography models be used to explain a spatial structure characterised by a few highly innovative cities and many cities with little innovative activity? The answer is no. The new core-periphery models offer little potential to explain long-term growth in high-tech regions because they incorporate only pecuniary externalities [i.e. savings in transportation costs] and exclude technological externalities. The strong presence of knowledge externalities in high technology regions, and their importance for high tech firms' innovations is well documented in the literature. Indeed, our empirical analysis gives some support to the hypothesis that knowledge externalities, not pecuniary externalities, are the key to distinguishing location preferences between high-tech and non high-tech firms.

We recognise that most of our analysis has been dominated by the effort to understand the variables that affect the location of high-tech firms and the long-term growth of high-tech regions [the core]. What can we say, however, about growth in the periphery [non high-tech regions]? To begin to answer this question, we offer a simple comparison between two regions with contrasting innovation dynamics: Silicon Valley and the U.S.-Mexico border. Both regions have been very successful in developing *specialised* labour and services and in generating the kind of information needed by their local firms. In this regard, firms in both regions locate for *similar* reasons.

A striking difference, however, exists in terms of the relative need of firms for one input, knowledge. High-tech firms depend strongly on both kinds of *knowledge:* the kind that can be appropriated and the kind that cannot [knowledge externalities]. In contrast, the maquiladoras depend only on *information.* The difference between information and knowledge is that knowledge leads to innovations, while information leads only to more efficient production of the same [old] good. Differences in the needs for knowledge are reflected in the fact that the relative supply of skilled and unskilled labour is one of the most striking differences between these two kinds of region.

What about the economic growth of both regions? As anticipated in the new economic development theory, the amazing innovation dynamics of firms in Silicon Valley have generated a process of *cumulative growth* that has continuously increased the standard of living in this area (see Saxenian 1994). In contrast, the U.S.-Mexico border region is associated *not* with the *creation* of new innovative firms, but with the *assembly* of products invented in the United States and other industrialised countries and with the relocation of plants of firms already established in these countries. While it is true that these firms have created employment in the border area, the standard of living in the region has not risen continuously. In sum, the optimal regional policies are those leading to the development of innovative activities.

It seems clear from our analysis that high-tech and non high-tech firms locate in a region for different reasons – in particular, they demand different amounts of skilled labour. We also observed that, even in cities such as San Jose and Austin, both kinds of firms often coexist. Thus, we question whether the same kind of location divergence will be found even within cities. Our empirical analysis gives some support to the hypothesis that high-tech and non high-tech firms locate for different reasons, even when they locate in the same city. It was also shown that studies using SIC classifications to define high-tech have a strong bias. About 50 percent of firms in our sample did not comply with the high-tech criteria.

Future research should take into consideration several issues. First, it should differentiate between high-tech and non high-tech firms, as the abundance of one or the other in a region creates different potential for long-term regional growth. Second, future studies should take into consideration that agglomeration processes in high-tech firms are explained by knowledge externalities, not pecuniary externalities. Third, empirical studies on high technology should use establishment data, not industry data, which introduce strong bias in the results.

We recognise that our empirical analysis has several limitations. One reason is that the analysis was static, as it dealt with the relative importance of Marshallian and pecuniary externalities at a specific moment in time. This made it impossible to study the dynamic processes of sustainable development in high-tech regions discussed in Section 8.2. Our analysis therefore says little about the sustained process of agglomeration, and the role of externalities in this process. This kind of analysis requires a much more sophisticated study of dynamic processes. Moreover, our data and methodology imposed some constraints on our findings. For instance, flow data was not used to measure the importance of transportation costs in the location decisions of firms.[9] It should also be pointed out that the case of the U.S.-Mexico border region was underlined in the conclusions only to illustrate the importance of innovations and skilled labour in long-term growth. It was not our intention to compare the growth dynamics of the U.S.-Mexico border with that of U.S. high-tech regions. Finally, although we have not considered the effect of certain non-measurable variables, such as the evolutionary trajectories of institutions, due to the difficulty of incorporating them in models, we recognise that they may well play a key role in the growth of both high-tech and non high-tech regions.

Endnotes

1   Dixit and Stiglitz (1977) derive demand functions from preferences [utilities] that exhibit a love for variety. Thus, for a given level of consumer spending in a product and a given price for the available varieties, welfare [consumer utility] rises as the number of varieties increases. The increase in utility is conditioned by the degree to which consumers desire variety.

2   The concept of iceberg transportation costs, introduced by Samuelson (1954) in the international trade literature, assumes that a fraction of any good shipped simply 'melts away' in transit, so transport costs are in effect incurred in the good shipped.

3   As Bona and Santos (1997, p. 243) pointed out, 'While the arsenal of techniques applicable to the analysis of non-linear systems has recently grown enormously, it must be acknowledged that our collective hands are often tied when confronted with what appear to be simple non-linear problems'.

4   Where $\tau = 1/T < 1$, with T [= quantity dispatched /quantity received] >1.

5   The DTM includes more than 90 percent of manufacturing establishments in Texas that have more than 10 employees, and more than 50 percent of those with fewer than 10. Because we expect to find most innovative high-tech firms in establishments with at least 10 employees, the DTM is an excellent database for our study.

6   The following industries [by 4-digit SIC code] were surveyed in these metropolitan areas. Austin: 3544, 3672, 3679, 3823, 3674, 2834, 3571, and 3842. Fort Worth: 3728, 3533, 3535, 3679, 3069, 3544, 2899, 3721, 2834, 3674. Dallas: 3544, 3661, 3721, 3728, 3674, 3679. Houston: 3533, 3823, 2899, 3561, 3569, 3511, 2821, 2869, 3571. San Antonio: 3728, 3544, 3679, 2899, 3674, 3537, 2842, 3721, 3531, 2834.

7   TDM involves a sequence of mailings and follow-ups designed to increase the response rate. Although TDM involves a fourth mailing of a letter and replacement questionnaire to non-respondents by certified mail [49 days after the initial mailing], this step was not performed due to time and funding restrictions.

8   The term *maquiladora* is derived from the Spanish word for the amount of corn paid by a farmer to the miller to grind the corn. Similarly, the *maquiladora* industry uses inputs provided by the client and returns the output to the same client.

9   Manfred Fischer highlighted this important point during my presentation of this chapter at the workshop on *Knowledge, Complexity, and Innovation Systems*, held in Vienna, July 1-3, 2000.

PART C: Innovation, Knowledge and Regional Development

9 Inventive Knowledge and the Sources of New Technology: Regional Changes in Innovative Capacity in the United States

Luis Suarez-Villa
School of Social Ecology, University of California at Irvine

9.1 Introduction

The United States experienced very radical changes in the regional sourcing of invention and new technology during the twentieth century. In less than five decades, areas that were previously peripheral or undeveloped turned into the most important sources of new technology. This remarkable transformation, and the factors which supported it, need to be understood if we are to make any sense of the forces that drive regional change.

The radical change experienced by the United States can be characterised as a process of *regional inversion*, whereby the predominance of certain regions is overturned and previously lagging areas take their place. In the American case, those lagging areas became the most important sources of invention and new technology in a relatively short period of time. It is hard to find other development processes as interesting as regional inversion, particularly when this is tied to the roots of technological change.

In this day and age, being a source of new technology is of paramount importance. It is often claimed that a 'new economy' is arising, driven mostly by invention and technology. Many of the old heuristics of macroeconomic policy are losing their value, as changes attributed to technology work their way into every economic activity. Invention is at the very root of this remarkable process of change, since it provides the ideas and tools that are the source of new technologies. The creative talents and knowledge required by invention are therefore very important, and they are bound to become a vital element of any region's human capital base.

For the purposes of this chapter, invention will be defined as the discovery of new ideas, processes or tools whose novelty can be proven by, for example, earning a patent award. Invention is therefore considered to be at the root of technological change, since it generates new ideas that can become new technologies. Innovation, on the other hand, is defined as the application of inventions for some economically or socially useful purpose. This is an important difference between invention and innovation which is usually overlooked in the literature on technological change. Technology then is defined as the sum total of all inventions and innovations that have been put to some economic or social use.

This chapter will provide, first, an overview of the changes that have turned the American Sunbelt into the most important regional source of invention and new technology. A historical indicator of invention will be used to show the Sunbelt's dynamism and its contrast with the regions that were once the United States' most important sources of technology. A subsequent section will then consider the main factors which made the process of regional inversion possible. Emphasis will be placed on the *endogenous* character of those factors that have supported the rise of previously peripheral areas as the most important sources of invention. Those factors will be discussed from a broad, macro-level perspective that takes into account their institutional context and the role of public resources.

9.2   The Emergence of the Sunbelt

The emergence of a previously peripheral or undeveloped region as a nation's most important source of new technology, as has occurred in the United States, has no parallel in any other advanced nation. Neither in Europe nor Japan can one find lagging areas that converged with and overtook the richer or knowledge-intensive regions. Processes of *inversion* of the established regional order are relatively rare. In the few historical instances when they occurred, they were part of a transition to a new epoch.

In the short span of five decades, the Sunbelt became the most important source of American invention and new technology. A number of entirely new industries and activities were created there. Most contemporary advances in electronics, micro computing, telecommunications, biotechnology and information processing were either created or developed in this region. Entirely new businesses arose in those activities, overtaking many old and well-established industrial giants, to become the largest firms in their field.

The rising economic and political influence of the Sunbelt during the past forty years or so has been a significant topic of interest for some scholars. However, its linkage with invention has been almost completely ignored in the social science literature (see Suarez-Villa 1993). Part of this neglect stems from the biases introduced by economic analysis, which has never been able to deal adequately with invention. Activities involving creativity, uncertainly, risk or human decision-making have been traditionally shortchanged by most economic models dealing with economic change. The very assumptions and constructs of such models typically negate a realistic treatment or even recognition of the factors that underpin invention. It should therefore not be surprising that the geographical dimension of invention has also been severely neglected in the economic literature. Most treatment of the Sunbelt phenomenon has thus been left to political scientists and historians, who are usually less inclined to explore invention or technology.

The Sunbelt region comprises twenty states in the southern half of the United States. As could be expected from its vast territory, there are significant differences between those states (see, for example, Schulman 1990; Sale 1975). Three states have played very important roles in the emergence of the Sunbelt and can be considered to be the 'locomotives' of its dynamism. Those states are California, Texas and Florida. A second tier of states, Virginia, North Carolina, Colorado, Utah, Arizona and New Mexico have also seen significant activity in some areas of technology production, services or research. A third tier of Sunbelt states has lagged behind the others, but may well see some growth in coming years as technology activities diffuse to new areas [for an extensive discussion on the Sunbelt states see Suarez-Villa 2000a, Chapter 3].

California's role in the emergence of the Sunbelt as a source of new technology has been paramount, because of its leadership in invention and in new technology industries, services and technological education. Silicon Valley, in northern California, was the birthplace of the electronics and micro computing revolution (see, for example, Hanson 1982; Frieberger and Swaine 1984; Saxenian 1994; Norton 1999). Among all metropolitan areas, Southern California concentrates the largest number of inventors, and now has what is perhaps the most diversified high technology industrial base in the world (see, for example, Suarez-Villa and Walrod 1997; Sivitanidou 1999). Texas and, particularly, the Austin area, has become a very important high tech centre, concentrating many electronics, micro computing and information processing businesses (see, for example, Echeverri-Carroll and Brennan 1999; Norton 1999). Central Florida is seeing the development of a significant cluster of high tech producers. North Carolina has developed research parks where a significant number of firms has undertaken high tech production and research. Virginia has become a major national centre of software and information processing services, with many businesses in those activities linked to federal government agencies (Stough 2000).

The emergence of the Sunbelt as a major source of new technology is best understood by its contribution to invention. As already stated, invention lies at the root of all technological knowledge and creativity. It involves processes of discovery which are extremely uncertain and therefore very risky. Invention typically involves much trial and error, despite the fact that it often relies on existing knowledge. While in many cases, inventions are achieved by launching into the unknown and by probing new ideas, in others they are attained by recombining various aspects of existing knowledge.

Innovation, on the other hand, involves the *application* of inventions. Such applications can occur by improving, modifying or developing an invention. The objective of innovation is usually to arrive at some economically or socially useful end. The cumulative total of those economically or socially useful applications is what we typically refer to as technology. Technology therefore comprises the sum total of all inventions being applied in any given society. Most research and development [R&D] today largely involves innovation. Studies of R&D units undertaken over the past four decades have shown that between 65 and 90 percent of their activities are not related to invention, and usually do not lead to patenting (see Suarez-Villa 2000a, chapter 3, for a review of the literature). Most often, R&D involves the adaptation or application of already existing inventions, or their modification and development for commercial use. As such, R&D data are poor indicators of invention or inventive activity. In this article, therefore, R&D will not be associated with invention, nor will such data be used as indicators of invention.

Invention in the United States is primarily a local or regional activity carried out by private individuals, either independently or through firms. Government institutions and laboratories typically carry out very little invention, and almost all invention is therefore performed by the private sector. The geographical scope of invention in the United States is mostly very limited, since most collaboration between inventors tends to occur in close geographical proximity. In this sense, therefore, the collaborative interregional or even interstate links between inventors are very short-range, tending to be confined to the local areas and states where these reside. Only within some large firms is it possible to find long-range geographical linkages for inventive collaboration. However, such cases are not representative of the way most invention occurs in the United States, and even in many large firms the tendency is usually to cluster inventors at one location (see Suarez-Villa 2000a, Chapter 3).

In this chapter, invention patenting will be used as an indicator of invention. Patenting data are about the most reliable kind of historical statistic that can be obtained in the United States. Their recording has been very consistent over time, and the criteria used for making patent awards remained uniform throughout the twentieth century (see, for example, Schmookler 1966; Griliches 1990; Suarez-Villa 1990, 2000a). Every application for a patent award must undergo a rigorous and time-consuming scrutiny that evaluates the novelty of the idea for which the award is requested. Moreover, patent data are compiled geographically, by state, according to the location of the first-named inventor in every award [for an extensive discussion of the advantages and disadvantages of patenting data, and a review of the literature on this topic, see Suarez-Villa 2000a, Chapter 3].

Patent awards have a seventeen-year legal validity in the United States. Extensions of that period are extremely rare. In any given year, therefore, one must consider not only the new patent awards made in that year, but also the sum total of awards made during the previous sixteen years. A cumulative seventeen-year estimate of patent awards would thus represent the total stock of legally valid inventions in any given year. This particular characteristic of patent data has been missed in most all previous research on patenting and invention.

A moving indicator of the total stock of invention patents will be used to estimate how inventive activity has changed. The *innovative capacity* indicator will show the cumulative number of valid invention patents available for application in any given year [more details on this indicator can be found in Suarez-Villa 2000a, chapter 3]. This indicator was first introduced by the author in a 1990 publication (Suarez-Villa 1990). Since then, it has undergone considerable testing, review and application with both national and regional data. Fig. 9.1 shows an estimate of the national innovative capacity indicator between 1880 and 1995. The rapid rise of the indicator after the 1950s is mostly a result of both the contributions of the Sunbelt and of corporate invention. Individual invention remained practically constant after the 1940s, as inventive activities grew and became an integral component of corporate strategy (see Suarez-Villa 2000a, Chapter 2).

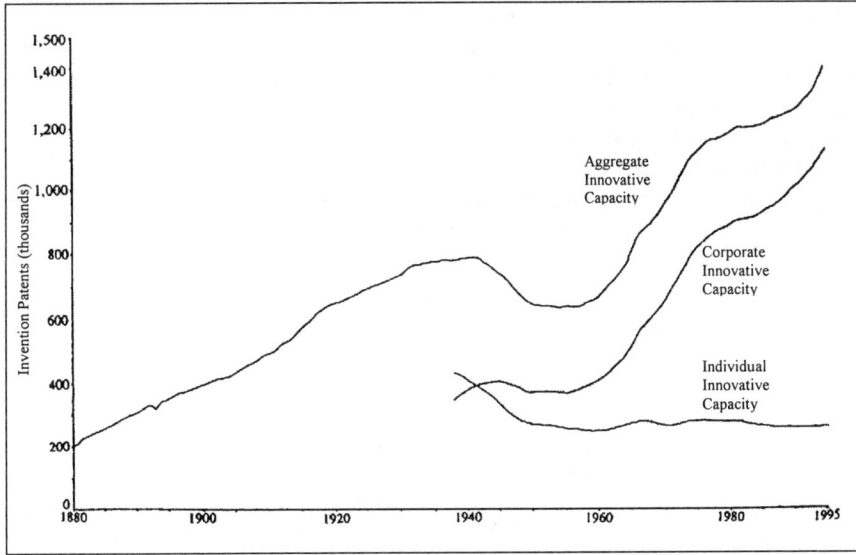

Data Source: U.S. Patent and Trademark Office

Fig. 9.1   Innovative capacity [1880-1995]

In many respects, the post-World War II decades were extremely important for American invention. The rise of American global corporations at a time when all other major industrial nations were still recovering from the effects of the war, provided many American companies with the resources to support invention in a major way (see, for example, Wilkins 1974; Chandler 1990). Reaching out globally at a time when few firms in other advanced nations could do so, gave many invention-rich American corporations an enormous advantage in the 1950s and 1960s. At the same time, increased market power and profits made it easier for such firms to deploy more resources for invention.

The main geographical sources of invention and new technology during the first half of the twentieth century were the north-eastern and mid-western regions. Those two regions were the richest and most industrialised in the United States. These two regions combined, although smaller in territorial area than the Sunbelt, had a much larger population, and they also comprised the largest and most important metropolises in the United States. In 1900, for example, the total population of the Sunbelt was only 72 percent that of the Northeast and Midwest combined. By 1950, the Sunbelt's population was still only 86 percent that of those two regions. It was not until the early 1980s that the Sunbelt's total population caught up with with their combined total. By the early 1990s its total population was 14 percent larger [U.S. Bureau of the Census 1950-1995]. Thus, despite the Sunbelt's much larger territory, its population size did not reach parity with that of the northeastern and mid-western regions until very recently.

Population size is, by and large, not very relevant for invention, however. There are plenty of regions around the world that have experienced substantial population growth without any increase in invention. Invention is a very elitistic activity undertaken by a very small number of highly talented individuals in the population. The correlation between population and inventive output or patenting is typically very poor, as might be expected from an activity that relies so much on individual creativity and that so very few individuals can undertake. Extremely few persons in any given population ever come up with an idea that can become an application for a patent, much less an actual patent award. Population numbers are therefore largely irrelevant for invention, even when the general level of technical skills is high.

All indications are that population migration from the Northeast and Midwest to the Sunbelt did not contribute to raising the latter's contribution to invention in any significant way. Although no historical migration data exists for inventors, it is generally considered that migration of inventors to the Sunbelt has been very limited over the past five decades and that the vast majority of the Sunbelt's inventive output was endogenously generated. To construct an indicator that considers population, such as patenting per inhabitant, would be highly misleading, since much of the Sunbelt's population growth has occurred through the immigration of low skilled individuals who would typically not be involved with invention. Such an indicator would severely distort the Sunbelt's contribution to invention, since it is well-known that both the Northeast and the Midwest have received much lower immigration than the Sunbelt. However, a more serious problem with such an indicator is the fact that reliable population data is collected in the United States only every ten years, as opposed to the annual reporting intervals for patenting data. Moreover, the way population data are reported would make it impossible to separate immigrants or even the less technically skilled from the general population to produce a compatible and reliable historical series. The historical data needed to separate those and other relevant categories of the population do not exist. For these reasons, the indicator provided in this contribution (and shown in Fig. 9.1 and Fig. 9.2) is considered to be more reliable than would one that includes population.

Throughout the first half of the twentieth century, the trend of the Sunbelt's innovative capacity resembled those of the Northeast and Midwest (see Fig. 9.2). An important break occurred during the late 1940s and early 1950s when the Sunbelt's innovative capacity began to rise, while that of the other two regions remained stagnant. The emergence of the Sunbelt as a source of invention was underpinned by the rise of entirely new economic sectors and activities that relied on a continuous stream of new inventions. Major invention-rich firms in those sectors and activities were created and grew in the Sunbelt. In general, the relocation of major firms from the Northeast or Midwest to the Sunbelt was not significant. Thus, from the standpoint of technology-rich firm creation, the rise of

the Sunbelt as a major source of invention and new technology can be considered
to be endogenously driven (see Suarez-Villa 2000a, Chapter 3).

Evidence supporting the rising post-war Sunbelt trend, shown in Fig. 9.2, can
be found in almost all technology-related aspects. For example, the number of
Sunbelt-based serial publications dealing with technology rose rapidly after the
early 1950s [Institute of Scientific Information 1950-1995]. Similarly, the number
of Sunbelt-based members of prestigious scientific organisations, such as the
National Academy of Sciences, grew rapidly. In the late 1940s, the membership of
the Academy has been mostly concentrated in the Northeast and the Midwest
[National Academy of Sciences 1987]. By the mid 1980s, the Sunbelt's
membership in that organization had reached parity with that of the Northeast and
Midwest. Similarly, many Sunbelt universities grew in quality to be among the
most highly ranked in many fields of science and technology.

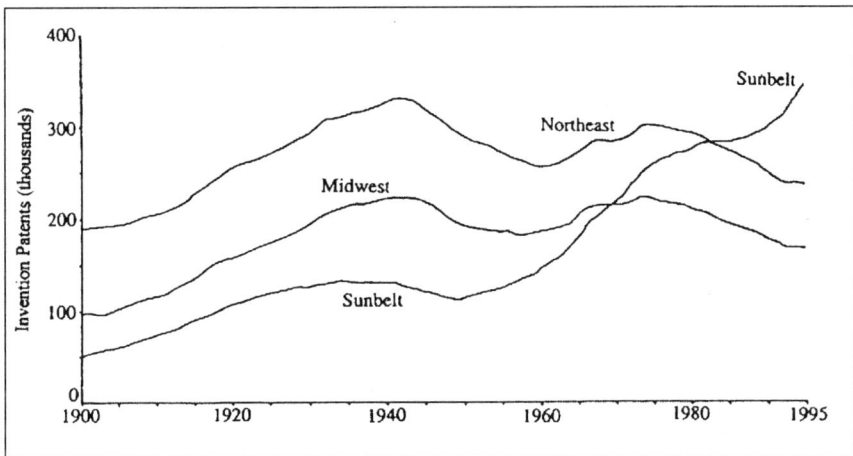

Data Source: U.S. Patent and Trademark Office

Fig. 9.2   Regional innovative capacity [1900-1995]

The rising importance of the Sunbelt as a source of invention is also reflected
in the distribution of annual patent awards (see Suarez-Villa 1993, 2000a; Varga
1999). Fig. 9.3 shows the rapid convergence between the Sunbelt and what
previously were the United States' most important regional sources of invention.
By 1995, the proportion of the Sunbelt was equal to that of the Northeast and
Midwest combined. The change shown in the graph is remarkable when one
considers that for almost the entire first half of the twentieth century the Sunbelt's
proportion of total patenting was practically constant and very low.

Equally remarkable is the fact that the two richest and most technologically
advanced regions, which traditionally concentrated the most highly skilled human
capital, experienced a sustained decline in patenting. The decline of the Northeast
and the Midwest was not based on any real deterioration of their technological

L. Suarez-Villa

institutions since, for example, the quality and recognition of their many universities was retained or even increased after the 1940s. Rather, systemic factors involving support for invention and technological creativity apparently benefitted the Sunbelt more than those two regions. These factors were sustained over the postwar decades by massive amounts of resources, which had a strong geographical bias. They were also rooted in deep social and economic beliefs related to access, opportunity and the role of government.

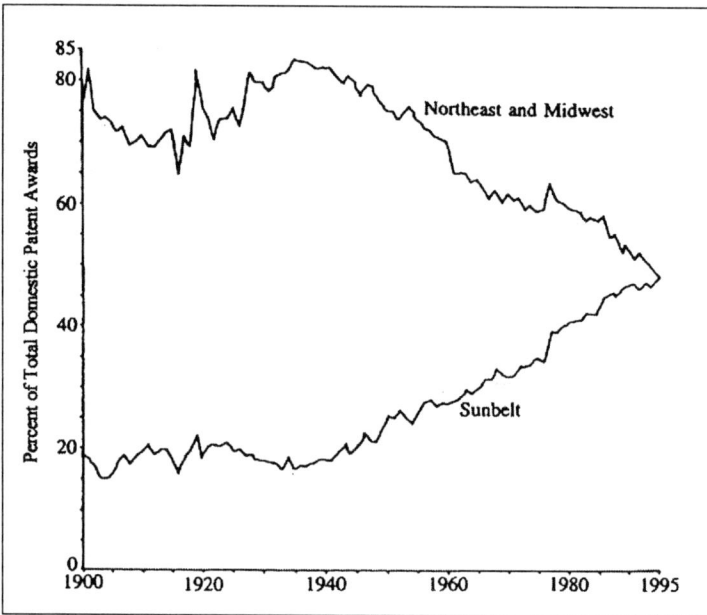

Data Source: U.S. Patent and Trademark Office

Fig. 9.3   Regional patenting distribution [1900-1995]

9.3   The Factors of Inversion

The factors which supported the emergence of the Sunbelt as the most important source of invention and new technology are grounded in the American institutional context. This means that they are, to a large extent, a product of the values and processes which have shaped how the allocation of public resources occurs, who benefits from such allocations, and how responsibility is distributed among the various levels of government. It has required the expenditure of enormous resources over long periods of time to expand access to technical knowledge. Similarly, to expand the opportunities for private capital to exploit in supporting invention has required substantial public sector commitments. Devolution and decentralisation have also been important in determining how the government has targeted the resources needed to support invention. Public resources and the way they have been allocated have therefore been a major influence in the *regional inversion* process of the last five decades.

One of the most important factors contributing to the rise of the Sunbelt was public infrastructure. Massive amounts of public resources have been spent since the late 1940s to provide many Sunbelt states with the sort of infrastructure that could support human capital development, particularly in the areas of science and technology (see Suarez-Villa 2000a, chapter 4; Suarez-Villa and Hasnath 1993). Much public spending also went to support infrastructure that is less directly related to invention, but which is nevertheless essential for building up or exchanging technological knowledge. Such infrastructure allowed communications to occur more efficiently, and it supported other public resources which had a more direct bearing on invention.

The data shown in Fig. 9.4 illustrate the kind of change in public spending allocations that benefitted the Sunbelt. Before the 1970s, the largest proportion of infrastructural spending was concentrated on the north-eastern and mid-western regions, particularly for public educational infrastructure (see U.S. Bureau of the Census 1975, 1981; American Public Works Association 1976). A sea change in the regional distribution of such expenditures occurred in the mid-1970s, as the proportion allocated to the Northeast and Midwest sank dramatically. The contrast after 1973 could not be more obvious. The significant aspect is not only the speed with which the shift in regional distribution occurred, but the fact that the Sunbelt's proportion permanently gained predominance.

The institutional and political influences behind the change shown in Fig. 9.4 are too complex to analyse here, but they nevertheless deserve to be mentioned. One change which had an important long term effect in the allocation of public resources was the growth of population in many Sunbelt states. In the American political system representation is largely based on population numbers, and areas that experience faster population growth can expect to accumulate more political power in the federal Congress. Such power eventually increases the allocation of public resources to the areas that gain population. Thus, faster population growth increased the proportional representation of the Sunbelt in the federal Congress, leading to greater allocation of public resources to many of its states.

Another institutional aspect which needs to be considered is the enormous fragmentation of political authority. Devolution is an important characteristic of the American political system. There are no federal schools or universities, for example, and responsibility for public education at all levels is largely left up to state and local governments. A great deal of competition for resources and personnel occurs between the states and also between local governments. This condition also applies to virtually all other aspects over which state and local authorities have responsibility. From road construction to the police, competition drives much of the resource allocation process. Having a growing population can therefore become an important advantage, not only because of the greater political power it confers, but also because it is a source of increased local and state tax revenues. Since public education and infrastructure tend to be financed from such revenues, there is a tendency for areas that gain population to have more resources to spend.

The kind of competitive attitude which pervades most state and local government activities, tends to limit inter-jurisdictional cooperation. It is quite common, for example, for municipalities within metropolitan areas to look no further than their municipal boundaries in almost every operational aspect, and to contribute nothing to the sustenance of the larger metropolitan area to which they belong (see, for example, Suarez-Villa 2000b). This sort of jurisdictional isolationism is a product of the competitive nature of American institutions at all levels. Although it is virtually impossible to demonstrate because of the lack of adequate data, this sort of pervasive competition may account for the radical and rapid regional shift seen in Fig. 9.4. Once a region gains in population and political clout, it is a foregone conclusion that an increasing amount of public resources will become available to it.

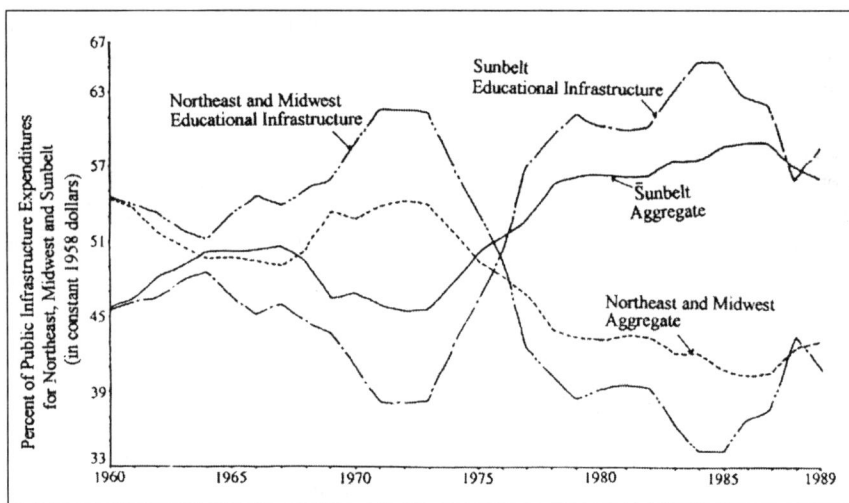

Data Source: U.S. Department of Commerce

Fig. 9.4   Distribution of public physical infrastructure by region [1960-1989]

The educational infrastructure data shown in Fig. 9.4 encompassed all types of public educational construction (see U.S. Bureau of the Census 1981, 1950-1995). It is therefore a gross indicator of human capital investment. Unfortunately, more specific data on the various types of educational infrastructure built, such as laboratories, research facilities and the like, do not exist. One reason for the lack of such data is the fragmentation of authority found in the American context, which has prevented the kind of coordination that is required to gather data in a more comprehensive and complete way.

It must also be noted that the educational data in Fig. 9.4 only represents public educational construction. No data on construction is available for the many private universities, institutes and schools in the United States. The contribution of these private institutions to American technological education and invention has been important. Private institutions have also been important in raising the level of technological and scientific knowledge in many Sunbelt states. Unfortunately, the lack of data on the private educational sector leaves this aspect of regional change incomplete.

A factor which is strongly related to the Sunbelt's rising profile in educational infrastructure was the *massification* of public education (see Suarez-Villa 2000a, Chapter 1). Starting in the late 1940s, a great deal of resources were devoted to opening access to higher education to almost everyone. At that point, access to higher education started to be considered a right or an 'entitlement' by most Americans. This situation was in deep contrast to the situation in many European nations, where only a small percentage of the secondary school population had access to university or college-level education. Needless to say, the enormous expansion of access to higher education from the 1950s to the 1970s exposed many young people to technology and science to an extent that had never occurred before. Many of these were the first in their family to ever attain university-level education. The new 'entitlement' affected Sunbelt states more than any part of the United States, mainly because their educational institutions had lagged behind those of the Northeast and Midwest until that time (see, for example, Graham and Diamond 1997; Geiger 1993).

The massification of access to education had one important characteristic. This is the fact that expanded access was not achieved at the expense of quality. Merit standards were maintained despite the opening up of higher education to almost all. How quality standards were sustained in the face of much expanded access is something that deserves some attention. It was achieved by creating a *division of labour* in public higher education, with some institutions retaining very high levels of quality while others retained open access by adopting much lower standards (see Suarez-Villa 2000a, Chapter 1).

Some Sunbelt states, such as California, created several tiers of public universities to implement a division of labour which expanded access. In California, three tiers of public higher education institutions were developed. The first tier, the University of California system, with nine campuses, is research-oriented and focuses more on graduate study and doctoral work. It competes effectively with the top private institutions in the nation, and admits only the top twelve percent of secondary school graduates within California (see, for example, Stadtman 1970; Smelser and Almond 1974). The second tier, the California State University system, with twenty campuses, is much more open and emphasises undergraduate education. Exposure to research is typically limited and doctoral programmes are not offered. The third tier, the Community College system, is the most accessible of all with over one hundred campuses. This provides two-year college diplomas and remedial education. Other Sunbelt states implemented similar or simpler hierarchies to provide wider access to higher education.

What made the California experiment interesting is that it was a planned, systematic effort to implement a division of labour that would open access to higher education to virtually everyone without sacrificing quality or merit (see, for example, OECD 1990; Smelser and Almond 1974). Other Sunbelt states also created a division of labour in public higher education, but most lacked the sort of long-range plan and discipline developed by California. In many other Sunbelt states, the tiers of the hierarchy developed informally without a clear delimitation of functions, leading to some confusion. One outcome of such confusion was that, unlike in the California system, older institutions tended to receive most of the public funding, while newer universities experienced many difficulties and shortfalls. Overall, however, the division of labour that was implemented throughout the Sunbelt did open up access remarkably, despite all the difficulties that could be expected from such a massive public effort.

The results of the massification of education, particularly in the Sunbelt, began to be felt in a relatively short time. By the mid-1970s, for example, the United States had the highest average number of years of higher education of any advanced nation. Similarly, the total number of years of education was also significantly higher than that of any other nation. An increasing number of foreign students then began to be admitted to American public universities as the 'baby boom' generation advanced past its school years. Major beneficiaries of such expansion in access, and of the subsequent increase in foreign student enrollments, were the science and technology fields. By the early 1980s the United States was the world leader in the number of per capita scientific articles published (Institute of Scientific Education 1950-1995). American universities also led the world in the value or amount of scientific and technological 'hardware' or equipment installed at its public universities. Because of their rapid growth and the enormous amount of resources they absorbed, Sunbelt universities were largely responsible for those advances (see Graham and Diamond 1997).

It is important to point out that the massification of education also created substantial disparities between institutions and between states. Unlike in Europe, for example, where efforts are made to avoid disparities between institutions, the division of labour in public higher education created enormous differences in quality in the United States. The fragmentation and devolution that are so typical of the American institutional context also contributed to those disparities. Individual states designed and followed their own plans and funding formulas without any central or national coordination. Thus, for example, some public universities in the Sunbelt became highly competitive with the top institutions in the nation, while other public universities within the same states languished and became repositories of the less qualified. The result of such disparities in institutional quality was an enormous difference in the value of diplomas between public universities and between states.

It is unknown to what extent such disparities affected science and technology education. However, it is possible that their impact on those fields was much less than for other areas, mainly because the less qualified student population has historically avoided those fields. Science and technology curricula are typically perceived to be the hardest, requiring substantial preparation and skills even for marginal achievers. Thus, the poorer institutions tended to avoid offering diplomas or substantial curricula in those fields, mostly because of low demand but also because of the more expensive hardware and facilities that they required.

The third factor supporting the rise of the Sunbelt as a major source of invention was venture capital. This factor affected corporate invention in the Sunbelt enormously. Recalling the data in Fig. 9.1, corporate invention was extremely important for the rising innovative capacity trend of the postwar decades. Venture capital made it possible for most new firms in invention-rich activities to be created and developed. Without adequate sources of venture capital both the rise of the Sunbelt and of corporate invention would have been jeopardised (see Suarez-Villa 2000a, Chapter 3).

The supply of venture capital was fundamental both to invention and to the high technology revolution. Unfortunately, venture capital was usually not available to the earlier technology firm start-ups. Initially, many of today's largest invention-rich companies were started with capital from their owners' personal savings. That was the case, for example, for Hewlett Packard and Apple Computer, both of which were started in the homes of their founders with their own personal financing (see, for example, Frieberger and Swaine 1984; Hanson 1982). At the time of the creation of these companies, venture capitalists viewed them as extremely risky and were unwilling to support them. It was not until the economic viability of their inventions and innovations was proven that they received financing for their activities. Unfortunately, there was a considerable loss of time and opportunities until such funding was obtained. Had these and other early high tech start-ups been able to secure adequate venture capital, their inventive activities might have yielded greater results during the crucial start-up phase.

As new high technology firms proved the value of their inventions, it became easier for start-ups to obtain venture capital financing. This development was extremely important for the Sunbelt, since most new high technology firms in the United States were started in the region. The traditional concentration of venture capitalists had been in the Northeast and Midwest, with New York City concentrating the largest number of those firms. In many cases, investment banks undertook venture capital financing, but they were rather conservative about their targets and tended to shy away from financing start-ups in new sectors. Also, most of the new technology activities were not well understood by the traditional venture capitalists, such as the large investment banks, and this made it much more difficult for new technology entrepreneurs to obtain financing.

As the high technology revolution advanced, Sunbelt start-ups found it easier to obtain venture capital. It is remarkable that in the short span of one decade most venture capital activity shifted from the eastern part of the United States to the west. Unfortunately, precise historical data on venture capital financing do not exist, but some recent information compiled by business consultants can provide some indication of the states that have become major venture capital recipients (see, for example, Suarez-Villa 2000a, Chapter 3; Bygrave and Timmons 1992; Hambrecht 1984). The data in Table 9.1 show California as the most important recipient of venture capital financing, with 39 percent of all funding and 41 percent of all financing agreements in the United States. In general, the Sunbelt states shown in the table accounted for over 56 percent of all funding and over 50 percent of all agreements. This was a far cry from the situation that prevailed only a decade before, when Sunbelt start-ups accounted for less than a quarter of all venture capital funding.

The rapid rise in Sunbelt venture capital financing of the 1990s and late 1980s made it possible for many *inventor-firms* to be created. Inventor-firms are companies which are owned and run by inventors, and which aim to fill a new market niche with a previously unknown product. Their owners may be engineers or scientists who set out to create a new firm out of one or several inventions to which they hold patents. Thus, their owners end up becoming scientist- or engineer-entrepreneurs. This contrasts with the traditional business model, where executives typically do not have much depth in the technical or scientific nature of their companies' inventions.

Table 9.1   Largest state recipients of venture capital investment [1996]

State (Region)	Venture Capital	
	% of Total Funds Invested	% of Total Agreements
California (Sunbelt)	39.0	40.8
Massachusetts (Northeast)	12.0	12.9
Texas (Sunbelt)	8.0	4.4
New York (Northeast)	7.9	4.0
Florida (Sunbelt)	5.0	2.1
Colorado (Sunbelt)	4.5	3.3
Illinois (Midwest)	3.5	2.6

Data Sources: OECD, Price Waterhouse LLP

The increasing availability of venture capital financing for the inventor-firms of the 1990s has made it possible for technology firms to dominate today's stock market. In the United States, every major stock market rally since the late 1980s has been led by technology firms. The value of technology shares today determines largely where the stock market will head and how strong it will be. A quick look at where the more important technology firms in the stock market are based will reveal that the vast majority have headquarters in Sunbelt states. Silicon Valley and southern California account for a large number of such firms, but other areas, such as Austin in Texas, are also becoming important headquarter locations. This situation is in deep contrast with the corporate giants of the industrial capitalist era, all of which were headquartered in north-eastern or mid-western states.

The rise of the Sunbelt as the United States' most important source of invention and new technologies has therefore depended largely on the development of infrastructure that could support invention-rich activities, on the massification of educational access which could provide inventive human capital, and on the greater availability of venture capital for new invention-rich firms. Three decades ago it would have been hard to imagine that entirely new sectors would rise up in a region that had always been peripheral to the American industrial context. Understanding the sources of the Sunbelt's technological leadership is perhaps the most interesting challenge that faces regional specialists in this new economic era.

9.4 Conclusions

The process of *regional inversion* that has turned the Sunbelt into the most important source of invention and new technology in the United States stands as one of the most interesting regional phenomena of the twentieth century. The magnitude and dynamism of this process of change have no parallel in any other advanced nation. It is important, therefore, to try to understand the major factors which have supported the rise of this once peripheral or undeveloped part of the United States.

Measuring inventive product through the *innovative capacity* indicator has highlighted the rapidity of the rise of the Sunbelt as a major source of technological creativity. The dynamism of the Sunbelt and the shift to corporate invention have underlain the growth of national innovative capacity during the past five decades. There is an important link between the rise of the Sunbelt and corporate invention, which has been fundamental in establishing the United States' global technological leadership. This link was the source of the new sectors which drove the high technology revolution of the late twentieth century.

The high technology revolution, which started out with electronics and computing and later advanced to telecommunications, information services and biotechnology, was largely a product of invention in the Sunbelt. All the major new centres of American high technology, such as Silicon Valley, southern California, eastern Texas, northern Virginia and the North Carolina research centres, were products of the factors which supported the new Sunbelt dynamic. Without the rise of the Sunbelt as a major source of new inventions, we would not be witnessing the emergence of the so-called 'new economy' and the technologies that drive it.

The factors which supported the rise of the Sunbelt for technology are macro-level and long term in scope and character. Infrastructural development, particularly for the kind of infrastructure that is more closely related to human capital development, emerged as a powerful transformational factor. The rapid expansion of public educational infrastructure was particularly important. The *massification* of educational access played a crucial part in expanding the technological knowledge of the population. Since most new inventions arise out of existing technology, expanding educational access to science and technology fields was fundamental for enhancing the Sunbelt's technological human capital. Venture capital was also a significant factor in the rise of the Sunbelt as a source of new technology, particularly due to the growing importance of corporate invention for the high technology revolution. Rapidly expanding access to venture capital was therefore crucial for the rise of the inventor-firms which created the inventions and businesses that drive today's 'new economy'.

The factors that supported the rise of the Sunbelt also promoted great disparities within the region. Some states, such as California, emerged as 'locomotives' of invention for the region while others lagged well behind in almost all technology- related aspects. Unequal dynamism and development have therefore also been a characteristic of the regional inversion process. Such inequities were compounded by the fact that major sectors of the 'new economy' emerged in the leading [or 'locomotive'] states of the Sunbelt. These disparities were also increased by the fragmentation and competition that is so characteristic of American institutions at all levels of authority.

Can the Sunbelt's regional inversion process be replicated elsewhere? Clearly, the rise of this region is inextricably linked with the American institutional structure and its emphasis on fragmentation, devolution and competition. It is hard to see how the factors which supported and promoted the rise of the Sunbelt as a major source of new technology could work in a different institutional context. Historical specificity and the broader institutional context may play more significant roles than anyone so far has given them credit for. However, it would be most interesting to see if the factors that supported the rise of the Sunbelt can produce similar results in a different institutional context. Unfortunately, at this time one can only speculate about the likely outcome of such possibilities. The Sunbelt so far remains unique among the regional development experiences of the late twentieth century.

10 Urban Innovation and Collective Learning: Theory and Evidence from Five Metropolitan Cities in Europe

Roberta Capello
Department of Economics, University of Molise, and Department of
Economics and Production, Politecnico Milano

10.1 Introduction

The tendency for innovation activity to cluster in large metropolitan areas is a widespread and well established phenomenon. Such areas are often regarded as 'centres of creativity' and have recently been referred to as 'islands of innovation', due to their capacity to induce economic progress and technological innovation (Davelaar and Nijkamp 1990; European Commission 1995; Hingel 1992; Simmie 1998 and forthcoming). The main explanation for their success is that they generate much greater agglomeration economies than elsewhere, so the metropolitan area is often conceived of as a breeding place for new activities. As has already been suggested (Glaeser et al. 1992), such a dynamic view of the city fits nicely with the recent approach to economic growth, which sees externalities [and particularly externalities associated with the stock of knowledge] as the 'engine of growth' (Romer 1986; Lucas 1988).

However, some doubts remain over the question of whether these agglomeration economies relate primarily to increases in the scale of activity in a *particular* industry [localisation economies] or more generally to the overall scale of activity in an area, hence affecting the productivity of *all* firms [urbanisation economies]. This question is still open and is one of the subjects of current theoretical and empirical debate, discussed in Henderson (1996), Parr (2000), Jacobs (1984), Krugman (1996b), Simmie and Hart (1999), Simmie and Sennet (1999) among others.

The aim of this chapter is to join the debate on the determinants of innovation in cities by examining the phenomenon through a rather new approach, based on the concept of the 'milieu innovateur'. The theory of the milieu innovateur has been extensively applied to the clustering of firms in industrial districts, being in part a dynamic interpretation of the industrial district model (Aydalot 1986;

Aydalot and Keeble 1988; Camagni 1991; Maillat, Quévit and Senn 1993; Ratti, Bramanti and Gordon 1997). An attempt has recently been made by the same group to apply the concept to the urban environment, and hence to demonstrate theoretically certain similarities between the city and the milieu.

At first glance, the two concepts seem rather difficult to compare. But if we go beyond the consideration of the physical element, some similarities between the milieu concept and the city do emerge. These are associated with elements of human 'relational' capital and collective learning, which represent the genetic elements of the milieu stemming from spatial interaction and proximity. These may be expected to be present also in the city (Camagni 1999).

The interpretation of the city as a place where milieu mechanisms can develop leads to an interesting conceptual question, i.e. whether the development of firms located in urban areas is influenced more by externalities stemming from their urban location [urbanisation economies] or by 'milieu economies'. This question represents a dynamic interpretation of the old debate on the relative effects of urbanisation and localisation economies on urban productivity.

The spatial relational approach through which we interpret the innovative capacity of firms in cities has not until now been subject to quantitative analysis. But thanks to the existence of a database on innovation in five metropolitan cities in Europe, created as part of a research project lead by Oxford Brookes University and financed by ESRC, we have been able to carry out an empirical investigation of the impact of dynamic agglomeration economies on the innovation capacities of cities [milieu economies versus dynamic urbanisation economies]. The five metropolitan areas involved in the study are Amsterdam, London, Milan, Paris and Stuttgart.

The structure of this chapter is as follows. The theoretical approach adopted for the study of the determinants of innovation in cities in explained in Section 10.2. We examine the role of spatial interactions, such as collective learning and relational capital, and the advantages of this approach in relation to more traditional ones. Section 10.3 presents the data, the methodology and the sample used, while the results of the empirical analysis are given in Sections 10.4, 10.5 and 10.6. A few conclusions are drawn in Section 10.7.

10.2 Learning in Cities

The importance attributed to the effect of territorial externalities on the performance of firms dates back to Marshall (1919) and is embedded both directly and indirectly in the work of geographers like Christaller (1933), economists like Lösch (1954), Hotelling (1929), Isard (1956), Myrdal (1959), Kaldor (1970) and more recently Mills (1970, 1993) and Henderson (1985, 1996), to mention just a few.

The concept of localisation economies, i.e. the external advantages associated with a single industry, dates back to the work of Weber (1929). Refinements and improvements in the field of location theory can be found in the works of Hoover (1937, 1948), Lösch (1954), Isard (1956), Koopmans (1957) Jacobs (1969) and Bos (1965), among others.

Table 10.1   Approaches to the efficiency of spatial agglomeration

Elements	Static Efficiency Approach	Dynamic Efficiency Approach	Spatial Relational Approach
Period of Theory's Development	1970s and 1980s	1990s	Late 1990s
Kind of Approach	Mostly empirical	Mostly empirical	Mostly conceptual
Efficiency Measure	Firms' productivity [static]	Firms' innovation Firms' growth [dynamic]	Firms' innovation [dynamic]
Preconditions for Firms' Efficiency	Agglomeration economies [static]	Stock of knowledge [static]	Collective learning [dynamic]
Level of Analysis	Sectoral or urban	Sectoral	Individual firm
Key Variables for Firms' Efficiency	Size of the sector Size of the city	Size of the sector Concentration of the sector	Presence of collective learning Presence of dynamic urbanisation economies
Methodology Used	Estimate of sectoral and urban production functions	Multiple regression analysis at the sectoral level	Multiple regression analyses at the firm level
Dependent Variables Used	Industrial productivity Urban productivity	Industrial productivity growth Number of patents or innovations achieved in each sector	Innovation achieved in each sector
Independent Variables Used	Size of the sector Size of the city	Specialisation index [lq] Industrial concentration index	Proxies for collective learning and dynamic urbanisation economies Specialisation index [lq] Size of firms
Main References	Hirsch [1968] Alonso [1971] Mera [1973] Henderson [1974, 1988] Segal [1976] Shefer [1973] Sveiskauskas [1975] Carlino [1980] Mills [1970] Sveikauskas, Gowdy and Funk[1988] Malmberg, Malmberg and Lundquist [2000]	Griliches [1992] Ellison and Glaeser, [1999] Glaeser [1997] Glaeser et al. [1992] Porter [1990] von Hipple [1994] Jaffe [1989] Jaffe, Traltenberg and Henderson[1993] Feldman [1994a] Audretsch and Feldmann [1996] Feldmann and Audretsch [1999] Oerlemans, Meeus and Boekema [1998] Satterthwaite [1992] Beeson [1992]	Camagni [1999] Crevoisier and Camagni [2000]

Since these early theoretical approaches, a vast literature has grown up, analysing the types of advantages stemming from agglomeration and testing their effects through empirical estimates. The debate on spatial economies can be classed into three main approaches (see Table 10.1):
- the static efficiency approach, developed primarily during the 1970s and 1980s;
- the dynamic efficiency approach, developed mainly during the 1990s;
- the spatial relational approach, developed in the late 1990s.

The Static Approach

During the seventies and the early eighties, agglomeration economies were considered the main determinants of urban productivity. The main question at that time was whether advantages of scale relate primarily to increases in the scale of activity in a particular industry, with all benefits accruing primarily to that industry [localisation economies], or whether they relate more generally to the overall scale of activity in an area, thereby affecting the productivity of all firms [urbanisation economies].

Many empirical studies were carried out, trying to capture the role of urban size or sector size on factor productivity in different ways, including:
- estimating an urban production function aggregated for all sectors. In this case the aim was to verify the existence of a multiplicative parameter stemming from the size of the city. Many studies reached common findings, which suggested that when a city doubles in size, factor productivity increases by 4-6 percent. These results provided *prima facie* evidence of the impact of urbanisation economies on factor productivity, although some doubts remained about whether small cities in fact operate with the same production function as large cities, as implicitly assumed by this methodology;
- estimating an urban production function disaggregated for the different sectors. In this case, it was shown that the size of the sector determines the increases in factor productivity, but some doubts remain here too over the methodology, which by definition ignores the fact that the mix of sectors may itself be the sources of urban advantages;
- analysing wage and income differentials between cities of different sizes. It is argued that large cities provide a greater advantage in terms of income and wages, although it could equally well be claimed that the wage and income differentials favouring large cities do not reflect greater efficiency but compensation for the disadvantages associated with size, such as pollution and congestion.

All of these approaches have a common feature - that of interpreting static efficiency, like factor productivity, through static elements, such as urbanisation and localisation economies. No final conclusion has so far been reached.

The Dynamic Approach

The first important step forward was the introduction of a dynamic approach, which instead of considering only static efficiency introduced dynamic elements, interpreting urban growth and innovation through the presence of spatial economies. In this strand of literature, urbanisation and localisation economies were considered not only to affect productivity [static efficiency], but also urban growth [dynamic efficiency] (Beeson 1992; Satterthwaite 1992) and innovation (Feldmann and Audretsch 1999). In particular, the debate focused on whether specialised or diversified knowledge spillovers provide a better explanation of innovation activities and technological change in spatially concentrated production systems.

Within this approach, three main externalities have been identified as determinants of urban innovation (Glaeser et al. 1992):

- *the Marshall-Arrow-Romer* externality. In this case, the externality concerns knowledge spillovers between firms within an industry. It is stated that the concentration of a given industry in a particular city helps knowledge spillovers among firms and therefore the growth of the industry as well as the city. Moreover, monopoly is considered to be better for growth than local competition, because the latter restricts the flows of ideas and means that externalities are internalised by the innovator;
- *the Porter externality.* Porter (1990) argues that knowledge spillovers in specialised, geographically concentrated industries stimulate innovation and therefore growth. He argues, however, that it is local competition, as opposed to local monopoly, which fosters the pursuit and rapid adoption of innovation;
- *the Jabobs externality.* Jacobs (1969, 1984) claims that the externality derives from the existence of a large number of industrial sectors in the city. Interaction between them generates a sort of cross-fertilisation of ideas through the transfer of knowledge among sectors. This approach, like the previous one, claims that local competition enhances the pursuit of innovation.

Many studies have been carried out to establish which of these views provides the best interpretation of the real world. The methodology generally adopted is to estimate regression models to test for the conditions in which cities grow fastest. The level of geographic specialisation and competition is also examined in an attempt to determine which, if any, of these externalities are important for growth. Most of the results seem to provide support for the diversity thesis.

Considering a cross-section of city-industries, Glaeser et al. (1992) found that industries grow more slowly in cities where they are heavily over-represented and where the level of local competition is high. By the same token, through an econometric analysis of 700 firms in the United States, Feldmann and Audretsch (1999) found that diversity across complementary economic activities sharing a common science base is more conducive to innovation than is specialisation. In addition, their results indicate that a degree of local competition for new ideas within a city is more favourable to innovative activity than is local monopoly.

A feature of this kind of approach is that it interprets a dynamic output, e.g. the degree of innovation in a city, or city growth, with static determinants, i.e. the stock of [diversified or specialised] knowledge. By contrast, the approach presented below provides a conceptual view of urban advantages and urban growth which interprets innovation through *dynamic* spatial externalities.

The Spatial-Relational Approach

In the late nineties, a new way of analysing urban location advantages was proposed, which we refer to as the 'spatial-relational approach'. This new approach compared the urban environment with the concept of the milieu. An *innovative milieu* is defined as a set of relationships occurring in a given area which encompasses a production system, various economic and social actors, a specific culture and a 'representation' system, generating a dynamic process of collective learning (Camagni 1991).

Despite the many differences between the concepts of the milieu and the city, especially if the latter is considered primarily in a physical sense, there are also similarities. In particular, they share a common genetic principle - that of agglomeration - through which they develop elements, processes and effects, ranging from the development of a common identity and sense of belonging to the socialised production of human capital and know-how (Camagni 1999). These elements, processes and effects are fundamental to both the innovative milieu and urban growth. The city, in this sense, is conceived in two dimensions:

- a relational dimension, since it is seen as a place where a set of territorial and social relationships take place;
- a dynamic dimension, since it is seen as a learning system.

These two dimensions represent the two generic elements of a learning process - synergy and continuity over time - suggesting a dynamic interpretation to city growth. According to this theory, the kind of learning mechanism which enhances urban innovative creativity is *collective learning*, i.e. learning which takes place in a *socialised* way, through creative knowledge which accumulates outside the single firm, but within the local area, as a sort of 'club good'. There is no [or low] rivalry in its use, and very limited exclusion of external agents from taking advantage of it. In this sense, collective learning supplies typical club goods *à la* Buchanan (1965) from which club externalities may be exploited. Collective learning is the territorial counterpart of learning processes happening inside the firm. It is seen as the vehicle for knowledge transmission, both in a time and spatial dimension. In the former, the transfer of knowledge is guaranteed by an element of continuity, in the latter by the interaction among agents (Capello 1999a, 1999b).

Collective learning may thus be defined as a dynamic and cumulative process of knowledge production, transfer and appropriation, taking place thanks to the interactive mechanisms typical of an area with a strong sense of belonging and strong relational synergies. The channels through which collective learning takes place are in fact thought to be (Camagni 1995; Capello 1999b):

- *high mobility of specialised labour and low external mobility*: this labour market structure guarantees cross-fertilisation processes for firms and professional upgrading for individuals. Local know-how grows through a collective and

socialised process, subject however to the risk of isolation and locking-in, unless external energy is also captured through selected external linkages;
- *stable linkages with suppliers and customers*: stable input-output relationships generate an exchange of codified and tacit knowledge with suppliers and customers. This accumulates over time and creates patterns of incremental innovation which feed a specific technological trajectory;
- *intense innovative interactions with suppliers and customers and mechanisms of local spin-off*. Local milieus provide both the social and market preconditions for this phenomenon to take place. From the social point of view, high trust and a common sense of social and cultural belonging make this process acceptable[1].

In the channels through which collective learning takes place, two main conditions are at work:
- geographical proximity, i.e. agglomeration;
- 'relational proximity', encompassing the linkages that occur due to the economic integration of firms, socio-cultural homogeneity of the local population and intense public/private co-operation and partnership[2].

Although agglomeration is a generic element of the urban environment, relational proximity in the form of synergy, governance and sense of belonging, is not by definition present in the city. The degree to which this exists and influences the innovative capacity of a city determines two different archetypes (Camagni 1999):
- the *city as a milieu*, where relational proximity gives rise to an urban context organised as a milieu, and where the engines for growth of the local area are collective learning processes and a common 'vision' for the evolution of the local area;
- the *urban production milieu*, where a network of informal or selected linkages develops within the urban context around a specialisation sector or 'filière' on the basis of collective learning mechanisms and common sectoral identity. In this case, urban evolution in specialised sectors may well depend on milieu economies and relational capital, while in non-specialised sectors traditional dynamic urbanisation economies, i.e. learning from co-operation with scientific urban institutions [universities, large public and private research centres], may be the main sources of growth.

The opposite archetype to the above is the 'pure city'. In this case, the main characteristics are strong sectoral de-specialisation, significant physical agglomeration economies [like the presence of advanced infrastructure], private services for different markets, and social heterogeneity of cultures. In the pure city, learning mechanisms for firms stem from the presence of so-called 'creativity centres', i.e. universities, and large private or public research centres, which create the sort of scientific atmosphere needed for knowledge acquisition.

In this respect, the spatial-relational approach has some similarities with the older theories. Here too we are interested in testing empirically whether the innovative capacities of firms in cities depend primarily on milieu economies, or on dynamic urbanisation economies. The difference between this and the so-called 'dynamic approach' is that the determinants of innovative activities are measured in terms of *dynamic elements*, and in particular:
- milieu economies, i.e. collective learning mechanisms;

- dynamic urbanisation economies, i.e. mechanisms of learning stemming from the interaction of firms with diversified business activities and scientific research centres. These channels of knowledge acquisition have nothing in common with the collective learning processes envisaged in the milieu. In fact, they stem neither from sectoral concentration, nor from socio-cultural homogeneity, but from the spatial concentration of activities in an urban setting.

The aims of the empirical part of the present study are:

- to test for the existence of milieu behaviour in firms in the metropolitan regions examined and determine whether it is more reasonable to speak of an urban milieu or an urban production milieu (see Section 10.4);
- to test whether milieu economies are more conducive to innovation than dynamic urbanisation economies (see Sections 10.5 and 10.6).

10.3 Data, Sample Characteristics and Methodology

The empirical analysis is based on a database built up as part of an ESRC research project led by Oxford Brookes University and was carried out by a research group composed of national subcontractors, one for each case study city, namely Amsterdam, London, Milan, Paris and Stuttgart. In each 'metropolitan city' [NUTS 3 level], firms in different sectors were interviewed about their innovation activity.

The database contains 159 observations, roughly equally distributed among the five cities. The choice of firms was made on the basis of two criteria: (a) firms which were winners of the BRITE Award, to guarantee as far as possible a similar choice throughout the five cities; (b) a random choice of other firms, when few BRITE Award firms were present.

The firms interviewed belong to both high-tech and low-tech sectors, but there was a higher share in the latter. They included small, medium and large firms, with an equal distribution among the cities. There were firms from both private and public sectors, although the private sector was far more strongly represented in the analysis [88.7 percent of the total sample]. In all cities there was a high proportion of firms involved in product innovation, while only a third declared that they had developed process innovations. The main characteristics of the sample are summarised in Table 10.2.

A common questionnaire was submitted to all firms, with the intention to collect information on:

- the type of innovation developed;
- the geographical location of customers, suppliers and competitors;
- forms of co-operation developed as support for innovation;
- sources of information used for the innovation activity;
- sources of knowledge on which innovation was based;
- the importance of location factors in innovation activities.

Table 10.2   Characteristics of sample firms

	Amsterdam	London	Milan	Paris	Stuttgart	Total Sample
Absolute Number of Observations	26	33	35	33	32	159
Share of Observations Among Cities						
Small Firms (≤ 49 employees)	14.7	23.5	27.9	17.6	16.2	100
Medium Firms (≤ 249 employees)	25.0	18.8	25.0	15.6	15.6	100
Large Firms (>249 employees)	10.7	17.9	14.3	28.6	28.6	100
Private Firms	12.1	21.3	24.1	19.9	22.7	100
Public Firms	0.0	33.3	16.7	50.0	0.0	100
High-Tech Firms	26.5	15.7	15.7	9.6	6.9	100
Low-Tech Firms	4.2	23.6	30.6	34.7	6.9	100
Product Innovations	17.4	24.8	21.1	17.4	19.3	100
Process Innovations	14.0	12.0	24.0	28.0	22.0	100
Share of Observations within Each City						
Small Firms (≤ 49 employees)	41.7	50.0	54.3	36.4	34.4	43.6
Medium Firms (≤ 249 employees)	33.3	18.8	22.9	15.2	15.6	20.5
Large Firms (>249 employees)	25.0	31.3	22.9	48.5	50.0	35.9
Total	*100.0*	*100.0*	*100.0*	*100.0*	*100.0*	100.0
Private Firms	100.0	93.8	97.1	90.3	100.0	95.9
Public Firms	0.0	6.3	2.9	9.7	0.0	4.1
Total	*100.0*	*100.0*	*100.0*	*100.0*	*100.0*	100.0
High-Tech Firms	88.0	43.3	37.1	24.2	84.4	53.5
Low-Tech Firms	12.0	56.7	62.9	75.8	15.6	46.5
Total	*100.0*	*100.0*	*100.0*	*100.0*	*100.0*	100.0
Product Innovations	73.1	81.8	65.7	57.6	65.6	68.6
Process Innovations	26.9	18.2	34.3	42.2	34.4	31.4
Total	*100.0*	*100.0*	*100.0*	*100.0*	*100.0*	100.0

Employment data for each city was collected at NUTS 3 level and at the national level for a NACE two digit sectoral disaggregation. This data allowed the creation of a specialisation index [a location quotient] for each sector, which was used as an index of sectoral specialisation.

The database refers to a single year and cross-sectional analyses were made. The dynamic element was expressed in the empirical analysis by using indicators of flows rather than of stocks. The innovative capacity of firms [their dynamic efficiency] and the learning channels through which they acquire new knowledge were used as indicators of incremental changes in the static level of firms' efficiency and the cumulated stock of knowledge. In particular, the channels through which firms were assumed to acquire innovation were innovative co-operation with suppliers and customers, long-standing relationships with local suppliers and customers, and proximity to a vast information source through co-operation with scientific research centres.

The questionnaire covered all the above aspects, and hence collected information on:

- whether the type of innovation developed by firms was new to the sector, the world, or only the firm;
- the relative importance of local suppliers, customers and competitors;
- the sources of knowledge on which firms base their innovation activity [to determine whether it was internal to the firm, internal to the sector present in the area, or internal to the city];
- whether sources of information useful for innovation activities were internal to the firm, to the specific sector in the city, or to the city as a whole;
- sources of useful co-operation for the innovative activity [with firms of the same group, with competitors and suppliers external to the firm, with innovative local suppliers, with innovative local customers, with universities and large research centres, or with other firms];
- traditional location factors, in terms of: (a) urbanisation economies, e.g. the presence of infrastructure provision, general business services, financial capital, or general information useful for innovation, and (b) milieu economies, e.g. the presence of local suppliers, customers and competitors stimulating innovative activity, a scientific atmosphere, trust and informal co-operation provided by the presence of friends and ex-colleagues in the innovative activity.

The local labour market, theoretically representing an important channel for local knowledge acquisition, was not included in the questionnaire. Further empirical analysis is needed to develop this perspective. Most questions provided discrete information on the degree of appreciation of the different sources of information, knowledge, co-operation and location advantages of each firm. Factor analysis[3] was used to transform these into continuous variables and reduce their number. The factors obtained are presented in Table A-10.1 of the Appendix. They represent the variables used in the empirical analysis described below, which is based on two different methodologies:

- cluster analysis, to detect the innovative behaviour of firms in the cities examined (see Section 10.4);
- multiple regression analysis, to study the determinants of innovation in the five cities (see Sections 10.5 and 10.6).

A word of caution is necessary when making international comparisons, since the different institutional contexts, different political cultures and different types of innovation system may influence the results of the analysis. No real remedy exists to resolve this methodological problem - the only solution is to take care in interpreting the results.

10.4 Innovative Behaviour of Firms in Metropolitan Cities

The first theoretical element to be tested empirically was the existence and importance of a 'milieu' in the metropolitan regions analysed. This required an analysis of the innovation behaviour of firms, with the aim of finding out whether the milieu externalities [in the form of collective learning] influence the innovative activity of firms. In the cluster analysis, a wide spectrum of possible variables explaining the innovative behaviour of firms, aggregated through a factor analysis, was used: type of innovation, sources of information, knowledge, or co-operation for the innovation activity, and the appreciated locational advantages for the innovation activity. The analysis was able to group firms according to the structural characteristics of their innovation behaviour. Tables 10.3a and 10.3b show the results obtained: four different types of innovative behaviour emerged, characterised by the size of the firm, and by specialisation of the sectors in which each firm operates.

The first cluster was typified by the behaviour of *small firms in specialised sectors*, and was made up of 94 observations, nearly 60 percent of the sample. In this cluster, a typical milieu economy and networking behaviour prevails, characterised by:

- innovative local suppliers [a channel through which collective learning takes place] as one of the main sources of knowledge for innovative activity;
- innovative local customers and suppliers as the main sources of co-operation, together with co-operation with other firms, underlining the importance of local economic interactions and networking mechanisms;
- locational preferences which reflected a 'milieu' approach. These included the existence of an industrial atmosphere created by the presence of ex-colleagues and friends, as well as the proximity of suppliers and customers. However, the firms also appreciate proximity to infrastructure and business services, which are more related to urban location.

A second cluster was typified by the behaviour of *small firms in non-specialised sectors*, and was made up of 14 observations [8.8 percent of the sample]. It is interesting that this group of firms behaves in a completely different way from the previous cluster. The firms generally seem to represent small branches of large companies, and choose an urban location for various reasons, including:

- to control the final market [proximity to customers];
- to control specific suppliers [proximity to suppliers];
- to take advantage of a location near a large urban area [proximity to business services and consultancies];
- to take advantage of an advanced scientific environment [proximity to R&D centres].

The interaction of this group of firms with local actors and local institutions is so weak that it is hard to envisage any territorial embeddedness, or any kind of spatial interaction among local economic actors:

- the main sources of knowledge are customers external to the area;
- the most appreciated channels for co-operation are also external customers and suppliers, or other firms of the same group;
- the locational advantages are associated with traditional urbanisation economies.

Table 10.3a Results of the cluster analysis: Firms in specialised sectors

	Small Firms (≤99 employees)		**Large Firms (>99 employees)**	
Specialised Sectors (ql > sample mean)	*Small Firms in Specialised Sectors (94 observations=59.1 percent of the sample)*		*Large Firms in Specialised Sectors (45 observations=28.3 percent of the sample)*	
	Market Size:		*Market Size:*	
	• National	0.11	• Non-European	-0.44
	• European	0.16	• Non-international	-0.24
	• International	0.03		
	Innovation:		*Innovation:*	
	• Imitative	0.05	• Breakthrough	0.53
	Sources of Knowledge:		*Sources of Knowledge:*	
	• Local innovative suppliers	0.05	• External suppliers	0.32
	• Consultancy services	0.004	• Ex-colleagues	0.26
			• Scientific research centres	0.06
	Sources of Co-operation:		*Sources of Co-operation:*	
	• Innovative local customers	0.06	• External suppliers	0.10
	• Innovative local suppliers	0.22	• Innovative local suppliers	0.22
	• Other firms	0.05	• Innovative R&D centres	0.40
			• Other firms	0.74
	Sources of Information:		*Sources of Information:*	
	• Scientific journals	0.59	• Internal	0.38
			• R&D centres	0.40
			• Technological information	0.59
			• Scientific journals	0.99
	Locational Advantages:		*Locational Advantages:*	
	• Presence of ex-colleagues and Friends	0.03	• Presence of ex-colleagues and friends	0.49
	• Proximity to infrastructure	0.03	• Proximity to suppliers and customers	0.45
	• Proximity to services to firms	0.06	• Proximity to information	0.02
	• Proximity to suppliers and Customers	0.01	• Proximity to high-quality public services	0.21
			• Proximity to R&D centres	2.80

Notes: Values = deviance from the sample mean

A third group consists of large firms in specialised sectors, which represent nearly 9 percent of the sample [14 observations]. Although they behave as typical large firms, appreciating their urban location and taking advantage of the scientific environment of the large metropolises. They also seem to appreciate 'milieu economies' resulting from the high specialisation and concentration of the sector in which they operate.

Table 10.3b   Results of the cluster analysis: Firms in non-specialised sectors

	Small Firms (<99 employees)		Large Firms (>99 employees)	
	Small Firms in Non-Specialised Sectors *(14 observations=8.8 percent* *of the sample)*		*Large Firms in Non-Specialised Sectors* *(6 observations=3.8 percent* *of the sample)*	
	Market Size:		*Market Size:*	
	• Local and regional	0.23	• Local and regional	0.02
	Innovation:		*Innovation:*	
	• Breakthrough	0.26	• No particular innovation	
	Sources of Knowledge:		*Sources of Knowledge:*	
	• External customers	0.29	• External customers	0.28
	• Consultancy services	0.17	• Ex-colleagues	0.07
			• Scientific research centres	0.05
			• Qualified labour market	0.09
	Sources of Co-operation:		*Sources of Co-operation:*	
	• External suppliers	0.08	• R&D centres	0.01
	• Other firms of group, suppliers	0.18	• Other firms of group	0.40
			• Innovative R&D centres	0.40
			• Other firms	0.74
	Soruces of Information:		*Sources of Information:*	
	• Ex-colleagues	0.07	• Ex-colleagues	0.01
			• Internal information	0.03
			• R&D centres	0.07
	Locational Advantages:		*Locational Advantages:*	
	• Proximity to R&D centres	1.65	• Proximity to competitors	0.15
	• Proximity to customers and suppliers	0.18	• High life quality standard	0.14
	• Proximity to infrastructure	0.11		
	• Proximity to services to firms	0.14		

(left margin label: Non-specialised sectors (ql < sample mean))

Notes: Values = deviance from the sample mean

The sources of knowledge and strategic information sources for the innovative activity of the third cluster are again typical of large firms:
- the main sources of knowledge are external suppliers and scientific research centres;
- internal information is the primary source of information;
- the scientific environment in which firms operate plays a key role in their innovative activity. One of the main sources of knowledge being R&D centres, which are also seen as locational advantages;
- the presence of highly qualified public services, such as schools, hospitals and public facilities [considered by previous studies as one of the main reasons for a metropolitan location of multinationals].

The importance of milieu economies for large specialised firms derives from elements such as:
- the appreciation of proximity to customers and suppliers;

- co-operation with innovative local suppliers [a traditional collective learning channel] as a way of feeding the firm's innovative activity.

The fourth and smallest cluster, accounting for six [3.8 percent] of the sample, is characterised by large firms in non-specialised sectors. The locational behaviour of these firms is influenced by their preference for centrality, due to the possibility of gaining:

- information from scientific research centres;
- knowledge from co-operation with scientific research centres;
- the highly qualified labour market.

The sources of development and creative activity of the firms in this cluster stem not from the local environment, but:

- from knowledge internal to the firm or
- from external resources: external customers and suppliers, or co-operation with other firms of the group.

The reasons for the choice of a metropolitan location of these firms also seem to be related to:

- the high quality of life in such areas, as previously mentioned in the case of large specialised firms;
- the possibility of exercises control over competitors and market share.

The definition of these four different behaviours provides two important results. The first conclusion is that a milieu kind of behaviour can also exist in urban areas. It has been seen that some firms take advantage of the interaction with local economic actors, co-operation with suppliers and customers to stimulate their innovative activity. It can be argued that these firms appreciate the existence of socialised knowledge mechanisms which feed their innovative capacity and stimulate innovative behaviour[4]. The second rather interesting result is that firms favouring this kind of spatial economy are characterised by their *size* and the *degree of specialisation* of the sector in which they operate.

As far as the size of the firm is concerned, small firms seem in general to appreciate milieu economies more than large ones, which are more likely to draw advantage from dynamic urbanisation economies [the highly qualified labour market and co-operation with research centres]. However, when the degree of sectoral specialisation is taken into consideration, another perspective emerges. Large specialised firms tend to feed their innovative activity with local specialised knowledge and seem to appreciate not only urbanisation economies, but also milieu economies which stem from the high degree of specialisation of the sector in which they operate. On the contrary, small firms in non-specialised sectors do not seem to appreciate milieu economies and tend to take advantage from their metropolitan location.

Fig. 10.1 summarises this important result, showing the importance of the interplay of the two elements mentioned above and depicting the behaviour of firms in different spatial economies. Two indices have been calculated, i.e. co-operation with research centres and co-operation with innovative suppliers, as proxies for dynamic urbanisation economies and milieu economies [i.e. collective learning] respectively.

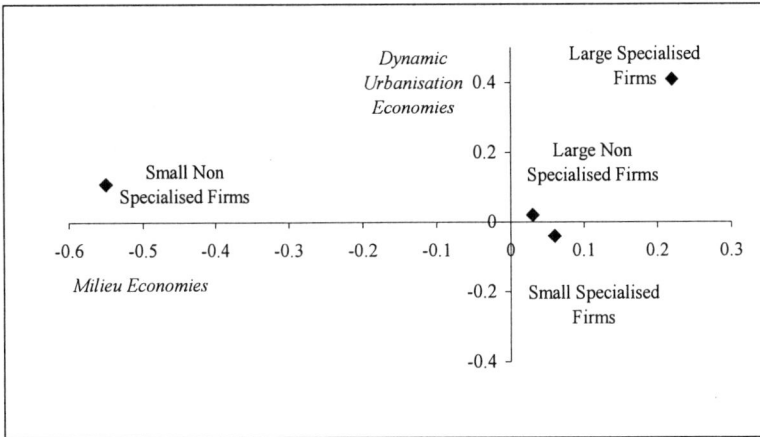

Fig. 10.1   Dynamic urbanisation economies versus milieu economies
for the four clusters

This gives rise to the following results:
- both non-specialised and specialised large firms take advantage of dynamic urbanisation economies;
- milieu economies are appreciated by both large and small firms operating in the more specialised sectors;
- small firms which operate in non-specialised sectors do not take advantage of milieu economies, but appreciate dynamic urbanisation economies in their innovative activities.

Milieu Economies, Internal and External Networking

The cluster analysis presented above shows that small specialised firms located in metropolitan cities take advantage of milieu economies for developing innovative activities. Another interesting suggestion put forward by the *milieu innovateur* theory is that, within the milieu, two kinds of co-operation processes are at work (Camagni 1991):
- a set of mainly informal, 'non-traded' relationships between customers and suppliers, private and public actors, and a set of tacit transfers of knowledge taking place through the individual chains of professional mobility and inter-firm imitation processes;
- a network of more formalised, mainly trans-territorial co-operation agreements between firms, collective agents and public institutions in the field of technological development, vocational and on-the-job training, infrastructure and service provision. These represent an organisational model somewhere between pure market and hierarchy (Gordon 1993).

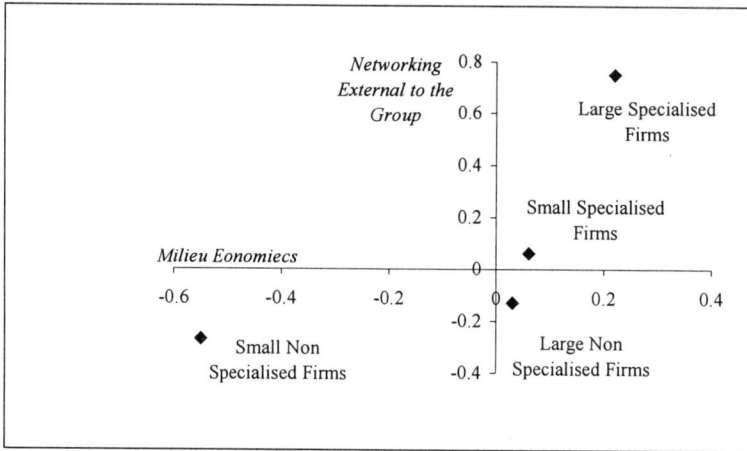

Fig. 10.2   Networking external to the area and milieu economies for the four clusters

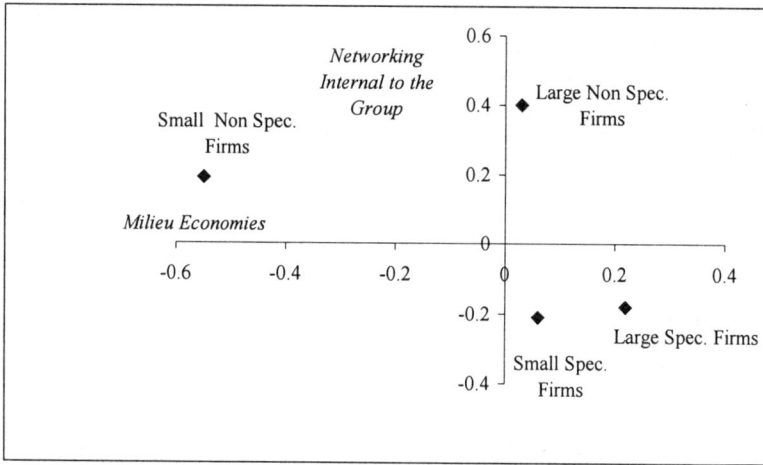

Fig. 10.3   Milieu economies and networking internal to the group for the four clusters

The former kind of relationship is in fact the 'glue' that creates the milieu effect. It is complemented by the latter, more formalised kind of relationship referred to as 'network relationships'. Both sets of relationship may be regarded as tools or 'operators' that help the [small] firm in its competitive struggle, enhancing its creativeness and reducing the dynamic uncertainty intrinsically embedded in innovation processes. In particular, the second kind of co-operative networking behaviour seems to be an efficient way for small firms to overcome the economically turbulent innovative phases, representing a way of obtaining information and knowledge from outside the area. We attempted to test this hypothesis in the case of our metropolitan firms. Two proxies were used, one for the existence of the milieu relationship [co-operation with innovative suppliers], the other for the network [co-operation with other firms]. They are presented in Fig. 10.2.

The results are quite interesting. It seems that specialised firms take advantage of both milieu economies and external networking, reflecting a kind of behaviour typical of innovative firms in milieu areas. On the contrary, the non-specialised firms, despite their small size, did not develop any interfirm innovative co-operation activity. They appeared to rely on networking with other firms of the same group for co-operation in innovative activities (see Fig. 10.3).

10.5 The Determinants of Innovation in Cities

The cluster analysis has provided some insight into innovative behaviour and spatial externalities of firms. In this part of the analysis, an attempt is made to define the determinants of innovation in cities, for firms of different size and in sectors with different degrees of specialisation.

The methodology used is based on a discrete choice modelling approach, more specifically a multinomial logit model. The choice of this methodology was driven by the essentially discrete nature of our database in relation to the innovation activity. The dependent variable is a dummy variable for the firms' innovative activity. Two types of model were estimated: for breakthrough innovation [firms replying that they develop innovation new to the world] and imitative innovation [firms replying that they develop innovation new to the firm]. It is in fact frequently mentioned in the literature that the quality of innovation depends on degree of interaction with the local territory (Oerlemans, Meens and Boeckema 1998). The variables used in the analysis of the innovation determinants were the following:

- company size [firms over 99 employees were used as a dummy variable for large firms];
- a location quotient [a continuous variable];
- company age [creation before or after 1970 used as a dummy discriminating a firm's age];
- indices of dynamic urbanisation economies, represented by proximity to information sources and competitors;
- indices of milieu economies, represented by proximity to local suppliers and customers, and innovative co-operation with local suppliers.

The results of the estimated logit models are presented in Table 10.4. As far imitative innovation is concerned, the results are quite interesting. The first model shows that imitative innovation is developed by firms operating in specialised sectors of the city. For these firms, the interaction with local suppliers [a traditional channel through which collective learning takes place] is very important for supporting innovative activity, while proximity to competitors or proximity to a large information base has a negative impact on their innovation capacity.

Table 10.4 Determinants of innovation activities in cities [Logit models]

	Model 1 Imitative Innovation	Model 2 Imitative Innovation	Model 3 Breakthrough Innovation	Model 4 Breakthrough Innovation
Constant	-3.60	-3.52	0.12	-0.38
	(27.9)	(20.01)	(0.27)	(1.77)
Firm Size		-1.37	0.98	
(1 = >99 employees)		(4.36)	(4.5)	
Location Quotient	0.92	0.85		
(continuous variable)	(7.04)	(5.79)		
Multinational Firms				1.34
(1=yes)				(7.07)
Milieu Economies:				
a) Cooperation with Local	0.71	0.73	-0.68	-0.70
Innovative Suppliers	(10.1)	(10.3)	(8.06)	(8.22)
(factor 3 of factor analysis b)				
b) Proximity of Suppliers and			-0.70	-0.81
Customers			(9.84)	(12.6)
(factor 4 of factor analysis f)				
Dynamic Urbanisation Economies:				
a) Proximity of Competitors	-0.87	-0.81		
(factor 7 of factor analysis f)	(6.70)	(5.64)		
b) Proximity of Information	-0.79	-0.67	0.40	0.36
(factor 8 of factor analysis f)	(7.00)	(4.68)	(3.46)	(2.90)
International Markets			0.41	0.42
(factor 1 of factor analysis e)			(3.03)	(3.01)
New Firms			1.77	1.54
(1=created after 1990)			(0.02)	(4.15)
Old Firms		0.99		
(1= created before 1970)		(2.81)		
Number of R&D Employees			0.80	0.84
(1=R&D employees > 20)			(2.88)	(3.46)
Goodness of Fit (chi-square)	143.3	216.8	170.4	149.8
Degrees of Freedom	4	6	8	8
Significance	0.00	0.00	0.00	0.00
Number of Observations	159	159	159	159

Notes: Wald test in brackets. Dependent variables: Imitative innovation = innovation new to the firm [1=yes]; Breakthrough innovation = innovation new to the world [1=yes].

Another interesting result emerges from model 2. Imitative innovation tends to be developed by relatively small firms [under 99 employees] operating in specialised sectors. For these firms, collective learning is a way of acquiring important knowledge for their innovative activity, while more dynamic urbanisation economies, like proximity to competitors or proximity to general information, have no strategic role. It also emerged that firms set up before 1970 were better able to exploit collective learning. This result is significant for three reasons:

- it underlines the importance of local interactions for firms operating in specialised sectors where the socialisation of knowledge and social interaction create a common knowledge base on which the innovative activity depends. This is especially so for small firms and those present in a given urban area for a number of years;
- local specialised knowledge and collective learning mechanisms are able to generate only incremental innovation and seem to be of little importance for breakthrough innovation processes;
- the Theil index, built as a proxy for the degree of specialisation of a city, did not as a whole provide statistically significant results. There is therefore no empirical support for the archetype of the city as a milieu.

Table 10.4 also provides interpretative models for breakthrough innovation processes [columns 3 and 4]. In this case, a contrasting picture emerges. Firms developing breakthrough innovation are mainly large firms [and mostly multinationals] operating in international markets which do not require collective learning mechanisms. On the contrary, the innovative activity of this kind of firm is favoured by an urban location due to proximity to a large information base. They are in general new firms, having located since 1990.

These results provide *prima facie* evidence that both the size of the firm and the degree of sectoral specialisation are discriminating factors affecting:

- the quality of innovation developed by firms;
- the dynamic spatial economies useful for innovative activities.

As far as this second result is concerned, it is interesting to discover the degree to which firms' size and sectoral specialisation play a role in the exploitation of spatial economies for a firm's innovative activities. This is the subject of the next section.

10.6 Dynamic Urbanisation Economies versus Milieu Economies in Innovative Behaviour: The Role of Firms' Size and Sectoral Specialisation

One of the main results of the above analysis is that dynamic urbanisation economies and milieu economies play a role in innovative activity, but that the influence depends on the size of the firm and its sectoral specialisation. Small firms operating in specialised sectors are more inclined to exploit milieu economies, while large firms tend to are favour dynamic urbanisation economies. In this part of the analysis we wish to measure:

- the impact of milieu economies and dynamic urbanisation economies on a firm's innovative capacity;
- how this impact varies according to a firm's size and the sectoral specialisation in which it operates.

For this purpose, the following two models were estimated:

$$I = \alpha_1 + \beta_1 \ln ql + v_1 \ln S + \varepsilon_1 due + \phi_1 me + \eta_1 (me\ ql) + \lambda_1 (me\ S) \qquad (10.1)$$

and

$$I = \alpha_2 + \beta_2 \ln ql + v_2 \ln S + \varepsilon_2 due + \phi_2 me + \eta_2 (due\ ql) + \lambda_2 (due\ S) \qquad (10.2)$$

where I denotes the innovation capacity of a firm, ql the location quotient of the sector in which the firm operates, S the size of the firm, due dynamic urbanisation economies, me milieu economies and α_1, α_2, β_1, β_2, v_1, v_2, ε_1, ε_2, ϕ_1, ϕ_2, η_1, η_2, λ_1 and λ_2 parameters.

The estimate of the first model allows us to capture the role of milieu economies on the firm's innovation activities and the way in which this role changes according to its size and the degree of sectoral specialisation. The second model captures the same effect for dynamic urbanisation economies. To measure this role, it is necessary to calculate the first derivative of innovation activities for dynamic urbanisation economies and milieu economies respectively:

$$\frac{\delta I}{\delta me} = \phi_1 + \eta_1\ ql + \lambda_1\ S \qquad (10.3)$$

and

$$\frac{\delta I}{\delta due} = \varepsilon_2 + \eta_2\ ql + \lambda_2\ S \qquad (10.4)$$

and to calculate how this varies with different values for firm size and the location quotient. The models were estimated using the following proxies:
- for size, we used the turnover of firms [in euros] [expressed in logarithmic terms]. Turnover was available only for 126 firms, limiting this part of the analysis to these 126 observations;
- for the specialisation index, we used the share of employment in one sector in a city compared with the same share of employment at the national level [location quotient, expressed in logarithmic terms];
- for the dynamic urbanisation economies, we used the co-operation with scientific research centres and universities strategic for the innovation activity [factor 5 of factor analysis b];
- for the milieu economies, we used the co-operation with local innovative suppliers for innovation [factor 3 of factor analysis b].

The results of the estimates of Equations (10.1) and (10.2) are presented in Table 10.5, and the results of Equations (10.3) and (10.4) displayed in Fig. 10.4. The results suggest that:

- imitative innovation activity [measured as the capacity of firms to introduce a new innovation] is developed by small firms operating in specialised industrial sectors of the city, taking advantage of milieu economies and, in particular, collective learning mechanisms (model 1, Table 10.5). Dynamic urbanisation economies do not provide any sort of help, and are even negatively correlated;
- the interaction between size, specialisation and agglomeration economies is statistically significant, with opposite signs: milieu economies are related negatively to the size of firms, and positively to the degree of sectoral specialisation (model 1), while dynamic urbanisation economies are positively linked to the size of the firm and negatively to the location quotient (model 2).

Table 10.5 Innovation, milieu economies and dynamic urbanisation economies [Linear regression models]

Independent Variables	Model 1	Model 2
Constant	1.65 (3.63)	-0.24 (-1.95)
Location Quotient [ln]	0.38 (3.22)	0.31 (2.38)
Turnover [ln]	-0.09 (-3.63)	-0.06 (-2.40)
Milieu Economies	0.97 (2.12)	0.21 (2.47)
Dynamic Urbanisation Economies	-0.17 (-2.20)	-0.60 (-1.70)
Service Firms (1=service firm)	-0.47 (-2.53)	
Milieu Economies * Turnover [ln]	-0.04 (1.70)	
Milieu Economies * Location Quotient [ln]	0.21 (1.79)	
Dynamic Urbanisation Economies * Turnover [ln]		0.03 (1.26)
Dynamic Urbanisation Economies * Location Quotient [ln]		-0.24 (-1.95)
Goodness of Fit (R-square)	0.24	0.20
Number of Observations	126	126

Notes: T-student in brackets. Dependent variable: Imitative innovation (factor 2 of factor analysis a) Milieu economies = Co-operation with local innovative suppliers for the innovation (factor 3 of factor analysis b) Dynamic urbanisation economies = Co-operation with scientific research centres and universities (factor 5 of factor analysis b).

Fig. 10.4 shows the results of Equations (10.3) and (10.4), calculated by keeping constant either the location quotient (Fig. 10.4 (a) and (c)) or the turnover

(Fig. 10.4 (b) and (d)), and measuring the values of the derivative for different levels of firm size (Fig. 10.4 (a) and (c)) and location quotient (Fig. 10.4 (b) and (d)). Fig. 10.4 (a) and Fig. 10.4 (b) show that the impact of dynamic urbanisation economies on innovative activities increases with a firm's size and decreases with the degree of sectoral specialisation. This means that larger firms exploit dynamic urbanisation economies more than small firms (Fig. 10.4 (a)) and that firms operating in highly specialised sectors do not draw particular advantage from their location in an urban setting (Fig. 10.4 (b)). The opposite results emerge in Fig. 10.4 (c) and (d), relating to the impact of milieu economies on the firm's innovative capacity. This appears to decrease with the firm's size (Fig. 10.4 (c)) and increase with the degree of specialisation (Fig. 10.4 (d)). Milieu economies are exploited more by small firms and, interestingly enough, by firms operating in specialised sectors. Small firms in specialised sectors therefore seem to recreate in an urban setting the sort of specialised industrial atmosphere typical of a milieu, giving rise to what could be called an 'urban production milieu'.

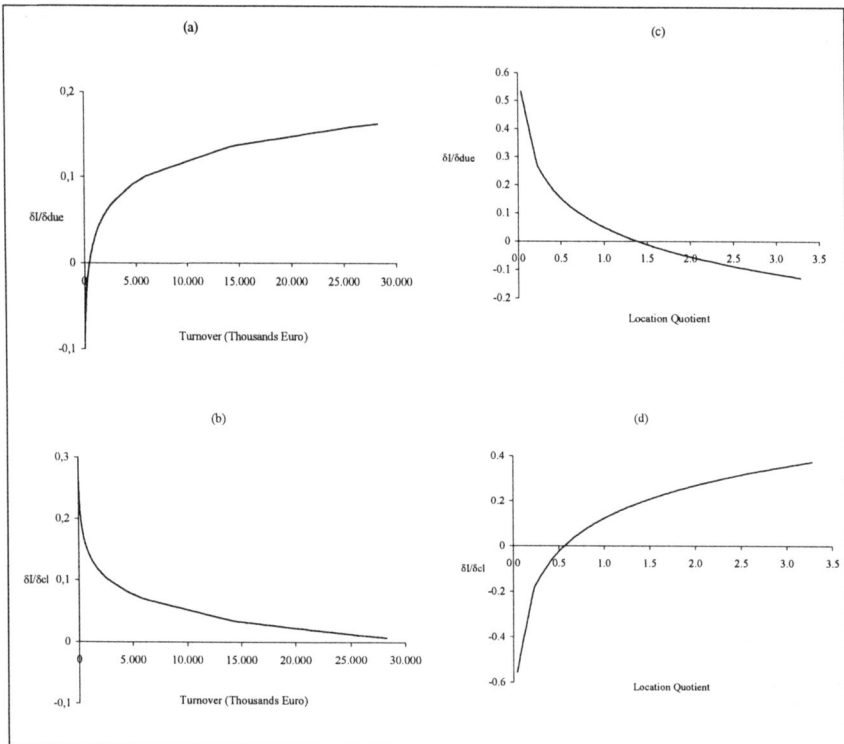

Fig. 10.4   Impact of dynamic agglomeration economies on firms innovation activities by firms' size and sectoral specialisation

10.7   Conclusions

This chapter has examined the capacity of the city to stimulate the innovative capacity of firms. The analysis was based on the theory of the *milieu innovateur*, which emphasises the role of relational proximity in favouring innovative processes. *Prima facie* evidence has been produced of the positive role played by milieu economies [expressed through the capacity of the city to produce knowledge in a socialised way, thanks to a strong and innovative interaction among economic actors] as opposed to dynamic urbanisation economies [i.e. the channels of knowledge acquisition typical of the large city, such as innovative interaction with universities and research centres].

An empirical analysis was carried out using a database for five European cities. International comparisons are fraught with difficulties deriving from differences in the institutional settings, political cultures and national innovation systems. These can lead to elements of ambiguity, however as long as caution is used in the interpretation of the results, such studies can be illuminating.

As far as we know, the present study represents a first attempt to provide a quantitative measure of collective learning in cities. It has produced some interesting findings, which help to provide new insights into the existence of 'milieu' effects in an urban setting. It therefore makes a useful contribution to the debate on the relative influence of dynamic urbanisation and dynamic localisation economies on the innovative activities of firms. These new insights derive from the conceptual approach used in the analysis. In order to test the importance of spatial externalities on the innovative capacity of the local area, the concept of the milieu was used, stressing the role of spatial synergies and local co-operation as channels through which local specialised knowledge can accumulate over time in a local area. Thanks to this conceptual approach, both the size and sectoral specialisation have been assessed as possible explanatory variables for spatial externalities. It is interesting however that neither the size of the cities nor their degree of specialisation appear to play an important role in the innovative capacity of the firms located there.

In this respect, the study provides evidence that the archetype of 'the city as an innovative milieu' is *not* supported by empirical evidence. On the contrary, there is evidence of the existence of an 'urban production milieu', since firms seem to exploit milieu economies in the form of collective learning. In fact, innovative co-operation with local suppliers and customers is one of the main determinants of firms' innovation activities. Moreover, the results suggest that the importance of localisation or urbanisation economies also depends on the size of the firm and the specialisation of the sector in which it operates. Small firms operating in specialised sectors, i.e. belonging to an industrial filière, take advantage of the traditional dynamic synergies typical of a milieu. Large firms, on the contrary, seem to prefer dynamic urbanisation economies, oriented towards the acquisition of knowledge stemming from their urban location. These results are supported by the quantitative analysis of the impact of dynamic urbanisation economies and milieu economies, and the effect of the size of firms and degree of sectoral specialisation.

These results may even explain why, after so long, empirical analyses have not yet reached a consensus. They have generally been applied at a sectoral or even aggregate urban level, ignoring the characteristics of single firms in terms of size and sectoral specialisation, two elements which are fundamental in explaining the role of dynamic agglomeration economies on innovation performance.

Acknowledgements. The project on which this chapter is based was led by James Simmie of Oxford Brookes University and financed by the ESRC. National subcontractors were responsible for the local enquires and local final reports: Roberta Capello, University of Molise and Politecnico of Milan, for the Milan area; Jeanine Cohen of University of Paris I, for the Paris area; Walter Manshanden of the University of Amsterdam for the Amsterdam area; James Simmie and James Sennet of Oxford Brookes University for the London area, and Simone Stramback of the University of Stuttgart, for the Stuttgart area. The results of the project will be published in Simmie 2001 (forthcoming). The author is grateful to Daniele Villa Veronelli for his support in data collection for the metropolitan area of Milan and in the elaboration of the whole database. Thanks are also due to Jeanine Cohen of University of Paris I, Erik Verhoef of the Free University of Amsterdam, James Sennet of Oxford Brookes University, Simone Stramback of the University of Stuttgart, Rolf Sternberg and Christine Tamasy of the University of Cologne for providing employment data at the required sectoral disaggregation and national and territorial level.

Endnotes

1   A vast literature exists on the social homogeneity of local districts. See among others, Bagnasco and Trigilia (1984); Becattini (1979, 1990). For an overall synthesis of local district theories see Rabellotti (1997); Bramanti and Maggioni (1997); Pietrobelli (1998).

2   The idea that territorial proximity is insufficient for milieu mechanisms has already been put forward by the French school on 'proximity'. See, among others, Bellet, Colletis and Lung (1993); Dupuy and Gilly (1995); Rallet (1993); Gilly and Torre (2000).

3   Factor analysis is a statistical technique used to identify the factors that can be used to represent relationships among sets of interrelated variables. The basic assumption is that underlying dimensions, or factors, can be used to explain complex phenomena. The goal of factor analysis is thus to identify not-directly-observable factors through a set of observable variables, therefore reducing the number without losing too much of their explanatory power.

4   A similar result was found in relation to the innovative behaviour of firms in the metropolitan area of Milan (see Capello 2001, forthcoming).

Appendix: Results of the Factor Analysis

Table A-10.1   Innovativeness of firms

Factors and Items	Factor Coefficients	Variance Explained	Labels
Factor 1 (innobth) *Innovation which is:*		34.3	*Breakthrough Innovation*
• New to the World	0.73		
• New to the Sector	-0.96		
Factor 2 (innovimi)		24.1	*Imitative Innovation*
• New to the Firm	0.98		
• New to the World	-0.62		
Factor 3 (innopg)		19.7	*Process Innovation*
• Process Innovation	0.82		
• Firm Size	-0.59		

Notes: Total variance explained: 78.2 percent
All variables assume 0-1 values (1=yes; 0=no).

Table A-10.2   Important co-operation linkages for innovation development

Factors and Items	Factor Coefficients	Variance Explained	Labels
Factor 1 (Copgruce) *Co-operation with:*		13.0	*Co-operation with Firms of the Same Group*
• Firms of the Same Group	0.67		
• Consultancy Services	0.53		
• External Customers	0.52		
Factor 2 (Copfores)		12.0	*Cooperation with Competitors and Suppliers External to the Firm*
• Competitors	0.77		
• Suppliers	0.73		
Factor 3 (Copflo)		11.0	*Cooperation with Local Innovative Suppliers*
• Local Suppliers	0.64		
• Innovative Suppliers	0.81		
Factor 4 (Copclilo)		10.0	*Cooperation with Local Innovative Customers*
• Local Customers	0.73		
• Innovative Customers	0.71		
Factor 5 (Coporg)		9.8	*Co-operation with Universities and Large Research Centres*
• Universities	0.61		
• Research Centres	0.79		
Factor 6 (Copimp)		9.4	*Co-operation with Other Firms*
• Other Firms	0.87		

Notes: Total variance explained: 66.4 percent
All variables assume 0-1 values (1= yes; 0= no).

Table A-10.3   Important sources of knowledge for innovation development

Factors and Items	Factor Coefficients	Variance Explained	Labels
Factor 1 (Conscien) *Sources of Knowledge:*		13.5	*Knowledge from Scientific Interactions*
• R&D Centres	0.75		
• R&D Employees	0.71		
Factor 2 (Coninform)		11.0	*Knowledge from Informal Sources*
• Ex-colleagues	0.84		
• Friends	0.87		
Factor 3 (Conatmur)		10.0	*Knowledge from Consultancy Firms and Technical Consultants*
• Consultancy Firms	0.79		
• Technical Consultants	0.63		
Factor 4 (Conmktll)		9.3	*Knowledge from the Local Labour Market*
• Internal Technicians Recruited from the Local Labour Market	0.68		
• Internal Managers Recruited from the Local Labour Market	0.71		
Factor 5 (Conclies)		8.9	*Knowledge from External Customers*
• Customers External to the Local Area	0.82		
Factor 6 (Conflo)		7.8	*Knowledge from Local Suppliers*
• Local Suppliers	0.92		
Factor 7 (Confores)		7.3	*Knowledge from External Suppliers*
• Suppliers External to the Local Area	0.85		

Notes: Total variance explained: 68 percent
All variables assume 0-1 values (1= yes; 0= no).

Table A-10.4   Strategic sources of information for the innovation

Factors and Items	Factor Coefficients	Variance Explained	Labels
Factor 1 (infscien) *Sources of Information:*		16.8	*Scientific Information*
• Government	0.72		
• Universities and Other HEIs	0.59		
• Private Non-Profit Research Centres	0.76		
Factor 2 (infpub)		16.4	*Information from the Press*
• Books	0.84		
• Journals	0.85		
Factor 3 (Infint)		14.7	*Information Internal to the Firm*
• Within the Firm	0.73		
• Other Firms within the Group	0.81		
Factor 4 (inftec)		13.0	*Technological Information*
• Research and Technology Organisations	0.86		
Factor 5 (infinf)		9.4	*Informal Information*
• Friends and Ex-colleagues	0.90		

Notes: Total variance explained: 72.9 percent
All variables assume values in a range from 1-5 (1=low importance; 5=very important).

Table A-10.5   Location of suppliers, competitors and customers

Factors and Items	Factor Coefficients	Variance Explained	Labels
Factor 1 (mktinter)		16.2	*International Markets*
• USA Competitors	0.48		
• Japanese Competitors	0.66		
• USA Customers	0.81		
• Japanese Customers	0.80		
Factor 2 (mktloreg)		12.2	*Local and Regional Markets*
• Local Customers	0.66		
• Regional Customers	0.72		
Factor 3 (mktnaz)		11.6	*National Markets*
• National Competitors	0.75		
• National Customers	0.84		
Factor 4 (mktflo)		9.9	*Local Suppliers*
• Local Suppliers	0.75		
Factor 5 (europmkt)		9.8	*European Markets*
• European Customers	0.79		
• European Competitors	0.75		

Notes: Total variance explained: 59.8 percent
All variables assume values in a range from 1-5 (1=low importance; 5=very important).

Table A-10.6   Quality of the local environment

Factors and Items	Factor Coefficients	Variance Explained	Labels
Factor 1 (proquaser)		46.5	*Proximity of Qualified Population Services*
• Proximity to Good Schools	0.79		
• Proximity to Good Leisure Facilities	0.85		
• Proximity to Good Private Services	0.88		
Factor 2 (loquavi)		32.9	*High Quality of Local Environment*
• Availability of Good Housing	0.78		
• Good Natural Environment	0.92		

Notes: Total variance explained: 72.9 percent
All variables were valued from 1-5 (1=low importance; 5=very important).

Table A-10.7   Locational advantages

Factors and Items	Factor Coefficients	Variance Explained	Labels
Factor 1 (lovacodi)		19.2	*Land and Labour Cost*
• Cost of Labour	0.76		*Advantages*
• Availability of Suitable Premises	0.84		
• Cost of Premises	0.85		
• Low Traffic Congestion	0.67		
Factor 2 (loprosi)		10.8	*Proximity to Business Services*
• Proximity of Business Services	0.75		
• Access to Private General Business Services	0.82		
• Access to Specialised Private Business Services	0.82		
Factor 3 (loinf)		9.8	*Proximity to Infrastructures*
• Good Rail Connections	0.84		
• Good Access to National Road Network	0.80		
• Good Access to Major Airport	0.83		
Factor 4 (proclifo)		7.5	*Proximity of Suppliers and Customers*
• Proximity of Customers	0.69		
• Proximity of Suppliers	0.83		
Factor 5 (loproorg)		6.4	*Proximity to Scientific Research Centres and Universities*
• Contribution from Technological Research Centres	0.78		
Factor 6 (loamiex)		5.1	*Proximity to Friends and Ex-colleagues*
• Presence of Ex-colleagues	0.81		
• Presence of Friends	0.82		
Factor 7 (loprocomp)		4.5	*Proximity of Competitors*
• Proximity of Competitors	0.67		
Factor 8 (loproinfor)		4.3	*Proximity of Information*
• Proximity to Information	0.64		

Notes: Total variance explained: 67.9 percent
All variables assume values in a range from 1-5 (1=low importance; 5=very important).

11 Distributed Knowledge in Complex Engineering Project Networks: Implications for Regional Innovation Systems

Neil Alderman
Centre for Urban and Regional Development Studies, University of Newcastle

11.1 Introduction

There is widespread agreement that we are witnessing a transition to a new type of economy fuelled not so much by the production of goods and services, but by the generation and circulation of knowledge (Florida 1995). Competitive advantage is increasingly seen to stem less from traditional capabilities in manufacturing, assembly, distribution or service, and more from the possession of unique capabilities based on know-how. In this so-called 'knowledge economy', knowledge has become a key component of conventional manufacturing activity. Correspondingly, the idea that knowledge has become the critical resource needing to be managed has taken root within the management literature (e.g. von Krogh, Roos and Kleine 1998) and has spawned all manner of approaches and perspectives. Knowledge management is now regarded by some to be the latest management 'fad' (Scarbrough and Swan 1999).

Associated with this shift in emphasis in the process of competition are changes in the way that major customers procure substantial engineered products or systems (McLoughlin et al. 2000). Increasingly, requirements are expressed in terms of complete turnkey systems, often with associated facilities management contracts over the lifetime of the equipment. Such projects have been observed in the fields of power generation, railway rolling stock, materials handling for ports, road construction, floating production facilities for offshore oil and gas exploitation, and are an intrinsic feature of the UK Government's Private Finance Initiative. The traditional manufacturing organisation is faced with a need to acquire new capabilities and new knowledge to enable it to supply such requirements. In order to do this the manufacturer is compelled to enter into new network relationships with a range of supply chain organisations and other partners.

This bringing together of different facets of knowledge to engender innovation is encapsulated in the 'systems of innovation' approach to the understanding of the innovation process (Edquist 1997). Stemming from an analysis of the factors influencing innovation at the national level within national innovation systems (Freeman 1995; Nelson 1993a; Lundvall 1992b), it recognises the importance of national institutions, embedded learning and evolutionary change in explaining innovation. At the same time, a number of commentators have argued that, in face of globalisation of the production process and the increasing complexity of technology and other facets of modern economic activity, the region as a locus or repository for the knowledge that such activity is dependent upon has regained its importance (Porter 1990; Cooke and Morgan 1998).

The conjoining of these perspectives gives rise to the concept of the regional innovation system (Braczyk, Cooke and Heidenreich 1998; Fischer 2000), consisting of an agglomeration of a range of components required by the new economic activity, operating at a scale of activity that is at least sub-national in extent. Examples of such territories have been shown to have developed the capacity to promote and support the innovation process in their own right (Cooke, Uranga and Etxebarria 1997). Policy makers at the regional level have seized on the possibilities offered by these ideas in attempting to promote 'clusters' of activity with the potential for innovation and economic growth through the generation and use of knowledge within the region (Lagendijk and Cornford 2000).

The purpose of this chapter is to argue that many of the components of such regional innovation systems are predicated on conditions pertaining to high volume manufacturing systems, exemplified by automotives or electronics. These systems possess a degree of temporal stability, thereby permitting the institutional structures and mechanisms for promoting the generation and use of knowledge to be established and become embedded within the region. The universality of this high volume manufacturing model has not been overtly questioned, but given the continued importance to the economy of much low volume manufacturing and project-based activity, particularly relating to complex capital products (Hobday 1998), it is necessary to question whether the promotion of localised clusters of low volume project-based activity through a regional innovation system approach is possible.

The low volume context of much capital goods manufacture, which is typical of the complex one-off engineering projects alluded to above, has been largely neglected in both theoretical and empirical analysis. In this chapter the particular characteristics of complex engineering projects are outlined and a framework for analysing such projects, termed the 'project value system', is described. This is then related to the regional innovation system concept to identify the extent to which a project's value system might map onto a regional innovation system. Using illustrations from a number of detailed empirical case studies, we then consider the ways in which the knowledge necessary for the successful delivery of complex engineering projects is distributed through a range of different network actors. It is suggested that changes in the nature of the market place for complex engineering projects are creating the conditions whereby knowledge is becoming both organisationally and geographically dispersed. This has implications not only for the organisation but also for the promotion and development of regional systems of innovation, which in many instances are seen to be largely irrelevant.

11.2 Complex Engineering Project Networks

In recent years the nature of many engineering projects has subtly yet substantially changed to create a greater degree of complexity, which demands learning and the acquisition and management of new sources of knowledge. This is particularly so in the capital goods sector, comprising those products, simple or complex, that 'tend to be manufactured in low volumes for industrial producers in a range of industries ... to use in the production of other goods and services' (Maffin and Thwaites 1998, pp.1-2). Complex engineering projects involving capital goods are characterised by a low degree of repeatability, and the end product is often a complex structure or system embodying many components and individual capital items.

Furthermore, the customer for such projects is often seeking to take a much reduced share of the financial risk and requires the contractor to extend the scope of delivery to encompass responsibility for some or all of the operation and maintenance of the equipment or system provided. In extreme cases, this may extend to the disposal of the product at the end of its life. Companies may, therefore, find themselves integrating not simply the component systems of the project in order to deliver a complete 'product' [systems integration], but integrating the whole of the delivery of the project through its entire life cycle, leaving the customer concerned only with the output. This 'bundling' of products, services and systems, it has been argued, can potentially enhance customer value (Gann and Salter 1998) and is therefore attractive to capital goods companies. Many engineering companies are therefore extending their activity into a contracting and/or facilities management role. In some cases they are divesting themselves of their traditional manufacturing capability, which comes to form a smaller and smaller proportion of the total project budget.

For many capital goods producers operating in traditional markets [e.g. in mechanical engineering], technological change in the core product is very slow, as technologies are mature and develop only incrementally (Alderman 1999). Most major advances tend to occur in terms of external technologies, such as control systems and software, so technical innovation occurs mainly in the supply chain (Hutcheson, Pearson and Ball 1996). Key supply items therefore have to be sourced externally as competence in such technologies is very difficult, time-consuming and costly to develop in-house, unless it is likely to bring long-term advantage (Grant 1997). These additional competencies are drawn from the project network. The pooling of resources in this way can be an efficient solution

(Håkansson 1987; Cooke and Morgan 1993) as it enables a sharing of risk and uncertainty, which is an important consideration in complex capital projects (Hansen and Rush 1998).

The benefits of networking have been widely recognised in the high volume manufacturing context where the possibilities for guaranteeing long-term supply arrangements make such relationships attractive to all parties (Lamming 1993). However, considerable resources are required to establish these relationships and they have to be seen as part of a long term strategy (Tan, Kannan and Handfield 1998; Harris et al. 1999). In the low volume capital goods context, this can be problematic, because the supply chains associated with specific projects are often ephemeral, created to meet the needs of a particular client (Alderman et al. 1998a). Long term relations, essential to building trust and goodwill (Granovetter 1985) are difficult to foster in this environment. Moreover, the capital goods producer is often sandwiched between a powerful customer and equally powerful suppliers. It is in part the volume of production demanded that confers some of this power on the supplier. The degree to which the benefits of networking may be realised will depend on the degree to which power is centralised or shared in the network and the degree and manner in which it is exercised (Harris et al. 1999).

11.3 Regional Innovation Systems and Project Value Systems

It can be seen, then, that in the low volume, project-based engineering context, the theoretical and empirical conditions for network forms of relationship are somewhat problematic. The distribution of knowledge in such systems is a crucial aspect of this. To consider the role of knowledge in complex engineering projects we bring together the two concepts of the regional innovation system and the project or value added system.

The value added system concept was developed to help articulate the contingencies and constraints operating on innovation within capital goods production (Alderman et al. 1998b)[1]. It derives from the work of Porter (1985) and subsequent developments by the MIT 'Management in the 1990s' project (see MacDonald 1991). It was Porter who articulated the notion of value chains and value systems. The latter consist of a series of value chains, linking suppliers, the producer, distributors and buyers. Porter's model was essentially a linear one. Our approach has extended these concepts to encompass a system of actors and their interrelationships that contribute to the ultimate delivery of value to a customer in the form of a product or project.

For example, individual firms or other organisations operate within a system that consists of a range of other actors. These include customers, suppliers, regulators, distributors, financiers and so on, who have an influence on the processes making up a complex engineering project. The project value added system consists of the set of actors who collectively influence the nature of the activities of the organisation delivering the project. The system will be shaped and influenced according to the way the actors within the system react to broader drivers on the system [e.g. economic conditions] and to the way the relations between these actors are structured.

The days when a vertically integrated company produced everything necessary for the delivery of a particular product to its customers are long gone. The trends, noted above, towards externalisation and the outsourcing of technology and other expertise mean that the value incorporated in an engineering project is no longer exclusive to the producer (Johnston and Lawrence 1988; Normann and Ramirez 1993). The complete project may require a range of innovations, technological, organisational, financial and so forth, which have to be assembled by the organisation delivering the project through the management of the project network. Understanding where this value originates requires the company to take a view of the complete system that influences the derivation of customer value:

'the only true source of competitive advantage is the ability to conceive the entire value-creating system and make it work' (Normann and Ramirez 1993, p.69).

The elements that make up a value added system are represented in Fig. 11.1. Briefly, we define the primary actors or players as those that directly add or obtain value from the project that forms the basis of the value added system. These include:

- the *'project integrator'* (Rudolph 1998; McLoughlin et al. 2000) responsible for the delivery of the total project package;
- *producers or prime contractors*: i.e. those designing, manufacturing, assembling or constructing a product or specific components or subsystems required by the project integrator;
- *clients or customers*: those commissioning the project;
- *end users*: those ultimately purchasing and using the end product or benefiting from the ultimate output from the project [e.g. rail passengers]; and
- *suppliers*: companies comprising the supply chain for the producers, prime contractors or project integrator, providing raw materials, capital goods, IT systems, technical know-how and so on, who in turn have their own lower tier suppliers.

Secondary actors or players are those who have a more indirect impact on the value embodied in the project. Their role will vary in importance depending on the context. Secondary actors include:

- the *engineering scientists* developing new technologies and theoretical principles, who may be found in the in-house R&D labs of the producing or integrating organisations, in Universities, or in other public or private research organisations;
- *financiers, regulators, insurers* etc., who may set boundaries on what is possible for the project integrator, and who may decide whether a project is considered 'bankable' [2],
- *other organisations* or agents, who may act as intermediaries, or who need to be brought into the network in some way in order to deliver the project [for instance, in the rail industry privatisation has created a raft of such organisations such as Railtrack, the Rolling Stock Companies - ROSCOS - and others];

- *consultants*, who may perform aspects of the design activities in the system or provide services relating to quality accreditation and approval, and other such actors; and
- *other stakeholders* who may desire to appropriate value, such as shareholders in the business (Kay 1993), but might also include employee representatives or union organisations, or consumer organisations [e.g. rail user groups].

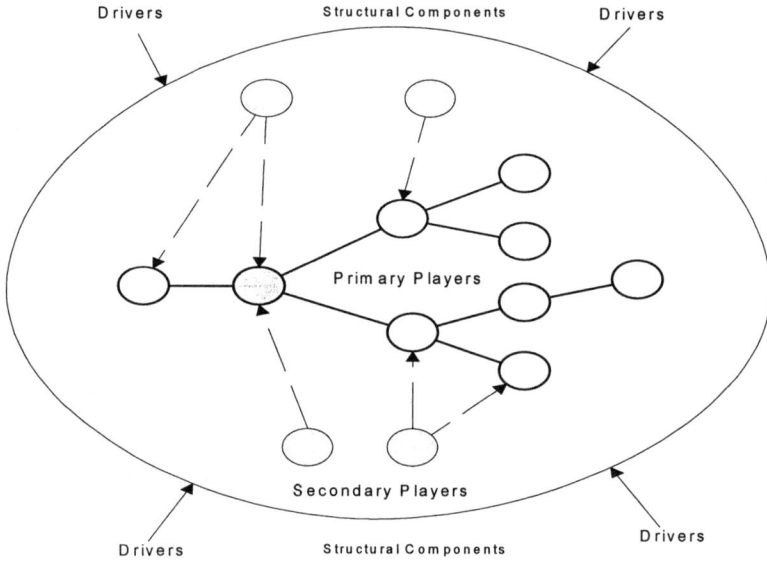

Source: Maffin et al. (1997)

Fig. 11.1   Elements of value added systems

The primary and secondary actors are considered to be part of the network of interest in so far as they have an influence on the ability of the organisation responsible for the delivery of the project to create and appropriate value.

The form of each project's value added system is conditioned by a series of components that impinge on each actor and structure the relationships between actors within the system to a greater or lesser degree. We refer to them as structural components because, in the short to medium term at least, they have to be taken as givens by any particular actor in the network. At a generic level these components include:

- the *pre-existing state of technology* and the *science base* [in the rail industry, this might include the current state of the art in tilting train technology, which does not reside with UK-owned companies];

- the *institutional framework* - in terms of legislation, regulation, finance etc, which can act as an important stimulant to or constraint on innovation;
- *human resources*, both within the organisation and within the [local] labour market; and
- *logistics* and *information systems*, which provide mechanisms for linking network actors in terms of goods and knowledge respectively [particularly relevant in projects that require the integration of a complex set of inputs from geographically dispersed sources of supply at a geographically remote location].

Both the actors and the structural components will be influenced and modified by the need to respond to a range of broader economic 'drivers' influencing each project value system. Such drivers include:

- *changing market requirements* [for example, demands for time compression, requirements for whole life project management, or provision of financial packages];
- *fundamental* and *invasive technological improvements*, which can change the basis of competition;
- *new sources of competition*, particularly since a move into markets requiring new capabilities will expose companies to competition from organisations already operating in such areas; and
- *regulation/de-regulation* and *environmental issues*, which impact on many areas of capital good procurement and supply.

The project value system bears considerable similarity to a system of innovation (see Fischer 2000, Fig. 3, for example). The innovation system, at whatever scale, sectoral or spatial, forms part of the wider system environment of the individual project, since innovations may be both endogenous and exogenous to the project value system. There is a fairly obvious conceptual overlap between the project-based value added system and an innovation system, because both are examples of open social systems (Flood and Carson 1993) where interaction with the environment is important for the production of new knowledge and new technologies (Cooke 1998, p.11). Moreover, many of the components of an innovation system are present in the value added system in the form of the various secondary actors and institutions that support the innovation process, thereby influencing the value that is created for the customer. The concept of value in the value added system goes beyond simple notions of cost, as in Porter (1985), to incorporate other features of projects that customers are looking for, such as technology, delivery time, through-life support and so on. The ability to deliver these components of value depends to a large degree on the acquisition of knowledge, or at least the management of this knowledge within the project network, by the lead organization.

The project value system is essentially an aspatial model in that the boundaries of the system are not geographically determined or fixed. In a complex engineering project it is hypothesized that the innovation system that supports it will extend across national boundaries. In their discussion of technological systems, Carlsson and Stankiewicz (1991) argue that spatial boundaries depend upon [amongst other things] technological and market requirements. Translating this idea into a project context implies that the system boundary will be determined by the scope and requirements of the individual project or the value requirements of the client or customer. There is thus no prior logic to suggest that a complex engineering project should be organised on any particular spatial basis.

The project value system is also aspatial in the sense that there is no over-riding requirement for spatial proximity on the part of the actors within the system. However, spatial proximity is argued to be an advantage, because collaboration within the network should be easier when partners are geographically close (Cooke and Morgan 1993; Yeung 1994). This proximity imperative is implicit in the concept of the regional innovation system.

In a regional innovation system, therefore, the key feature is a close networking of firms, possibly in a geographical cluster, supported by a set of institutions and actors that provide knowledge and other inputs, such as technology transfer and vocational training, to create a system for advanced technology production (Cooke 1998). Importantly, the regional innovation system combines the economic with associated governance structures (Cooke and Morgan 1998). Central to the concept is the idea that economic performance depends on the interaction of economic actors such as firms and research institutes 'with each other and with the government sector in *knowledge production and distribution*' (Gregerson and Johnson 1997, p. 482, emphasis added). Thus, in a regional [or national] innovation system, it is the generation and use of knowledge that is the key factor.

The key question here is whether or not value added systems and regional innovation systems map onto each other in any meaningful way. Whilst one would not necessarily expect to find a one-to-one mapping, it is easy, nevertheless, to conceive how in a high volume manufacturing context there may be considerable overlap. Here, for instance, an original equipment manufacturer [OEM], such as a major vehicle assembler, attracts a satellite supply chain into the local area and the local support infrastructure gears up to provide specialist services, training and other forms of support dedicated to that industry sector and the needs of the local OEM in particular. Certainly, such systems will become better developed in terms of innovation if the OEM locates its R&D activity in the region and the local institutes of higher education [HEIs] have specialized in relevant areas of technology[3].

In such circumstances the regional innovation system is sustained for the very reason that the network relationships between the firms are sustained: there are long term agreements and partnerships based on high volumes of production over a several year period. This permits the necessary investment in innovation and supporting infrastructure. As Cooke (1998) puts it: 'the 'systemness' ... derives from the relatively stable and regular flows of information among the members of the regional innovation community' (p.16). In the low volume context these conditions do not hold, because the client [equivalent to the OEM in the high volume model] changes from project to project, and the network of suppliers supporting the equipment producer also changes to some degree with each new project.

It is therefore questionable whether, in a low volume manufacturing context, such a regional innovation system could exist in the same way. One possible exception to this is the well-documented case of 'Motor Sport Valley' in the south east of Britain (Henry and Pinch 2000). Racing cars are not manufactured in high volumes, but there is a clear agglomeration of activity in the region with an identifiable 'knowledge community'. However, this system operates under the influence of very different factors to the sorts of engineering projects we are concerned with here. Sponsorship and advertising are dominant features of the economics of motor racing. Moreover, whilst a racing car is certainly a complex product, it has a relatively short life-span and the industry does not exhibit the same kinds of financing arrangements or facilities management contracts that are an increasing feature of other engineering industries.

In the next section, a number of detailed case studies are used to illustrate empirically the extent to which there is an overlap between the project value system and the regional innovation system in the context of complex engineering projects. The limitations of regional innovation systems from a low volume, project-based perspective are highlighted.

11.4 Case Studies

Four case studies will be outlined in this section. Two of the cases were studied as part of a project to develop the value added system concept described above[4]. A third example is an engineering design and project management organisation based in the North of England, which serves sectors such as iron and steel, water, and process plant. It was interviewed during the research definition phase for an as yet unfunded project. A final example is drawn from public domain documentation of a supply and maintain contract for new railway rolling stock.

Case 1: Mechanical Handling Equipment for the Offshore Industry

This case draws on a study of a company we shall call Mecho Ltd. It is a long established and relatively small unit within the materials handling division of a global corporation. Until comparatively recently Mecho manufactured its own products, but has now effectively outsourced all production and assembly activity to specialise in design and project management in relation to the supply of mechanical handling equipment to the marine, dock and offshore industries. Demands for components and sub-products or assemblies to meet the design requirements are placed on a number of suppliers and, for the project in question, assembly of the total product was undertaken by another subsidiary within the same division of Mecho's parent company.

This project involved the design, construction and delivery of a substantial piece of mechanical handling equipment for an international client operating offshore, which assumed the product would embody the latest technologies. Because Mecho has outsourced its manufacturing capability, it was required to put together a complex supply chain in order to carry out the project. Furthermore, many of the latest advances in technology are found within key components, the know-how for which resides with the suppliers.

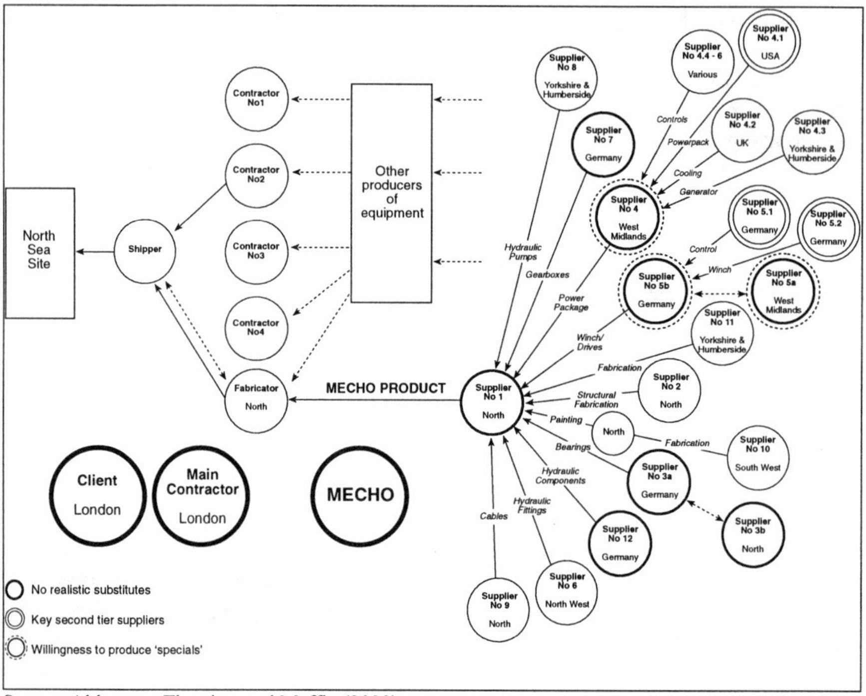

Source: Alderman, Thwaites and Maffin (2000)

Fig. 11.2   Critical suppliers in the Mecho supply chain

Fig. 11.2 shows that the geographic pattern of Mecho's supply chain was highly dispersed. In common with many capital goods we have investigated, most of the critical technologies are sourced from North America and Europe, notably Germany. Critically, Mecho Ltd is able to exert little influence over these suppliers. Within the UK only one first-tier supplier (No.10) is located in reasonable proximity to Mecho itself and this is largely for historical reasons[5]. It might appear that there is a degree of clustering in the north of England around the location of the assembler (i.e. Supplier Nos. 1, 9, 3b, 2), but this does not provide the basis for a regional [sub]system, because, on closer inspection, many of the UK sources of supply are themselves reliant on foreign inputs of major components, or even the total package in the case of supplier 3b.

In this project, the geography of information sources and knowledge flows is largely parallel to the flow of goods and services. With the exception of non-graduate labour and training and minor aspects of technical information, Mecho cannot be said to be embedded in its local area. Moreover, even at the national level, many of the support mechanisms available to its competitors on the continent are absent in the UK. Thus, there is no longer a Trade Association in the UK to represent Mecho's activities. Moreover, the setting of standards for the industry is led by the German standards institutions and Mecho finds that it has less influence on such matters than some of its European competitors. This has a bearing on innovation, since industry standards pose requirements and constraints on the way the company's products, and those of its suppliers, have to be developed. While information flows between all participants in a network of this type are crucial to the success of the project, as far as Mecho's competitiveness and client satisfaction is concerned, the key elements of value in the system stem from information and knowledge links with a select group of foreign sources.

Case 2: A Greenfield Factory Development

The second case study revolves around a firm of civil and structural engineering consultants we shall refer to as Integrator plc. The company's principal activities are related to industrial, quasi-public sector and commercial developments, including the design and construction management of major projects for local, national and international clients.

This project was concerned with the delivery of a new manufacturing facility to a location in the north of England on behalf of a foreign inward investor. Essentially, this was a construction project valued at over £30 million, with Integrator plc primarily concerned with the provision of a building[6]. Integrator's Client was the driving force in this particular project network. The Client was a transnational corporation manufacturing final goods for mass consumer markets. It wished to commission the 'fast-track' construction of a new manufacturing facility in Europe. Clearly, it did not wish to have to manage the myriad different contractors needed for a complex project of this nature and so Integrator plc was selected to provide this integration role. This involved a complex process of site selection, facility design, construction and commissioning for operation. Although cost was a major consideration, other key value elements for the Client were lead time and project management capability.

The Client had a good idea of what it wanted by way of a European manufacturing facility [a building with bays for manufacturing, packaging, warehousing and offices with flexibility to accommodate expansion]. Once the decision had been reached to locate on a specific site in the North of England, these concepts needed to be translated into practical detail and constructed within a local regulatory and operational framework. This meant that the project system involved a range of other agencies from utilities providers, local councils and planning authorities and the local development agency, to Government Ministries [e.g. Agriculture, Fisheries and Foods] and the National Rivers Authority [in relation to waste water and site run-off]. Failure to meet the project timetable by any of these agencies could have had serious ramifications, so the management of relationships with external organisations was a critical feature of the project.

In order to bid for the contract, Integrator deemed it necessary to bring together a range of capabilities and knowledge in a consortium consisting of three other firms [Architect Ltd, M&E Ltd and QS Ltd.[7]]. The skills of the other consortium members complemented those of Integrator, which lay in civil and structural design, and project management. Integrator acted as team leader. All parties were well known to each other, or recommended by one of the others, and all had worked previously for the Client. The other consortium members were all part of multi-locational service companies, but this was not a locally-oriented consortium. Each of the partners had specialised offices based in different regions, reflecting the fact that in such organisations there is a natural tendency for specific offices to develop expertise in different specialisms of their profession. Consequently, Integrator went where the relevant expertise was, and not necessarily to the local office. This seems to contradict the regional innovation system idea at least for this type of project activity.

The actual construction work was contracted to a separate firm, Constructor plc, which undertook some of the work itself, but subcontracted the rest. It was responsible for the co-ordination and management of these subcontractors. Constructor was selected from a long list of competitors as a result of its innovativeness and willingness to adopt the spirit of co-operation required by the Client. The ability to embed the project within the locality proved limited, however, despite a desire on the part of the Client to use local sub-contractors. Most were found to be too small for such a large project and local involvement was therefore limited to relatively minor inputs and activities.

A number of innovations were incorporated in the facility in terms of the arrangements of the buildings, provided by Integrator plc, which improved on the layout and design of the Client's pre-existing facilities, and also architectural innovation, provided by Architect Ltd, in the form of a novel atrium and roof design. The only significant input to the innovative elements of the project from elsewhere in the supply chain came via the cladding supplier, one of the construction subcontractors. Again, the source of these innovations was external to the region within which the project was located and totally independent of the rest of the local support infrastructure.

Case 3: Process Plant Design

The third case concerns the issues facing a firm of process plant designers operating primarily in export markets. This company perceives its core competencies (Hamel and Prahalad 1994) to lie with engineering design and project management. The company is facing increased complexity in the engineering projects it is engaged in as a result of the growing importance of finance as a central component of what is valued by the client. The knowledge and ability to put together an appropriate financing package is becoming an increasingly important order-winning criterion. It was suggested that for some projects the financial benefits may stem not from the design or supply, but from the manipulation of the finance itself.

The company is also too small to undertake its own R&D and therefore relies on major breakthroughs in process technology emerging from the manufacturers [e.g. steel companies] who have the resources to try new things. Successful technologies are then licensed in. For example, at the time of the interview, the company was selling Japanese and South Korean technologies. These technologies are incorporated into the process plant designs that the company is specialised in, provided for clients who often have little or no knowledge of the processes involved. The company is consequently dependent upon a global search process in relation to new innovations and not linked in to any kind of regional or local innovation system.

This company is another example of an engineering concern that has shed its manufacturing capability. One reason for this is the requirement for local content on the part of foreign clients and their governments. The supplier base is therefore international and changes from project to project as the geographical locus of each project shifts. This has a number of implications. In some localities there is a cost penalty arising from the need for extra project management in order to ensure that manufacturing activity which is local [to the client] is undertaken to the required quality and safety standards.

Possibly of greater long term concern is the fact that, in this situation, the in-house design team becomes steadily more distanced from the manufacturing process. This has two potential effects. First is the fact that there is a gradual loss of knowledge based on the principles of 'learning by doing' and 'design for manufacture' as the feedback between manufacturing and design dwindles. Secondly, this knowledge is also lost to the region, thereby weakening whatever regional innovation system may have existed, leaving gaps in the knowledge base required for complex engineering projects of this type[8].

Case 4: The Virgin Voyager Contract

A good example of the geographically dispersed nature of knowledge and capability in a complex engineering project is provided by the UK railway industry and the case of the procurement of new diesel-electric multiple units [DEMUs] from the Canadian company, Bombardier, by Virgin Trains for the Cross-Country franchise[9].

The complexity of the project is attributable to a number of factors. Firstly, the contract is not just for the supply of new trains [and in fact a very large number, even by rail industry standards [10]], but also includes a maintenance package for the new trains lasting the duration of the franchise [another 12 years]. This, then, is a good example of a traditional manufacturer being forced to move into new areas of expertise and knowledge. Moreover, the maintenance contract is let on a fixed price basis, so that it is in the manufacturer's interests to achieve whatever savings are possible in terms of maintenance costs, since this will feed directly into the bottom line. Design for maintenance has therefore become a much higher priority than on a traditional supply contract, where the costs of maintenance are borne fully by the customer after the normal warranty period.

Secondly, the project is technically complex, despite most of the key technologies being tried and tested. A large number of suppliers are responsible for key components of the trains, the tilting technology has to be incorporated and the whole integrated with the UK rail infrastructure - no mean task, given the age, condition and variability of the latter!

Thirdly, the project is organisationally complex. The value systems associated with rail industry projects are particularly extensive. In addition to the principal client-supplier relationship, both parties draw on a range of design consultancies for both conceptual and detail design activity. The whole project has required an external financing package, in this case £350 million provided by GL Trains Ltd[11]. Other agents with an interest in the project include Railtrack, which is the infrastructure owner with responsibility for approving the 'safety case' for all new trains[12], the Shadow Strategic Rail Authority, responsible for the award and monitoring of the operating franchises, and the Office of the Rail Regulator. All of these actors have an influence on innovation within the project, and all provide particular knowledge inputs, whether financial, technical or operational, that have to be coordinated and managed. Whilst it is perhaps possible to conceive of a national or sectoral system for innovation within the UK rail industry, this is not reflected in a regional innovation system for any of the UK train manufacturers[13].

The knowledge and capability of Bombardier itself is also geographically distributed. In order to satisfy the terms of the contract, 'Bombardier has set up a network involving four factories in north west Europe ... the most important plant in the later stages of the work is [the] Wakefield factory ... [but] Bombardier is also drawing on the skills of its established factories at Brugge and Manage in Belgium and Crespin in north east France' (Modern Railways, April 2000, p. 35).

Thus, it can be seen that for a complex project such as this, there is a dispersion of knowledge and skill within the corporation which is also geographically distributed across national boundaries. In order to manage this distributed network, Bombardier have had to establish a telecommunications network to link designers at each site with a common computer aided design [CAD] platform. The Voyager trains are reported to represent a new venture for the Wakefield factory and the company has therefore relocated staff from the UK to its sister factory in Brugge for lengthy periods in order that they gain 'hands-on experience' before the vehicles reach Wakefield for the final assembly tasks.

Train manufacturers increasingly are outsourcing much of the basic production work and focusing on their core competence of overall design and assembly, essentially an integration task. Table 11.1 lists the key suppliers to Bombardier and reveals the international flavour of the supply network. Moreover, it is apparent that suppliers of critical components within the UK are also highly dispersed, from Chippenham in the South West to Gateshead in the North East. There are, of course many other suppliers in a project like this. Many of these will inevitably be local to the assembler, particularly those concerned with more basic fabrication tasks, and, of course, each key supplier has its own supply chain in turn. The main point being made here, however, is that the core technologies in the product, which are the main areas for innovation, are not clustered on a regional basis, but distributed quite widely.

Furthermore, whilst the train manufacturers may have their preferred suppliers, other circumstances might dictate that the supply base changes from one project to the next. A major consideration in the Voyager project, as reported in Modern Railways, would appear to be the need to satisfy the safety acceptance procedures, which has promoted the application of technologies that are already tried and tested in the UK rail context. Such considerations may therefore lead to the choice of one supplier in preference to another.

Table 11.1 Key product suppliers to Bombardier

Company	Product	Location
ALSTOM	Electric traction	UK, North West
Cummins	Diesel engine	USA/ UK, North East
Covrad	Radiator	UK, W. Midlands
Eminox	Exhaust system	UK, E. Midlands
Faiveley	Heating/air conditioning Doors Data recorder	France
Westinghouse[14]	Brakes	UK, South West
INBIS	Shop/servery, galley	UK, E. Midlands/ Yorks & Humberside
Temoinsa	Toilets	Spain
Groupo Antolin Loire	Seats	France
Dellner	Coupler	Sweden
Joyce Loebl	Visual passenger information system	UK, North East
Televic	Audio passenger information system	Belgium

Source: Abbott (*Modern Railways*, p. 40) supplemented by original research

11.5 Conclusions

The basic argument advanced in this contribution is that current thinking about the nature and importance of regional innovation systems implicitly assumes what we have termed a 'high volume manufacturing model'. In this model, stable network relationships between major producers or assemblers and their local [and non-local] suppliers can be established, with an associated regional infrastructure geared to delivering appropriate training and development, financial, technical and other business support on a continuous basis. On the basis of our investigations into the capital goods industry and complex engineering projects, we can perhaps posit that low volume manufacturing or project-based activities are not conducive to the formation of such regional innovation systems, owing to the fact that relationships tend to be project-based, intermittent or ephemeral. The long-term commitments needed to create a viable innovation system are difficult to justify and achieve under these circumstances.

Furthermore, it has been argued that the increasing complexity of projects in low volume engineering reduces the importance or relevance of a regionally-organised system of technical support and knowledge production, as the range of capabilities required by those organisations attempting 'project integration' increases and activities become more and more specialised. Companies therefore have to search out these capabilities, often on a global basis, because they are unlikely to be found together in the same locational context. Providing an appropriate institutional support infrastructure under such circumstances is proving highly problematic for local agencies.

The case studies presented above illustrate the way that the 'project value system' is extended, as engineering projects become more complex and require the enrollment of a wider range of actors. These value systems are found to be geographically dispersed and do not map conveniently onto the regional innovation systems that might historically have supported manufacturers supplying such equipment. It is tempting to suggest that the project value system is simply another manifestation of a sectoral innovation system (Breschi and Malerba 1997), but this is not the whole story. The value components that contribute to the outcome of the project are ultimately specific to that project and are likely to change from one project to the next. The structural components of the value system are more stable and aspects of these may be regional [the local labour market], sectoral [industry associations, international standards and accreditation systems etc.] or national [Government support schemes, tax regimes etc.]. The sectoral innovation system is just one part of the wider system environment within which the project operates. Moreover, value may be derived in other ways than through technical innovation *per se*, implying that the system of innovation is perhaps a necessary, but not sufficient, component of the project value system.

A key conclusion is therefore that, in the low volume context of complex engineering projects, the precise composition of the relevant innovation system is project specific and, as a consequence, possesses a unique geography. Proximity and regionally agglomerated actors may or may not be important and, even if they are, there is no guarantee that this will remain the case in subsequent projects. The case studies illustrate how the project network is built up from components of a system that is international in extent, some parts of which may be sectoral, others regional or national.

As we have noted elsewhere in the context of the engineering product development process, it is the ability to tap into national and international knowledge networks that is critical, rather than the creation of local networks (Alderman 1999). This is arguably more pertinent in the case of large-scale complex engineering projects of the type described in this paper. In the case studies cited, the sources of innovation have primarily been external to the region of the manufacturer or integrator of the project. Further support for this is provided by evidence from the Aberdeen offshore oil complex, a classic low volume industry engaged in major complex projects, which found there to be '...little evidence of firms using the various local or regional 'institutional' agencies for more innovative activities such as new product development or product diversification...' (Cumbers 2000, p.380). Critical supply chains are typically international in flavour and even basic manufacturing tasks, where important knowledge can emerge from learning by doing, is being outsourced by the manufacturers of complex capital goods to low labour cost locations abroad, as in the case of the process plant designer described above.

Other empirical research (e.g. Wiig and Wood 1997) has reinforced the suggestion that 'individual firm strategies and networks actually work against the formation of visibly integrated regional innovation systems' (p.95). This process is exacerbated over time as companies outsource more activity, particularly the manufacturing and assembly of complex products. As a result, the regional innovation systems within which these companies might have operated in the past are also potentially weakened, as key components of the knowledge required for complex projects are lost to the region. Through this process, engineering firms become 'disengaged' from their local environment as they enter or create project networks that are geographically dispersed.

In looking at relatively mature industries, such as those comprising the bulk of capital goods, it is not surprising to find this dispersion of expertise and knowledge. As industries mature and competition intensifies, it is normal to observe comparatively high levels of exit from the industry, thereby reducing the competitive advantage that may have been provided by a 'cluster' of firms undertaking such activity (Tichy 1998). This naturally tends to lead to isolated geographical pockets of expertise distributed on a global basis. Through this process we should expect to find former patterns of clustering and historically constituted regional innovation systems losing their coherence and atrophying. Indeed, in our most extreme example of the mechanical handling company, we find that not only have all vestiges of a regional innovation system disappeared [if one ever existed], but that arguably there is not even a national innovation system in the UK to support the mechanical handling sector of capital goods production. Consequently, most developments in standards and legislation relating to mechanical handling technologies emerge from outside the UK.

The construction of the project network by the project integrator therefore represents the principal mechanism by which geographically dispersed sources of expertise and knowledge are brought together. Locations of knowledge are linked through the project network on a predominantly temporary basis. Although longer term partnerships and relations with preferred suppliers may cement these linkages and render them permanent or semi-permanent, the case study evidence suggests many will remain fluid, responding to the requirements of individual projects on an *ad hoc* basis.

The current predilection of development agencies to pursue cluster strategies (Steiner 1998; Lagendijk 1999c), which are intrinsically related to the idea of the regional innovation system, appears limited, if not misplaced, in relation to low volume engineering. Just as the organisation and management of complex projects differs from the organisation and management of high volume manufacturing activity, so the development of support for innovation in a complex project environment also needs a different way of thinking and approach. The knowledge required for complex engineering projects may have to be assembled at a particular geographical location [as in the Integrator plc. case] but tends to draw on international networks. Resources have to be mobile rather than regionally embedded.

Complex engineering projects are beginning to require management skills that go beyond normal capabilities. This is exacerbated by the frequent need to cross international boundaries, with differences of language, business culture and social norms impinging on the conduct of the project. For regions with a strong engineering base, such as the North East of England, there is perhaps a need not so much for a developed regional innovation system to support technological change, but for the development of human resources that are suited to the new circumstances under which the region's engineering firms are having to compete.

Acknowledgements. The financial support of the EPSRC [Grant No. GR/K 95512] and the ESRC [Grant No. L700257003] for aspects of the work reported here is gratefully acknowledged. David Maffin, Alfred Thwaites, Roger Vaughan and the author jointly carried out the empirical case studies referred to in this paper. In addition to the two latter colleagues, Chris Ivory, Ian McLoughlin and Paul Braiden are also engaged in the current study of knowledge management in long-term engineering projects. The present chapter will have been influenced through discussions with them. The stimulating comments and suggestions of participants at the Vienna Workshop were also very helpful. However, the author alone is responsible for the views expressed. Finally, the contribution of a large number of respondents in the empirical case studies is gratefully acknowledged. Confidentiality requirements prevent them from being identified, but without their considerable input, in-depth research of this nature would not have been possible.

Endnotes

1   The term 'value added system' is used as a generic descriptor for any such system, while the term 'project value system' is used when referring to a specific project.
2   It was recently reported that in the railway industry's 'first major example of alliancing … [between] … Railtrack, Balfour Beatty Rail and Westinghouse Signals … a bespoke form of insurance policy had to be drawn up for the alliance', because the project structure was so novel (Abbott 2000, p.12).
3   It is no accident that Nissan located its new R&D facility near Cranfield, close to the West Midlands automotive industry, rather than with its assembly facility in Sunderland in the North East of England.
4   More details have been presented elsewhere (Alderman et al. 1997; Alderman, Thwaites and Maffin 2000).
5   Mecho appears to maintain a closer relationship with its more local suppliers, but in this case supplier 10 is concerned with fabrication which, whilst important, is not a significant source of innovation.
6   The other main aspect of the project concerned the provision of the process equipment inside the factory, but this was subject to commercial secrecy and not the subject of our detailed study.
7   These organisations covered architectural design, mechanical and electrical design and quantity surveying/cost estimating.
8   In practice, individual companies can overcome this, because they can go outside the region for that knowledge, but the region itself arguably is less able to do so.
9   The factual information relating to this case study is drawn from the public domain rather than original empirical fieldwork, specifically the April 2000 edition of Modern Railways (Abbott 2000). We are currently studying another major train manufacturer, ALSTOM, and its contract with Virgin to build the high speed tilting trains for the West

Coast Main Line (McLoughlin et al. 2000). This material is subject to confidentiality constraints at the present time. However, there are many similarities in terms of the overall structure and complexity of the project, as Virgin have clearly applied the same philosophy in each case.

10 Bombardier are supplying 352 vehicles to Virgin: a mix of tilting and non-tilting DEMUs.

11 GL Trains is a specialist company jointly owned by an American leasing company and Lombard, a subsidiary of Natwest Bank.

12 This process is supported by a range of other actors, safety assessors, testing authorities and acceptance bodies, including Her Majesty's Railway Inspectorate.

13 Bombardier manufacture at Wakefield in West Yorkshire. Other important centres include Adtranz at Derby in the East Midlands and ALSTOM at Birmingham in the West Midlands.

14 Recently acquired by Knorr Bremse of Germany.

12 Endogenous Technological Change, Entrepreneurship and Regional Growth

Zoltán J. Ács
Robert G. Merrick School of Business, Department of Economics and
Finance, University of Baltimore

12.1 Introduction

This chapter builds on recent research by Ács and Varga (1999) which looked into the question why some regions grow faster than others. [1] In this previous project, three distinct strands of literature were examined, each with a long and distinguished history: New Economic Geography (Krugman 1991b), New Growth Theory (Romer 1990), and the New Economics of Innovation (Nelson 1993a). The aspects investigated were the unit of analysis, how endogenous growth was modelled, and the interactions between the actors and institutions in innovation processes. The authors searched for insights that would help develop a clear analytical framework which integrates economic growth, spatial interdependencies and the creation of new technology as an explicit production process to formulate production-oriented regional policies (Nijkamp and Poot 1997).

Each of the three approaches naturally revealed strengths and weaknesses. In the case of Krugman's theory on economic concentration, the main contribution is not so much in the individual elements, but in the way the system is put together. His model provides a case for the traditional economic treatment of spatial issues. However, while the model is strong in representing the characterisation of specific combinations of the three parameters favouring geographical concentration, it is weak in modelling the growth process. *Cumulative causation effects induced by forward and backward linkages do not seem to be sufficient to trace the economic growth of individual regions.*

One of the principal assumptions in the theory of endogenous growth is that the total stock of knowledge is freely accessible to anyone engaged in creating new sets of technological knowledge. However, *this assumption is not verified in the growing literature of geographic knowledge spillovers*[2]. New technological knowledge [the most valuable type of knowledge in innovation] is usually present in a tacit form, so its accessibility is bounded by geographic proximity and/or by the nature and extent of the interactions among actors of an innovation system.

In the systems approach, innovation is viewed as ubiquitous phenomenon (here innovation is conceived in a much broader way than just technical innovation). In all parts of the economy, at all times, there are ongoing processes of learning, searching and exploring which result in new products, new techniques, new forms of organisation and new markets. The first step in recognising innovation as a ubiquitous phenomenon is therefore to focus upon its gradual and cumulative aspects. Such a perspective gives rise to simple hypotheses about the dependence of future innovation on the past. In this context, an innovation may be regarded as a new use of pre-existing possibilities and components. *Almost all innovations reflect existing knowledge combined in new ways.*

The concept of a 'national' innovation system can be questioned. Krugman (1995) has suggested that as globalisation spreads, and as economies become less constrained by national frontiers, they become more geographically specialised. Important elements of the innovation process tend to be regional rather than national. Some of the largest corporations are weakening their ties to their home country and spreading their innovation activities to source different regional systems of innovation. *Regional networks of firms are creating new forms of learning and production.* These changes are important and offer an alternative to national systems of innovation.

Systemic shortcomings were also found in the literature in the understanding of regional economic growth. Perhaps the most important missing ingredient is the role of entrepreneurship in the innovation process. None of the above strands of literature answers the question, *'How is the stock of knowledge in society discovered and exploited?'* Entrepreneurial discovery and the competitive market process approach, which has emerged in modern Austrian economics during the past quarter century, has developed out of elements derived from Mises and from Hayek. From Mises, we have learned to see the market as an entrepreneurially driven process, and from Hayek gained an appreciation of the role of knowledge and its enhancement through market interaction. These elements have been welded into an integrated theoretical framework, but both are missing from the systems approach to innovation (Nelson 1993a) and endogenous growth (Romer 1990).

The aim of this chapter is to arrive at better understanding of the relationship between endogenous technological change and entrepreneurial discovery in economic growth. It is argued that in order for growth to take place, existing knowledge needs to be turned by entrepreneurs into future goods and services. In the next section, we investigate endogenous technological change to understand its basic contribution to economic growth. Section 12.3 outlines the process of entrepreneurial discovery and traces the connection between knowledge and entrepreneurship, Section 12.4 presents empirical evidence on the existence of entrepreneurship at the regional level, while Section 12.5 examines the role of public policy in promoting regional growth.

12.2 Endogenous Technical Change

In the eighteenth century, the avant-garde theories of Adam Smith concerning the wealth of nations provided a cheery alternative to the dismal science of Thomas Malthus. However, for over a century, from Marx to Jorgenson, the prospects of diminishing returns were central to our understanding of economic growth. Today, the New Growth Theory provides a more optimistic alternative to the conventional wisdom that counsels diminished expectations when it comes to future growth. Above all, advances in technology, and interdependencies between new ideas and new investment ultimately save the day, yielding brighter prospects for long-run prosperity.

We begin with a short survey of endogenous growth theory to bring out its strengths and weaknesses for regional analysis[3]. This section is not intended as a detailed survey of the endogenous economic growth literature[4]. It focuses only on those aspects of crucial importance to the survey. The distinguishing feature of endogenous economic growth theory as compared to the neoclassical growth model the fact that it models technological change as the result of profit motivated investments in knowledge creation by private economic agents. The novel formulation of technological knowledge in Romer (1990) is the key to this new and rapidly evolving field of economic growth theory. According to this formulation, technological knowledge is a non-rival, partially excludable good. Such formulation of technological knowledge as a key factor in the production function results in a departure from the constant returns to scale, perfectly competitive world of the neoclassical growth theory.

Central to the neoclassical theory of economic growth as formulated in Solow (1956) is the production function. Assuming that capital does not depreciate, the labour force does not grow and technology does not change over time (Helpman 1992) the production function has the form of:

$$Y = F(K, L) \tag{12.1}$$

where Y represents aggregate production, K the capital stock and L the labour force. $F(.)$ is the constant returns to scale production function. It is assumed that the capital stock grows without bounds. However, the growth rate of per-capita income is bounded. The growth rate of per capita income is therefore:

$$g = s\, F_K(K, L) \tag{12.2}$$

where g is the growth rate of per capita income, s is the savings rate and F_K is the marginal product of capital.

Equation (12.2) states that per-capita income grows as long as the marginal product of capital exceeds zero. However, assuming constant growth in the capital stock, per-capita income approaches zero. Relaxing the assumptions of stable labour force and no depreciation of capital does not change essentially the main point of the model. The condition for sustained per-capita income growth in the long run is that, resulting from continuous capital accumulation, the marginal product of capital should not decrease below a positive lower bound.

Development in the state of technology is essential to offset the effect of reduced capital accumulation causing per capita income to decline in the neoclassical model of economic growth. Introducing technological progress in the production function it takes the form:

$$Y = F(A, K, L) \tag{12.3}$$

where A stands for the state of technology. Assuming that A increases, it will increase the marginal product of capital which will lead to a higher per capita income. As a result, in a steady state, the rate of technical development equals the rate of capital accumulation.

The essential role of technological progress in economic growth has been emphasised above. However, technological development remains unexplained in the neoclassical theory of economic growth. As a public good, it is considered exogenously determined, although (as shown in Solow 1957 and Maddison 1987) the major portion of economic growth can be attributed to technological change. Capital accumulation, the main concern in the neoclassical model, explains only a small fraction of it.

Early attempts in the literature to endogenise technological progress include Arrow (1962), who introduced 'learning by doing' into technological development, Lucas (1988), who modelled human capital as the determinant factor in technical change, and Romer (1986), who explicitly included research in the production function. In Arrow's formulation:

$$Y_i = A(K) F(K_i, L_i) \tag{12.4}$$

the state of technology depends on the aggregate capital stock in the economy. Subscript i denotes individual firms. According to Lucas' model of endogenous technological change, spillovers result from human capital accumulation rather than the accumulation of physical capital increasing the technological level in the economy:

$$Y_i = A(H) F(K_i, L_i) \tag{12.5}$$

where H stands for the general level of human capital in the economy. In Romer (1986), it is assumed that spillovers from private research efforts lead to an increase in the public stock of knowledge. This could be written as:

$$Y_i = A(R) \, F(R_i, \, K_i, \, L_i) \tag{12.6}$$

where R_i stands for the results of private research and development efforts by firm i and R denotes the aggregate stock of research results in the economy.

As stated by Romer (1990), the major conceptual problem with the formulation of endogenous growth in Equations (12.4) – (12.6) is that the entire stock of technological knowledge is considered to be a public good. In reality, as daily evidence suggests, new technological knowledge can be partially excluded [at least for a limited length of time] by means of patenting. Not until the formulation of monopolistic competition by Dixit and Stiglitz (1977), applied in the dynamic context by Judd (1985), was economic growth modelled within an imperfectly competitive market. Judd's approach was combined with 'learning by doing' (Romer 1990) to create the first model of endogenously determined technical change with imperfectly competing firms.

In this model, any firm developing new technological knowledge has some market power and earns monopoly profits on its discoveries. The 'New Theory of Economic Growth', builds on the above view of the available stock of technological knowledge, as well as formulating the economy within a framework of imperfect competition. At the core of New Growth Theory is the concept of technological knowledge as a non-rival, partially excludable good, as opposed to the neoclassical view of knowledge as an entirely public good. Knowledge is considered a non-rival good because it can be used by one agent without limiting its use by others, and partially excludable because it is possible to a certain extent to prevent its use by others. Knowledge can, for example, be made partially excludable by the patent system, and commercial secrecy.

Knowledge can enter into production in two ways. Firstly, newly developed technological knowledge is 'invested' by the firm in the development of its production. In this role, knowledge can be protected from direct imitation by others through patenting. However, this new set of knowledge may spill over to other researchers through the study of the patent documentation (Romer 1990). It may in this way stimulate the production of further inventions in the research sector. This second indirect role of knowledge can be formalised as:

$$dA = G \, (H, \, A) \tag{12.7}$$

where H stands for human capital used in research and development, A is the total stock of technological knowledge available at a certain point in time, whereas dA is the change in technological knowledge resulted in private efforts to invest in research and development.

Human capital creates new knowledge - at the same time, the productivity of human capital depends on the total stock of available knowledge (A). The greater A, the higher the productivity of H and the less expensive it is to create new technological knowledge. In the words of Grossman and Helpman (1991, p.18), 'The technological spillovers that result from commercial research may add to a pool of public knowledge, thereby lowering the cost to later generations of achieving a technological breakthrough of some given magnitude. Such cost reductions can offset any tendency for the private returns to invention to fall as a result of increases in the number of competing technologies.'

A major assumption in the theory of endogenous growth is that the total stock of knowledge, A in Equation (12.7), is freely accessible to anyone engaged in research. However, this assumption is not borne out in the growing literature on knowledge spillovers. New technological knowledge usually exists in tacit form and its accessibility is bounded by geographic proximity and/or the extent of the interactions among actors of an innovation system.

Similar to the case of relaxing the neoclassical assumption of equal availability of technological opportunities in all countries of the world (Romer 1994) a relaxation of the assumption that the term A in Equation (12.7) is evenly distributed across space within countries seems to be also necessary. The non-excludable part of the total stock of knowledge seems rather to be correctly classified if it is assumed to have two portions: a perfectly accessible part consisting of already established knowledge elements [obtainable via scientific publications, patent applications etc.] and a novel, tacit element, accessible by interactions among actors in the innovation system. While the first part is available without restrictions, accessibility of the second one is bounded by the nature of interactions among actors in a system of innovation[5].

12.3 Entrepreneurial Discovery

Even if the total stock of knowledge were freely available, knowledge about is existence would not necessarily be. Hayek (1945) made the important point that one of the key features of a market economy is the partitioning of knowledge among individuals. In other words, knowledge is not at everyone's disposal, and that no two individuals share the same knowledge or information about the economy. Thus, only a few people may know about a new invention, a particular scarcity or a resource lying fallow. This knowledge is typically idiosyncratic because it is acquired through each individual's own channels including their job, social relationships, and daily life. It is this specific knowledge, obtained through a particular 'knowledge corridor' that may lead to some profit-making insight.

The fact that different economic agents do not have access to the same observations, interpretations or experience has two fundamental implications for entrepreneurship. Firstly, it is this uneven dispersion of information which creates opportunities for discovering new goods and services. Secondly, however, the very same dispersion may cause hurdles to the discovery, creation and exploitation of opportunities, because of the absence of reliable knowledge about markets for future goods and services. It is therefore necessary to understand (i) how opportunities for the creation of new goods and services arise in a market economy and (ii) how individual differences in the capacity to overcome these hurdles come about. Thus, to understand entrepreneurial ability, it is important 'to understand how opportunities to bring into existence 'future' goods and services are discovered, created, and exploited, by whom and with what consequences (Shane and Venkataraman 2000).'

How do opportunities arise in the economy? The first premise is that in most societies, markets are inefficient most of the time, thus providing opportunities for enterprising individuals to exploit. This is clearly illustrated in the work of Kirzner (1997) who assumes that markets are usually in disequilibrium. A second premise suggests that even if markets are in equilibrium, human enterprise combined with the lure of profits and the advance of knowledge and technology will eventually destroy the equilibrium. This is familiar as Schumpeter's 'creative destruction'. These two premises are based on the underlying assumption that change is a fact of life. The result of this natural process is both a continuous supply of lucrative opportunities to enhance personal wealth, and a continuous supply of enterprising individuals seeking such opportunities.

There are at least four classes of opportunity. The first arises through the inefficiencies in existing markets, either due to information asymmetries among market participants or to the limitations of technology in satisfying certain known but unfulfilled market needs. The second is the emergence of significant changes in social, political, demographic and economic forces that are largely outside the control of individual agents. The third source is through inventions and discoveries that produce new knowledge - dA in Equation (12.7). A fourth source of opportunity is the accumulated stock of knowledge, A, that exists in society.

It is of course one thing for opportunities to exist, but an entirely different matter for them to be discovered and exploited. Even new technology needs to have opportunities for its exploitation. 'Opportunity discovery' is a function of the distribution of knowledge in society. Opportunities rarely present themselves in neat packages - they almost always have to be discovered or recognised. Thus, the coming together of opportunity and enterprising individuals is critical to understanding entrepreneurship.

The role of specific knowledge in motivating the search for profitable opportunities is critical to our understanding of what triggers some individuals and not others. The possession of useful knowledge varies among individuals. These differences matter as they influence the decision to search for and exploit an opportunity, and the success of the exploitation process.

Specific knowledge alone is not a sufficient condition for successful enterprise. The ability to make the connection between specific knowledge and a commercial opportunity requires a set of skills, aptitudes, insight and circumstances that is neither uniformly nor widely distributed in the population. Thus two people with the same knowledge may put it to very different uses. One may have an insight, but not necessarily be able to profit from it. The incentive, capacity and specific behaviour needed to profit from useful knowledge or insight vary among individuals and explain diversity in the exercise of enterprise.

Bringing new products and markets into existence usually involves an element of risk. By definition, entrepreneurship requires making investments today without knowing what the returns will be tomorrow. There is a fundamental uncertainty that cannot be insured against or diversified away (Knight 1921b). Individuals vary in their perception of such downside risk, and in their aptitude and capacity to deal with it. The important thing is that individuals process and interpret the same statistical information differently, and these variations have a significant impact on the decision to undertake a new enterprise and the success of the endeavour.

While idiosyncratic insight and the ability to convert knowledge to commercial profit leads to successful enterprise, these same qualities also cause problems for an entrepreneur. The creation of new products and markets implies that much of the information required by potential stakeholders – such as future technology, price, tastes, supply networks, distributor networks and strategy – is not available, or not reliable. Reliable information will exist only when the market has been successfully created. Potential stakeholders thus have to rely on the entrepreneur for information, but without the benefit of the entrepreneur's special insight. In almost all projects entrepreneurs have more information about the true qualities of the project than any other party. Because of this information asymmetry, neither buyers nor suppliers may be willing to make the necessary investment or enter into formal co-operative arrangements to develop the business.

Despite the absence of current markets for future goods and services, and in spite of the moral hazards when dealing with investors, suppliers and customer markets for future goods and services, the simple fact is that some individuals do succeed in creating new markets and products. Thus, the ability to overcome the problems involved varies among individuals. Entrepreneurs are often able to shift considerable risk to other stakeholders. They are in fact funded by venture capitalists to discover new knowledge to create future goods and services (Venkataraman, 1977). The companies founded by these entrepreneurs are characterised by rapid growth and value creation.

A critical decision for the entrepreneur is how to organise relationships with resource suppliers in order to foster the development of a new business. Stated differently, when there are several possible institutional arrangements for creating a future product or service [such as a new firm, a franchise or license arrangement, a joint venture, or a simple contractual agreement], why do entrepreneurs choose a particular mode? Moreover, what are the consequences of this choice on the distribution of risks and rewards among the various stakeholders? The usual assumption about the execution of entrepreneurial activity has been that most [if not all] new business creation occurs within a hierarchical framework, either as novel start-ups or as new entities within an existing corporate body.

We have shown that entrepreneurial capacity plays an important role in the discovery of knowledge and the turning of that knowledge into future goods and services. The manifestation of this discovery process is fierce competition, and the outcome is the phenomenon of the rapidly growing firm or establishment. These rapidly growing establishments occupy a special place between the small and medium-sized firms [SMEs] and large firms. We now turn to an examination of entrepreneurial activity at the regional level, measured through the distribution of rapidly growing establishments.

12.4 Regional Growth

This section presents empirical evidence on the geographical distribution in the United States of rapidly growing establishments. It is assumed that a high proportion of rapidly growing establishments indicates the presence of successful entrepreneurs and that these successful entrepreneurs have taken advantage of sources of knowledge that are available, at least to some extent, regionally[6].

The data on high growth establishments have been calculated from the Longitudinal Establishment and Enterprise Microdata [LEEM][7]. The LEEM has annual data for all U.S. private sector [non-farm] businesses with employees. The current form of the LEEM archives facilitates the tracking of employment, payroll, firm affiliation and employment size for the over eleven million establishments that existed between 1989 and 1996[8]. The basic unit of the LEEM data is the business establishment [location or plant].[9] An establishment is a single physical location where business is conducted or where services or industrial operations are performed. The microdata describe each establishment for each year of its existence in terms of its employment, annual payroll, location [state, county, and metropolitan area], primary industry, and start-up year. Additional data for each establishment and year identify the firm [or enterprise] to which the establishment belongs, and the total employment of that firm[10].

A firm [enterprise or company] is the largest aggregation [across all industries] of legal business entities under common ownership or control. Establishments are owned by legal entities, which are typically corporations, partnerships, or sole proprietorships. Most firms are composed of a single legal entity that operates a single establishment - their establishment data and firm data are therefore identical and they are referred to as 'single unit' establishments or firms. The single unit businesses are frequently owner-operated. Only four percent of firms have more than one establishment, and they and their establishments are described as multi-location or multi-unit.

Labour Market Areas [LMA's] within the U.S. are defined according to the 1990 specification of Tolbert and Sizer (1990) for the Department of Agriculture. The United States includes the 50 states and the District of Columbia [territories are excluded]. There are 394 LMA's, all based on aggregations of counties, many of them cutting across state boundaries. We use the most recently specified state and county for each establishment in the LEEM, assuming that most of the few locations coding changes are corrections. Businesses that report operating statewide have been placed in the largest LMA in each state.

Industry codes are based on the most recently reported 4-digit SIC code for the original establishment in each firm. For most firms [single location firms] this is the only establishment. For most new multi-unit firms, the industry classification of the primary location is the same as that of secondary locations. We use the most recently reported SIC code, rather than the first reported SIC, because the precision and accuracy of the codes tends to increase over time. The Census often lists new establishments before detailed industry codes are available for them. We have defined six sectors in this study instead of one-digit sectors (see Table 12.1).

At the industry sector level, many of the counts of establishments in an LMA with high growth were not large enough to satisfy the new guidelines for disclosure clearance. To facilitate clearance, the data were converted to shares of the total number of establishments in the LMA, and these were converted to decile rankings, from zero for the lowest group, to nine for the highest group. About 40 LMA's are classified in each group, with United States average high growth shares for the sector shown on the top line of each table. To construct each decile column in a table, the rates or shares for that column are listed in ascending order. They have been divided into ten nearly even sized groups, and the appropriate decile rankings assigned to each member of each group.

Table 12.1   Classification of industries

Sector	Standard Industrial Classification	
Distributive	4000-5199	Transportation, communication, public utilities, and wholesale trade
Manufacturing	2000-3999	
Business Services	7300-7399 and 8700-8799	Including engineering, accounting, research and management services
Extractive	0700-1499	Agricultural services and mining
Retail Trade	5100-5299	
Local Market[11]	1500-1799 and 6000-8999	Excluding business services (construction,consumer and financial services)

Table 12.2 focuses on the share of businesses in each LMA that have recorded annual growth exceeding 15 percent per year during the first half of the 1990s. This business success measure is specified as the share of 1991 establishments that have grown at least 101 percent between 1991 and 1996, with a minimum of five new jobs:

> **Higrowth** – share of establishments with employment growth 1991-1996 of at least five employees [for those under five in 1991] and averaging at least 15 percent compounded annual growth [at least 101 percent increase from 1991 level].

Table 12.2 provides data aggregated for all industry sectors, for the United States as a whole and for each LMA. It gives the name of the largest town in the LMA, the State, the 1990 population of the LMA, the identification number, the total number of establishments in 1991, and the share of those establishments in 1996 that qualified as high growth establishments according to the definition

238   Z. Ács

given above. These have been listed in decreasing order according to the high growth share. While the US average was 4.69 percent, the LMA with the highest proportion of high growth establishments is Provo, Utah, with 7.96 percent.

Table 12.2 Share of high growth establishments by Labour Market Areas

Largest Town	State	Population 1990	Labour Market Areas	Establ 91 [number]	Higrowth [%]
	USA	248,669,873	USA	5,544,033	4.679
Provo	UT	280,740	360	4,170	7.962
St. George	UT	150,418	359	3,187	7.342
Farmington	NM	129,979	353	3,157	7.317
Austin	TX	922,307	312	20,915	7.239
Phoenix	AZ	2,313,258	350	50,608	7.064
Fayetteville	AR	267,534	303	5,517	6.924
Salt Lake City	UT	1,129,963	361	23,391	6.913
Fort Collins	CO	339,896	288	7,870	6.760
Flagstaff	AZ	242,271	354	6,037	6.609
Elkhart	IN	323,967	137	7,336	6.529
Las Vegas	NV	892,568	379	19,322	6.521
Atlanta	GA	2,725,351	91	69,279	6.465
Laramie	WY	136,503	287	5,898	6.443
Albuquerque	NM	640,537	349	14,240	6.334
Denver	CO	1,875,828	289	51,542	6.238
Boise City	ID	401,186	358	9,885	6.181
Colorado Springs	CO	448,210	284	9,790	6.129
Killeen	TX	268,822	329	3,747	6.085
Grand Juction	CO	155,236	352	4,319	5.997
Nashville	TN	996,401	56	24,458	5.990
Tucson	AZ	794,180	351	16,663	5.941
Portland	OR	1,515,310	388	39,656	5.923
Pensacola	FL	515,942	109	10,863	5.919
Wilmington	NC	278,374	15	6,805	5.878
Logan	UT	119,392	362	1,857	5.870
Raleigh	NC	1,089,423	17	25,768	5.848
Anchorage	AK	55,043	341	12,861	5.832
Grand Rapids	MI	1,108,630	122	22,999	5.822
Little Rock	AR	554,185	42	13,036	5.807
Dallas	TX	2,584,139	331	64,298	5.796
Minneapolis	MN	2,530,955	215	59,091	5.781
Fort Worth	TX	1,443,402	330	29,981	5.777
Baton Rouge	LA	709,562	35	13,449	5.718
Monett	MO	102,422	298	2,442	5.692
Gainesville	GA	190,941	94	4,103	5.679
Kennewick	WA	269,370	391	5,611	5.632
Bend	OR	159,537	392	4,608	5.621
Columbus	IN	146,039	143	3,014	5.541
Spokane	WA	568,021	389	13,837	5.514
Bryan	TX	169,826	318	2,996	5.507
Callup NM	AZ	199,935	355	2,331	5.491

Prepared May 2000 from the LEEM8996 file at Census Center for Economic Studies by Armington and Ács.

What is interesting about the results is that, as predicted by the theory above, entrepreneurial activity is distributed across the country. However, the distribution of activity is contained within a narrow range. For example, all regions have high growth establishments, and the difference between the highest and the lowest share of high growth firms is nowhere as skewed as the distribution of high tech employment, patents or industrial R&D (Anselin, Varga and Ács 1997).

Table 12.3 allows comparisons across business sectors for each LMA. For each LMA, the table lists the total number of establishments in each sector in 1991 and the decile ranking for that sector's share of high growth establishments between 1991-1996. The first four columns of Table 12.3 are similar to the first four columns of Table 12.2, showing (1) the biggest place in the LMA, (2) the state, (3) the identification number of the LMA and (4) the number of establishments. The first row shows the US average for each sector, and the rest ofthe entries are listed by LMA identification number.

The decile rankings for all industry sectors are shown in column five. This allows overall comparison of each LMA to the national average high growth share. The next twelve columns show paired data – the number of establishments and their decile ranking – for each of the six broad industry sectors defined above. For example, in Johnson City, Tenessee there were 538 establishments in business services in 1991. Their high growth share had a decile ranking of two, within that sector, relative to all other LMAs.

In order to facilitate comparison, we have created an extract from Table 12.3 showing the five LMA's with the largest share of high growth establishments, and the five LMA's with the smallest share of high growth establishments based on the higrowth ranking in Table 12.2. We have also included the U.S. averages in this extract. As shown in Table 12.4, the LMAs with the highest shares are all in the west, with Utah, New Mexico, Texas, and Arizona represented. The five LMAs with the lowest shares are more dispersed across the US, with Mississippi, South Dakota, New York, North Carolina, and Hawaii represented. The last line of this extract table gives the U.S. average shares. If these U.S. averages were expressed in decile ranking terms, they would be high 4s or low 5s, depending on the relative sizes of the strongest and weakest LMAs.

The LMAs with the highest share of high growth establishments have an overall decile ranking of 9. Similarly, the LMAs with the lowest share of high growth LMAs have a decile ranking of 0. However, within most of the LMAs there is some variation in ranking across sectors. The LMA with the highest share of high growth establishments, Provo, has a decile ranking is a nine for all sectors, indicating that all sectors are very strong relative to the national average.

Table 12.3 Share of high growth establishments by Labour Market Areas, comparisons across business sectors

Biggest Place	State	Labour Market Areas	All Sectors Estab91 [no.]	Higro[a] [%]	Business Services Estab91 [no.]	Higro[a] [%]	Distributive Estab91 [no.]	Higro[a] [%]	Extractive Estab91 [no.]	Higro[a] [%]	Local Market Estab91 [no.]	Higro[a] [%]	Manufacturing Estab91 [no.]	Higro[a] [%]	Retail Trade Estab91 [no.]	Higro[a] [%]
USA			5,544,033	4.696	455,852	6.198	689,972	5.573	101,973	5.686	2,541,040	4.447	359,304	7.034	1,395,892	3.483
Johnson	TN	1	9,839	1	538	2	1,107	3	121	3	4,543	1	669	5	2,861	1
Morristown	TN	2	3,853	7	149	8	323	1	53	8	1,727	5	312	7	1,289	9
Knoxville	TN	3	15,166	7	1,154	6	1,967	4	250	9	6,801	7	871	5	4,123	7
Wiston-Sale	NC	4	11,221	4	836	5	1,206	8	180	5	4,915	4	833	2	3,251	3
Greensboro	NC	5	21,345	5	1,447	4	2,774	6	291	4	9,185	4	2,112	4	5,536	3
North Wilkesb	NC	6	2,456	1	108	3	245	0	41	9	1,053	1	260	3	749	2
Spartanburg	SC	7	6,441	4	356	6	785	5	99	9	2,834	3	576	7	1,791	1
Gastonia	NC	8	7,457	5	372	5	741	0	82	8	3,351	7	880	4	2,031	2
Charlotte	NC	9	28,383	8	2,437	9	4,332	7	449	7	12,475	8	2,118	4	6,572	3
Morganton	NC	10	4,158	2	169	1	354	6	80	8	1,906	2	387	2	1,262	5
Hickory	NC	11	7,462	6	402	5	849	4	110	3	3,041	4	1,063	7	1,997	3
Asheville	NC	12	11,000	7	659	4	979	5	192	3	5,409	8	676	6	3,085	5
Florence	SC	13	12,091	6	593	2	1,125	6	183	7	5,459	7	694	2	4,037	8
Fayetteville	NC	14	9,448	5	473	1	925	6	152	7	4,409	7	569	6	2,920	4
Wilmington	NC	15	6,805	9	464	5	838	9	100	9	3,042	9	398	7	1,963	8
Rocky Mount	NC	16	4,303	6	226	7	506	4	78	5	1,907	4	261	7	1,325	6
Raleigh	NC	17	25,768	9	2,471	9	2,735	8	471	9	11,881	9	1,443	7	6,767	6
Goldsboro	NC	18	3,495	3	151	8	447	1	47	6	1,526	1	236	3	1,088	8
Greenville	NC	19	8,956	7	502	4	1,014	3	141	9	4,010	6	443	7	2,846	7
Virginia Beach	NC	20	20,291	4	1,579	6	2,241	5	303	9	9,956	6	755	4	5,457	7
Washington	NC	21	2,918	0	106	5	340	0	40	1	1,296	5	175	2	961	8
South Boston	VA	22	2,799	0	94	1	310	6	35	0	1,264	0	257	0	839	2
Lynchburg	VA	23	4,455	2	276	8	460	1	62	7	2,159	1	336	2	1,162	8
Richmond	VA	24	21,831	6	1,817	8	2,630	6	338	8	10,864	6	1,078	1	5,104	5
Newport News	VA	25	9,548	3	691	8	850	1	155	6	4,734	3	351	3	2,767	4
Roanoke Rapids	NC	26	2,550	0	69	5	342	0	37	0	1,123	0	207	0	772	0
Biloxi	MS	27	6,656	7	407	7	704	7	95	7	3,055	7	345	1	2,050	8

Prepared May 2000 from the LEEM8996 file at Census' Center for Economic Studies, by Armington and Ács.
1991 establishments and decile rankings of shares with high (>15%) average annual growth to 1996, (with percentage share for United States).
Note a   Higro (decile=9 is highest share, decile=0 is lowest share within sector)

Table 12.4   Labour Market Areas with the largest and smallest share of high-growth establishments

Rank			1990		All Sectors		Business Services		Distributive	
	Largest Town	State	Population	LMA*	Estab91 [no.]	Higro[a] [%]	Estab91 [no.]	Higro[a] [%]	Estab91 [no.]	Higro[a] [%]
1	Provo	UT	280,740	360	4,170	9	370	9	424	9
2	St.George	UT	150,418	359	3,187	9	147	8	393	9
3	Farmington	NM	129,979	353	3,157	9	218	4	365	9
4	Austin	TX	922,307	312	20,915	9	2,233	9	2,120	9
5	Phoenix	AZ	2,313,258	350	50,608	9	5,366	9	5,874	9
390	Greenville	MS	151,652	48	2,626	0	115	0	373	0
391	Aberdeen	SD	110,753	266	2,939	0	96	6	514	2
392	Amsterdam	NY	111,451	185	2,174	0	78	0	258	2
393	Roanoke Rapids	NC	133,384	26	2,550	0	69	5	342	0
394	Hilo	HI	120,317	356	3,226	0	194	0	366	0
	US average		248,669,872		5,544,033	4.679	455,852	6.198	689,972	5.573

Rank			Extractive		Local Market		Manufactures		Retail Trade	
	Largest Town	State	Estab91 [no.]	Higro[a] [%]	Estab91 [no.]	Higro[a] [%]	Estab91 [no.]	Higro[a] [%]	Estab91 [no.]	Higro[a] [%]
1	Provo	UT	70	9	1,929	9	326	9	1,051	9
2	St.George	UT	75	5	1,446	9	182	9	944	9
3	Farmington	NM	178	3	1,411	9	124	9	861	9
4	Austin	TX	526	8	9,965	9	1,022	9	5,049	9
5	Phoenix	AZ	1,163	9	23,351	9	3,020	9	11,834	9
390	Greenville	MS	57	0	1,127	0	131	3	823	0
391	Aberdeen	SD	44	4	1,331	0	107	0	851	0
392	Amsterdam	NY	21	4	984	0	213	0	620	0
393	Roanoke Rapids	NC	37	0	1,123	0	207	0	772	0
394	Hilo	HI	51	0	1,5586	0	108	1	921	0
	US average		101,973	5.69	2,541,040	4.447	359,304	7.034	1,395,892	3.483

Source: Derived from Table 12.2 and Table 12.3
        1991 establishments and decile rankings of high (>15% average annual to '96) growth shares
        (with percentage share for United States).
Note a   Higro (decile=9 is highest share, decile=0 is lowest share)

The LMA with the third highest share of high growth establishments is the one containing Farmington, New Mexico. While it has a decile ranking of only four in business services, and three in extractive industries, the ranking is nine for all other sectors. This means that although this LMA had an extremely high share of high growth businesses overall, one sector had roughly the national average and another was well below the national average. If we look at the number of establishments in each sector, we note that over five percent of the businesses in this LMA were in extractive industries, while nationally less than two percent are in this sector. Therefore, in spite of its overall strong performance, the above normal share of extractive businesses performed below average during the five-year period. It would be interesting to investigate whether sectors with a relatively strong presence generally tend to have a weaker growth performance. That is, in LMA's is there a positive association between high local sector strength and low relative performance?

The second set of LMA's in Table 12.4 includes those with the lowest share of high growth establishments. All five have a decile ranking of zero. However, there is some variation across industry sectors. For example, while the LMA containing Greenville MS has an overall decile ranking of zero, it ranks three in manufacturing. The LMA with Hilo HI had the lowest overall share of high growth businesses during this period, but its 108 manufacturing businesses had a decile ranking of one, indicating that their share with high growth firms was higher than that of a tenth of the other LMAs [those ranked zero].

Establishments in the two largest sectors, which both serve primarily local markets, had lower than average shares of high growth businesses, with 4.45 in the local market sector and 3.48 percent in retail trade. All the smaller sectors had substantially higher average shares of high growth establishments – business services 6.20 percent, distributive firma 5.57 percent, extractive 5.69 percent and manufacturing 7.03 percent. Although the manufacturing sector lost jobs overall during this period, it accounted for the highest share of high growth establishments in the U.S.

What emerges clearly from the empirical investigation is that the variation in regional growth is large but not enormous. If the number of high growth firms reflects strong enterprise qualities, then entrepreneurship would seem to be important for economic growth. What is important now is to focus in on a select policy tool that builds on the strength of endogenous technical change and compliments the innovation system approach (Edquist 1997).

12.5 Theory and Empirical Results

The purpose of this section is to bring together the theoretical discussion in Sections 12.2 and 12.3 with the empirical results. The fundamental thesis of New Growth Theory is that economic growth is non-diminishing because technological knowledge is a non-rival, partially excludable good.

The starting point for most theories of innovation is the firm. In such theories the process by which the firm generates technological change is assumed to be endogenous. For example, in the most prevalent model of technological change process in the literature, the knowledge production function, the firm engages in the pursuit of new knowledge as an input into the process of generating innovative activity. The most important source of new knowledge is generally considered to be R&D.

This model can however be questioned, because in many industries it is small firms which serve as the engine of innovation. This may seem surprising because the bulk of industrial R&D is undertaken in the larger corporations, in fact small enterprises account for only a minor share of in-house R&D. Thus the knowledge production function would suggest that innovative activity is more prevalent in large firms instead of small firms. However, many smaller firms also innovate, so where do the entrepreneurs get the innovating producing inputs, i.e. the knowledge?

One answer to this question is that although the model of the knowledge production function may be valid, the unit of observation – the level of the establishment or firm – may be less valid. Another strand of literature suggests that knowledge spills over from the firm or research institute which produces it to other firms which commercialise that knowledge. This view is supported by theoretical models which focus on the role played by knowledge spillovers between firms in generating increasing returns, and ultimately economic growth (Romer 1990).

Location theory suggests that geographical proximity is needed to transmit knowledge, especially tacit knowledge. This is supported by the observation that knowledge producing inputs are not distributed evenly throughout the economy. So knowledge spillovers tend to be localised within a geographic region. Over time there tends to be an accumulation of knowledge especially in the larger cities. The importance of geographical proximity for knowledge spillovers has been confirmed in a wave of recent empirical studies including Jaffe (1989), Ács, Audretsch and Feldman (1992, 1994); Jaffe, Trajtenberg and Henderson (1993); Anselin, Varga and Ács (1997, 2000). For a critical survey of the literature on spillovers see Karlsson and Manduchi (Chapter 6 in this book).

The mechanism by which new knowledge is spread is through entrepreneurial functions. It frequently occurs that a firm will commercialise knowledge that was discovered by another firm or institution. This shifts the emphasis from firms and institutions to the individuals who possess new economic or technical knowledge. When the lens shifts away from the firm to the entrepreneur as the unit of observation, the question we should ask is how entrepreneurs can gain the best returns from the new knowledge in their possession.

One measure of successful entrepreneurship is the number of rapidly growing establishments. We have assumed in this study that the existence of high growth establishments gives at least some indication of the presence of entrepreneurial capacity. Several interesting observations can be made about the data concerning businesses in the U.S. *Firstly*, the average share of high growth establishments across the U.S. as a whole is 4.6 percent. It was found that all Labour Market Areas have high growth firms. Even the lowest growth regions have over two percent of high growth establishments. This supports the idea that innovations are ubiquitous, and occur in high technology areas as well as non technology areas. *Second,* LMAs with a large share of high growth establishments have four times as many entrepreneurs as LMAs with a low share of high growth establishments.

Third, while we do not have data on knowledge inputs in the LMAs with a large share of high growth establishments, it appears that these inputs are found predominantly in the West and the Southwest of the United States. This supports the observation of Suarez-Villa (Chapter 9 in this book), that there has been a shift in technology leadership from the Northeast to the West. In fact, most of the LMAs with a high share of high growth establishments, for example Provo, Utah, Austin, Texas, Phoenix, Arizona, Salt Lake City, Utah, are in the West and are centres of technology. This chapter does not test the hypothesis that a higher share of rapidly growing establishments leads to more job growth. However, it is well established in the entrepreneurial literature that rapidly growing establishments account for the lion's share of new jobs in a region.

12.6 Policy Implications

In the introductory chapter of Nelson's *National Innovation Systems*, a central hypothesis was formulated about the 'new spirit of what might be called technonationalism', which combines a strong belief that the technological capabilities of a nation's firms are a key source of their competitive prowess with the belief that these capabilities are in a sense national, and can be built by national action (Nelson 1993a, p.3). While Richard Nelson and Nathan Rosenberg are careful to explain that one of the main concerns of their 15-country study was to establish 'whether and in what ways the concept of a 'national' system made any sense today,' they also add that de facto 'national governments act as if it did' (Nelson 1993a, p.5).

Before we get too involved in the policy implications of innovation systems, it is important to look at the main lessons of the New Growth Theory, of which there are three[12]. The first is that no process is more central to the New Growth Theory than investment in research and development. While in many countries both government and the private sector fund R&D, the policy question is whether public investment actually stimulates private investment. In fact, there are two questions here. Firstly, assuming that R&D does contribute to economic growth – and all the evidence suggests that it does – are there technical reasons to suppose that private sector firms would under invest in R&D? Secondly, is there solid evidence that that public sector investment in R&D, especially in universities, stimulates subsequent private sector investment? The answer to both questions is yes. The social rate of return to R&D exceeds the private rate, and R&D is more productive in larger cities than smaller ones. Governments that fail to fund R&D therefore undermine the crucial role of R&D in economic growth.

Lesson Number One: Governments should increase investment in R&D at regional level

Since technological progress results from human inventiveness, the accumulation of human capital (H) in Equation (12.7) *serves a double purpose*. It augments the value of labour in the production of goods and services and, at the same time, promotes the discovery of new technology. Hence, there are potentially prodigious social returns to human capital that are not captured in the standard cost-benefit framework used for determining investment levels. We tend to underinvest in human capital because we ignore these particular social returns. Increased public investment in education and training will therefore more than pay for itself, since it will underwrite future technological discovery, which will in turn raise national growth rates (Bluestone and Harrison 2000).

Lesson Number Two: Governments should increase expenditure on human capital at all levels

The third lesson is that if the discovery of new knowledge is important, then entrepreneurial capacity and market processes are central to the production of knowledge spillovers. The institutional process that supports entrepreneurship is important for economic growth in general and regional growth in particular.

In the systems approach, innovation is viewed as a ubiquitous phenomenon. In all sectors of the economy and at all times, we may expect to find ongoing processes of learning, searching and exploring which result in new products, new techniques, new forms of organisation and new markets. The first step in recognising this to focus on the gradual and cumulative aspects of knowledge acquisition. *Such a perspective gives rise to simple hypotheses about the dependence of future innovation on the past accumulation of knowledge.* In this context, an innovation may be regarded as a new use of pre-existing possibilities and components. Almost all innovations reflect existing knowledge combined in new ways.

The connections between institutions and innovation are ubiquitous and exist at many levels. They exist at the level of the firm, where institutions affect the relations between R&D, production, and marketing – relations that strongly influence innovation. They exist at the level of the market – the relationship between firms themselves and between firms and households. Relations between firms and government agencies with respect to technology policy form the third level. Institutions provide three general levels of functions: to reduce uncertainty by providing information, to manage conflict and cooperation, and to provide incentives. How exactly do institutions do this? Recent literature has suggested that one of the most successful initiatives to foster rapidly growing firms and help them exploit the knowledge base, is the development of networked 'incubators' (Martin 1997; Murray 1993; Spitzer 1988).

In the past incubators were conceived as non-profit bodies and most of their sponsors were also non-profit. There is mixed evidence of the success of these organisations. They were supposed to create a positive environment for start-ups, create local employment, and contribute to the community and diversification of the local economy. However, during the past decade a new business model of incubation has been developed that specifically focuses on fostering the growth of new technology-based firms (Petzinger 1999). The new model relies on exploiting the accumulated regional knowledge base and relies on entrepreneurial discovery. These incubators are privately funded and are like a business. They have a short window for success, usually about 18 months, will assist firms to go public, and recycle human resources.

These incubators provide a high level of services to their clients, including strategic counsel and marketplace vision, human resource recruiting, funding, licensing and partnership negotiations, business and other funding preparation. They are also built on networks. According to a recent article in the *Harvard Business Review* (2000, p.76)[13] 'When properly designed, networked incubators combine the best of two worlds – the scale and scope of large, established spirit of small venture-capital backed firms – all while providing unique networking benefits. Because of this combination, we believe that networked incubators represent a fundamentally new organizational model that is especially well suited for creating value and wealth in the new economy.' These services are provided in exchange for a share of equity in the new companies that the incubator owns. However, the model also provides for reciprocal ownership by the tenants in the incubator.

Business incubators represent one of the hottest growth industries of the new millennium. Most incubators offer facilities in which multiple startups have ready access to office space, support services and hands-on help from experienced managers. By matching promising technologies and business models with capital, services and experience, incubators greatly increase the odds for survival and success of a start-up.

Lesson Number Three: Networked incubators may be one of the most effective institutions to foster entrepreneurial led regional growth

Endnotes

1   Several books have appeared, simultaneously and independently, trying to identify the underlying processes and interconnections that govern regional innovation (Braczyk, Cooke and Heidenreich 1998; de la Mothe and Pacquet 1999; Ratti, Bramanti and Gordon 1997; DeBresson 1996; Ács 2000). Although these books take different approaches, rely on different methodologies, use different data, and define the unit of analysis differently, they all suggest that there is something fundamental at work at the regional level. While they are all interesting, and illuminate pieces of the regional innovation puzzle, neither singularly, nor in concert, do they answer the bigger question as to why some regions are more innovative than others and therefore grow faster.
2   Jaffe 1989, Ács, Audretsch and Feldman 1991, 1994; Glaeser et al. 1992; Anselin, Varga and Ács 1997, 2000; Varga 1998) and innovation systems (e.g. Saxenian 1994; Braczyk, Cooke and Heidenreich 1998; Fischer and Varga 2000; Oinas and Malecki 1999; Sternberg 1999; Ács 2000).
3   This section draws heavily on Ács and Varga (1999).
4   For such surveys see for example Grossman and Helpman 1991; Helpman 1992; Romer 1994; Barro and Sala-i-Martin 1995; Nijkamp and Poot 1997; Aghion and Howitt 1998.
5   If we are concerned with the regional distribution of A, then regional systems of innovation are the proper unit of analysis.
6   This section draws on work with Catherine Armington at the Center for Economic Studies.
7   The SUSB data and their Longitudinal Pointer File were constructed by Census under contract to the Office of Advocacy of the U.S. Small Business Administration. For their documentation of the SUSB files, see Armington (1998).
8   This file was constructed by the Bureau of the Census from its Statistics of U.S. Business [SUSB] files, which were developed from the economic micro data underlying the County Business Patterns. These annual data were linked using the Longitudinal Pointer File, which facilitates tracking establishments over time, even when they change ownership and identification number.
9   For evidence on regional variation on new firm formation see Armington and Ács (2000).

10 The LEEM data cover all private sector businesses with employees, with the exception of those in agricultural production, railroads, and private households. This is the same universe covered in the annual County Business Patterns publications, but establishments with a positive payroll during a year but no employment in March of the same year are not counted for that year.

11 There is a small number [10,000 to 16,000] of new firms each year for which no industry code is ever available. Most of these are small and short-lived. These have been added to the local market category, which is by far the largest of our sectors.

12 A fourth implication of the New Growth Theory is that given the fixed and finite costs of new discoveries, the larger the size of the market the faster the rate of growth. Therefore, tapping into unexplored domestic markets and into foreign markets for trade can raise the rate of growth by boosting the incentive for technological discovery and therefore boost output itself (Ács and Morck 1999).

13 The complete study [The State of the Incubator Industry] can be downloaded from the web at www.hbsp.edu/hbr/incubator.

PART D: Modelling Complexities

13 Modelling of Knowledge, Capital Formation, and Innovation Behaviour within Micro-Based Profit Oriented and Correlated Decision Processes

Günter Haag and Philipp Liedl
Steinbeis Transfer Centre, Applied Systems Analysis

13.1 Introduction

In the neoclassical theory of economic growth, innovations are viewed as being exogenous. Technological progress is treated as a shift parameter in production functions. This view cannot however explain the causes of technical progress, and is unsuited to the study of the impact of a varying rate of technical change (Jungmittag, Blind and Grupp 1999). Furthermore it does not consider interactions between firms using different technologies. The assumption of economic equilibrium in the neoclassical approach is inconsistent with Schumpeter's concept of the creative entrepreneur (Schumpeter 1934) because, once equilibrium is reached, no agent of the economic system can leave, due to optimal decision-making. Technological spillover effects are interpreted as having a negative effect on the market, because the spread a new technology decreases the incentive to invest in R&D.

The work presented in this chapter was guided by the shortcomings mentioned above and the fact that R&D-expenditure has been identified as one of the most important determinants for the development of markets. Market results are not only determined by the ratio of prices and marginal costs, but also the rate of technical change and the variety of products (Audretsch 1996). During the last decades several new approaches, which go beyond the neoclassical growth theory, have been developed to describe the effect of knowledge and innovations on the success of firms. Some of them will be summarised in the following sections.

The first attempt to treat technological progress as an endogenous process was made by Arrow (1962). He regarded technological progress as an experience resulting from learning-by-doing within firms. In this sense technological progress is a by-product of investment, so the innovation process is not depicted directly (Pyka 1999).

An important recent approach to the modelling of innovations uses game theory to simulate the behaviour of the innovation system (Audretsch 1996). A number of studies on license agreements and the adoption of innovations have been carried out using the game theoretical approach. Examples are the work of von Hippel (1989) and Witt (1995), who both use the prisoner's dilemma to describe the conflicting situation in which a firm has to decide whether to exchange knowledge with another firm or not.

In their fundamental work, Nelson and Winter (1982) suggested another way of considering innovations. They used an evolutionary model to describe changes in the behaviour of firms with respect to the development of new techniques. They call all regular and predictable behavioural patterns of firms *routines*. Changes in those routines are made by *searching* for new rules. Stochastic elements are used to model uncertainty of the outcome of research activities.

In this context Pyka (1999) emphasised the importance of interactions between firms. In his model, firms participate in so-called informal networks (von Hippel 1989) whose members exchange new techniques. This leads to the view of a collective innovation process where all economic agents can benefit from the new techniques and technologies developed by other members of the network. One important condition for these 'technology absorbing' firms is the existence of available knowledge they already possess enabling the use of the new technique themselves. In this sense spillovers are interpreted as having a positive effect for firms which are members of the informal network, since this saves R&D investments for parallel- or multi-development of one and the same technique.

It is the purpose of this contribution to provide a decision-based framework for the complex and interwoven processes of production and innovation, labour force and capital formation as well as knowledge production and knowledge transfer processes within and between industries, including the role and efficiency of transfer institutions. The Master Equation framework is used to model and simulate the decision behaviour of firms [entrepreneurs] in formulating their overall business strategies.

The Master Equation approach is a fundamental tool for describing the dynamics of probability distributions related to a multiplicity of interacting economic agents. This approach has its roots in statistical physics and was first applied to sociology and economics by Weidlich and Haag (1983). The Master Equation was applied by Cantner and Pyka (1998) to describe probability distributions of a large number of agents either to express technological changes or changes in behavioural patterns of firms, i.e. whether to cooperate with other firms or not (Pyka 1999) or to develop new products or not (Woeckener 1993).

The interaction of knowledge creation with standard capital formation in an economy is considered in Andersson (1981) and Andersson and Matsinen (1980). However, considerable difficulties were met in estimating the stock of knowledge and its depreciation rates in a reasonable and acceptable way. Furthermore, the R&D undertaken in the production process, as well as the transfer of know-how from research institutions, universities, consultants or transfer institutions like the Steinbeis Institutes are essential elements of the process of developing new products or production processes. This means that knowledge has – at least partially – the character of a public good, in that it may participate as an input in the production process without being 'used up'. On the other hand, knowledge acquisition, information retrieval, knowledge transfers etc. requires time, effort and costs. Consequently these processes have to be considered within the open space of firms' strategic decisions.

In this contribution we follow the approach of Haag (1989) and Müller and Haag (1996). The classical input factor, labour, is 'enhanced' by transfers of technologies from the economic and scientific system. The accumulation of labour and knowledge-transfers is called *know-do* (Müller and Haag 1996). The concept of know-do is that research co-operation within the economic system requires a specific amount of working hours by firms, and co-patenting activities with a university demands a certain amount of time in the scientific area. So the efforts spent in knowledge transfers can be measured in terms of working hours.

Outsourcing the development of new technologies gives firms the advantage of being able to use these technologies without needing to employ their own workers on the research process. This clearly represent a positive effect of know-how transfer. On the other hand, developments made by competing firms may have negative impact on the firm, e.g. if the innovation at one firm replaces the innovation of another firm. In other words, technological spillovers may have positive and negative effects.

Before we start with the development of our macro-economic model we have to define what is meant by the term *innovation*. We will use the definition of the OECD and Eurostat given in the Oslo-Manual (OECD / Eurostat 1997, pp. 47-49):

'*Technological product* and *process* [TPP] *innovations* comprise implemented technologically new products and processes and significant technological improvements in products and processes. A TPP innovation has been implemented if it has been introduced on the market [product innovation] or used within a production process [process innovation]. TPP innovations involve a series of scientific, technological, organizational, financial and commercial activities. The TPP innovating firm is one that has implemented technologically new or significantly technologically improved products or processes during the period under review.'

The term *product innovation* includes new products whose technological characteristics or intended uses differ significantly from those of existing products. But it also includes technologically improved products which provide a significant enhancement or upgrading of the performance of an existing product (OECD/Eurostat 1997). A *process innovation* is the adoption of technologically new or significantly improved production methods, including product delivery. These methods may be used to produce or deliver technologically new or improved products, which cannot be produced or delivered using conventional production methods, or simply to increase the production or delivery efficiency of existing products. (OECD/Eurostat 1997).

At this point we need to distinguish between *innovations* and *inventions*. Only inventions implemented in connection with products or processes are referred to as innovations. Furthermore, it is important in this context to distinguish between *knowledge* and *information*. The term *knowledge* is commonly used to describe research results such as tacit knowledge that are difficult to codify (Nelson and Winter 1982) and hence difficult to transfer between participants of a network of innovating firms (Teece 1998, see also Chapter 6 in this volume). To describe data that is easy to codify and thus easy to transfer, the term *information* is used. Thus, information involves all messages and routinised data that is not tacit and is not too complex to codify or to store (Kobayashi, Sunao and Yoshikawa 1993).

This chapter is organised as follows. In Section 13.2 the macro-model of the nested innovation process will be introduced, then in Section 13.3 some simulations of interacting firms will be presented. The variety of possible solution patterns, including limit cycles and deterministic chaos. This may lead to dangerous trajectories of firms between being exhausted or in a growing phase. To round up the discussion, some concluding remarks are made in Section 13.4.

13.2 Modelling Nested Innovation Processes

In this section we firstly develop an appropriate complex macro-model of the nested innovation processes of a network of firms which avoids some of the shortcomings of earlier knowledge-based concepts (Haag 1989, Müller and Haag 1996).

As empirical studies show (Fischer and Menschik 1994), there are internal and external impulses for firms to innovate. Internal impulses come from the firm's own research and development and from management decisions. External impulses arise from observation of competitors and their products, from trade fairs or product shows and congresses, and from contacts with universities and research centres. So an appropriate model for simulating innovation processes in a firm should include the impacts of other firms and influences from the scientific system as well as their own R&D activities. The term *scientific system* includes universities, research institutes, technology consulting agencies and technology transfer centres like the Steinbeis Institutes.

In order to describe the impact of different transfer activities on the production of a specific firm, the notion of *spillover effects* [inter-firm as well as intra-firm spillovers and transfers between the scientific system and the firms] is commonly used. Firstly we examine the impact of spillover effects on the evolution of inter-linked firms or companies. We therefore consider a network of interacting firms belonging to one or several sectors. This network is illustrated in Fig. 13.1, together with the linkages to the scientific system. Competition between the various firms in the market is not considered in this introductory phase. Hence, we limit the interactions of the firms to spillover effects only. These spillover effects may have both positive and negative impacts on the economic development of a firm. Positive spillover effects can be identified if, for example, a product or process innovation in one sector has an positive impact on other sectors. For example, developments in the semiconductor industry have led to a reduction in the size of electronic components used by the computer industry for their own products. This reduction was necessary to allow the computer industry to produce small personal computers for private consumers.

On the other hand, negative spillover effects may be observed if an innovation in one specific sector has a negative impact on the development of products or processes in another sector. Silverberg and Lehnert (1994) give an example of a negative effect of this kind in the primary energy sector. During the 20th century, wood and coal were replaced by oil and nuclear power. This meant that all the machinery associated with the transform of the former energy sources into electric power became obsolete and had to be replaced.

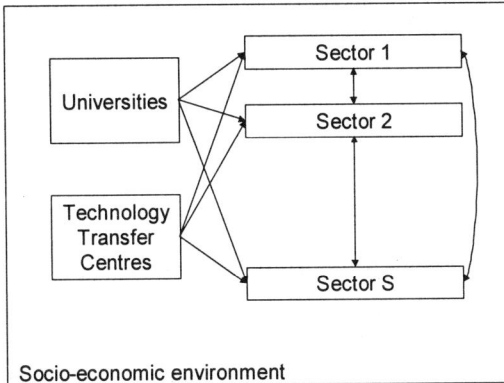

Fig. 13.1   Nested system of economic agents

Firms belonging to a given sector may in principle interact both with other firms in the same sector or firms in other sectors as illustrated in Fig. 13.2. Within a sector the competition for the most advanced products [high-tech products] is responsible for spillover effects with respect to product innovations (Pyka 1999). Process innovations are introduced mainly to reduce costs of production and to enhance the quality of products (Fischer and Menschik 1994).

An example of a spillover effect having positive and negative impacts within a sector is the imitation of innovations of one firm by others. On one hand, for the innovating firm, having first introduced the product on the market, the spillover effect is negative. The imitating firms, on the other hand, benefit from this innovation and experience a positive spillover effect. Such processes of innovation diffusion are described among others by Pyka (1999) and Maier (1998). An example of a negative spillover effect of this kind was the complete replacement of the video-recording system Betamax, introduced by Sony in 1975, by the VHS-system JVC (Arthur 1989, Woeckener 1995). Co-operation between two firms may however lead to positive spillover effects for both (von Hippel 1988, Pyka 1999). Due to the fact, that spillover effects are occurring between firms of different sectors as well as between firms of the same sector, it is sufficient to introduce only one multiple index in order to describe the firms' position within the innovation network.

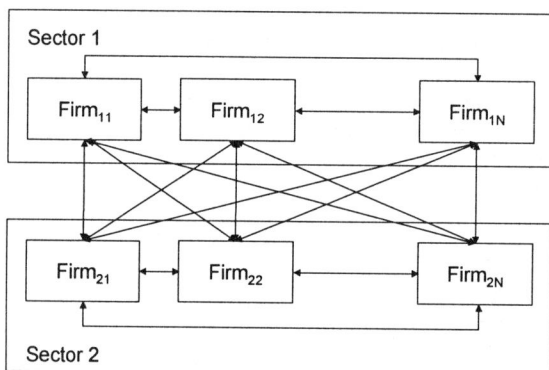

Fig. 13.2   Interdependencies between firms belonging to two industrial sectors

Modelling of the Macro-Economic Network of Interlinked Firms

In the following, the model will be formulated in mathematical terms. The focus is mainly on the supply side of the market. Besides considering the spillover effects in the innovation sector described above and the innovation efforts of firms, production and consequently economic development depend on the classical factors capital and labour. It is assumed that decisions on investments are made by the firms with respect to profit maximisation. However, in describing the production of individual firms, taking into account the impact of spillover effects, we do not follow the concepts of Lucas (1988) or Romer (1986) who take the human capital of a society, and the results of R&D of the whole economy into account for spillover effects. As mentioned by Romer (1990), a major problem of these models is that the entire stock of knowledge is regarded as a public good.

We assign each firm i a production function of the form $Q_i = f(K_i, L_i, I_i)$. So we have a distinction in input factors between the capital stock K_i of the firm i, labour force L_i and the impact of innovations I_i. We have chosen a modified Cobb-Douglas production function[1]:

$$Q_i = a\, K_i^{a_1} \left(L_i + b\, I_i\right)^{a_2} \tag{13.1}$$

This means I_i is that part of a mixture of knowledge, innovation and information used by firm i in its production process, whereby the parameter b describes the efficiency of using this mixture for production. Thus, I_i covers knowledge and information with public good character as well as without. Spillover effects will be modelled within I_i and not directly in the production function. For simplicity, it is assumed that only one product is produced. Moreover we assume that the elasticities of production α_1 and α_2 as well as the scaling factor a are the same for all firms. The variable for innovation activity I_i contains the impact of innovations on the firms output developed by firm i itself and innovations from other firms and from the scientific system. Therefore, positive or negative effect on firm i due to spillover effects may be observed.

The assignment of innovations to the classical production factor labour has been described by Müller and Haag (1996). Because of the difficulty of estimating the impact of innovations a new production factor called 'know-do', D_i , was introduced. Besides intra-firm labour, all inputs such as inter-firm innovation transfers and inputs from universities or transfer institutes can be measured in terms of working hours. Introducing know-do into our model we have:

$$D_i = L_i + b\,I_i \qquad\qquad\qquad (13.2)$$

In case of $I_i = 0$, i.e. if firm i is not innovating and there are no innovation or technology transfers from outside the firm, the chosen production function (13.1) becomes the neoclassical Cobb-Douglas production function. In Fig. 13.3 we give a schematic overview of the model in case of one firm.

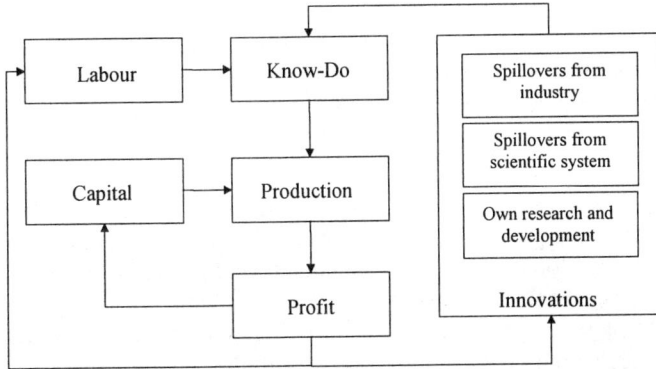

Fig. 13.3   Graphical description of the innovation process in case of one firm

As already noted, we focus mainly on the supply side of the market. However, the limited demand of consumers as well as limited resources like the area useable for production will be indirectly considered by introducing a fixed maximal production capacity $Q_{max,i}$. A simple formulation for the effective production $Q_{eff,i}$ modified in this way reads[2]

$$Q_{eff_i} = \frac{1}{\dfrac{1}{Q_i} + \dfrac{1}{Q_{max\,i}}} \quad or \quad \frac{1}{Q_{eff_i}} = \frac{1}{Q_i} + \frac{1}{Q_{max_i}} \tag{13.3}$$

The maximal capacity of production prevents production increasing to unrealistic high values. According to Fig. 13.4, in the case of very high values of Q_i, $Q_{eff,i}$ becomes approximately equal to $Q_{max,\,i}$. If the production Q_i is much lower than the maximal capacity $Q_{max,\,i}$ equation (13.3) becomes equal to (13.1).

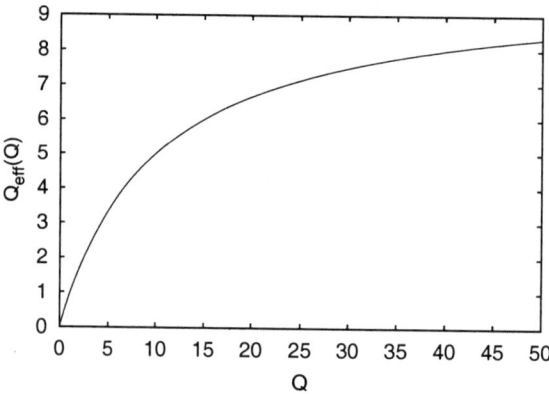

Fig. 13.4   Effective production Q_{eff} in dependency of Q with a maximal production capacity of $Q_{max} = 10.0$

Continuing in the development of our model, we assume that firm i can use a fraction $s_i Q_{eff,\,i}$ of its effective production for non-material investments in the three different fields of activity: for increasing the capital stock K_i, investment in the labour force L_i and investment in innovations I_i, where s_i indicates the rate of savings of firm i. Savings are regarded as being constant over time in a first approximation. Of course, the rate of savings has to fulfill the condition $0 \le s_i \le 1$. The fraction $s_i Q_{eff,\,i}$ will be further divided into a part $\mu_i^K s_i Q_{eff,\,i}$ that is used for capital accumulation, a part $\mu_i^L s_i Q_{eff,\,i}$ that is used for labour formation and a third part $\mu_i^I s_i Q_{eff,\,i}$, used for the formation of innovations. For the ratios of investment μ_i^K, μ_i^L, μ_i^I the following condition has to be satisfied:

$$\mu_i^K + \mu_i^L + \mu_i^I = 1 \tag{13.4}$$

The change in time of the capital used for production by firm i is modelled, according to Müller and Haag (1996), via:

$$\frac{dK_i}{dt} = \mu_i^K s_i \, Q_{eff_i} - \delta_i K_i \qquad (13.5)$$

where δ_i is the rate of decrease of the capital stock.

Contrary to standard neoclassical theory of growth (Solow 1956; Gandolfo 1980) we do not regard labour as being proportional to population. We are not examining the whole economy, but the behaviour of individual firms, therefore we do not assume a direct link between population growth and labour force in the firms under consideration. Rather, the demand for labour in the firms will depend besides the wages on their productivity and the overall economic situation. We thus model the labour force endogenously. Let the number of workers N_i employed by firm i be given by:

$$\frac{dN_i}{dt} = \mu_i^L s_i \, Q_{eff_i} \frac{1}{c_{Ni}} - v_i \, N_i \qquad (13.6)$$

where v_i is the rate of decrease of labour force due to natural decreasing factors. This means that investments in the labour force are related to production. If we translate the number of workers into labour hours h worked in one production period, we receive the total labour hours used for production L_i in one period:

$$L_i = h \, N_i \qquad (13.7)$$

For the costs of one worker during his employment c_{N_i} we obtain:

$$c_{N_i} = w_i \, h \, n \qquad (13.8)$$

where w_i is the hourly wage rate and n is the number of production periods a worker is employed. With (13.7) and (13.8) we receive from (13.6) the change in the labour force over time:

$$\frac{dL_i}{dt} = \mu_i^L s_i \, Q_{eff_i} \frac{1}{w_i n} - v_i \, L_i \qquad (13.9)$$

If we formally reduce the length of the employment contracts to the length of one production period, i.e. $n = 1$, equation (13.9) yields:

$$\frac{dI_i}{dt} = \mu_i^L s_i Q_{eff_i} - v_i L_i \tag{13.10}$$

In this approach we have assumed that the labour market always provides enough workers to the firms. Of course this is realistic in sectors of high unemployment, but not in sectors where skilled labour is rare. A similar approach was adopted by Zhang (1990) and Cigno (1982) who modelled population growth endogenously, depending on production. We have therefore regarded labour, as in neoclassical theory, as being proportional to the population.

We now return to the modelling of innovations. These may be product innovations as well as process innovations. Innovations may occur in the continuous form of incremental improvements of products or processes as well as a discrete form, in the case of radically new products or processes (Freeman 1988). As explained above, following the concept of Müller and Haag (1996), we measure innovations I_i in terms of working hours spent on the development of new products or processes. This concept allows us to regard all innovations as being continuous. We have to consider efforts in research and development within a firm as well as spillover effects from other firms and from the scientific system. The rate of change in innovations is modelled in the following way:

$$\frac{dI_i}{dt} = \mu_i^I s_i Q_{eff\,i} \left(\mu_i^{industry} \sum_{j=1}^{N} g_{ij} D_j + \mu_i^{science} \sum_{k=1}^{M} g_{ik}^{science} D_k^{science} \right) f(I_i) - \gamma_i \tag{13.11}$$

where γ_i is the rate of decrease in innovation activities and accounts for the obsolescence of older innovations, N the overall number of firms and M the number of scientific institutions taken into account.

The investments $\mu_i^I s_i Q_{eff\,i}$ in innovations are subdivided into a fixed fraction $\mu_i^{industry}$, which is the part of investment going into co-operation with other firms ($i \neq j$) and into own research and development ($i = j$), and another part $\mu_i^{science}$, which corresponds to the fraction of investment going into co-operation with the scientific system. Therefore, it must hold that:

$$\mu_i^{industry} + \mu_i^{science} = 1 \tag{13.12}$$

We regard the amount of know-do D of firms and scientific facilities as being responsible for spillover effects. Firm i has access to the labour force of other firms, e.g. by outsourcing the development of new technologies. This means that investment in the own labour force can be reduced.

The factors g_{ij} and $g_{ik}^{science}$ in (13.11) are interaction coefficients describing the strength of the spillover effects from other firms and the scientific system. By using adequately chosen interaction coefficients, the different scenarios of interrelated firms or sectors can be described. A simple example is a firm which invests only in its own R&D. In this case we have $g_{ii} > 0$, $g_{ij} = 0$, $g_{ij}^{science} = 0$. For a firm using innovations of other firms without investing in its own R&D, the coefficients would be $g_{ii} = 0$, $g_{ij} > 0$, $g_{ij}^{science} = 0$.

The transferability of knowledge and information can be characterized by the value of the interaction coefficients g_{ij} and $g_{ik}^{science}$. This approach to modelling spillover effects differs from that presented by Karlsson and Manduchi in Chapter 6 in this book. Thay have taken regional aspects, i.e. the distance between firms, in order to model the impact of spillovers. However, in principle, one could assume that the interaction coefficients g_{ij} and $g_{ik}^{science}$ also depend on distance.

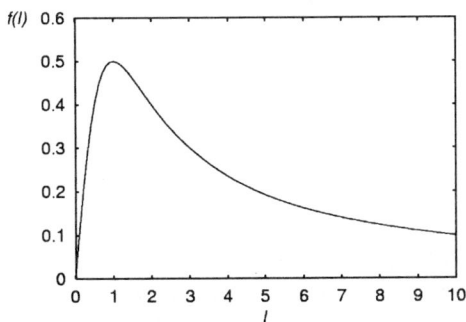

Fig. 13.5   Spillover function $f(I_i)$

The spillover function $f(I_i)$ describes the impact of spillover effects on firm i which depends on its own knowledge stock. Following Pyka (1999), a firm with few knowledge-accumulating activities can gain only little advantage from innovations produced by other firms because of lack of experience. On the other hand, a firm positioned at the technological frontier also has difficulty in increasing its knowledge stock through co-operation with others. So by increasing I_i, the spillover function $f(I_i)$ will increase at first and then with larger values of I_i, it will decrease. We will assume the following spillover function:

$$f(I_i) = \frac{I_i / I_0}{1 + (I_i / I_0)^2} \qquad (13.13)$$

The shape of this function is shown in Fig. 13.5. Of course different functional forms of $f(I_i)$ can be used. However, the general results are not affected by the specific form of (13.13).

We next take a look at the problem of modelling the decision of firm i to split up the total non-material investments $s_i Q_{eff,i}$ into the three parts: capital stock, labour force and innovation activity. We assume profit maximizing rules as criteria used in the decision-making process The profit π_i of firm i is defined as returns from production $Q_{eff,i}$ reduced by the costs for capital and labour:

$$\pi_i = Q_{eff_i} - r\,K_i - w_i\,L_i \tag{13.14}$$

For the sake of simplicity, wages w_i and the rate of interest r are viewed as being constant. Without loss of generality the price of the product can be set to one. Therefore, r and w_i are measured in units of the good produced (Zhang 1999). Each firm i will try to maximise its profits with respect to its investments in K, L and I. The marginal profits:

$$u_i^{(m)} = \frac{\partial \pi_i}{\partial m} \tag{13.15}$$

can be interpreted as utility to invest (Müller and Haag 1996), where $m = K, L, I$.

The probability of changing the type of investment from n to m is then determined by the differences $u_i^{(m)} - u_i^{(n \neq m)}$. It is appropriate to model the dynamics of the decision behaviour of the firms via the Master Equation framework in order to include the statistical effects of uncertainty in the decision process (Haag 1989). This leads to the evolution of the investment ratios $\mu_i^{(m)}$ over time:

$$\frac{d\mu_i^{(m)}}{dt} = \varepsilon_i \sum_n f^{(nm)} \mu_i^{(n)} \exp\left[\lambda_i \left(u_i^{(m)} - u_i^{(n)}\right)\right] \tag{13.16}$$
$$- \varepsilon_i \sum_n f^{(mn)} \mu_i^{(m)} \exp\left[\lambda_i \left(u_i^{(n)} - u_i^{(m)}\right)\right]$$

where ε_i is the speed of adjustment. The parameter λ_i describes the intensity of response due to differences in those marginal profitabilities. The parameters $f^{(nm)}$ consider possible barrier effects due to insufficient information between the different investment types. Since we are dealing with the decision behaviour of firms, it seems to be justified to assume $f^{(mn)} = f^{(nm)} = 1$.

The outcome of the Master Equation approach (13.16) guarantees that the values of $\mu_i^{(m)}$ are in a range of $0 \leq \mu_i^{(m)} \leq 1$. Furthermore this approach satisfies the normalization condition

$$\sum_m \mu_i^{(m)} = 1 \tag{13.17}$$

at all times. The stationary solution of (13.16) can easily be determined:

$$\mu_{stat_i}^{(m)} = \frac{\exp\left(2\,u_i^{(m)}\right)}{\sum_n \exp\left(2\,u_i^{(n)}\right)} \tag{13.18}$$

Equation (13.18) also represents the outcome of a 'random utility' model. Therefore the notion of 'utility to invest' $u_i^{(m)}$ seems to be justified. However, since the stationary solution depends on the marginal profitabilities (13.15), and so on the performance of the firms, it depends among other factors on the different time scales of the economic system, whether or not the investment ratios $\mu_i^{(m)}$ approach an almost stable equilibrium state or end up in a dynamic mode [limit cycle or chaotic state].

It is worth emphasising that the decisions of firms are not made on the basis of perfect foresight as in neoclassical theory. Due to the interlinked network of firms and time-lagged impacts of production decisions, there is uncertainty about the outcome of the investment decision, which can be observed in practice and provides an important point of critique on the 'omniscience' of agents in neoclassical theory (Nelson and Winter 1982, Pyka 1999, Silverberg 1997). In our approach, uncertainties in the decision process are taken into account via the Master Equation. Thus, by considering the dynamics of different strategies of investment behaviour by individual firms together with heuristically founded dynamic equations of motion for the capital stock, the labour force and the innovation activities of firms and production, a complex system of an interlinked network of firms is obtained. In the next section, this system is simulated by means of some specific examples.

13.3 Simulations for Selected Examples and Scenarios

The following simulations of the dynamics of the system of inter-linked firms is based upon the differential equations (13.5), (13.10), (13.11), (13.16). The values of the model parameters are chosen for demonstration purpose only, considering some plausibility arguments. In particular, the rate of decrease in the labour force ν_i is much higher than the rate of decrease of the capital stock δ_i, because the length of the employment contracts has been reduced to the length of one production period.

The time variable in the simulations has to be regarded as scaled time. The time periods on which the patterns observed in the simulations take place depend very much on the values of the rates of decrease δ_i, ν_i, γ_i and the interaction coefficients g_{ij}, $g_{ik}^{science}$. The unit of capital stock K_i is measured in scaled currency, whereas the unit of labour force L_i and innovation activity I_i is measured in scaled working hours.

Simulations with Fixed Ratios of Investment

We will start by considering a very simple system of two firms interacting with one scientific institution. In this first example, the ratios of investment $\mu_i^{(m)}$ are constant over time, so the dynamics of the system is studied only with respect to the development of capital stock K, labour force dynamics L and innovation activities I. It is assumed that the scientific institution provides a constant amount of know-do, $D_I^{science}$, to firm 1. Each simulation is carried out for a scaled time period of $\tau=1000$. The parameters of the model are listed in Table 13.1.

Table 13.1 Parameters for the simulations with fixed ratios of investment

Global Parameters			Firm Specific Parameters				Firm 1	Firm 2
			Firm 1	Firm 2				
a	1.0	$Q_{max, i}$	10.0	10.0	g_{i1}		0.00	0.78
b	1.0	s_i	0.5	0.5	g_{i2}		-0.60	0.00
α_1	0.5	w_i	0.4	0.4	$g_{i1}^{science}$		Var*	0.0
α_2	0.5	δ_i	0.1	0.1	μ_i^K		0.33	0.33
r	0.1	ν_i	1.0	1.0	μ_i^L		0.33	0.33
$D^{science}$	2.3	γ_i	0.1	0.1	μ_i^I		0.34	0.34
I_0	1.0				$\mu_i^{industry}$		0.5	1.0
					$\mu_i^{science}$		0.5	0.0

* This parameter will vary.

Own innovation efforts are not considered for either of the two firms ($g_{ii} = 0$). It is aasumed that firm 1 receives innovations only from the scientific system, whereas firm 2 uses the innovative potential of firm 1 for innovating activities ($g_{21} > 0$). This results in a negative interaction coefficient g_{12} and thus a negative spillover effect for firm 1. Fig. 13.6 shows these interdependencies in a schematic form.

The choice of signs of the interaction coefficients g_{ij} with respect to the field of innovations corresponds to a predator-prey situation in biology, e.g. in Lotka-Volterra-systems. However, as the economic system also depends on changes in the labour force and capital stock, the analogy is rather incomplete. Nevertheless Lotka-Volterra-systems have been used in economics to model business cycles explicitly, e.g. by Gabisch and Lorenz (1987) and Nijkamp and Reggiani (1998).

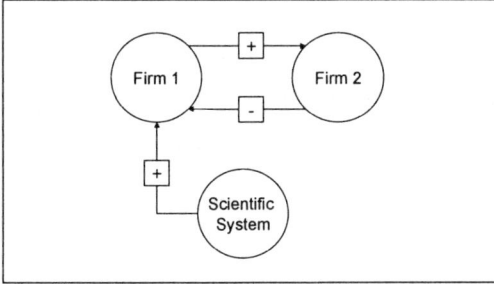

Fig. 13.6 Interdependencies of the agents involved in the first simulation

We first investigate the behaviour of firms for different values of the interaction coefficient $g_{11}^{science}$. In other words, the effect of different transfer activities between the scientific system and firm 1 is considered. The initial values for the input factors of both firms used in the simulations are listed in Table 13.2.

Table 13.2 Initial values for the simulations with different values of $g_{11}^{science}$

	$K_i(t=0)$	$L_i(t=0)$	$I_i(t=0)$
Firm 1	1.0	0.5	1.0
Firm 2	1.0	0.5	1.0

Fig. 13.7 (a) – (e) show the results of the corresponding simulations. Special attention should be paid to the evolution of the labour force and innovation activities. For the first two values of the interaction coefficients, $g_{11}^{science} = 0.4$ and $g_{11}^{science} = 0.9$, the support from the scientific system is insufficient and does not keep both firms alive. In both cases, the labour force and innovation activity, and therefore also production, decrease towards zero. Due to the lack of own research and development activities, and the fact that the ratios of different investment activities are constant in this particular example, the firms continue to invest in innovations, and this development cannot be stopped. In the second case, where $g_{11}^{science} = 0.9$, it is evident that the labour force of firm 2 is increasing rapidly due to positive spillover effects from firm 1.

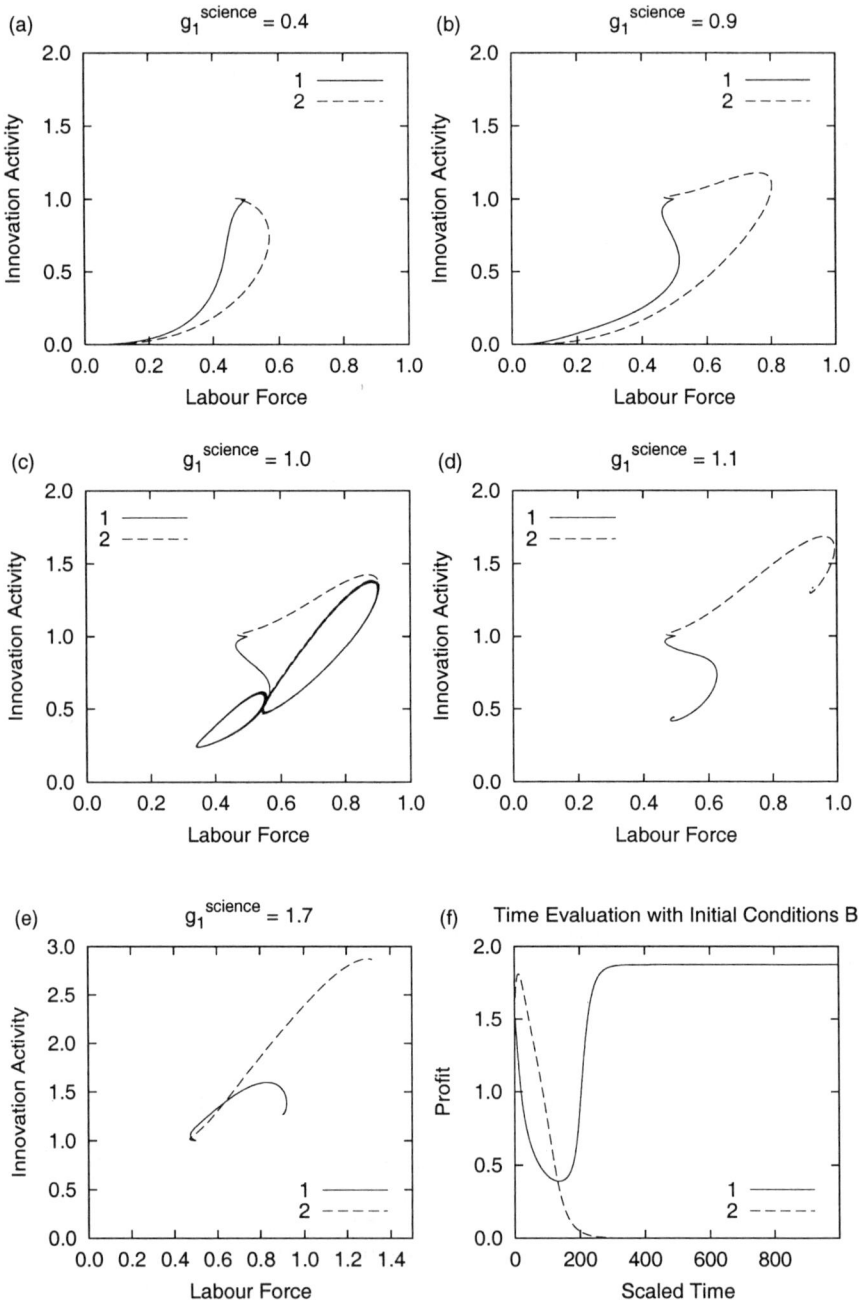

Fig. 13.7   Computer simulations: (a) – (e) labour force and innovation activity for different values of $g_1^{science}$; (f) time evaluation with $g_1^{science} = 1.0$ and initial condition B

A limit-cycle appears if the support of the scientific system reaches a critical threshold value ($g_{11}{}^{science}$ = 1.0), as shown in Fig. 13.7 (c). Both firms are producing profitably, so the labour force and innovation activity for each firm increase and decrease cyclically. The capital stock and profit obtained show the same temporal behaviour as the labour force and innovation activities. It becomes clear that the suppression of firm 1 by firm 2 [which is surely not intended by firm 2, because its success depends mainly on the success of firm 1) leads to the collapse of firm 2 itself. This collapse enables firm 1 to recover and the cycle starts again.

It is surprisinf that there is a further increase in labour force, although innovation activity is already decreasing. On the other hand, the labour force continues to decrease for a while, when innovation activity increases again. At higher values of $g_{11}{}^{science}$ ($g_{11}{}^{science}$ = 1.1 and $g_{11}{}^{science}$ = 1.7), Fig. 13.7 (d) and (e), the trajectories end up at a point of stable equilibrium where both firms can coexist.

In the following, we focus on the case $g_{11}{}^{science}$ = 1.0. In other words, we assume that the spillover effects from the scientific system are large enough for the occurrence of a limit-cycle. We will examine the system with respect to different initial values for capital stock, labour force and innovation activity. For every initial condition listed in Table 13.3 the evolution of the system is calculated for a period of 1,000 time steps.

Table 13.3   Initial conditions for the simulation with $g_{11}{}^{science}$ = 1.0

Initial Condition	$K_1(t=0)$	$L_1(t=0)$	$I_1(t=0)$	$K_2(t=0$	$L_2(t=0)$	$I_2(t=0)$
A	3.60	0.90	1.25	5.30	1.31	2.85
B	3.50	0.90	1.50	2.50	0.40	1.50
C	0.30	0.02	0.35	0.30	0.02	0.35
D	0.30	0.02	0.20	0.30	0.02	0.20
E	1.00	0.50	1.00	1.00	0.50	1.00

The results are illustrated in Fig. 13.8. Let us first take a look at the relationships between capital stock and labour force for firm 1 in Fig. 13.8 (a). For the initial conditions B and C, the system ends up at a stable equilibrium point in the upper right corner of the figure. When this point is reached, the ratio K/L remains constant. This stability point corresponds to the results obtained in neoclassical models, according to which a stable equilibrium point exists for the ratio K/L (Zhang 1990, Gandolfo 1980). For the initial condition A, K is approximately proportional to L, so K/L is nearly constant. In case of the limit-cycle, the ratio of capital and labour changes drastically at the turining points. Between these points K/L is again approximately constant.

In this simulation, the values of the stable equilibrium point at which the system ends up for $g_{11}{}^{science}$ = 1.7, according to the previous simulation in Fig. 13.7 (e), are taken as the initial condition A. Here it becomes clear that a sudden change of one parameter, e.g. from $g_{11}{}^{science}$ = 1.7 to $g_{11}{}^{science}$ = 1.0, can cause the evolution of the firms to become unstable again. Both firms will collapse if, as in this simulation, they do not change their strategy, e.g. change the ratios of investment.

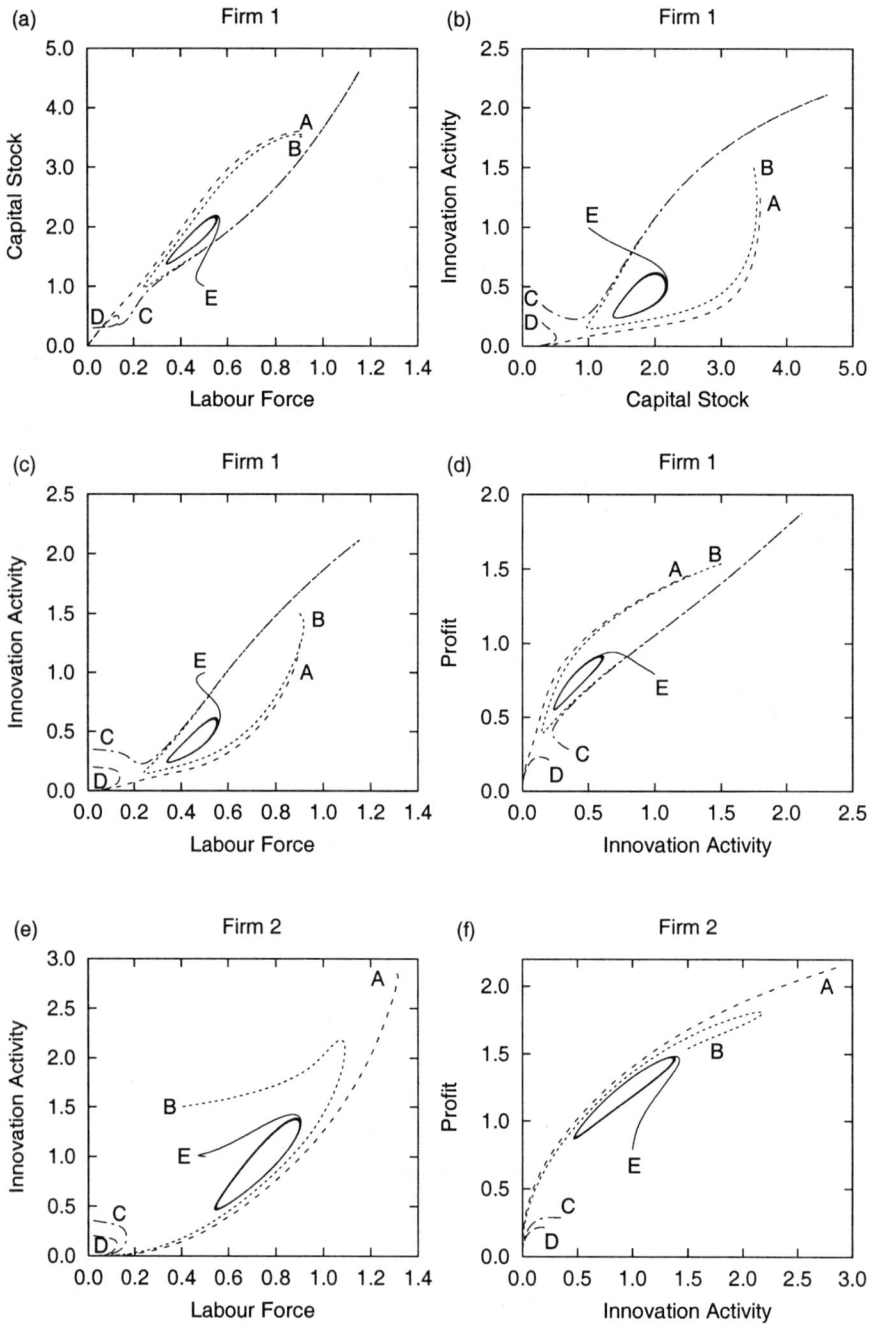

Fig. 13.8   Computer simulations: (a) – (f) evaluation with different initial conditions and $g_1^{science} = 1.0$

Under initial condition B, firm 1 first evolves in direction of bankruptcy, as seen in Fig. 13.8 (a) – (d). Firm 2 is causing this development due to its negative spillover effect on firm 1. This development continues until the profit of firm 2 has declined to values below that of firm 1. This is depicted in Fig. 13.7 (f), where the temporal development of the profits of the two firms are shown. For firm 2, at the point of time when the profit of firm 1 reaches its minimum, it is to late for a recovery. Firm 1 can then evolve without undue negative impacts from firm 2, and its profit reaches the stable equilibrium point determined by the maximal production capacity $Q_{max, i}$ as introduced in Section 13.2.

For the initial conditions C and D it is shown how a small difference in the initial conditions for R&D can affect the evolution of both firms. These differences, in the case of firm 1 determine its survival on the market, as illustrated in Fig. 13.8 (d). Under initial condition C, firm 1 initially increases its capital stock and labour force, whereas its innovation activity decreases slightly (Fig. 13.8 (b), (c)). At larger values of K_1 and L_1 innovation activity starts to increase and the system ends up in the same stable equilibrium point as under initial condition B.

Initial condition E corresponds to the initial condition for which the system ends up in a stable limit cycle, shown in Fig. 13.7 (c). The profit of both firms oscillates between the two extremes and depends on their innovation activities. This can be seen in Fig. 13.8 (d) and (f). In this case, not only the profit of firm 2, but also the capital stock, labour force and innovation activity reach higher levels than for firm 1.

From the shape of the trajectories in Fig. 13.8, one can conclude that for the set of parameters used in this simulation, a stable equilibrium point in the upper right corner of Fig 13.8 (a) – (d) exists, as well as a stable limit-cycle and a saddle point which lies in the bottom left corner. This saddle point is a critical point which decides upon the success of firm 1 on the market.

Simulations of Decision-Making Strategies

We will now study the dynamic behaviour of three decision-making firms that have different co-operation strategies. Decision-making in this context refers to changes in the ratios of investment over time. Again there is one scientific institution, which provides a constant amount of know-do $D_1^{science}$ over the time period considered. The simulations are carried out for a scaled time period of $\tau = 400$. The values of the model parameters are listed in Table 13.4.

As already noted in Section 13.2 the parameters $f^{(nm)}$ are set equal to 1.0. An overview of the interactions between the firms resulting from the parameters g_{ij} and $g_{il}^{science}$ are shown in Fig. 13.9. Firm 1 co-operates with firm 2, which cooperates in turn with firm 3. But firm 3 does not want to co-operate with firm 1. On the contrary, it competes with firm 1 in the field of know-do. The initial values of the simulations are listed in Table 13.5.

Table 13.4   Parameters of the simulation with changing ratios of investment

Global Parameters			Firm 1	Firm 2	Firm 3		Firm 1	Firm 2	Firm 3
								Firm Specific Parameters	
a	1.0	$Q_{max,\,i}$	10.0	10.0	10.0	g_{i1}	0.00	0.30	0.00
b	1.0	s_i	0.5	0.5	0.5	g_{i2}	0.00	0.00	0.20
α_1	0.5	w_i	0.1	0.1	0.1	g_{i3}	-0.80	0.00	0.00
α_2	0.5	δ_i	0.05	0.05	0.05	$g_{i1}{}^{science}$	1.0	0.0	0.0
r	0.1	v_i	1.0	1.0	1.0	$\mu_i{}^{industry}$	0.5	1.0	1.0
$D^{science}$	4.0	γ_i	0.1	0.1	0.1	$\mu_i{}^{science}$	0.5	0.0	0.0
I_0	1.0					ε_i	1.0	1.0	1.0
						λ_i	10.0	10.0	10.0

Table 13.5   Initial values for the simulation of the three firms

	$K_i(t{=}0)$	$L_i(t{=}0)$	$I_i(t{=}0)$	$\mu_i{}^K(t{=}0)$	$\mu_i{}^L(t{=}0)$	$\mu_i{}^I(t{=}0)$
Firm 1	1.0	0.5	0.5	0.33	0.33	0.34
Firm 2	1.0	0.5	0.5	0.33	0.33	0.34
Firm 3	1.0	0.5	0.5	0.33	0.33	0.34

In Fig. 3.9 the results of the simulations are depicted. The market share m_i of firm i is calculated according to:

$$m_i = \frac{Q_{eff_i}}{\sum_j Q_{eff_j}} \tag{13.19}$$

which carries the assumption that all firms belong to the same industrial sector.

First of all it is noticeable that the support from the scientific system leads to a large increase of innovation activity in firm 1, as shown in Fig. 13.10 (e). Maxima and minima of the capital stock and the labour force are delayed compared to the innovation behaviour. The increase in innovation activity causes an increase in profit, and therefore an increase in capital stock and labour force due to increasing investments.

A noticeable time-delay occurs in the starting point of the innovation cycle in the three firms. This phenomenon can be interpreted as a process of diffusion of innovations. Firm 1 is the first to innovate. Firm 2 follows with a further time delay because it has to collect enough know-do concerning the new innovation before it can implement it. Delays in the diffusion process of innovations are often mentioned in the literature, see for example Freeman (1988), Maier (1998) or Silverberg (1991). A remarkable feature is that the maximum innovation activity of firm 2 lies at higher values than for firm 1. It seems that the innovation can be improved upon by firm 2. The innovation activity reaches firm 3 with a considerable delay, because firm 3 is not profiting directly from firm 1. The maximum in innovation activity of firm 3 does not reach the values of the maxima of firms 1 and 2. The reason for this lies in the fact that the interaction coefficient g_{32} is smaller than g_{21}. On the other hand, innovation activities in firm 1 and 2 are already declining at this point in time, so firm 3 gets less of the innovation transfer from firm 2. The decline of innovation activity in firms 1 and 2 is caused by firm 3, due to its negative spillover effect on firm 1, as well as the increasing importance of the term $-\gamma I_i$ in (13.11) for higher values of I_i. As time goes by, further innovation cycles with the same pattern but decreasing amplitude can be observed, until the system reaches a stable equilibrium point.

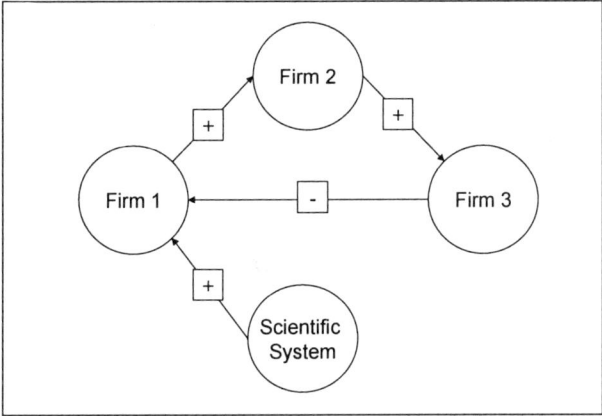

Fig. 13.9 Interdependencies of the agents in the simulation

From the behaviour of strategic investment of firms in Fig. 13.10. (b) and (d), it becomes obvious that a change in the ratios of investment effects the development of the corresponding variable (K, L, I) several time steps later. The firms cannot instantly adjust their investment behaviour to match the economic situation. The percentages of investment in the different fields of innovation, which are not shown here, behave similarly than the ratio of investment in labour force. Together they evolve anti-cyclically to the ratios of investment going into capital formation. Because of the relatively high wages compared to the rate of interest, more investment goes into capital formation than into the labour force.

An analysis of market shares shows that the market share of firm 2 evolves far more smoothly than the market shares of the other two firms. Firm 1 is able to increase its market share as long as firm 2 has no significant innovation effort. Firm 1 rules the market in this phase because it was the first firm to introduce a new innovation on the market. At the point when innovation activity in firm 1 reaches its maximum, firm 1 has already lost a considerable part of its market share to firm 2 and firm 3. Firm 3 gains a considerable market share several time periods later, after it too has introduced the innovation. By this time the innovation activity of firm 1 has reached its minimum.

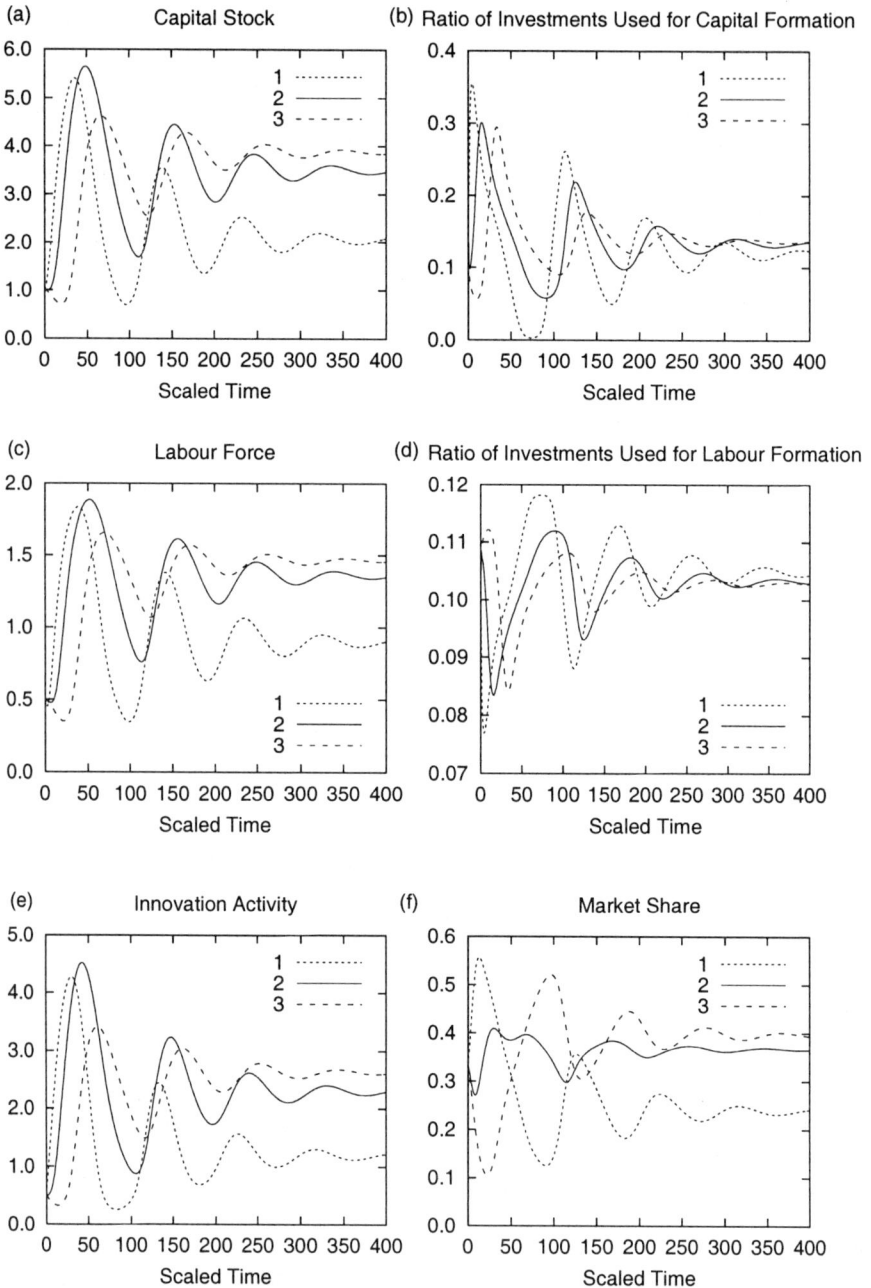

Fig 13.10   Computer simulations: (a) – (f) evaluation of the dynamics of three different firms over time

The gain in market share of firm 2, which adopts the innovation early compared to firm 3, is not as large as for firm 3. A similar observation was made by Silverberg (1991). He found that early adopters tend to lose out in terms of market share and profitability, whereas adopters that introduce the innovation later, but not too late, can gain greater advantage from it. Moving towards equilibrium, the market shares of firms 2 and 3 level out at higher values than the firm 1, although it is the only firm benefiting from the scientific system directly.

The accumulated production of the three firms, not illustrated here, shows a cyclical development with maxima at times of high overall innovation activity in all firms and minima at times of low overall innovation activity. This behaviour is reasonable, since if new innovations enter the market, the market-potential as a whole can be increased due to increasing demand for the new or improved products. On the other hand, the demand becomes saturated over time, if no new innovation follows.

13.4 Conclusions

The economic success of interlinked firms belonging to different sectors, and the existence of stable points of equilibrium depend largely on the signs of spillover effects between the firms operating in the market and the impact of the scientific system. For special constellations, the occurrence of a stable limit-cycle can be observed. In such a situation, firms are permanently fighting for dominance of the market.

One important result of the simulations above is the fact that large increases in innovation activities [innovation pushes] can be caused by spillover effects without the existence of random events. So besides random events, which play an important role in the context of the development of new products (Ebeling 1992), an increase in know-do/innovation success can occur through the concentration of the labour force and knowledge base [know-do] of several firms and the scientific system.

The simulations above confirm the result of Silverberg (1991) that the timing of adoption of an innovation has an important impact on the development of future market share. If innovations are adopted too early, a lower market share is gained than when the adoption takes place at the optimal time, determined by the current stock of knowledge.

Our investsigation has focused mainly on the supply side of the market. Modelling the demand side endogenously would provide a more realistic and comprehensive description of the economic system, for example by assuming a limit to the maximal production capacity. But the price to be paid in terms of increased modelling complexity would be very high. In order to obtain a detailed insight into the complex interlinked activities of firms behaving under profit maximisation conditions, it would seem worth studying selected examples of economic networks of firms. In particular, the role of know-do supporting institutions should be further investigated. Of course, the proposed modelling framework has its shortcomings, but with respect to the insights gained into the complex decision-making processes of co-operating and competing firms, future research in this direction is justified.

Endnotes

1   The Cobb-Douglas production function is used only for illustrative purpose.
2   Instead of using effective production, the same results can be obtained interpreting $s_i Q_{eff,i}$ as realised non-material investment Y_i, where $1/Y_i = 1/s_i Q_i + 1/Y_{max_i}$.

14 Communication and Self-Organisation in Complex Systems: A Basic Approach

Frank Schweitzer and Jörg Zimmermann
Real World Computing Partnership – Theoretical Foundation GMD
Laboratory

14.1 Introduction

The emergence of complex behaviour in systems consisting of interacting elements is among the most fascinating phenomena of our world. Examples can be found in almost every field of today's scientific interest, ranging from coherent pattern formation in physical and chemical systems (Feistel and Ebeling 1989; Cladis and Palffy-Muhoray 1995), to the motion of swarms of animals in biology (DeAngelis and Gross 1992) and the behaviour of social groups (Weidlich 1991; Vallacher and Nowak 1994). In the social and life sciences, it has generally been held that the evolution of social systems is determined by numerous factors –cultural, sociological, economic, political, and ecological, etc. However, in recent years, the development of the interdisciplinary 'science of complexity' has led to the insight that complex dynamic processes may also result from simple interactions. Moreover, at a certain level of abstraction, one can find many common features between complex structures in very different fields (Schweitzer 1997a, b).

Recent progress in the understanding of non-equilibrium phenomena in complex systems has initiated a great deal of analysis, modelling and simulation of 'living' systems by means of methods derived from statistical physics. This holds not only for biological systems (Parisi, Müller and Zimmermann 1998), but also for social and economic systems (Dendrinos and Sonis 1990; Weidlich 1991; Lewenstein, Nowak and Latané 1992, Helbing 1995; Kacperski and Holyst 1996; Allen 1998). Already in the early seventies, physicists and regional scientists had realised that these methods could help to understand social phenomena, such as opinion formation, migration, and settlement formation (Weidlich 1972, Weidlich and Haag 1983).

Very recently, physicists have also focussed their interdisciplinary interests to particular economic processes, such as trading, market dynamics, decision processes, economic agglomeration, or company growth (Bruckner et al. 1994;

Levy, Levy and Solomon 1995, Galam 1997, Schweitzer 1998, Lee et al. 1998). The joint efforts of many research groups spread over the world eventually lead to the establishment of *econophysics* - a young and fast growing field, the potential importance of which can be hardly overestimated (Mantegna and Stanley 2000, Schweitzer and Helbing, 2000). Even with the analysis of financial time series as its current focus, econophysics is meant to be a more comprehensive enterprise. Basically, it focusses on the question how and to what extent methods from statistical physics can be used for the analysis, modelling, simulation, and optimisation of economic systems.

In order to make this enterprise a successful one, a broad and openminded dialog is needed between physics, economics and the social sciences. This dialog should help to overcome the gap between these different disciplines (i) by providing methods from the natural sciences, which could be adapted to solving problems in social or economic fields, and (ii) by increasing among natural scientists the sensitivity for problems in the fields of economics and the social sciences.

The contribution of this chapter to the discussion is twofold: in the first part, we intend to address general problems involved in defining and simulating complex systems which are also of relevance in an economic context. The second part deals with a basic model which aims to simulate communication and self-organisation in an agent system via the exchange of information. By 'basic' we mean that we focus only on particular interactions among the agents, and make no attempt to model a specific socio-economic system most realistically. Instead, we concentrate on the spatial co-ordination of decisions among the agents dependent on the various types of information received. We consider certain important features, such as the exchange of information with a finite velocity, the existence of a memory, and the local heterogeneity of available information.

14.2 Complex Systems and Self-Organisation

Despite many efforts, there is no commonly accepted definition of a complex system (Ebeling, Freund and Schweitzer 1998). Heuristic approaches basically focus on the interaction between ['microscopic'] subsystems and the emergence of new qualities at the ['macroscopic'] system level. The following are two of the many possible definitions:

- 'Complex systems are systems with multiple interacting components whose behaviour cannot be simply inferred from the behaviour of the components' [New England Complex Systems Institute].
- 'By complex system, is meant a system comprised of a [usually large] number of [usually strongly] interacting entities, processes, or agents, the understanding of which requires the development, or the use of, new scientific tools, non-linear models, out-of equilibrium descriptions and computer simulations' [Journal *Advances in Complex Systems*].

The latter description already raises the question as to which kind of scientific methodologies or tools are required to investigate complex systems. It is now commonly accepted that *computer simulations* will play a major role in this enterprise, as a third methodology in addition to formal theories and empirical studies [experiments].

Among the simulation approaches developed within the last twenty years, the *multi-agent* approach seems to be most promising and versatile. While agent models were originally developed in the *Artificial Life Community* (Maes 1991; Meyer and Wilson 1991), it has emerged that they could well be suitable for use in a number of scientific fields, ranging from ecology to engineering (DeAngelis and Gross 1992; Lam and Naroditsky 1992), including economics and the social sciences (Andersen, Arrow and Pines 1988; Troitzsch et al. 1996; Hegselmann, Mueller and Troitzsch 1996; Arthur, Durlauf and Lane 1997; Silverberg 1997; Schweitzer and Silverberg 1998). However, agent-based models are not restricted to the social and life sciences, they are also useful in natural sciences in cases where continuous approximations are less appropriate. Here, the discrete approaches to structure formation range from lattice gas models in hydrodynamics to stochastic cellular automata and models of active walkers or active Brownian particles (Boon 1992; Crutchfield and Hanson, 1993; Lam 1995; Schimansky-Geier et al. 1995, Schweitzer 1997e).

The advantage of an agent-based approach is that it is applicable also in cases where only a small number of actors [particles, agents] govern the future evolution. Here deterministic approaches or mean-field equations are not adequate to describe the behaviour of a complex system. Instead, the influence of history, i.e. irreversibility, path dependence, and the occurrence of random events/ stochastic fluctuations play a considerable role.

In general, agents are regarded as relatively autonomous entities which may represent local processes, individuals, species, agglomerates, chemical components, firms, etc. These entities are governed by a set of rules which determine how they interact with each other. Which rule applies to which specific case may depend on local variables. These can be influenced in turn by the [inter]action of the agents. In this contribution, an *agent* is taken to be a subunit with an 'intermediate' complexity. This means that he/she is not a 'physical' particle only reacting to external forces but, on the other hand, does not have the same complex capabilities as the whole system. The agent is characterised by a range of 'activities' which will depend on the kind of system the model is applied to – some examples are given later in this chapter.

A *multi-agent system* [MAS] may consist then of a *large number* of agents, which can be also of *different types*. The complex behaviour of the multi-agent system as a whole basically depends (i) on the complexity of the agent [i.e. the range of possible actions], (ii) on the complexity of the interaction. The latter is perhaps even more important, since it has been shown in physical systems, for example, that a rich variety of structures can emerge even from the interaction of *simple* entities. The interactions between agents may occur on *different spatial and temporal scales*. That means, in addition to local or spatially restricted interactions which occur only at specific locations or if agents are close to each other, that we also have to consider global interactions involving all agents. In addition, the time scale of interactions is significant. Whereas some interactions occur quite frequently, i.e. on a shorter time scale, others become effective only over a long period. A third distinction is that between direct and indirect interactions. The latter occurs for example if agents use a common resource that can be exhausted in the course of time. In this way, the actions of all agents are indirectly coupled via the resource, and its current availability provides some information about the cumulative activity of others.

As a result of the different interactions, we can observe different kinds of collective dynamics and the *emergence of new system properties* not readily predicted from the basic equations. This process is often referred to as *self-organisation,* i.e. 'the process by which individual subunits achieve, through their cooperative interactions, states characterized by new, emergent properties transcending the properties of their constitutive parts' (Biebricher, Nicolis and Schuster 1995).

However, whether or not these emergent properties occur depends not only on the properties of the agents and their interactions, but also on the external conditions, such as global boundary conditions, or the influx/outflux of resources [matter, energy, information]. A description which attempts to include these conditions is given by the following heuristic definition: 'self-organisation is defined as spontaneous formation, evolution and differentiation of complex order structures forming in non-linear dynamic systems by way of feedback mechanisms involving the elements of the systems, when these systems have passed a critical distance from the statical equilibrium as a result of the influx of unspecific energy, matter or information' (SFB230 1994).

In this sense, a self-organised structure can be considered as the opposite of a hierarchical structure, which basically proceeds *from top down to bottom.* Here, structures are *originated* bottom up, leading to an emerging hierarchy, where the structure of the 'higher' level appears as a new quality of the system (Haken 1978; Darley 1994). For the prediction of these global qualities from local interactions there are fundamental limitations which are considered for example in chaos theory. Stochastic fluctuations provide the opportunity for unlikely events to occur, and these in turn affect the history of the system. This means that the properties of complex systems cannot be determined by a hierarchy of conditions, since the system creates its complexity in the course of evolution with respect to global constraints. Considering that the boundary conditions too may evolve and new degrees of freedom appear, co-evolutionary processes become important, and the evolution may occur at a new level.

14.3 Complex Versus Minimalistic Agents

In order to gain insight into the interplay between microscopic interactions and macroscopic features in complex systems, it is important to find a level of description which makes it possible to consider specific features of the system and reflect the origin of new qualities, but is not overwhelmed with microscopic details. As pointed out above, the multi-agent approach may provide a suitable tool for describing and simulating complex systems. However, this raises the question of how to design the agent's features appropriately. There is no general answer to this, since the agent design depends largely on the system being considered. Nevertheless, some general observations apply.

Let us take the example of agent-based computational economics, which aims to describe and simulate the economic interaction between 'agents', which can be either firms or individuals (Föllmer 1974; Holland and Miller 1991; Lane 1992; Arthur 1993; Epstein and Axtell 1996; Kirman 1993). One of the standard paradigms of neoclassical economic theory, the *rational agent* model is based on two main assumptions: i) that the agent has *complete knowledge* of all possible actions and of their outcomes, or at least a known probability distribution of outcomes, and ii) that the agent knows that all other agents know exactly what he/she knows and are equally rational, i.e. the *common knowledge assumption* (Silverberg and Verspagen 1994). In this particular form, the rational agent is just one example of a complex agent with either knowledge-based or behaviour-based rules (Maes 1991), performing complex actions, such as rational choices or BDI [belief-desire-intention] (Müller, Wooldridge and Jennings 1997).

The complex agent on the other hand is capable of specialisation, learning, genetic evolution, etc. Different problems are involved in the design of such a complex agent – two will be mentioned shortly. One problem often ignored in neoclassical analysis concerns the *information flow* in the system. The common knowledge assumption implicitly demands an infinitely fast, loss-free and error-free distribution of information throughout the whole system, but does not give a hint of how this can be achieved. A more realistic assumption would be based on a heterogeneous, time-delayed, incomplete and noise-affected information distribution, but this would require the explicit modelling of the information flow between the agents.

A second problem concerns the combinatorial explosion of the state space. The freedom to define rules and interactions for the agents provides an interesting development, but each additional rule expands the state space of possible solutions for the agent system. Already for 1,000 agents with 10 rules, the state space contains about 10^{13} possibilities. Hence almost every desirable result could be produced from such a simulation model and the freedom could very soon turn out to be a pitfall.

In fact, due to the complexiy of their simulation, many of the current multi-agent tools, for example SWARM, lack the possibility of investigating the influence of specific interactions and parameters systematically and in depth. Instead of incorporating only as much detail as is *necessary* to produce a certain emergent behaviour, they put in as much detail as *possible*, and thus reduce the chance of understanding *how* emergent behaviour occurs and *what* it depends on.

Quite a different approach is offered by *minimalistic agent design*, which is used in this paper. A minimalistic agent acts on the *simplest possible set of rules,* without deliberative actions. The minimalistic agent model is based on a large number of 'identical' agents, and the focus is mainly on *co-operative interaction* instead of autonomous action. Of course, this approach too involves a certain trade-off. Some features which might be considered important for a specific system have to be dropped in order to investigate in detail a particular kind of interaction. Instead of describing the whole system realistically, the minimalistic approach focuses on particular dynamic effects within the system dynamics. The advantage is that it also provides numerous quantitative methods to investigate the influence of certain parameters or quantities. For example, bifurcations, the structure of the attractors, conditions for stable non-equilibrium states, etc. can be investigated by means of advanced methods borrowed from statistical physics, thus providing a clear idea of the role of particular interaction features.

Like any other agent-based approach, minimalistic agent models are based on a specific kind of reductionism which should be addressed from the viewpoint of the philosophy of science. Self-organisation theory is often interpreted as a holistic approach which overcomes classical reductionism, especially in the natural sciences. However, self-organisation itself is a phenomenon that implies a certain perspective, i.e. the observation depends on the specific level of description, or on the focus of the 'observer' (Niedersen and Schweitzer 1993). In this sense, self-organisation theory is aithetical [from the Greek meaning of 'aisthetos' or perceivable] (Schweitzer 1994). The particular level of perception used for self-organisation has been denoted *mesoscopy* (Schweitzer 1997c). It differs from the microscopic level of perception, which focuses primarily on the smallest entities or elements, as well as from the macroscopic level, which focuses on the system as a whole. Instead, mesoscopy concentrates on elements which are complex enough to allow an interaction that eventually results in emergent properties or complexity at the macroscopic scale. These elements are the 'agents' in the sense denoted above: they provide an 'intermediate complexity' and are capable of a certain level of activity, i.e. they do not just passively respond to external forces, but are actively involved, for example in non-linear feedback processes.

Thus, in the following discussion we have to bear in mind the degree of 'reductions' implied in the agent-based models regarding the system elements and their interactions. If we wish to develop a generalised self-organisation theory, we have to understand carefully the nature of these reductions, especially in the social and life sciences (Hegselmann, Mueller and Troitzsch 1996). Self-organisation in social systems has to deal with the mental reflections and purposeful actions of agents, who create their own reality. While we are convinced that the basic dynamics of self-organisation provide analogies between structure formation processes in very different fields regardless of the elements involved, we should not forget the differences between these elements, especially between humans and physical particles. Thus, a deeper understanding of self-organisation, complex dynamics and emergence in socio-economic systems must also include greater insight into these reductions.

14.4 An Information-Theoretic Approach

In the following, we characterise the minimalistic agent approach in terms of an information-theoretic description. This will allow us to describe the interaction of agents as a generalised form of 'communication', based on the exchange of information. To this end, we need to distinguish between three different kinds of information: functional, structural and pragmatic information (Ebeling, Freund and Schweitzer 1998; Schweitzer 1997d).

Functional information denotes the ability of the agent to process external information [data] received. It can be regarded as an *algorithm* specific to the agent. This algorithm is applied to 'data' in a very general sense, which can be also denoted as *structural information* because it is closely related to the [physical] structure of the system. The DNA is an example of structural information in a biological context. As a complex *structure*, it contains a mass of [structural] information in a coded form, which can be selectively activated according to different circumstances. Another example of structural information would be a book or a user manual, written in the letters of a particular alphabet.

Structural information is meaningless, since it does not contain *semantic* aspects, only *syntactic* aspects. Hence, the content of structural information can be analysed for instance by means of different physical measures [e.g. conditional or dynamic entropies, transinformation etc.] (Ebeling, Freund and Schweitzer 1998). Functional information on the other hand is related to the semantic aspects of information; it reflects the contextual relations of the agent. It is the purpose of functional information to *activate* and to *interpret* the existing structural information with respect to the agent, thus creating a new form of information called *pragmatic information*. This is a type of operation-relevant information which allows the agent to act. In the examples above, cells are able by means of specific 'functional' equipment to extract different [pragmatic] information from the genetic code, which then allows them to evolve differently e.g. in morphogenesis. In the example of the user manual, the reader [agent] is able by means of specific functional information, i.e. an algorithm to process the letters, to extract useful 'pragmatic' information from the text, which allows him/her to act accordingly. If the functional information [algorithm] does not match the structural information [data], i.e. if the manual is written in Chinese and the reader can only process Roman letters, then pragmatic information will not emerge from this process – even though the structural and functional information is still there. With respect to pragmatic information, we can express this relationship as follows: *it is the purpose of functional information to transfer structural into pragmatic information.*

In order to characterise the minimalistic agent model in terms of a generalised communication approach based on the exchange of information, we have to identify the kind of functional and structural information used in the system, and then have to investigate what kind of pragmatic information may emerge. As explained above, the functional information may be a [simple] algorithm which can be repeated by each agent. For example, we may assume that at every time step the agent is able (i) to read data, (ii) to write data, and (iii) to process the data currently read [e.g. to compare their value]. The data read and written are structural information stored on a *blackboard* external to the agent. Like a normal blackboard, it has the role of a communication medium (Veit and Richter 2000). Because of this, the communication among the agents may be regarded as *indirect communication*; however, the involvement of a medium always seems to be necessary, even in 'direct' oral communication.

The emergence of *pragmatic* information for a specific agent will of course depend on (i) the functional information, i.e. the 'algorithm' for processing a specific piece of structural information, and (ii) the availability of this structural information, i.e. the access to the respective blackboard at a particular time or place. For different applications, we may consider various possibilities of restricting the access to the blackboard both in space and time. In this way, the communication between agents can be modelled as *local* or *global*. On the other hand, we may also assume that there are different *spatially distributed* blackboards in the system, modelling a spatially heterogeneous distribution of [structural] information.

In addition, we may consider *exchange processes* between different blackboards, for instance for the reconciliation of data. The data originally stored on a particular blackboard may then also propagate to other blackboards in the course of time. This way, we observe a rather complex interaction dynamics determined by two quite different space and time-dependent processes: (i) changes of blackboards caused by the agents, (ii) changes of blackboards caused by eigendynamics of the structural information. The latter may also involve dynamic elements such as (i) a finite life time of the data stored which models the existence of a memory, (ii) an exchange of data in the system with a finite velocity, (iii) the spatial heterogeneity of available structural information.

As one possibility for considering these features within a minimalistic agent model, we have introduced the concept of a *spatio-temporal communication field* (Schweitzer and Holyst 2000) which models the eigendynamics of an array of spatially distributed blackboards. This communication field is external to the agents, but is created by the various types of data [structural information] consecutively produced by them. For example, agents contribute to the communication field by making choices, by consuming resources, by producing some output, e.g. economic production or simply as waste or thermal radiation which heats up the atmosphere. The distribution of the different kinds of structural information is spatially heterogeneous and time dependent. However, it may affect other agents in different regions of the system, provided they notice this information and are capable of extracting some meaning from it, which may then influence their own decisions or output.

This concept has proved applicable to a variety of models. We should like to mention just two examples here. In a model of economic agglomeration, the communication field has been regarded as a spatially heterogeneous, time dependent *wage field*, from which agents can extract meaningful information for their migration decisions. The model describes the emergence of economic centres from a homogeneous distribution of productivity (Schweitzer 1998). Another example deals with spatial self-organisation in urban growth (Schweitzer and

Steinbrink 1997). In order to find suitable places for aggregation, 'growth units' [specific urban agents] may use the information of an urban attraction field which has been created by the existing urban aggregation. In this way it provides an indirect communication between different types of urban agent.

To be more specific, we wish to examine in the following section a simple model of agents communicating by means of a two-component communication field, which contains information for their local decisions.

14.5 Basic Model of Communicating Agents

Let us consider a rather simple toy model of agent interaction. Suppose, we have a 2-dimensional spatial system with a total area A, where a community of N agents exists. In general, N can be changed by birth and death processes, but A is to be assumed fixed. Each agent i is assigned two individual parameters: its position in space, r_i, which should be a continuous variable, and its current 'opinion', θ_i [with respect to a definite aspect or problem]. The latter is a parameter with a discrete value representing an *internal degree of freedom* [which is a rather general view of 'opinion'].

To be more specific, let us discuss the following example. Imagine a certain problem, for instance the separate disposal of recycling material. Each agent in the system needs to decide whether he/she will collaborate in the recycling campaign or refuse to do so. There are therefore two [opposite] opinions, i.e. $\theta_i \in \{+1, -1\}$. From the classical economic perspective, each agents' decision to collaborate or not will depend on an estimate of their *utility*, i.e. what they may gain compared to the effort involved. Here, however, we neglect any question of utility and simply assume that the agent is more likely do what others do with respect to the specific problem, i.e. he/she will decide to collaborate in the recycling campaign if most of his/her neighbours do so, and vice versa. This kind of contagious behaviour in decision-making processes is well known in many different fields, from fashion demand to the choice of movies (Weidlich 1991; Helbing 1995; Lane and Vescovini 1996; Solomon et al. 2000).

This problem raises a question about the interaction between agents at different locations, i.e. how is agent i at position r_i affected by the decisions of other agents at closer or far distant locations? In a checkerboard world, commonly denoted as cellular automaton, a common assumption is to consider only the influence of agents which are at the [four or eight] nearest neighbour sites, or at the second-nearest neighbour sites (Schelling 1969; Sakoda 1971, Hegselmann and Flache 1998). On the contrary, in a mean-field approximation, all agents are considered equally influencial via a mean field which affects each agent at the same time and in the same manner.

Our approach is different from those described above, as we consider a continuous space and a gradual, time-delayed interaction between all agents. We assume that agent i at position r_i is not directly affected by the decisions of other agents, but receives information about their decisions indirectly via a *communication field* generated by the agents with the different opinions. This field is assumed a scalar *multi-component spatio-temporal field* $h_\theta(r,t)$, which obeys the following equation:

$$\frac{\partial}{\partial t} h_\theta (r,\, t) = \sum_{i=1}^{N} s_i\, \delta_{\theta,\theta_i}\, \delta(r - r_i) - k_\theta\, h_\theta(r,t) + D_\theta\, \Delta h_\theta(r,t) \tag{14.1}$$

All agents contributes permanently to this field with their personal 'strength' or influence, s_i. Here δ_{θ,θ_i} is the Kronecker Delta indicating that the agents contribute only to the field component which matches their opinion θ_i. $\delta(r-r_i)$ means Dirac's Delta function used for continuous variables, which indicates that the agents contribute to the field only at their current position, r_i.

The *structural information* generated this way has a certain life span $1/k_\theta$, further it can spread throughout the system by a diffusion-like process, where D_θ represents the diffusion constant for information exchange. We have to take into account that there are two different opinions in the system, hence the communication field should also consist of two components, $\theta = \{-1, +1\}$, each representing one opinion. Note that the parameters describing the communication field, s_i, k_θ, D_θ do not necessarily have to be the same for the two opinions.

Equation (14.1) for the communication field h_θ (r,t) is a partial differential equation which is continuous in space and time. In a *discretised* version, it describes a spatial array of two different kinds of blackboards, each storing the contributions s_i ['data', structural information] produced by the agents of a particular opinion in a particular spatial domain $x + \Delta x, y + \Delta y$. These blackboards are updated in time intervals Δt and have an eigendynamic determined by the exchange and the life time of the 'data' stored. This eigendynamic can be used to reflect some important features of communication in social systems:

- the existence of a *memory*, which reflects the past history of actions. In our model, this memory exists as an external memory, the lifetime of which is determined by the decay rate of the structural information, k_θ;
- an *exchange of information* in the community with a finite velocity. In our model, this exchange is described by a diffusion-like process with the exchange constant D_θ. This implies that the structural information will eventually reach every agent in the system, but of course at different times;
- the influence of *spatial distances* between agents. Thus, the information generated by a specific agent will affect agents located closer earlier than distant agents, and with a larger weight.

The communication field h_θ (r,t) influences the agent's decisions as follows. At a certain location r_i , the agent i with opinion θ_i is affected by two kinds of information: the information h_θ $(r_i ,\ t)$ resulting from agents who share his/her opinion, and the information $h_{-\theta}(r_i,\ t)$ resulting from the opponents. The diffusion constants D_θ determine how fast he/she will receive any information, and the decay rate k_θ determines, how long a generated information will exist. The agent, who is dependent on the information received locally, has two opportunities: he/she can *change his/her opinion* or keep it. A possible equation for the transition rate of a change of opinion reads (Schweitzer and Holyst 2000):

$$w\left(-\theta_i \mid \theta_i\right) = \eta \exp\left\{-\frac{h_\theta(r_i,t) - h_{-\theta}(r_i,t)}{T}\right\} \qquad (14.2)$$

The probability of changing opinion θ_i is rather small if the local field $h_\theta(r_i,t)$, which is related to the support of opinion θ_i, overcomes the local influence of the opposite opinion. Here, η defines the time scale of the transitions. The scaling parameter T may be interpreted as a 'social temperature' (Kacperski and Holyst 1996) describing a degree of *randomness* in the behaviour of the agents, but also their average volatility (Bahr and Passerini 1998).

In order to summarise our model, we note the non-linear feedback between the agents and the communication field as shown in Fig. 14.1. The agents generate the field, which in turn influences their further decisions. In terms of synergetics, the field plays the role of an order parameter which couples the individual actions and this way initiates coherent behaviour within the agent community.

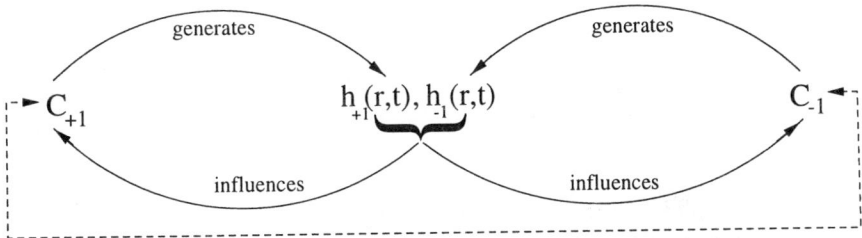

Fig. 14.1   Circular causation between the agents, C_{-1}, C_{+1}, and the two-component communication field, $h_\theta(r,t)$

For N = constant, the community of agents may be described by the multivariate distribution function $P(\underline{\theta},\underline{r},t) = P(\theta_1,r_1,...., \theta_N,r_N,t)$ which gives the probability of finding N agents with the opinions $\theta_1,...., \theta_N$ at positions $r_1,....,r_N$ on the surface A at time t. The time dependent change of $P(\underline{\theta},\underline{r},t)$ can then be described by a *Master Equation* which considers any possible transition within the opinion distribution $\underline{\theta}$ (the formal details are skipped here, for a more complete explanation see Schweitzer and Zimmermann 2000). The Master Equation, together with equations (14.1) and (14.2) forms a complete description of our system which depends on the parameters describing the agent density (N,A) and the components of the communication field $(s_i, k_\theta, D_\theta)$. In order to find possible solutions of the Master Equation we will use computer simulations, and in particular apply the stochastic simulation technique. The results are presented in the following sections.

14.6 The Mean-Field Approach

Before investigating the spatially distributed system, we shall discuss a mean-field approximation in order to obtain some insight into the complex dynamics of the agent system. This case, which has been discussed in greater detail by Schweitzer and Holyst (2000), may have some practical relevance for communities existing in small systems with small distances between the agents. In particular, in such small communities a very fast exchange of information may hold, i.e. spatial heterogeneities in the communication field are equalised immediately. In terms of the blackboard interpretation, this means that all agents have access to the same [two] blackboards independent of their locations. Thus, in this section, the discussion can be restricted to subpopulations with a certain opinion rather than to agents at particular locations.

Let us define the share x_θ of a subpopulation θ and the respective mean density $\bar{n}_\theta$ in a system of size A consisting of N agents:

$$x_\theta(t) = \frac{N_\theta(t)}{N} \quad \text{and} \quad \bar{n}_\theta(t) = \frac{N_\theta}{A} \tag{14.3}$$

where the total number of agents sharing opinion θ at time t fulfills the condition

$$\sum_\theta N_\theta(t) = N_{+1}(t) + N_{-1}(t) = N = const. \quad \text{and} \quad x_{+1}(t) = 1 - x_{-1}(t) \tag{14.4}$$

The dynamics of the system is then determined by the equations for the subpopulation $x_\theta(t)$, which are coupled via the equations for the two-component communication field $h_\theta(r,t)$. The stationary states of the dynamics follow from the conditions $\dot{x}_\theta = 0$, $\dot{h}_\theta = 0$. For the two field components we find with the assumption that agents with the same opinion θ will have the same influence $s_i \rightarrow s_\theta$ and with $\bar{n} = N/A$ (Schweitzer and Holyst 2000):

$$\overline{h}_{+1}^{\,stat} = \frac{s_{+1}}{k_{+1}}\,\overline{n}x_{+1}\,; \quad \overline{h}_{-1}^{\,stat} = \frac{s_{-1}}{k_{-1}}\,\overline{n}\,(1-x_{+1}) \tag{14.5}$$

Let us for the moment assume that the parameters of both field components are identical, i.e. $s_{+1} = s_{-1} \equiv s$, $k_{+1} = k_{-1} \equiv k$, a more complex case will be discussed below. Then, we find for the stationary values of x_θ in the case $\theta = +1$ (Schweitzer and Holyst 2000):

$$(1-x_{+1})\exp\!\left[\kappa\,x_{+1}\right] = x_{+1}\exp\!\left[\kappa\,(1-x_{+1})\right] \tag{14.6}$$

Here, the *bifurcation parameter*

$$\kappa = \frac{2s\overline{n}}{kT} \tag{14.7}$$

includes the specific internal *conditions* within the community, such as the population density, the social climate, the individual strength of opinions, and the life time of the information generated.

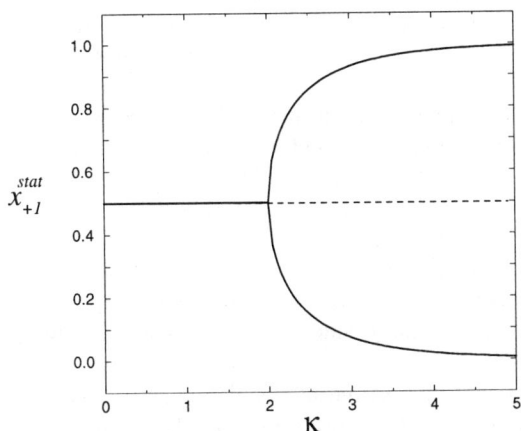

Fig. 14.2   Stationary solutions for x_{+1} (Equation (14.6)) for different values of κ

Schweitzer and Holyst (2000) found that, depending on the value of κ, different stationary values for the fraction of the subpopulations exist (see Fig. 14.2). For $\kappa < 2$, $x_{+1} = 0.5$ is the only stationary solution, which represents a stable community where both opposite opinions have the same influence. However, for κ > 2, the equal distribution of opinions becomes unstable, and a separation process towards a preferred opinion is obtained, where $x_{\pm 1}$ = 0.5 plays the role of a separation line. Two stable solutions are found where both opinions coexist with different shares. This can be seen in Fig. 14.2 where the bifurcation at the critical value κ^c = 2 is clearly visible. Hence, each subpopulation can exist either as a *majority* or as a *minority* within the community. In a deterministic approach, which of these two possible situations emerges depends on the initial fraction of the subpopulation. For initial values of x_{+1} below the separatrix, 0.5, the minority status is most likely to be the stable situation (Schweitzer and Holyst 2000). In the stochastic approach considered here, the emergence of a possible minority/majority relation will also depend on the fluctuation during the early stage of the evolution of the agent system as shown below.

From the condition $\kappa = 2$ we can derive a *critical population size*

$$N^c = k \, A \, T \, / \, s \qquad\qquad (14.8)$$

where for larger populations an equal fraction of opposite opinions is certainly unstable. If we consider a *growing community* with fast communication, then both contradicting opinions are balanced, as long as the population number is small. However, for $N > N^c$, that is after a certain population growth, the community tends towards one of these opinions, thus necessarily separating into a majority and a minority. Which of these opinions dominates depends on small fluctuations in the bifurcation point. Fig. 14.3 shows a particular result obtained from computer simulation of 400 agents who at $t = 0$ are randomly assigned one of the opinions $\{+1, -1\}$.

As indicated in Fig. 14.3, there is a latent period in the beginning *before* the minority/majority relation emerges, i.e. during this period it is not clear which of the two subpopulations will gain the majority status. This initial time lag t^* is needed to establish the communication field which plays the role of an *order parameter* known from synergetics (Haken 1978). Consequently, for $t \geq t^*$, a transition from the unstable equal distribution between both opinions toward a majority/minority relation is clearly visible in Fig. 14.3. The time period required to eventually establish this situation is quite short, since the case discussed here has a very fast exchange of information.

The results of a computer simulation of the relative subpopulation sizes x_{+1} (o) and x_{-1} ($\lozenge$) versus time t for a community of $N = 400$ agents are shown in Fig. 14.3. Parameters values are: $A = 400$, $s = 0.1$, $k = 0.1$, $T = 0.75$, i.e. $\kappa = 2.66$. Initially, each agent has been randomly assigned the opinion +1 or -1. The dashed lines indicate the inital equal distribution ($x_\theta = 0.5$) and the minority and majority sizes ($x_\sigma = \{0.115; 0.885\}$) which follow from Equation (14.6).

Finally, we wish to point out that the symmetry between the two opinions can be broken due to external influences on the agents. Schweitzer and Holyst (2000) have considered two similar cases: (i) the existence of a *strong leader* in the community, who possesses a strength s_l which is much larger than the usual strength s of the other individuals, (ii) the existence of an external field, which may result from government policy, mass media, etc. which support a certain opinion with a strength s_m. The additional influence $s_{ext} := \{s_l /A, \ s_m /A\}$ mainly effects the mean communication field due to an extra contribution, normalised by the system size A. It was found within the mean-field approach that at a critical value of s_{ext}, the possibility of a minority status completely vanishes. Hence, for a certain supercritical external support, the supported subpopulation will grow towards a majority, regardless of its initial population size, with no chance for the opposite opinion to be established. In the real world, this situation quite often occurs in communities with one strong political or religious leader ['fundamentalistic dictatorships'], or in communities driven by external forces, such as financial or military power ['banana republics'].

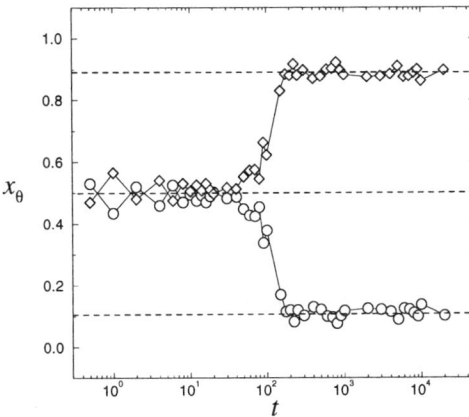

Fig. 14.3   Computer simulation of subpopulation sizes

14.7 Spatial Information Distribution

The previous section has shown within a mean-field approach the emergence of a minority/majority relation in the agents community. With respect to the example of the recycling campaign adressed previously, it means that *either* most of the agents decide to collaborate *or* most of them decide to refuse to collaborate. If we start from an unbiased initial distribution, i.e. an equal distribution between both opinions, then there is no easy way to break the symmetry towards a preferred opinion, except when an external bias is taken into account.

We now investigate a possibility of breaking the symmetry by means of different information distribution. This requires considering the spatial dimension of the system explicitly. Let us start with the previous example of $N = 400$ agents randomly distributed in a system of size A with random initial opinions $+1$ ($\Diamond$) and -1 (o). The snapshots seen in Fig. 14.4 are taken at three different times: (a) $t = 10^0$, (b) $t = 10^2$, (c) $t = 10^4$. For the parameters and initial conditions see Fig. 14.3, additionally $D = 0.06$. Agents receive information about the opinions of other agents by means of the two-component communication field h_θ (r,t), Equation (14.1), which now explicitly considers space and therefore the 'diffusion' of information. The two-dimensional system is treated here as a torus, i.e. we assume periodic boundary conditions.

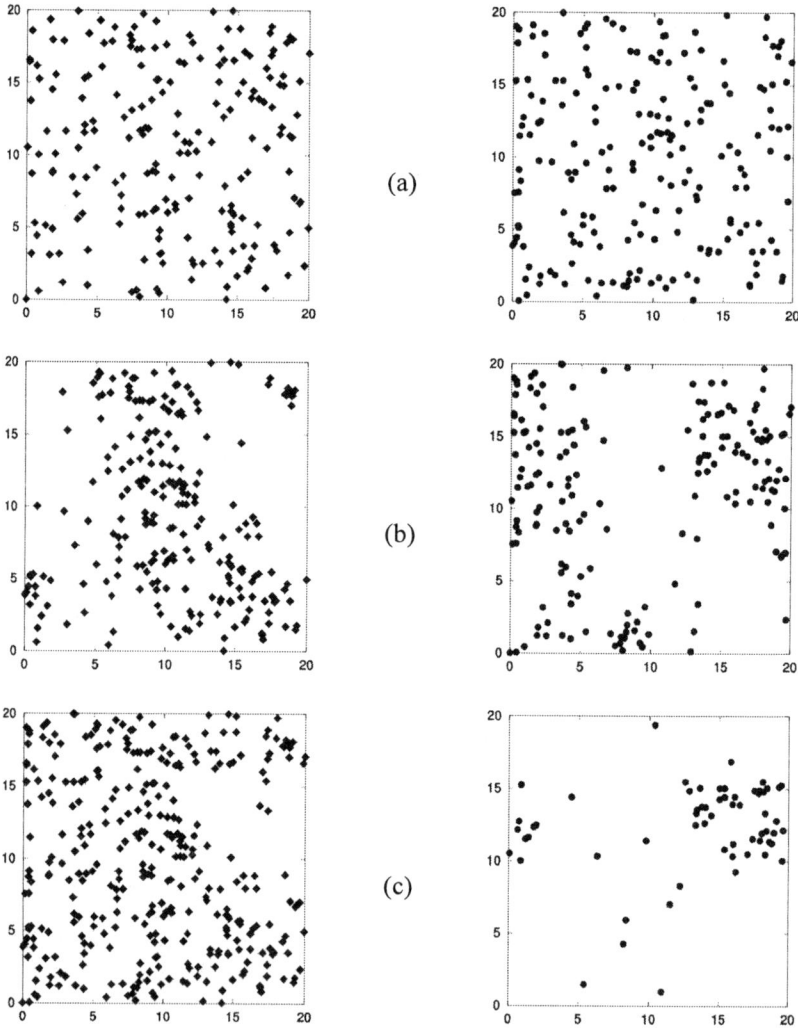

Fig. 14.4   Computer simulations of the spatial distribution of agents

As a first example, we assume that the parameters decribing the communication field are again the same for both components, i.e. $s_{+1} = s_{-1} \equiv s$, $k_{+1} = k_{-1} \equiv k$, $D_{+1} = D_{-1} \equiv D$. While Fig. 14.4 shows three snapshots of the spatial distribution of the agent's opinion, Fig. 14.5 shows the respective evolution of the subpopulation shares. Once again we find the emergence of a majority/ minority relation – this time however, on a larger time scale compared to Fig. 14.3, detemined by the information diffusion, expressed in terms of D. However, the initial latent time lag t^* for the emergence of the majority/minority relation is about the same, which is needed again to establish the communication field.

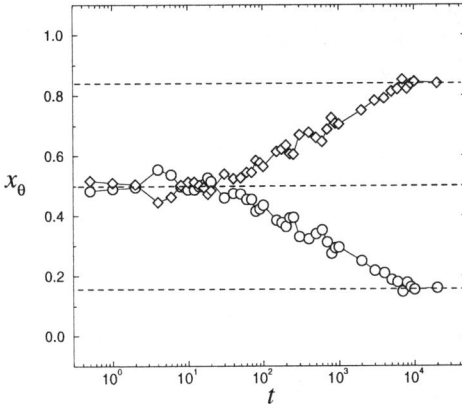

Fig. 14.5   Relative subpopulation sizes x_{+1} (◊) and x_{-1} (o) versus time for computer simulation shown in Fig. 14.4

As illustrated by the snapshots in Fig. 14.4, the minority and majority subpopulations organises itself in space in such a way that both are *separated*. Thus, despite the existence of a global majority, we find regions in the system which are dominated by the minority. From this we can conclude that there exists a *spatial coordination of decisions*, i.e. agents who share the same opinion are spatially concentrated in particular regions. With respect to the example of the recycling campaign, this means that those agents who collaborate [or refuse to collaborate in the opposite case], are mostly found in a spatial domain of a like-minded neighbourhood. This result might recall the famous simulations of segregation in social systems (Schelling 1969; Sakoda 1971; Hegselmann and Flache 1998), however, we should like to point out that in our case the agents do not *migrate* toward supportive places, they *adapt* to the dominant opinion of their neighbourhood.

The spatial distribution of the majority and the minority is also reflected in the different components of the communication field, as shown in Fig. 14.6. We find that the maxima of both components are of about equal value. However, whereas the information generated by the majority is roughly spread over the whole system, the information generated by the minority eventually concentrates only in specific regions dominated by them.

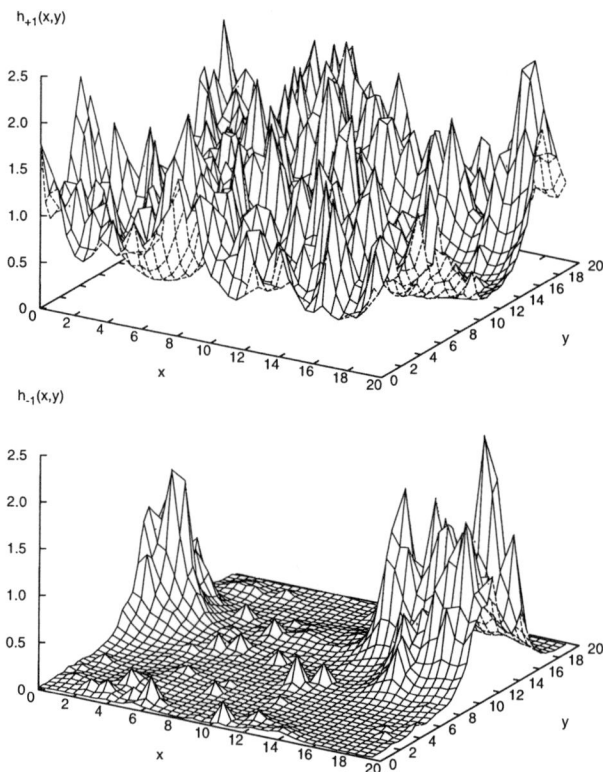

Fig. 14.6   Spatial distribution of the two-component communication field, [top] $h_{+1}(r,t)$, [bottom] $h_{-1}(r,t)$ at time $t = 10^4$, which refers to the spatial agent distribution of Fig. 14.4(c)

So far we have noticed the importance of fluctuations during the initial time lag t^*, that decide which of the two possible opinions will appear as the majority opinion. For the spatial co-ordination of decisions we may now exploit the different properties of the information exchange in the system, expressed in terms of the parameters s_θ, k_θ, D_θ of the communication field. For instance, we may assume that the information generated by one of the subpopulations is distributed *faster* in the system than the information generated by the other one. Alternatively, we may consider different life times for the different components of the communication field. However, in order to model a faster exchange of information, it is not sufficient to simply increase the value of D_θ. We need to consider its effect on the local values of the communication field in more detail. A closer inspection of Equation (14.1) shows that faster communication in the system via faster diffusion of the generated information also lowers the information available at the agent's position (Schweitzer and Zimmermann, 2000). This might be considered as a drawback in modelling information exchange by means of reaction-diffusion equations. Obviously, the field $h_\theta(r,t)$ obeys certain boundary conditions and conservation laws which do not hold for information 'per se'. In particular, the local value of available information is not lowered if this information spreads out faster, but the local value of the 'communication field' obeying Equation (14.1) does.

In order to compensate for the unwanted effect of a local decrease of $h_\theta(r,t)$, we have to choose the parameters s_θ, k_θ, D_θ in such a way that the ratios:

$$\frac{k_\theta}{s_\theta} = \beta \quad \text{and} \quad \frac{D_\theta}{s_\theta} = \gamma \tag{14.9}$$

both need to be constant for the two components $\theta = \{+1, -1\}$. In this case, Equation (14.1) for the dynamics of the multi-component communication field can be rewritten as:

$$\frac{\partial}{\partial \tau} h_\theta(r, \tau) = \sum_{i=1}^{N} \delta_{\theta,\theta_i} \delta(r - r_i) - \beta h_\theta(r, \tau) + \gamma \Delta h_\theta(r, \tau) \tag{14.10}$$

where the time scale τ is now defined as $\tau = t(D_\theta/\gamma)$. If both parameters β and γ are kept constant, Equation (14.10) means that the dynamics of the respective component of the communication field occurs on a different time scale τ, dependent on the value of D_θ. In terms of the blackboard interpretation, this means that the array of blackboards containing the information about a particular opinion θ will be updated more [or less] frequently than the blackboard array representing the opposite opinion. An increase in the diffusion constant D_θ then models the information exchange on a faster time scale, as expected, without affecting the stationary distribution resulting from Equation (14.10).

Computer simulations of the evolution of the subpopulations for the case of different information diffusion are shown in Fig. 14.7. We find once again the emergence of a majority/minority relation – but this time the subpopulation (-1) with the faster diffusing communication field is more likely to become the majority in the system. We have also found that the minority is no longer concentrated in particular regions, but only randomly distributed, so there is no longer a co-ordination of decisions on the side of the minority.

From various runs of the computer simulations we can deduce the following general conclusions regarding the influence of the ratio $d = D_{+1}/D_{-1}$, under the assumption that β and γ are kept constant (see Schweitzer and Zimmermann 2000):

- For $d = 1$, both subpopulations have an equal chance of becoming the majority in the system. With an increasing difference in the values of D_{+1} and D_{-1}, the subpopulation with the faster [more 'efficient'] communication is more likely to become the majority.
- With an increasing difference between D_{+1} and D_{-1}, the initial time lag, when the decision about which subpopulation becomes the majority is still pending, decreases (see Fig. 14.3 and Fig. 14.7). This reduces the likelihood of early fluctuations to break the symmetry towards one of the subpopulations.
- With an increasing difference between D_{+1} and D_{-1}, the size of the respective minority will decrease (this can be also seen by comparing Fig. 14.3 and Fig. 14.7). A smaller minority will also have a smaller chance of organising itself in space, forming regions of co-ordinated decisions. Furthermore, due to the shorter initial time lag, they will also not have the time to establish their own communication field.

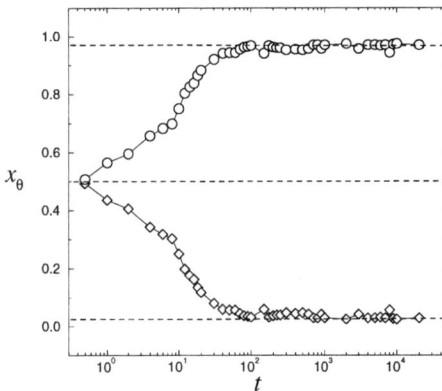

Fig. 14.7   Relative subpopulation sizes x_{+1} ($\Diamond$) and x_{-1} (o) versus time t for the computer simulation with different information diffusion. Parameters: $D_{+1} = 0.02$, $D_{-1} = 0.06$, $\beta = 1$, $\gamma = 0.6$

In order to summarise our simulations, we should like to link the discussion to the different kinds of information introduced in Section 14.4. At the individual or microscopic level, we have the genuine *local* decisions of each agent which result from an interplay between the functional and the structural information. The latter describes the data stored in the different arrays of blackboards, which are modelled here as a two-component communication field $h_d(r, t)$. Functional information refers to the ability of the agent to read data from and to write data on the blackboards or the communication field, respectively. The functional information also considers the contextual situation of the agent, i.e. whether he/she is able to get access to a particular blackboard dependent on his/her current position. In addition, it describes how these data are processed by the agent – in this particular example, how their values are compared.

As the result of this interplay, *pragmatic* information emerges and allows the agent to make the decision whether to collaborate or not. This pragmatic information is individual, since it exists only for a particular agent, at a particular position and a particular time. Already the next time step may change the whole situation: depending on the structural information read, the agent may make a quite different decision. Thus pragmatic information is not an invariant of the dynamics, it has to be consecutively generated by each agent. The local and independent decisions of the different agents are coupled via the communication field, which is generated jointly by the agents, but also feeds back to their decisions. This kind of non-linear feedback between local actions and non-local coupling suddenly results in global repercussions: the random distribution of agents with a particular opinion changes to become an ordered state at the macro scale. Thus, the emergence of a spatial co-ordination of decisions can be regarded as a transition from locally-independent decisions to the globally-coordinated decision of the agents.

14.8 Conclusions

Self-organisation and the emergence of new properties at the collective level play an important role in socio-economic dynamics. Despite this commonly accepted view, a number of conceptual problems associated with defining and simulating complex socio-economic systems still exist. In the first part of this chapter, we have addressed some of these issues. Since complex systems usually consist of a large number of interacting components, multi-agent models can play a valuable role in exploring and simulating their dynamic behaviour. However, the dependence of emergent system properties on specific agent's interactions is sometimes hard to investigate systematically because of the rather complex design of multi-agent systems. Therefore, we have proposed a 'minimalistic' agent approach which focuses only on particular interactions, with no attempt to model a socio-economic system more realistically. The minimalistic agent design can be seen as a stepping stone strategy where more sophistication can be added gradually on the path to a deeper understanding of complex phenomena. At first, the restriction to a simple set of interaction rules necessarily involves reductions, but on the other hand it opens the way to the application of quantitative methods developed in the field of physics for the analysis of interacting systems. Various promising examples for this transfer of methods can be found in the fields of quantitative sociology (Weidlich 1991, 2000; Helbing 1995) and econophysics (Mantegna and Stanley 2000; Schweitzer and Helbing 2000).

In our approach, the basic interaction between the agents can be described as a generalized form of *communication*. Each agent produces/releases structural information [generalized form of 'data'] over the course of time which is stored externally on blackboards. Agents are also able to 'read' the structural information stored, provided they possesses an algorithm denoted as functional information for processing these data. This way, pragmatic or action-relevant information can emerge from the interplay between functional and structural information. That means the agent is enabled to perform a specific task, e.g. to make a decision.

In order to give an example of the minimalistic agent design, in the second part of the chapter we investigated the *spatial coordination of decisions* within a multi-agent system. Each agent at his/her current location has to decide whether he/she wants to collaborate in a campaign or not. Unlike the classical economic approaches, this decision is not based on the calculation of utilities, but simply on the decision of other agents. This raises the problem of communication, i.e. how an agent at a particular location receives information about the decisions of other agents. The spatio-temporal distribution of information in the system is described by means of a two-component communication field which couples the actions of the different agents and in this way plays the role of an order parameter. We find both analytically and by means of stochastic computer simulations, that for some critical parameters, such as the population density a majority/minority relation appears, i.e. a majority of the agents either decides to collaborate or not to collaborate. Furthermore, in relation to the spatial extension of the system, we find that both the majority and the minority organise themselves into particular spatial domains. This means that we find a spatial co-ordination of particular decisions mediated by the communication among the agents.

Both the appearence of the majority/minority and the spatial concentration of these like-minded subpopulations are emergent properties of the multi-agent system. In addition to the spatial self-organisation [emergence of spatial domains], we can observe self-organisation in the state space of possible decisions [majority/minority relation]. This dynamic process can be influenced by means of the communication among the different subpopulations. We found from our computer simulations that the subpopulation with the more efficient communication [i.e. structural information distributed faster, and the blackboard arrays updated more frequently] have a much better chance to become the majority in the system. This also allows an interpretation in the socio-economic context: if the decision between competing opinions about a given subject is still pending and is not determined by the private utilities of agents, the speed of distribution of relevant information may decide the success of a given opinion.

Our study has focused on the communication between agents from a somewhat 'minimalistic' point of view – but it is worth noting that in contrast to other approaches widely used in economics, we have not embarked on common knowledge assumptions or rational decisions. On the contrary, we have emphasised important questions such as the heterogeneous distribution of information, effects of local decisions, or the effectiveness of communication among different subpopulations. Besides the possibility of obtaining quantitative results [such as the critical population size], our minimalistic agent model also provides a better understanding of the process of self-organisation, because the complex dynamics emerge from readily understandable interactions between relatively unsophisticated agents.

15 Agents, Interactions, and Co-Evolutionary Learning

David F. Batten
Temaplan Group, Applied Systems Analysis for Industry and Government

15.1 Introduction

Innovation breeds new products, new processes and new organisational forms. In other words, it breeds changes to the status quo. Yet innovative change is not the sole province of the creative individual. More often than not, fundamental changes to a society or economy result from the collective behaviour of groups of interacting agents. When tacit knowledge is shared, agents can behave in a myriad of different ways. How does this heterogeneous microworld of individual behaviours generate the global macroscopic regularities of society? Much of the creative capacity of such tacit collectives is hidden within a *virtual* system of accumulated interactions. If we choose to isolate the agents, then these virtual parts disappear. If we choose to aggregate the agents, then the virtual parts disappear. It's the virtual parts of an interactive society that we must discover.

To probe this virtual domain, where groups of heterogeneous agents can produce surprising emergent outcomes at the collective level, we must find new ways of doing social science. Traditional, closed-form models fail in such open-ended environments. Because the behaviour of each agent determines the outcome, but the outcome history, in turn, shapes the behaviour of each agent, learning is *co-evolutionary*. If we want to understand how such a self-reinforcing society may evolve over time, we must develop learning tools that are able to produce emergent behaviour. We must give up the analytical [top-down] perspective and turn to synthetic [bottom-up] methods.

One promising synthetic approach is known as *agent-based simulation*. Agent-based simulation does not simplify an economy, but incorporates as much detail as is necessary to produce emergent behaviour. We can view it as a way of doing thought experiments on complex systems. Some pioneering attempts at agent-based modelling in the social sciences were carried out *by hand* a generation ago by Harvard's Thomas Schelling. In many respects, Schelling anticipated various themes that are currently at the forefront of agent-based simulation, social complexity, and economic evolution. For example, he realised that macroscopic order can emerge from situations where numerous agents interact, even with changing patterns of interactions between them. Furthermore, he understood that studies of such situations can quickly get too difficult for explicit mathematical solutions. Regrettably, he lacked the computational capacity to reveal the full consequences of his pioneering ideas.

Today's agent-based experiments *in silico* have three fundamental ingredients: boundedly-rational agents, an environment in which they 'live', and a set of rules. Although these artificial agents are intelligent and adaptive, the processes of inference, learning and discovery they can embrace inside the computer remain rather simplistic. Learning agents can be arbitrarily intelligent, but unless they can know and react to other agents' learning methods, they cannot know if their own learning processes are efficient. Agents can only discover the efficacy of their own learning methods by testing them against others. Thus the main aim of this chapter is to explore the scope for creating an adaptive, open-ended learning environment inside the confines of a computer.

The chapter[1] begins with a short review of Schelling's pioneering efforts without the aid of the computer. This is followed by a small sample of studies of locally interacting agents *in silico*, whose behaviour generators produce novel, large-scale effects. We cannot 'solve' such agent-based simulations, but instead we must 'evolve' them. The collective outcomes are uncertain at the outset, being sensitive to small variations in one or two key parameters. Individual beliefs become endogenous, competing within a larger ecology of all agents' beliefs. Furthermore, this ecology of beliefs co-evolves over time, as agents adapt to the choices of other agents, the collective outcomes and the state of their environment. Thus the path of innovative change is uncertain. The contribution concludes by suggesting that we develop more realistic co-evolutionary learning mechanisms in agent-based experiments that attempt to grow artificial societies. Such mechanisms can be seen as the driving forces behind systems of innovation.

15.2   Schelling's Self-Organising Neighbourhoods

Much of the excitement emanating from the work on complexity is associated with the process of self-organisation. To see this process at work in a socio-economic context, we can turn to the innovative attempts at agent-based modelling carried out a generation ago by Schelling (1969, 1978)[2]. His classic ideas on complexity and self-organisation were summed up in a deceptively simple account of how neighbourhoods in a city could become segregated. In Schelling's model, there are two classes of agents. He thought of them as blacks and whites, but they could be any two classes of individuals that have some difficulty in getting along together – e.g. boys and girls, smokers and non-smokers, butchers and vegetarian restaurants. For our purpose, a chessboard can play the role of a 'simplified city'.

Think of the sixty-four squares as a grid of potential house locations, although the principles hold just as convincingly over much larger [and irregularly-shaped] spatial domains.

Each agent cares about the class of his immediate neighbours, defined as the occupants of the eight abutting squares of the chessboard. Although there are eight squares abutting each square, less than eight of the squares may be filled with neighbours. Preferences are honed more by a fear of being isolated rather than from a liking for neighbours of the same class. It's pretty obvious that such preferences will lead to a segregated city if each agent demands that a majority of his neighbours be the same class as himself. The novelty of Schelling's work was that he showed how much milder preferences, views that seem compatible with an integrated structure, typically lead to a highly segregated city – once the interdependent nature of any changes are taken into account. Consider the following simple rule:

an individual who has one neighbour will only try to move if that neighbour is a different type; one with two neighbours wants at least one of them to be the same type; one with three to five neighbours wants at least two to be his or her type; and one with six to eight neighbours wants at least three of them to be like him or her[3].

At the individual level, this rule of neighbourhood formation is only mildly colour-conscious or culture-sensitive. With such preferences, an integrated residential pattern can be found that satisfies everybody. The familiar checkerboard layout, where most individuals have four neighbours of each class, does the trick – as long as we leave the corners vacant.

Nobody can move in such a layout, except to a corner. There are no other vacant cells. But nobody wants to move anyway. Since it's an integrated equilibrium structure, there's no incentive to change it. But what if a few people are forced to move. For example, what if three neighbours – who work together at a nearby office – are relocated by their company. They must sell up and move to another city. Will the integrated equilibrium remain? Let's try to find out.

After they leave, the neighbourhood layout looks like the board shown in Fig. 15.1. The departing workmates vacated the squares located at coordinates C4, D3 and E2. Once they move out, however, other nearby neighbours of the same type suddenly feel too isolated. For example, residents at D1 and F1 discover that only one of their four neighbours is the same class as them. Thus they decide to move to locations where the neighbourhood rule is satisfied again, say A1 and H8.

A self-reinforcing pattern of interdependencies becomes evident. Another resident can become unhappy because the departing resident tips the balance in his neighbourhood too far against his own class, or because his arrival in a new location tips the balance there too far against agents of the other class. Surprisingly, our integrated equilibrium begins to unravel. An unsatisfied individual at C2 moves to C4, leaving another at G2 with nowhere to go. G2 has no alternative but to move out completely, precipitating a chain reaction of moves in response to his decision. Residents at F3, H3, G4, H5, E4, F5 and G6 all follow suit. Despite the fact that agents only have mild preferences against being too much in the minority, some of them are forced to move out and pockets of segregation begin to appear on our chessboard city (see Fig. 15.2).[4]

There are now 49 agents residing in the city. Let's trigger some more change by removing another nine of them using a random number generator, then picking five empty squares at random and filling them with a new type of agent on a 50/50 basis. The pattern shown in Fig. 15.1 is unstable with respect to some random shuffling, and tends to unravel even further. Fig. 15.3 shows the result after our random number generator has done the job.

Some other residents will now be unhappy with their locations and will move [or move again]. Seemingly simple moves provoke self-reinforcing responses. Thus, a new chain reaction of moves and countermoves is set in motion. To simulate this chain reaction on a computer, the order in which people move, and the way they choose their new location, would need to be specified. As we are doing this by hand on a chessboard, we can watch the structure evolve. When it finally settles down, our randomly generated series of moves leads to the layout in Fig. 15.4.[5]

What a surprise! Even though the individuals in our city are tolerant enough to accept an integrated pattern, they still end up highly segregated. Even though their concerns are local, the whole city gets reorganised into homogeneous residential zones. Surprisingly, short-range interactions can produce large-scale structure. Our chessboard city has engaged in a process of self-organisation. Large-scale order has emerged from a disordered initial state. Segregation may not be our favourite form of order, but it's order nevertheless. It's also meaningful, because all city-dwellers in Fig. 15.4 are now content.

This large-scale order emerges because the integrated pattern (shown in Fig. 15.1) is *unstable*. Scramble it a little and you start a chain reaction of moves that leads to segregation. We could say that you get *order from instability*. This is another of the hallmarks of self-organisation.

Schelling tuned his rules very carefully. He specified that an individual would be satisfied as long as at least 37.5 percent of his or her neighbours were of the same class. If that figure had been a fraction lower, say 33.3 percent, then only two residents in Fig. 15.1 – those located at positions D1 and F1 – would have wanted to move. Once they had moved – say to A1 and H1 – then everyone else in the city would have been satisfied. In other words, the original integrated equilibrium would have remained stable. There's an important message here about emergence. As Holland (1998) has suggested, *emergent properties of agents' interactions are bound up in the selection of rules or mechanisms that specify the model*. In this case, a small change in class consciousness – the migration rule – can result in a large change in the ensuing number of moves. There's a small range over which the degree of segregation is by no means obvious. Once class consciousness gets too strong, however, a highly segregated pattern appears immediately.

The sudden and unexpected appearance of highly segregated areas is indicative of a qualitative change in the aggregate pattern of behaviour. We might say that the location pattern has 'flipped' into a qualitatively different state. In fact this non-linear change is indicative of something like a *phase transition*. At first the integrated equilibrium remains fairly stable to slight increases in class consciousness. Then, rather suddenly, the number of moves skyrockets dramatically. Although we cannot be sure that the whole city ever reaches a state of self-organised criticality, various avalanches of change [in the form of clusters of migration of different sizes] can occur. Global order emerges from the growing scale of local interactions.

	A	B	C	D	E	F	G	H
1		#	O	#	O	#	O	
2	#	O	#	O		O	#	O
3	O	#	O		O	#	O	#
4	#	O		O	#	O	#	O
5	O	#	O	#	O	#	O	#
6	#	O	#	O	#	O	#	O
7	O	#	O	#	O	#	O	#
8		O	#	O	#	O	#	

Fig. 15.1 An almost-integrated pattern of residential location

	A	B	C	D	E	F	G	H
1	#	#	O		O		O	
2	#	O		O		O		O
3	O	#	O		O		O	
4	#	O	#	O		O		O
5	O	#	O	#	O		O	
6	#	O	#	O	#	O		O
7	O	#	O	#	O	#	O	#
8		O	#	O	#	O	#	#

Fig. 15.2 The pattern after the chain reaction of moves

	A	B	C	D	E	F	G	H
1		#	O		O		O	
2	#	O				O	#	O
3	O	#	O				O	
4	#	O		O		O	O	O
5		#	O	#	O		O	#
6	#	O	#	O	#	O		O
7	O		O	#	O	#	O	#
8		O	#		#	O	#	#

Fig. 15.3 Perturbing the pattern even further

	A	B	C	D	E	F	G	H
1	#	#	#	#	O	O	O	
2	#	#	#			O	O	O
3	#	#			O	O	O	
4	#	#	#	O	O	O	O	O
5	#	#	O		O		O	
6	#	O		O				O
7	O		O		O		O	#
8		O				O	#	#

Fig. 15.4 A segregated city

The idea that local interactions can produce global structure – via non-equilibrium phase transitions – came from the pioneering work of some physicists and chemists studying self-organisation in physical systems.[6] Yet Schelling's model permits us to see exactly how the process works in an economy. To some extent, his model oversimplifies reality. The tendency is to divide a city into vast # and 0 areas. What really happens is that the chain reaction of moving households dies out at some point, leaving the city locked into various # and 0 domains of different sizes. The resulting classes of individuals are not simply two-dimensional, they are n-dimensional. So much so that it's sometimes difficult to discern the true class or 'colours' of all your neighbours. Despite these drawbacks, Schelling's insights were well ahead of their time and the rich dynamics contained therein are quite extraordinary.

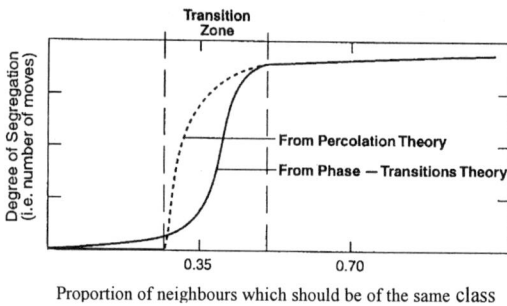

Fig. 15.5 A phase transition on the degree of segregation as class consciousness changes

Such a model helps us to understand why distinctions between weakly-interactive and strongly-interactive patterns of interaction are so important in an economy. The transition zone between these two states provides basic clues about how the behaviour of a group of agents, or even a whole society, may undergo unexpected change. When patterns of interaction between agents become sufficiently dense, just as they did in Schelling's model, a qualitatively different kind of collective behaviour arises. Something unexpected happens.

15.3 Simulation via Cellular Automata

Schelling's chessboard city, together with his rules determining moves, resemble a two-dimensional cellular automaton [CA]. CA were originally put into practice by von Neumann (1966) to mimic the behaviour of complex, spatially extended structures. Because they're really cellular computers, today they're being put to use as simulators, designed to help with time-consuming calculations by taking advantage of fast parallel processing. In this section, we discuss a few examples of agent-based simulation via CA, especially those that have been used to sharpen our intuition about socio-economic behaviour in cities.

Simulation games that deal with urban problems are gaining in popularity. The success of software packages such as SIMCITY are proof of that. In this section, we'll restrict ourselves to a few simulations using cellular automata [CA], because this breed of simulation boasts two advantages over other simulation techniques. First, it's explicitly dynamic. Second, it links macrostructure to microbehaviour. In the same issue of the *Journal of Mathematical Sociology* that published Schelling's famous article, there's a lesser-known article by Sakoda (1971) entitled 'The Checkerboard Model of Social Interaction'. After describing a similar model to Schelling's, Sakoda stresses that the main purpose of cell-based modelling is not a predictive one, but clarification of concepts and 'insight into basic principles of behaviour.' It's these insights that make CA and checkerboard modelling promising when it comes to deepening our primitive understanding of innovation and socio-economic dynamics.

A two-dimensional CA consists of the following: (1) a two-dimensional grid; (2) at each grid site, there's a cell which is in one of a *finite* number of possible states; (3) time advances in *discrete* steps; (4) cells change their states according to *local* rules, so that the state of a cell in the next period depends upon the states of neighbouring cells in past periods; (5) the transition rules are mostly *deterministic*, although non-deterministic rules are also possible; (6) the system is *homogeneous* in the sense that the set of possible states is the same for each cell and the same transition rule applies to each cell; (7) the updating procedure usually consists of applying the transition rule *synchronously* or selecting cells *randomly*.[7]

The number of different transition states in a CA can quickly go through the roof. Consider a two-dimensional CA with just two possible cell states, a neighbourhood of one cell and its four orthogonally adjacent neighbours, and with only the last period having an influence on the next period. In such a seemingly simple case, the number of different transition states is $2^{32} = 4,294,967,296$! No wonder we need a computer to implement a CA-based approach to simulation. The fortunate thing is that the kinds of problems tackled successfully in some of the physical sciences using CA just happen to be amongst the most urgent, unsolved problems in the social sciences. Economics is a case in point. Table 15.1 provides a comparative overview.

Table 15.1   Similarities between CAs and socio-economic dynamics

	Cellular Automata	**Socio-Economic Dynamics**
Basic Elements	Cells are the basic units or 'atoms' of a CA	Individual agents are the basic units of an economy
Possible States	Cells assume one of a set of alternative states	Agents form mental models which enable them to make choices from alternatives
Interdependence	The state of a cell affects the state of its closest neighbours	The choices made by each agent affect the choices of other agents
Applications and Tasks	Modelling the emergence of order, macro outcomes explained by micro rules, and the path dependence of dynamic processes	Important tasks include: understanding the emergence of order, macro to micro relationships, and economic dynamics

Using the simplest CA, it's easy to show that complex global patterns can emerge from the application of local rules. Schelling's patterns of segregation are an example of global emergence, and emergence is one of the things that makes CAs so intriguing. In a world where global outcomes fuse in subtle and diverse ways with local action, CAs look like a methodological paradigm for the 21st century.[8] They're the source of, and inspiration for, major developments in complex adaptive systems. The promising new field of Artificial Life is one of the more obvious examples. The message is that many classes of dynamics can be simulated through CA.

Perhaps the greatest attraction of a CA-based approach to socio-economic dynamics is the equal weight given to the importance of space, time and system attributes. When Sakoda and Schelling published their checkerboard articles, however, they didn't mention the CA concept. Yet CA and socio-dynamic checkerboard models have much in common (as Table 15.1 testifies), like grid structure and local neighbourhoods. Strictly speaking, Schelling's model is not a pure CA, since it allows agents to migrate from one cell to another. Checkerboard models focus primarily on 'sorting and mixing,' i.e. agents searching for and moving to attractive locations in space. They don't just concentrate on cells changing their state at a given site [like CAs], but on changing their site as well. For simulations over geographical space, this point is important. We must distinguish between models that allow individuals to move – *migration* models – and those that do not – *steady site* models (Hegselmann 1996). Moving an agent to an empty cell in his neighbourhood can be treated as the application of a rule by which an ocupied cell and a neighbouring empty one exchange states.

Another important feature of CAs is the definition of neighbourhoods. Two kinds are popular in two-dimensional CA: the *von Neumann* neighbourhood [four neighbouring cells north, south, east and west of the cell in question] and the *Moore* neighbourhood [with the same four cells plus those which are NW, NE, SE and SW]. More distant neighbours may have an influence on state changes, but it's assumed in strict CA that the temporal dynamics will take care of these effects. In other words, growth and decline imply spatial diffusion. Fortunately, a halfway house exists, embracing some CA principles, but also relaxing the neighbourhood definition. These are the so-called cell-space [CS] models introduced by Albin (1975).

Although many CA applications reported in the urban modelling literature – such as the work pioneered by Michael Batty and Roger White – relax the neighbourhood effect to allow for action-at-a-distance, this is not in the spirit of strict CA. Perhaps the most important challenge for this class of simulation model is the specification of the nexus between changes at the physical and the human levels. Two key systems are of interest: (1) the relatively slow developmental changes that take place across networks of housing and infrastructure constructed in cities, and (2) the relatively rapid behavioural changes that agents can implement by altering their own mental models and choices. On the surface at least, one might argue that CA transition rules should be based on agents' local behaviour. However, real urban 'cells' like houses, roads and green areas are more spatial in character, and are governed by a broader set of co-evolutionary forces. The real challenge is to address the tangible and intangible changes interdependently.

Since CA employ repetitive application of fixed rules, we should expect them to generate self-similar patterns. Indeed, many do produce such patterns. If Schelling had used computer simulation to explore a much larger chessboard city, self-similar patterns of segregation may even have been visible in his results. Being akin to periodicity on a logarithmic scale, such self-similar patterns would conform to power laws. The footprints of power laws can be found everywhere. They turn up in the frequency distribution of many catastrophic events – like floods, forest fires and earthquakes. They are also thought to be responsible for the music most listeners like best – a succession of notes that's neither too predictable nor too surprising. In each case, the activity going on is relatively predictable for quite long periods. Suddenly this quiescent state is interrupted by brief and tumultuous periods of major activity, roaming and changing everything along the way. Such punctuations are another hallmark of self-organised criticality.

Recent studies have demonstrated the self-organising properties of some agent-based, urban simulation models. For example, a simple heuristic CA model, called *City*, was developed to study sociospatial segregation in a similar spirit to Schelling's work (see Portugali 2000). *City*'s territory is a two-dimensional square lattice of cells, each of which may be regarded as a house or a *place*. Individuals [persons, families or households] occupy or leave various places, thereby generating the migration dynamics and sociospatial structure of the *City*. Residents and place-hunters base their decisions on preferences about the types of individuals in neighbouring places. Model results display self-organisation, local instabilities, captivity, and other interesting phenomena. Later versions feature two-levels: a population level composed of individuals with cultural and economic properties, and a housing-stock level consisting of a two-dimensional lattice of cells. Immigrants and inhabitants interact with each other and the system of cells [houses], and this interaction gives rise to migration dynamics, changes in the properties of individuals, and changes in the properties of cells.

Another interesting agent-based model goes by the name of *SIMPOP*. It was developed by a French group of geographers, with the aim of unearthing a set of rules which transform systems of cities over time (see Sanders et al. 1997). They experiment with the effects of various hypotheses using a grid of hexagonal cells. Settlements are characterised by types of economic functions. The general evolutionary patterns which emerge from their work are consistent with the author's arguments put forward elsewhere (see, e.g. Batten 2000). For example, the universality of power laws and the rank-size rule is demonstrated under a variety of initial conditions. *SIMPOP* also suggests that transitions between different urban regimes are a necessary characteristic of urban evolution.

One of the strengths of agent-based simulation is its ability to generate simple and complex regimes of behaviour. CA-based systems can simplify and complexify life in its various forms. Cities and their market economies are perfect illustrations of this. Urban economies typically switch from being weakly-interactive to states in which they're *too-strongly-interactive*. If the pendulum swings too far too quickly, collective outcomes can be counterproductive. Stock market booms and crashes are typical examples. So are various forms of congestion and pollution. These types of oscillatory behaviour are important facets of urban and economic evolution, and will be discussed more fully in the next two sections.

15.4 Growing an Innovative Silicon Society

In this section, we discuss an interesting, two-level, silicon world, the joint brainchild of Epstein and Axtell (1996). Their aim was to 'grow' a social order, complete with all sorts of innovative capabilities, from scratch. They did this by creating an ever-changing environment and a set of agents who interact with each other and their environment according to a set of behavioural rules. History is said to be an experience that's only run once. Clearly Epstein and Axtell don't hold with that view. Their idea is that an entire society – like an economy complete with its own production, trade and culture – could be 'recreated' from the inter-actions among the agents. As Epstein suggests: 'You don't solve it, you evolve it.' They call the laboratory in which they conduct their simulation experiments a *CompuTerrarium*, and the landscape which the interacting agents inhabit a *Sugarscape*.[9] Let's take a closer look at how socio-economic life develops in this artificial world.

The action takes place on a grid of fifty-by-fifty cells. But the landscape denoted by this grid is not blank, as it is on a typical CA. On it is scattered this silicon world's only resource: sugar. In order to survive, the entities that inhabit this sweetened landscape must find and eat the sugar. The entities themselves are not just cells that are turned on or off, mimicking life or death. Each is an agent that's imbued with a variety of attributes and abilities. Epstein and Axtell call these *internal states* and *behavioural rules*. Some states are fixed for the agent's life, while others change through interaction with other agents or with the environment. For example, an agent's sex, metabolic rate, and vision are hard-wired for life. But individual preferences, wealth, cultural identity and health can all change as agents move around and interact.

Although every interacting agent appears on the grid as a coloured dot, each may be quite different. Some are far-sighted, spotting sugar from afar. Others are thrifty, burning the sugar they eat so slowly that each meal lasts an eternity. Still others are short-sighted or wasteful. Rapacious consumers eat their sugar too quickly. The obvious advantage of this heterogeneity is that it's capable of mimicking [albeit simplistically] the rich diversity of human populations in terms of their preferences and physiological needs. Any agent that can't find enough sugar to sustain its search must face that ultimate equilibrium state: it simply dies!

Sugarscape resembles a traditional CA in its retention of rules. There are rules of behaviour for the agents and for the environmental sites [i.e. the cells] which they occupy. Rules are kept simple, and may be no more than the commonsense ones for survival and reproduction. For example, a simple movement rule might be: *Look around as far as you can – find the nearest location containing sugar – go there – eat as much as you need to maintain your metabolism – save the rest.* Epstein and Axtell speak of this as an agent-environment rule. A rule for reproduction might be: *Breed only if you've accumulated sufficient energy and sugar.* Also, there are rules governing socio-economic behaviour, such as: *Retain your current cultural identity [e.g. consumer preferences] unless you see that you're surrounded by many agents of a different kind – if you are, change your identity to fit in with your neighbours or try to find a culture like your own.* This rule smacks of Schelling's segregation model, because it highlights co-evolutionary possibilities among nearby neighbours.

The *CompuTerrarium* leaps into action when hundreds of agents are unleashed randomly onto the grid. Coloured dots distinguish agents who can spy sugar easily from more myopic agents. Naturally, all the agents rush towards the sugar. The latter may be piled into two or more huge heaps or scattered more evenly throughout the landscape. Strikingly, many agents tend to 'stick' to their own terrace, adjacent to their 'birthplace.' Because natural selection tends to favour those agents with good eyesight and a low metabolic rate, they survive and prosper at the expense of the short-sighted, rapacious consumers. In short, the ecological principle of carrying capacity quickly becomes evident. Soon the landscape is covered entirely with red dots [high-vision agents].

Even with relatively simple rules, fascinating things happen as soon as the agents begin to interact on the Sugarscape. For example, when seasons are introduced and sugar concentrations change periodically over time, high-vision agents migrate. But low vision, low metabolism agents prefer to hibernate. Agents with low vision and high-metabolism usually die, because they're selected against.

All of the time, the surviving artificial agents are accumulating wealth [i.e. sugar]. Thus there's an emergent wealth distribution on the Sugarscape. Herein lies the first topic of particular interest to us. Will the overall wealth be distributed equally, or will agents self-organise into a Pareto distribution? In other words, will equity prevail or will the ubiquity of power laws prevail again? Although quite symmetrical at the start, the wealth histogram on the Sugarscape ends up highly skewed. Because such skewed distributions turn up under a wide range of agent and environment conditions, they resemble an emergent structure – a stable macroscopic pattern induced by the local interaction of agents. Self-organisation seems to be at work as usual, and the power law prevails once again!

Although these few examples are a useful way of illustrating the variety of artificial life evolvable on the Sugarscape, they hardly herald an impending revolution in our understanding of how an economy changes. For that we must expand the behavioural repertoire of our agents, allowing us to study more complex socio-economic phenomena. Epstein and Axtell (1996, pp. 91-137) have made a start on this expansion. When a second commodity, spice, is added to the landscape, a primitive trading economy emerges. By portraying trade as welfare-improving barter between agents, they implement a trading rule of the form: *Look around for a neighbour with a commodity you desire, bargain with that neighbour until you agree on a mutually acceptable price, then make an exchange if both of you will be better off* (Epstein and Axtell 1996, chapter 4).

Surprisingly, this primitive exchange economy allows us to test the credentials of that classical theory of market behaviour: the efficient market hypothesis. The first stage of the test involves imbuing agents with attributes consistent with neoclassical economic wisdom – homogeneous preferences and infinite lifespans for processing information. Under these conditions, an equilibrium price is approached. But this equilibrium is not the general equilibrium price of neoclassical theory. It's *statistical* in nature. Furthermore, the resulting resource allocations, though locally optimal, don't deliver the expected global optimum.

There remain additional gains from trade that the agents can't extract. What we find is that two competing processes – exchange and production – yield an economy that's perpetually out of equilibrium.

Once we imbue agents with human qualities – like finite lives, the ability to reproduce sexually, and the ability to change preferences, the trading price never settles down to a single level. It keeps swinging between highs and lows, very much like price oscillations in real markets. Basically, it appears to be a random distribution. But it turns out that there's structure after all! Although the seemingly-random price fluctuations continue indefinitely, the fluctuations appear to be variations from an identifiable price level. This particular price just happens to be the same equilibrium level as the one attained under those all-too-unrealistic assumptions underpinning the efficient market hypothesis. Thus we gain the distinct impression that any equilibrium state associated with the efficient market hypothesis is nothing more than a limiting case among a rich panorama of possible states that may arise in the marketplace. If the agents are not textbook agents – if they look a little bit human – there is no reason to assume markets will perform the way economic textbooks tell us they should.

How is the distribution of wealth affected by trade? It turns out that the overall effect of trade is to further skew the Sugarscape's distribution of wealth. By increasing the carrying capacity, and allowing more agents to survive, it also magnifies differences in wealth. Trade increases the interactions between agents, thereby strengthening the power law fit even further. Experiments with a wider set of choice possibilities endorsed the view that it's devilishly difficult to find conditions under which a society's wealth ends up being evenly distributed. We may conclude that there's a definite tradeoff between economic equality and economic performance. This bears a striking qualitative similarity to findings in various economies around the world.

There's so much more one could say about the socio-economic laboratory constructed by Epstein and Axtell. Many other innovative possibilities – such as the emergence of cultural groups, webs of economic intercourse, social clusters, institutional structures, and disease – can all be scrutinized under the Sugarscape microscope. The pair are now working to extend the Sugarscape in order to capture the way of human life in the late twentieth century. Thus far, the agents have sex but there are no families, no cities, no firms and no government. Over the next few years, they hope to produce conditions under which all of these things emerge spontaneously. We can liken their work to a study of innovation from the primitive to the post-industrial. As life on the Sugarscape is in its infancy at present, who's to say what might happen in time?

Sugarscape is an important example of agent-based simulation for several reasons. First, although economists and other social scientists study society, they do so in isolation. Economists, geographers, psychologists, sociologists and archaeologists rarely interact meaningfully or pool the knowledge they've accumulated. Regional scientists are more transdisciplinary, but the current organisation of university departments further endorses these divides. Yet life on the Sugarscape brings all these narrow views together, broadening our understanding in a meaningful way. Second, Sugarscape activities are interactive and dynamic. Thus it's far more process-dependent than classical models. Third,

Sugarscape recognises and preserves differences in culture and skills that human populations exhibit. Finally, for the first time in history, the social sciences have the opportunity to conduct and repeat experiments and test hypotheses to do with unexpected innovations in social and economic behaviour. Sugarscape typifies this new way of doing social science as we enter the unprecedented era of agent-based simulation. If your business is understanding socio-economic behaviour, it's an excellent starting point for rule-based experiments.

There is a rapidly expanding literature in the field of agent-based computational economics. Each simulation attempts to model artificial economic societies in depth. Though space precludes a broader review of this exciting field, most of these simulation models show that complex behaviour need not have complex origins. Some complex outcomes arise when agents are imbued with relatively simple predictors or behaviour generators. Other emergent behaviour is attributable to predictors which differ in terms of the time horizons over which they're applied. Since it's hard to work backwards from a complex outcome to its generator(s), but far simpler to create many different generators and thereby synthesise complex behaviour, a promising approach to the study of self-organisation in complex socio-economic systems is to undertake a general study of the kinds of collective outcomes that can emerge from different sets of predictors as behaviour generators (see Rasmussen and Barrett, 1995; Barrett, Thord and Reidys 1998). As we've already stressed, work of this kind must be done by simulation experiments.

15.5 Co-Evolutionary Learning

The real difficulty with our economy is that each of us is part of the very thing that we're desperately trying to understand. This has the hallmark of a systems problem. But it's not a classical systems problem, like how a clock 'tells the time' or how a car 'moves'.[10] Clocks and cars are structurally complex, but behaviourally simple. Their behavioural simplicity transcends the structural complexity of all their intricate parts. An economy, however, is behaviourally complex. Because some key 'parts' are human agents, they're observers as well as participants, learning from their experiences as well as contributing to the collective outcome. What people believe affects what happens to the economy *and* what happens to the economy affects what people believe. Thus any serious study of emergence must confront *adaptive learning* (see Holland 1998; Batten 2000).

Whenever agents learn from, and react to, the moves of other agents, predicting the collective outcome is only possible if the economy is linear. If learning is only weakly-interactive, the behaviour of the whole economy is just the sum of the behaviour of its constituent parts. For learning to be adaptive, however, the stimulus situations must themselves be steadily evolving rather than merely repeating. This requires strongly-interactive conditions. The existence of a recursive, non-linear feedback loop is the signature of co-evolutionary learning. People learn and adapt in response to their *recent* experiences. Each agent's decision affects other agents, and thus the collective outcome as a whole; and this collective outcome, in turn, influences the agents' future beliefs and decisions. In other words, the behaviour of the whole is greater than the sum of its parts.

Unexpected outcomes trigger avalanches of uncertainty, causing each agent to modify his view of the world. How the world looks to each of us depends on the kind of 'glasses' we're wearing. As Kant suggested, nobody can have certain knowledge of things 'in themselves.' Each of us only knows how things appear to us. If we're only privy to part of the information about an economy, then there are clear limits to what we can know. Each agent's mind sets these limits. When we ask questions about this economy, we're asking about a totality of which we're but a small part. We can never know such an economy completely; nor can we see into the minds of all its agents and their idiosyncrasies.

The key to understanding adaptive behaviour lies with explanation rather than prediction. When economic agents interact, when they must think about what other agents may or may not be thinking, their collective behaviour can take a variety of forms. Sometimes it might look chaotic, sometimes it might appear to be ordered, but more often than not it will lie somewhere in between. At one end of the spectrum, chaotic behaviour would correspond to *rapidly* changing models of other agents' beliefs. If beliefs change too quickly, however, there may be no clear pattern at all. Such a volatile state could simply appear to be random. At the other end of the spectrum, ordered behaviour could emerge, but only if the ocean of beliefs happens to converge onto a mutually consistent set of models of one another.

For most of the time, however, we'd expect mental models of each other's beliefs to lie somewhere in between these two extremes, tending to change, poised ready to unleash avalanches of small and large changes throughout a system of interacting agents. Why should we expect this? Given more data, we would expect each agent to improve his ability to generalise about the other agents' behaviour by constructing *more complex* models of their behaviour. These more complex models would also be more sensitive to small alterations in the other agents' behaviour. Thus as agents develop more complex models to predict better, the co-evolving system of agents tends to be driven away from the ordered regime toward the chaotic regime. Near the chaotic regime, however, such complexity and changeability would leave each agent with very little reliable data about the other agents' behaviour. Thus they would be forced to simplify, to build *less complex* models of the other agents' behaviour. These less complex models are less sensitive to the behaviour of others and live in calmer oceans.

Because it's impossible to formulate a closed-form model under such volatile conditions, traditional economic models fail in this environment. A typical closed-form model of an economy is an attempt to gain understanding of [parts of] that economy through a simplified representation of [parts of] it. If the agents in that economy are adaptive, however, the set of unknown predictors is too large and variable to be simplified in a meaningful way. In John Holland's jargon, the full set of predictors forms an *ecology*. If we want to understand how this ecology might evolve over time, we're forced to resort to simulation experiments.

Simulation doesn't simplify that economy, but incorporates as much detail as is necessary to produce emergent behaviour. There's simply no other way of accommodating such a large, ever-changing population of active predictors.

The defining characteristic of a complex adaptive system [like an economy] is that some of its global behaviours cannot be predicted readily from knowledge of the underlying interactions (Darley 1995). Instead, this kind of behaviour is emergent. An emergent phenomenon is defined as collective behaviour which doesn't seem to have any clear explanation in terms of its microscopic parts. What does emergence tell us? It tells us that an economic system of interacting agents can spontaneously develop collective properties that are not at all obvious from our limited knowledge of each of the agents themselves. These statistical regularities are large-scale features that emerge purely from the microdynamics. They signify order despite change. Sometimes, this order takes the form of self-similarity at different scales.

For emergent socio-economic phenomena, the optimal means of 'modelling' is agent-based simulation (Darley 1995; Batten 2000). Simulation itself is really a blend of modelling and computation. It's basically the art of using computers to calculate the interactions among separate algorithmic representations. In an economy, for example, the algorithmic representations might be agents [firms] engaged in the business of trade between nations. If developed in an appropriate fashion, at the very least a simulation should demonstrate sufficient understanding of the original economic system so as to be able to reproduce its behaviour. But surely if we understand something very well, we shouldn't need to perform such a simulation. Sadly, the complexities within an economy may preclude such a deeper understanding.

Our astonishment at the fact that we seem unable to predict emergent properties doesn't stem from any inability to understand, but from those inherent properties of the system attributable to the accumulation of interactions. As an economic system becomes more emergent, the propagation of information through accumulated interaction will blur the boundaries of any analysis that we try to perform. All useful predictive knowledge is contained in the accumulation of interactions. Thus we must resort to simulation.

15.6 Discovering Artificial Economics

Brian Arthur is a Sante Fe Institute economist who has tried the agent-based simulation approach to financial markets. Together with John Holland, Blake LeBaron, Richard Palmer and Paul Tayler, Arthur created an artificial stock market inside his computer, inhabited by 'investors' who are individual, artificially-intelligent programs that can reason inductively.[11] In this *market-within-a-machine*, artificial investors act like economic statisticians. They're constantly testing and discarding expectational hypotheses of how the market works and which way prices will move. As all of us know, however, there's simply no way any individual investor can tell what tomorrow's prices will be in the stock market.

There are plenty of clues around, of course. For example, a popular guide to the state of prices the next day is the value of tomorrow's stock index in the futures market. If that value is above today's closing value, it means that the bulk of investors expect tomorrow's prices to rise. But there are literally hundreds of different hypotheses about tomorrow's state of play. Here's a couple of other possibilities:

IF today's price is higher than its average in the last 100 days,
THEN predict that tomorrow's price will be three percent higher than today's.
OR
IF today's price breaks the latest trendline upwards,
THEN predict that next week's price will be five percent lower than this week's.

Some investors may keep several models in mind, others may retain only one at a time. In the Prediction Company's artificial stock market, each agent adopts his 'most reliable' model – the one that performs best in the market's current state. Naturally enough, different expectational models may perform better than others under different conditions. Thus investors must retain and adopt a suite of models for their buy and sell decisions. Eventually, however, the poorer performing models are discarded. The silicon agents use a genetic algorithm to produce new forecasting models from time to time.

Learning in this silicon world comes from two sources: discovering 'new' expectational models and identifying the one that performs best from among your current set.[12] Prices form endogenously from the bids and offers of the silicon agents, and thus ultimately from their beliefs. Such expectational models are akin to Pigou's 'changes in men's attitudes of mind,' and display some feedback effects inherent in his theory (see Pigou 1927). For example, if enough traders in the market happen to adopt similar expectational models, positive feedback can turn such models into self-fulfilling prophecies.[13] The agent-based experiments conducted by the Prediction Company have typically involved about 100 artificial investors each armed with 60 expectational models. As this pool of 6,000 expectational models co-evolves over time, expectations turn out to be mutually reinforcing or mutually negating. Temporary price bubbles and crashes arise, of the very kind that Pigou attributed to excesses of human optimism or pessimism. These more volatile states may be attributed to the spontaneous emergence of self-fulfilling prophecies.

A key aspect of agent-based simulations are their internal dynamics. Expectations come and go in an ocean of beliefs that form a co-evolving ecology. How do the beliefs of fundamentalists fare in this silicon world? Do technical trading beliefs ever gain a firm footing? The results so far suggest that both views are upheld, but under different conditions.[14] If a majority of investors believe in the fundamentalist model, the resulting prices will validate it; and deviant predictions that arise by mutation in the population of expectational models will be rendered inaccurate. Thus they can never get a solid foothold in the market. But if the initial expectations happen to be randomly distributed uniformly about the fundamentalist ones, trend-following beliefs that appear by chance have enough density to become self-reinforcing in the ecology of beliefs. Then the use of past prices to forecast future ones becomes an emergent property.

In this mutated regime, no stationary equilibrium seems to be reached. As Epstein and Axtell found with Sugarscape, the market keeps evolving continuously. If initially successful agents are 'frozen' for a while, then injected back into the market much later, they do no better than average. The market seems to be impatient, moving on and discovering new strategies that replace earlier ones. There's no evidence of market 'moods,' but there is evidence of GARCH.[15] The presence of GARCH means that there are periods of persistent high volatility in the price series, followed randomly by periods of persistent low volatility. Such phenomena make no sense under an efficient market hypothesis. But in an evolutionary marketplace, prices might continue in a stable pattern for quite some time, until new expectations are discovered that exploit that pattern. Then there'll be very rapid expectational changes. These transform the market itself, causing avalanches of further change. Once again, there's evidence of punctuated equilibria and self-organised criticality. Perhaps that see-sawing action we observe in markets is symptomatic of a system driving itself to and from the edge of chaos![16]

If it does behave in such a way, this would be further evidence that markets undergo phase transitions. Observable states oscillate between necessity and chance, between the deterministic and the near-chaotic, between the simple and the complex. In summing up, Arthur (1995, p. 25) states: 'We can conclude that given sufficient homogeneity of beliefs, the standard equilibrium of the literature is upheld. The market in a sense in this regime is essentially 'dead.' As the dial of heterogeneity of initial beliefs is turned up, the market undergoes a phase transition and 'comes to life.' It develops a rich psychology and displays phenomena regarded as anomalies in the standard theory but observed in real markets. The inductive, ecology-of-expectations model we have outlined is by definition an *adaptive linear network*.[17] In its heterogeneous mode it displays complex, pattern-forming, non-stationary behaviour. We could therefore rename the two regimes or phases *simple* and *complex*. There's growing evidence suggesting that actual financial markets live within the complex regime.'[18]

It seems that market participants are involved in an incessant game of co-evolutionary learning. Agent-based simulation experiments like the Sante Fe Artificial Stock Market offer a keyhole through which we can gain useful insights into adaptive behaviour. Similar studies by others have also shown that heterogeneous behaviour on the part of participants can provide opportunities for making consistent profits, that participants with stable bankrolls appear to have an advantage over those who don't, and that small perturbations can sometimes drastically alter the behaviour of the participants.[19] As empirical evidence mounts against the view that markets are efficient, new explanatory approaches like that of the adaptive, boundedly-rational investor will gain more credibility. Scaling principles, power laws, and simulation experiments will play an increasingly important part in this new behavioural revolution. Behavioural experiments in such silicon worlds may herald a new kind of economic dynamics, an experimental economics that relies heavily on agent-based simulation. We might even call it *Artificial Economics* (see Batten 2000). This agent-based approach to the social sciences is just beginning to get underway.

15.7 Some Final Thoughts

What are the principal advantages and disadvantages of agent-based simulation? The main advantage is that the system's dynamics is generated by way of the simulation itself. Interactions between agents can accumulate, multiple pathways can be recognised, and emergent properties can be revealed; without making ad hoc assumptions about these properties. Disadvantages are the extremely high computational demands and the fact that it may not always lead to a better understanding of the mechanisms that caused the dynamics. Agent-based simulation reveals, but doesn't necessarily explain, the inherent dynamics. Thus it can help to sharpen our intuition about complex dynamic behaviour.

The time seems ripe for a synthetic approach to economics. Instead of taking economies apart, piece-by-piece, this new field of Artificial Economics must attempt to put them together in a co-evolutionary environment. We might even find that a synthetic approach leads us beyond known phenomena: beyond *economic-life-as-we-know-it* and into the less familiar world of *economic-life-as-it-could-be*. It would address the problem of creating diverse behaviour generators, which is partly psychological and partly computational. Like nature, an economy is fundamentally parallel. Thus we must recapture economic life as if it's *fundamentally and massively parallel*.[20] If our models are to be true to economic life, they must also be highly distributed and massively parallel.

There are some exciting experiments underway which attempt to replicate the rich diversity of socio-economic behaviour inside the computer. A wide range of important collective phenomena can be made to emerge from the spatio-temporal interaction of autonomous adaptive agents operating under simple local rules. Some of these open-ended experiments have been mentioned in this paper. Their common feature is that the main behaviours of interest are properties of the interactions between agents, rather than the agents themselves. When such multifaceted agents are released into an environment in which [and *with* which] they interact, the resulting society will – unavoidably – couple demography, economics, cultural change, conflict, and public policy. All these spheres of *social* life will emerge – and merge – naturally and without top-down specification, from the purely local interactions of the individual agents.[21]

A society depends on non-linear interactions between multifaceted human agents for its very existence. Innovations are contained in these *virtual parts* of a society. If we choose to isolate the agents, then the virtual parts disappear. If we choose to aggregate the agents, then the virtual parts disappear. Bottom-up simulation offers an exciting new way of discovering the virtual parts and exploring innovation. The fixed coefficients of aggregate models become dynamic, emergent entities in bottom-up models. Despite their lack of visibility, it's the virtual parts of an economy that the new 'Artificial Economics' must seek. In this quest, agent-based simulation may be the key experimental tool and the computer can be seen as the essential scientific laboratory.

Endnotes

1   This contribution reviews some of the agent-based approaches discussed in Batten (2000) on a selective basis.

2   Krugman suggests that the first chapter of Schelling's (1978) book is 'surely the best essay on what economic analysis is about, on the nature of economic reasoning, that has ever been written'. (Krugman 1996a, p.16). The two chapters on 'sorting and mixing' provide an excellent, non-mathematical introduction to the idea of self-organisation in cities.

3   An equivalent way of stating this rule is that each individual is satisfied as long as at least three-eighths of his or her neighbours are of his or her class.

4   Because of the limited computational power available a generation ago, Schelling discovered pockets of segregation by moving coins around on top of a table decorated with suitable grid paper. As he notes: 'Some vivid dynamics can be generated by any reader with a half hour to spare, a roll of pennies, a roll of dimes, a tabletop, a large sheet of paper and a spirit of scientific enquiry or, lacking that spirit, a fondness for games.' (see Schelling 1978, p. 147).

5   This layout is one of a number of possibilities, since the order in which individuals move remains unspecified. The final outcome will also be sensitive to the initial conditions (as depicted in Fig. 15.1). As Schelling noted, repeating the experiment several times will produce slightly different configurations, but an emergent pattern of segregation will be obvious each time.

6   The notion of phase transitions has its roots in the physical sciences, but it's relevance to economic evolution has been recognised recently. In the social sciences, phase transitions are difficult to grasp because the qualitative changes are hard to see. Far more transparent is the effect of temperature changes on water. As a liquid, water is a state of matter in which the molecules move in all directions, mostly without recognising each other. When we lower its temperature below freezing point, however, it changes to a crystal lattice – a new solid phase of matter. Suddenly, its properties are no longer identical in all directions. The translational symmetry characterising the liquid has been broken. This type of change is known as an equilibrium phase transition. Recent advances in systems theory, especially studies led by Hermann Haken on the one hand, and Ilya Prigogine and the Brussels School of Thermodynamacists on the other, have discovered a new class of phase transitions – one in which the lowering of temperature is replaced by the progressively intensifying application of non-equilibrium constraints. It's non-equilibrium phase transitions that are associated with synergetics and processes of self-organisation. See, for example, Haken (1978) or Nicolis and Prigogine (1977, 1989).

7   Although John von Neumann and Stanislaw Ulam were the first to introduce the CA concept about fifty years ago, it's pretty safe to say that John Conway popularised the concept through his invention of the game of 'Life'. In Life, cells come alive [i.e. 'turn on'], stay alive [i.e. 'stay on'] or die [i.e. 'turn off'], depending on the states of neighbouring cells. Although Life is the best known CA, it's perhaps the least applicable to real configurations.

8   In their introduction to a special issue of the journal Environment and Planning B devoted to urban systems as CA, Batty, Couclelis and Eichen (1997) make this suggestion. The reader is directed to this issue for an overview of ways in which urban dynamics can be simulated through CA.

9   Readable summaries of this metaphoric world of artificial life can be found in Casti (1997, chapter 4) and Ward (1999, chapter 2).

10 As Cohen and Stewart (1994, p. 169) have noted, 'You can dissect axles and gears out of a car but you will never dissect out a tiny piece of motion'.
11 The Sante Fe Artificial Stock Market has existed in various forms since 1989. Like most artificial markets, it can be modified, tested and studied in a variety of ways. For glimpses into this new silicon world, and its methods of mimicking the marketplace and its gyrations, see Arthur (1995) or Arthur et al. (1997).
12 'New' expectational models are mostly recombinations of existing hypotheses that work better.
13 In a series of interesting studies – typified by De Long et al. (1990) or Farmer (1993) – it has been shown analytically that expectations can be self-fulfilling. Thus we may conclude that positive feedback loops, or Pigovian herd effects, do have a significant role in shaping the market's co-evolutionary patterns.
14 In proposing his general theory of reflexivity, George Soros suggested that 'since far-from-equilibrium conditions arise only intermittently, economic theory is only intermittently false...' There are long fallow periods when the movements in financial markets do not seem to follow a reflexive tune but rather resemble the random walks mandated by the efficient market theory;' (see Soros 1994, p. 9).
15 GARCH = Generalized AutoRegressive Conditional Hederoscedastic behaviour.
16 Peter Allen has pointed out that an adaptive trading strategy is one that can give good results despite the fact that we cannot know the future, because there are different possible futures. When discernable trends become apparent, the strategy must be able to react to this. By taking such actions, however, the strategy will change what subsequently occurs in reality. This co-evolutionary behaviour implies that markets will always drive themselves to the 'edge of predictability;' in other words, to the edge of chaos.
17 For a precise definition of an adaptive linear network, see Holland (1988).
18 See Arthur (1995, p. 25).
19 Such a set of simulation experiments can be found in de la Maza and Yuret (1995).
20 Massively parallel 'architecture' means that living systems consist of many millions of parts, each one of which has its own behavioural repertoire.
21 See Epstein and Axtell (1996, p. 158).

16 Major Actors in the Innovation Diffusion Process

Michael Sonis
Department of Geography, Bar-Ilan University

16.1 Introduction: Innovation Diffusion as a Way to Complexity through the Process of Complication

This chapter provides an explanation of the quintessential role of innovation diffusion as a part of the process of complication, i.e. the deepening of complexity in evolving complex systems. The main feature of the evolution of a complex system is the emergence of new properties which did not exist previously and which add new information to the system. Here we should stress the difference between invention and innovation. While invention involves the appearance of new information, innovation implies the spread of information [new or old] within the system. Innovation diffusion is the mechanism of complication, i.e. of evolution of complexity. The evolution of complexity [complication] of the physical universe included in the past at least three quintessential events: the Big Bang, which flooded the universe with radiation and, a billion years later, the darkening of the firmament, because of the appearance of atoms and the creation of stars, black holes and galaxies. Since then the rate of complication of the physical universe has become very low, while the complication of biological, ecological and especially social reality has continued at an accelerated rate.

The purpose of this contribution is to describe innovation diffusion in socio-ecological and socio-economic systems. In these systems innovation diffusion is generated by the choice of alternative innovations: the innovation is the subject of individual choice within the collective. The innovation diffusion of some social phenomena is a complicated process, whose description and understanding require a unifying multidisciplinary view. The present work is a reappraisal and reinterpretation of numerous papers written by the author during last two decades. In this study different disciplinary approaches have been used to analyse innovation diffusion. The achieve a unifying view means applying the duality principle: a transfer of ideas from the depth of understanding of one approach to the depth of understanding of another. Methodologically this duality means that, despite the different interpretations, the mathematical models generated through the various approaches are analytically similar. Each approach is associated with a different methodological base relating to the behaviour of 'social man'. In this study, the mathematical description of the complex behaviour of 'social man' in choice processes within the collective is based on four different approaches, which give the same mathematical form to the innovation diffusion process in real space-time. These approaches are (i) empirical regularities of the choice process – the S-shaped change in the portion of adopters of alternative innovations; (ii) the first principles of parsimonious human behaviour as collective beings, (iii) the competitive behaviour of social elites in the form of variation principles and (iv) the 'lock in' captivity phenomenon in the behaviour of social elites. These different approaches reflect the behaviour of four types actor involved in the innovation diffusion process, spreading the information within society in space-time.

16.2 Major Actors in the Innovation Diffusion Process

It is assumed in this study that four major types of actor participate in the dynamic process of innovation diffusion: (i) a set of *alternative competitive innovations* representing the emerging properties of new alternatives, spreading within a given territorial unit; (ii) different groups of innovation *adopters* (iii) different groups of choice-makers, *innovators and innovating elites,* i.e. the systems supporting, producing and spreading the innovations; and (iv) *an active territorial environment* adjusting the innovations to the structure of socio-economic hierarchical territorial organisation. We shall now analyse the social content of the behaviour of each group of actors and construct mathematical models to represent their behaviour.

Ecological Approach

Empirical regularities of innovation spread: competition between adoption and non-adoption. Innovations can be considered as separate entities with distinctive behaviour: they spread in space-time within society, competing for a population of adopters, and coexist or disappear; so the ecological analogy with the behaviour of animal associations competing for the same food supply is a meaningful metaphor. An empirically measurable parameter is the *tempo of spread* – the most commonly recognised form for the innovation diffusion process being the logistic S-shaped curve of the innovation adoption share (see Brown 1981). The S-shaped growth of the relative portion $y(t)$ of adopters of the innovation at time t can be presented by the following curve:

$$y(t) = \frac{1}{1 + C e^{-at}}$$ (16.1)

where $C = (1-y_0)/y_0$ is the ratio of adopters to non-adopters at time zero and a is the tempo of innovation spread. This curve describes the S-shaped growth of the portion of adopters of innovation, from the very small share of 'early' adopters to the portion of 'late' adopters, through exponential growth replaced by asymptotic change near the saturation level of adoption. Equation (16.1) is a solution for the Verhulst logistic differential equation (Verhulst 1838):

$$\frac{dy}{dt} = a y (1 - y)$$ (16.2)

which presents a contagious-type of innovation diffusion process. Potential innovation users become adopters as a result of direct or indirect contact with existing adopters. Here $y(1-y)$ *is* proportional to the maximal amount of such contacts, and the parameter a is included to measure the amount of contact leading to actual innovation adoption. So it is possible to reinterpret the parameter a as a measure of the effectiveness of transfer of information about the innovation between the population of adopters and non-adopters (see Casetti 1969; Sonis 1981).

To represent innovation spread as competition between adoption and non-adoption, let us consider the relative portions of adopters and non-adopters $y_1 = y$, $y_2 = 1-y$. Then the Verhulst equation (16.1) will have the following form:

$$\frac{d \ln y_1}{dt} = a y_2; \quad \frac{d \ln y_2}{dt} = a y_1 \quad \text{and} \quad y_1 + y_2 = 1$$ (16.3)

or in vector form:

$$\frac{d}{dt} \ln \begin{pmatrix} y_1 \\ y_2 \end{pmatrix} = \begin{pmatrix} 0 & a \\ -a & 0 \end{pmatrix} \begin{pmatrix} y_1 \\ y_2 \end{pmatrix} \quad \text{and} \quad y_1 + y_2 = 1$$ (16.4)

where the frequency vector describes the fractional share of innovation adopters within the population and the antisymmetric matrix [i.e. the first term on the right hand side of Equation (16.4)] represents the influence of adopters on non-adopters and vice versa. In symbolic form, the system (16.3) can be written as:

$$\frac{d}{dt}\ln Y = A\,Y \tag{16.5}$$

The co-influence parameter a in Equation (16.2) can be interpreted as a value of an antagonistic game between adoption and non-adoption (Sonis 1983a, b). This interpretation implies the competitive exclusion principle: if the adoption is a 'winner' in the antagonistic zero-sum game against non-adoption (i.e. $a > 0$), then in the long run everyone will adopt the innovation. This principle is one of the most important aspects of the behaviour of conservative multi-species ecological systems, which forbids the stable coexistence of two [or more] species with identical habits within an ecological niche with a limited food supply (Hardin 1961).

Multinomial generalisation of the Verhulst equation for the diffusion of competitive innovations – the multinomial Verhulst-Volterra system. The methodological basis for passing from the the case of adoption/non-adoption of a single innovation to the diffusion of a set of innovations is the understanding of innovation as a choice alternative. This means that the innovation diffusion is caused by the adopters, whose choice introduces the innovation to a definite place at a definite moment in time.

In this section we introduce the multinomial generalisation of the Verhulst Equation (16.1) for the case of a set of competitive innovations. The innovations are called a 'set of competitive innovations' if they are mutually exchangeable and mutually exclusive, i.e they have the same functional properties and the adoption of one of them excludes the simultaneous adoption of the other. Under the assumption that there is a random mixing of adopters and a process of information transfer, the diffusion of the set of competitive innovations can be described by the following multinomial generalisation of the Verhulst equation:

$$\frac{dy_i}{dt} = y_i \sum_{j=1}^{n} a_{ij}\, y_i = f_i(y_1, y_2, \ldots, y_n) \qquad \text{for } i = 1,2,\ldots,n \tag{16.6a}$$

$$\sum_{i=1}^{n} y_i = 1 \tag{16.6b}$$

where $y_i(t)$ is the portion of the users of the ith innovation alternative at time t, and a_{ij} are time independent coefficients of the influence of the ith innovation on the adoption of jth innovation. For consideration of the information transfer process generating the system (16.6) see Sonis (1983b). The system (16.6) can be also written in the form (16.5). This system of log-linear equations represents the Volterra relative ecological multi-species competition dynamics with zero 'self-growth' rate and unit 'average weight' of the species [where the condition of relative growth, see equation (16.6b)] replaces the famous Volterra condition of the constant 'value of life', see Volterra 1927, Hofbauer and Sigmund 1988). We will call system (16.6) the multinomial Verhulst-Volterra system.

The possibility of continuing the diffusion process from an arbitrary distribution of adopters between innovations and the conservation condition (16.6b) imply that the interaction matrix $A = (a_{ij})$ is antisymmetric, i.e.

$$a_{ij} + a_{ji} = 0 \quad \text{and} \quad a_{ii} = 0 \tag{16.7}$$

This antisymmetry can be interpreted as meaning that each pair i and j of competitive innovations are participating in a zero-sum antagonistic game with a value a_{ij}, so that the competitive exclusion principle holds: if $a_{ij} > 0$ for some fixed i and for each j, then only the ith innovation will survive in the long run. This will be proved later with the help of Lyapunov method of asymptotical stability.

The system (16.6) includes quadratic non-linearities, so usually it is impossible to find the explicit analytical form of the solution for the system of differential equations (16.6). But, as will be shown below, under the additional assumption of totally antagonistic competitive innovations, explicit solutions can be obtained.

Totally antagonistic competitive innovations and multinomial logistic growth. Competitive innovations are called totally antagonistic innovations if for each closed chain of innovations $i_1, i_2, ..., i_k, i_1$ the sum of the corresponding influence coefficients is equal to zero:

$$a_{i_1 i_2} + a_{i_2 i_3} + \cdots + a_{i_k i_1} = 0 \tag{16.8}$$

This requirement is equivalent to the properties:

$$a_{ij} = 0; \quad a_{ji} = -a_{ij} \quad \text{and} \quad a_{ij} = a_{1j} - a_{1i} \tag{16.9}$$

For the set of totally antagonistic innovations the 'projective transformations':

$$z_1 = \frac{1}{y_1} \quad \text{and} \quad z_i = \frac{y_i}{y_1} \quad \text{for} \quad i = 2, 3, ..., n \tag{16.10}$$

convert the system (16.6) into the linear system:

$$\begin{cases} \dfrac{dz_1}{dt} = -\sum_{j=2}^{n} a_{1j} z_j \quad \text{and} \quad z_1(0) = \dfrac{1}{y_1(0)} \\ \dfrac{dz_i}{dt} = -a_{1i} z_i \quad \text{and} \quad C_i = z_i(0) = \dfrac{y_i(0)}{y_1(0)} \quad \text{for } i = 1, 2, ..., n \\ z_i = 1 + z_2 + z_3 + ... + z_n \end{cases}$$

(16.11)

That has the following solution:

$$\begin{cases} z_i(t) = C e^{-a_{1i}t} & \text{for } i = 1, 2, ..., n \\ z_1 = 1 + z_2 + z_3 + ... + z_n \end{cases}$$

(16.12)

Consequently, we obtain the following explicit solution of the multinomial totally antagonistic Verhulst-Volterra system (16.6):

$$y_1(t) = \left(1 + \sum_{j=2}^{n} C_j e^{-a_{1j}t}\right)^{-1}$$

$$y_i(t) = C_i e^{-a_{1i}t}\left(1 + \sum_{j=2}^{n} C_j e^{-a_{1j}t}\right)^{-1} \quad \text{for } i = 1, 2, ..., n$$

(16.13)

The requirement of totally antagonistic innovation (16.9) gives us a symmetric form of the multinomial logistic growth:

$$y_i(t) = \left(1 + \frac{1}{y_i(0)}\sum_{\substack{j=1 \\ j\neq i}}^{n} y_j(0) e^{-a_{ij}t}\right)^{-1} \quad \text{for } i = 1, 2, ..., n$$

(16.14)

In the case of clusters of totally antagonistic innovation alternatives, these formulae describe the competition between innovating elites, which materialises in the form of long-term 'creative gales of destruction' (Schumpeter 1934, 1939,

1950). Each such competition cycle is based on the exclusion of non-efficient innovation alternatives, and incorporates three types of behaviour: (1) the decline of the cluster of old alternatives, from a stabilised distribution covering all susceptible population, gradually diminishing to the level of preservation niches, or totally disappearing; (2) the growth of the cluster of new innovations, starting from incubator niches and gradually diffusing throughout the susceptible population through competition with other alternatives; and (3) the wave-like growth-decline of the 'satellite' innovation cluster, initially growing within the susceptible population at the expense of the intensified decline of the cluster of previous old alternatives and eventually losing its relative share through internal competition within the cluster and external competition with more efficient innovation alternatives. Moreover, at the end of the competition cycle, a stabilised distribution of innovations within the susceptible population is achieved. The next cycle will start with the emergence of new clusters of innovations, overcoming their threshold barriers and competing with both old and new innovation alternatives. Thus, each competition cycle includes short term and medium term cycles, and the succession of two or more cycles generates long term growth-decline waves (see Sonis 1992a).

Time-dependent totally antagonistic competition and harmonic logistic growth. In the case of time-dependent totally antagonistic competition between innovations, which satisfyies the conditions:

$$a_{ij} = a_{ij}(t), \ a_{ii}(t) = 0; \ a_{ji}(t) = -a_{ij}(t) \text{ and } a_{ij}(t) = a_{1j}(t) - a_{1i}(t) \tag{16.15}$$

analogously to (16.14), the following solution of (16.6) can be explicitly derived:

$$y_i(t) = \left(1 + \frac{1}{y_i(t_0)} \sum_{\substack{j=1 \\ j \neq i}}^{n} y_j(t_0) \exp\left(-\int_{t_0}^{t} a_{ij}(t)\, dt\right) n\right)^{-1} \quad \text{for } i = 1, 2, ..., n \tag{16.16}$$

In the case of one innovation with a time-dependent tempo of spread, Dodd referred to this solution as 'harmonic logistics' (Dodd 1956; also Rapoport 1963).

Equilibria of the multinomial Verhulst–Volterra system. Equilibria of the multinomial Verhulst-Volterra system (16.6) satisfy the following system of algebraic equations:

$$\begin{cases} y_i \sum_{j=1}^{n} a_{ij} y_j = 0 \qquad\qquad \text{for } i = 1, 2, ..., n \\ \sum_{i=1}^{n} y_i = 1 \end{cases} \tag{16.17}$$

The following description holds for all solutions of the system (16.17), i.e all possible equilibria:

(a) competitive exclusion equilibria:

$$e_k = (0, ..., 0, 1, 0, ..., 0)^T \qquad (16.18)$$

and (b) equilibria with r non-zero components $0 \le r \le n$. Without loss of generality one can assume that:

$$y_1, y_2, ..., y_r > 0, y_{r+1} = y_{r+2} = ... = 0 \qquad (16.19)$$

Consider now the matrix A_r that is derived from the interaction matrix $A = (a_{ij})$ by exclusion of the rows and columns $r+1$, $r+2$, ..., n. If det $A_r \ne 0$ then there is no equilibrium of the form (16.19); if det $A_r = 0$ then equilibrium of the form (16.19) exists and its components are:

$$y_s = \frac{A_{rs}}{\sum_{k=1}^{r} A_{rk}} \qquad \text{for } s = 1, 2, ..., r \qquad (16.20)$$

where A_{rs} are the co-factors of the component a_{rs} in the rth row of the matrix A_r.

It is important to stress that the antisymmetry conditions (16.7) imply automatically that for odd r det $A_r = 0$ and for even r the determinant of matrix A_r is positive. Thus, an equilibrium of the form (16.19) can only exists for odd $r = 2p+1$.

Lyapunov local stability analysis of competitive exclusion equilibrium. It is well known (see, for example, Korn and Korn 1961, chapter 9) that for an equilibrium state Y of a system of non-linear differential equations, such as (16.6), to be asymptotically stable, all eigenvalues of the Jacobi matrix J of the linear approximation of system (16.6) at equilibrium Y:

$$J_{|Y} = \left(\frac{df_i}{dy_j}_{|Y} \right) = (J_{ij}) \qquad (16.21)$$

must have negative real parts. In the case of our system (16.6), the components:

$$\frac{df_i}{dy_j} = \begin{cases} a_{ij}y_i & \text{for } i \neq j \\ \sum\limits_{\substack{k=1 \\ k \neq i}}^{n} a_{ik}y_k & \text{for } i = j \end{cases} \qquad (16.22)$$

Therefore, for the competitive exclusion equilibria $Y = e_k$ the components J_{ij} of the Jacobi matrix at equilibria e_k are:

$$J_{ij} = \begin{cases} a_{ik} & i \neq j \\ 0 & i = j \end{cases} \qquad (16.23)$$

and for $i \neq j$

$$J_{ij} = \begin{cases} 0 & i \neq k \\ a_{kj} & i = k \end{cases} \qquad (16.24)$$

Thus,

$$J = J_{|e_k} = \begin{pmatrix} a_{1k} & 0 & \cdots & 0 & \cdots & 0 \\ 0 & a_{2k} & \cdots & 0 & \cdots & 0 \\ \vdots & \vdots & & \vdots & & \vdots \\ a_{k1} & a_{k2} & \cdots & 0 & \cdots & a_{kn} \\ \vdots & \vdots & & \vdots & & \vdots \\ 0 & 0 & \cdots & 0 & \cdots & a_{nk} \end{pmatrix} \qquad (16.25)$$

The eigenvalues of this matrix are the elements of its main diagonal, i.e. the components of the k-th column of the interaction matrix A. So the competitive exclusion equilibrium $Y = e_k$ is asymptotically stable if and only if the k-th column of the interaction matrix A includes only strictly negative non-diagonal terms:

$$a_{ik} < 0 \text{ for } i \neq k; \ i = 1, 2, \dots, n \qquad (16.26)$$

or, because of antisymmetry of the interaction matrix A, the competitive exclusion equilibrium $Y = e_k$ is asymptotically stable if and only if the k-th row of the interaction matrix A includes only strictly positive non-diagonal terms:

$$a_{kj} > 0 \text{ for } j \neq k; \quad j = 1, 2, ..., n \qquad (16.27)$$

The interpretation of the conditions (16.26) is as follows: the game-theoretical interpretation of coefficients a_{ij} as pay-offs in antagonistic zero sum games among pairs of innovations lead to the following extension of the competitive exclusion principle: if the k-th innovation is a winner in all games against other innovations, the state e_k is a stable attractor, representing the final distribution of adopters:

$$\lim_{t \to +\infty} Y = e_k = e_+ \qquad (16.28)$$

The inversion of the time direction can give the initial frequency distribution $e_s = e_-$, which is asymptotically stable iff the elements of the s-th row of the matrix A are strictly negative:

$$a_{sj} < 0 \text{ for } j \neq s; \quad j = 1, 2, ..., n \qquad (16.29)$$

so that

$$\lim_{t \to -\infty} Y = e_s = e_- \qquad (16.30)$$

It is important to stress that, due to the antisymmetry, the interaction matrix A includes not more than one strictly positive or strictly negative row, i.e. if some asymptotically stable competitive exclusion equilibrium exists (for $t \to \pm\infty$), then it must be unique. Let us consider the matrix $sign\, A = (sign\, a_{ij})$ with components:

$$sign\, a_{ij} = \begin{cases} + & a_{ij} > 0 \\ 0 & a_{ij} = 0 \\ - & a_{ij} < 0 \end{cases} \qquad (16.31)$$

Thus, the qualitative picture of asymptotical stability of competitive exclusion equilibrium is:

$$
sign \ A = \begin{pmatrix} 0 & & & & & \\ & \vdots & & & & \\ + & \cdots & + & \cdots & 0 & \cdots & + & \cdots & + \\ & \vdots & & & & \\ - & \cdots & - & \cdots & - & \cdots & 0 & \cdots & - \\ & & & & \cdots & & \\ & & & & 0 & & \end{pmatrix} \quad \begin{aligned} & \lim_{t \to +\infty} Y = e_k = e_+ \quad \textit{final state} \\ \\ & \lim_{t \to -\infty} Y = e_s = e_- \quad \textit{initial state} \end{aligned}
$$

(16.32)

Remark: The periodic motion and the loss of dimensionality of solutions near the repeller Y of the type (16.19) is considered in Sonis (1992a). In this case the trace of the Jacobi matrix $J_{|Y}$ at Y is zero, so the eigenvalues of the Jacobi matrix at Y are zeros or purely imaginary complex numbers.

Linear Markov chain approximations near competitive exclusion equilibrium. As shown above, the approximate linearised Lyapunov differential equation in the vicinity of competitive exclusion equilibrium e_k is:

$$
\frac{d}{dt}(Y - e_k) = J_{|e_k}(Y - e_k)
$$

(16.33)

or, since $J_{|e_k} = 0$,

$$
\frac{dY}{dt} = JY
$$

(16.34)

As we know from the theory of ordinary differential equations (see, for example, Gantmacher 1990, chapter V.6) the solution of (16.34) is given by the formula $Y(t) = e^{Jt}Y(0)$. It is easy to see by induction that, since $a_{kj} = -a_{jk}$ the formula (16.25) gives:

$$
J^m = J^m_{|e_k} = \begin{pmatrix} a^m_{1k} & 0 & \cdots & 0 & \cdots & 0 \\ 0 & a^m_{2k} & \cdots & 0 & \cdots & 0 \\ \vdots & \vdots & & \vdots & & \vdots \\ -a^m_{1k} & -a^m_{2k} & \cdots & 0 & \cdots & -a^m_{nk} \\ \vdots & \vdots & & \vdots & & \vdots \\ 0 & 0 & \cdots & 0 & \cdots & a^m_{nk} \end{pmatrix}
$$

(16.35)

Therefore,

$$
\begin{aligned}
e^{Jt} &= 1 + J + \frac{1}{2} J^2 t^2 + \ldots + \frac{1}{m} J^m t^m + \ldots = \\
&= \begin{pmatrix}
e^{a_1 k t} & 0 & \cdots & 0 & \cdots & 0 \\
0 & e^{a_2 k t} & \cdots & 0 & \cdots & 0 \\
\vdots & \vdots & & \vdots & & \vdots \\
1 - e^{a_1 k t} & 1 - e^{a_2 k t} & \cdots & 1 & \cdots & 1 - e^{a_{nk} t} \\
\vdots & \vdots & & \vdots & & \vdots \\
0 & 0 & \cdots & 0 & \cdots & e^{a_{nk} t}
\end{pmatrix}
\end{aligned}
\tag{16.36}
$$

If the k-th column of the interaction matrix A includes only non-positive elements, then for each i and for $t>0$ the elements $e^{a_{ik} t}$ lie between 0 and 1, so the matrix e^{Jt} is an Markovian matrix. Thus, the approximative linearised Lyapunov differential equation (16.34) in the vicinity of competitive exclusion equilibrium e_k generates a continuous homogeneous Markov chain with a stochastic matrix $(e^{Jt})^T$.

16.3 The First Principle of Individual Choice within the Collective

In this section we consider the demand pull aspect of the innovation diffusion process, representing the choice behaviour of adopters. We recall that the driving force for the spread of innovation is the individual adopter's consideration of innovation as a choice alternative. So we describe the classical models of choice behaviour of economic man [Homo Oeconomicus], political man [Homo Politicus], and contrast these models of choice behaviour with that of social man [Homo Socialis]. We present here a vision of the innovation diffusion as a meso-level collective socio-ecological spatial dynamic choice process, and describe the relationships between the meso- and macro-level of cumulative socio-spatial collective choice behaviour. The conceptual basis of this vision is the gradual transfer from the concepts of Homo Oeconomicus and Homo Politicus to the concept of Homo Socialis.

The choice behaviour of Homo Oeconomicus. Homo Oeconomicus is a totally egoistic, rational omniscient creature who is supposed to accomplish a rational free choice between different innovation alternatives on the basis of the utility maximisation principle (Luce 1959; Dreze 1974). This means that Homo Oeconomicus has full information about the possible alternatives, knows their properties and diffusion capacity, and can evaluate their utilities rationally and precisely. He recognises the universal form of individual utility function, which includes the factors of rational expectations. Various extensions of this concept have been made on grounds such as bounded rationality, satisficing behaviour rules, habitat consumption, choice and search routines, disequilibrium adjustment processes, etc. It could be argued that all of these partially present the behaviour of Homo Socialis.

The choice behaviour of Homo Politicus. Homo Politicus is a rational omniscient creature who is supposed to accomplish a rational choice between different political alternatives on the basis of the maximisation of societal [collective] utility. Homo Politicus recognises all possible political alternatives, knows all the parameters and limitations of the societal utility function and is obliged to maximise it by individual choice.

The choice behaviour of Homo Socialis. Homo Socialis (the notion coined by Perroux 1964, see also Sonis 1992b) is a 'collective' being who cannot exist or survive without or outside society [the collective]. He does not possess full information about all possible innovation alternatives, does not know the utility properties of these innovations and has no knowledge about the form of his own utility function. Information about the innovations and their utility is obtained through the learning process. This learning process includes:

• Imitation of the choice behaviour of other adopters.

• Extraction of information about innovations and their utility through direct and indirect contacts [social interaction] with 'near-peers' [adopters or adoption units nearby who have adopted or rejected an innovation] located within an active, uncertain territorial environment, and through the mass media which presents 'ready-made' opinions and solutions, making the rational evaluation of innovations and their utilities difficult. Each person who deals with a given innovation becomes a 'specialist', whether rational, wise or not, heavily influencing the subjective mental evaluation of marginal spatio-temporal utilities [expectations of gains in the future or in other location]. So in the eyes of Homo Socialis, the tempo of innovation spread and public opinion about innovation are the indicators of the utility or disutility of innovation.

• 'Learning by using' the chosen innovation.

The subjective mental expectations of gains in the future or in other locations represent the main propensity of Homo Socialis towards parsimony of effort and expense. This propensity to thrift replaces for Homo Socialis the utility maximisation principle.

The adopter as a 'collective being' in terms of innovation choice. The central core of the Innovation Diffusion Theory includes the following hypothesis concerning the choice behaviour of Homo Socialis, based on a 'collective consciousness': the choice behaviour of Homo Socialis is a collective meso-level choice behaviour such that the relative changes in choice frequencies depend on the distribution of innovation alternatives between adopters of innovations. This hypothesis expresses essentially the viewpoint of social statistical mechanics by including consideration of the collective conscience of 'human molecules' arising from social interactions and the effects of mass media information.

Let us construct the mathematical form of this first principle of individual choice within the collective (Sonis 1992b). For this purpose, we consider a set of n competitive innovation alternatives, i.e. the full set of mutually exchangeable and mutually exclusive choice alternatives, a multidimensional space R of space-time parameters and the parameters of decision-relevant attributes of choice-makers and innovation choice alternatives. Let the frequency vector:

$$y(r) = \left(y_1(r), y_2(r), ..., y_n(r) \right) \quad \text{for } r \in R \tag{16.37}$$

represent the relative distribution of adopters at each point $r \in R$. The relative change in the frequency component $y_i(r)$ in direction s in R is:

$$\frac{\partial y_i(r)}{\partial s} / y_i(r) = \frac{\partial \ln y_i(r)}{\partial s} \tag{16.38}$$

where $\partial/\partial s$ is the directional derivative in the direction s in space R. The first principle of choice behaviour in the collective, formulated above, can be presented analytically in the form of a system of partial differential equations for all choice alternatives i and all directions s in the anisomorphic space R:

$$\begin{cases} \dfrac{\partial \ln y_i(r)}{\partial s} = f_{si}(r, y(r)) & \text{for } i = 1, 2, ..., n \\ \displaystyle\sum_{i=1}^{n} y_i(r) = 1 & \text{for } r \in R \end{cases} \tag{16.39}$$

The integrability conditions for this system mean that for each two arbitrary directions p and s in R, the following holds:

$$\frac{\partial f_{si}(r, y(r))}{\partial p} = \frac{\partial^2 \ln y_i(r)}{\partial p\, \partial s} = \frac{\partial^2 \ln y_i(r)}{\partial s\, \partial p} = \frac{\partial f_{pi}(r, y(r))}{\partial s} \tag{16.40}$$

As is well known from the theory of functions of many variables the conditions (16.40) mean that for each i there exists the scalar potential $V_i(r)$ such that:

$$\frac{\partial}{\partial s} V_i(r) = f_{si}(r, y(r)) \tag{16.41}$$

After substitution of (16.41) in (16.39), we obtain the multidimensional generalisation of the Verhulst-Volterra system for the space-time spread of totally antagonistic competitive innovations:

$$\begin{cases} \dfrac{\partial y_i(r)}{\partial s} = \sum_{j=1}^{n} y_i \, y_j \dfrac{\partial}{\partial s} \big[V_i(r) - V_j(r) \big] & \text{for } i = 1,2, \, ..., \, n \\[4mm] & \qquad\qquad\qquad\qquad\quad (16.42\text{a}) \\[2mm] & \qquad\qquad\qquad\qquad\quad (16.42\text{b}) \\[2mm] \sum_{i=1}^{n} y_i(r) = 1 & \text{for } r \in R \end{cases}$$

The system of equations (16.42) gives explicitly the description of the sociological process of transfer of individuals from one alternative to other. The products $y_i \, y_j$ represent the frequency of direct and non-direct contacts between adopters if innovation alternatives i and j; the potentials $V_i(r)$ reflecting public opinion' about the utilities of innovation i; and the expressions $a_{si} = \partial/\partial s \, V_i(r)$ present changes in these subjective, mental utilities in the direction s; the expressions:

$$a_{sij} = a_{si} - a_{sj} = \frac{\partial}{\partial s} V_i(r) - \frac{\partial}{\partial s} V_j(r) \qquad\qquad (16.43)$$

represent both the effectiveness of contacts between adopters of innovations i and j and the dynamic marginal utilities from the transfer from i to j, i.e. subjective mental expectations of the future gains from this transfer. The interaction matrix:

$$A = \left(a_{sij} \right) = \left(\frac{\partial}{\partial s} V_i(r) - \frac{\partial}{\partial s} V_j(r) \right) \qquad\qquad (16.44)$$

is antisymmetric and represents a 'totally antagonistic dynamic hypergame' which is the set of antagonistic zero-sum games between all possible pairs of competitive innovations with the value of game:

$$a_{sij} = \frac{\partial}{\partial s} V_i(r) - \frac{\partial}{\partial s} V_j(r) \qquad\qquad (16.45)$$

in each space/time direction s. Let us find an explicit solution for the Verhulst-Volterra multidimensional system of partial differential equations (16.42). Obviously:

$$\frac{\partial \ln(y_i / y_1)}{\partial s} = \frac{\partial}{\partial s} V_i(r) - \frac{\partial}{\partial s} V_1(r) \qquad\qquad (16.46)$$

So we can conclude that:

$$\frac{y_i}{y_1} = \frac{y_i(0)}{y_1(0)} \exp\left(\frac{\partial}{\partial s} V_i(r) - \frac{\partial}{\partial s} V_1(r)\right) \tag{16.47}$$

Using the conservation condition (16.42b), we obtain the following explicit solution of the Verhulst-Volterra multidimensional system of partial differential equations (16.42a):

$$y_i(r) = \frac{\exp[V_i(r) - V_i(0)]}{\sum\limits_{j=1}^{n} \frac{y_j(0)}{y_i(0)} \exp[V_j(r) - V_J(0)]} \qquad \text{for } i = 1, 2, ..., n \tag{16.48}$$

This solution for each i resembles the well-known multinomial Logit model with the deterministic part of the utility $V_i(r) - V_i(0)$ (see, for example, Heckman 1981). This resemblance gives additional support to the interpretation of scalar potentials $V_i(r)$ as utilities representing mental expectations of future gain from the choice of an innovation alternative.

16.4 Innovators and Innovating Elites

In this section we consider the supply push part of the innovation diffusion process, looking at the behaviour of different groups of choice-makers, i.e. innovators and innovating elites, consisting of the various systems supporting, producing and spreading innovations. In Schumpeter's economics of capitalistic development (Schumpeter 1950) innovators appear as entrepreneurs, in the Political Sciences innovators are seen as charismatic political leaders, in show business they are 'stars' and 'superstars', in the Gumilev theory of ethnogenesis, innovators are referred to as 'passionarii' (Gumilev 1994).

Innovators obtain information about the innovations and their utility through the learning process. This learning process includes:
- Imitation of the choice behaviour of other decision makers.
- Research and Development [R&D] which includes: (i) the extraction of information about innovations and their utility from inventors and experts and (ii) interaction with adopters of innovation.
- 'Learning by doing', i.e. the further adaptation of innovation to the desires of adopters.

The essential property of innovators is 'emotional intelligence' (Goleman 1995).

The variational principle of duality between supply push and demand pull: meso-level competition between social elites versus micro-level social contacts. We now show that the generalised Verhulst-Volterra mathematical formalism (16.5) can be presented in mathematical form with the help of the meso-level variational principle of collective choice behaviour which determines the balance between the resulting cumulative social spatio-temporal interactions among the population of adopters susceptible to the innovation and the cumulative equalisation of the power of elites supporting different choice alternatives. This balance governs the dynamic innovation choice process and constitutes the dynamic meso-level counterpart of the micro-economic individual utility maximisation principle.

Let us consider within the framework of a set of n competitive innnovations the cumulative portions of adopters of each innovation alternative at time t:

$$Y_i(t) = \int_0^t y_i(t)\, dt \qquad \text{for } i = 1, 2, ..., n; \ t \in (0,T) \qquad (16.49)$$

The following variational integral can be constructed (see Dendrinos and Sonis 1994):

$$Var\ Int = -\int_0^T \left(2\sum_{i=1}^n y_i \ln y_i + \sum_{i,j=1}^n a_{ij}\, y_i\, Y_j \right) dt \qquad (16.50)$$

This plays the role of a welfare function arising from interaction of adopters of different innovations and governs the dynamics of choice of individuals within society [the collective]. The first variation of this integral vanishes, giving the Euler system of second order differential equations:

$$Y_i''(t)/Y_i'(t) = \sum_{j=1}^n a_{ij}\, Y_j' \qquad \text{for } i = 1,2, ..., n; \ t \in (0, T) \qquad (16.51)$$

which after the substitutions $Y_i = y_i$; $Y_i'' = dy_i/dt$ gives the multinomial Verhulst-Volterra system:

$$
\begin{cases}
\dfrac{dy_i}{dt} = y_i \sum_{j=1}^{n} a_{ij}\, y_j & \text{for } i = 1,2,\,...,\,n;\; t \in (0,T) \\[2em]
\sum_{i=1}^{n} y_i = 1
\end{cases}
\tag{16.52}
$$

Thus, the multinomial Verhulst-Volterra system of innovation diffusion of competitive innovations is a solution for the fixed end points variational problem of stationarity of integral (16.52) The stationary value of this variational integral is

$$
Stationary\ Value\ of\ VarUnt = -\int_{0}^{T}\left(\sum_{i=1}^{n} y_i \ln y_i\right) dt
\tag{16.53}
$$

Thus, this stationary value turns out to be the cumulative entropy [over time T] of the distribution of innovations between adopters. This value can be interpreted as a measure of the results of the process of 'equalisation of the power of influence' of all innovation alternatives over time horizon T. Expressions (16.50) and (16.53) imply that in a real life diffusion process, the 'equalisation of power of influence' of social elites supporting the innovation spread coincides with the cumulative interactions between adopters causing the transfer from innovation to innovation:

$$
-\int_{0}^{T}\left(\sum_{i=1}^{n} y_i \ln y_i\right) dt = \int_{0}^{T}\sum_{i,j=1}^{n} a_{ij} y_j \left[\int_{0}^{t} y_i(\tau)\,d\tau\right] dt
\tag{16.54}
$$

Thus, the 'equalisation of power of influence' of social elites depends on the level of freedom of adopters to move from one alternative to another.

The manipulative power of elites: the Ten Commandments of aggressive intolerance. The influence of powerful elites on individual decision-making depends on the ability of the elite to attract and capture [lock in] the largest possible portion of adopters. Elites will do everything in their power to attract and capture adopters, using social manipulation, and information which distorts reality in the form of poetic history and poetic geography. They have the ability to influence the mass media because they recognise that Homo Socialis has insufficient understanding and the propensity to replace rationality by emotional intelligence. In modern times with the spread of cultural, religious and nationalistic fundamentalism, when the 'Cold War' era of ideological collisions has been replaced by cultural and ethnic clashes (Huntington 1996), there is an urgent need to understand properly the essence of elite behaviour.

As an example of the possible implications of the socio-ecological approach, let us formulise a model of agressive elite behaviour in the polemic form of a list of 'Ten Commandments' of aggressive intolerance. This ladder of increasing intolerance (Sonis 1997) consists of a series of public utterances of various cultural, religious and political leaders. Most of them will be familiar to the public at large:

- I am Us,
- We know the Truth, all the truth, we are speaking truth - therefore we are always right,
- All others are wrong,
- Whoever is not with us is against us, against the truth,
- Whoever is against us is an enemy, the enemy of truth,
- The enemy should be restrained,
- If you do not know the truth, we will teach you; if you do not want to, we will make you,
- Once an enemy - always an enemy,
- The enemy should surrender,
- If the enemy does not surrender, he must be destroyed.

16.5 Active Environment and Socio-Ecological Niches

The active environment is not involved directly in the innovation diffusion process, and does not produce new innovation alternatives, but intervenes indirectly in the spread of innovations. The external intervention of an active environment restricts the innovation choice behaviour of individuals [within the demand pull component] and changes the competitive abilities of elites [within the supply component] through the redistribution of innovation alternatives. An active socio-economic territorial environment smoothes out the extreme action of competitive exclusion of innovations and generates socio-economic territorial frameworks [niches], thereby fostering 'incubators' for new innovating elites and supporting the preservation of existing ones. Thus, the 'creative gales of destruction' are supported by the action of external interventions, such as governmental support of regional development, the implementation of local and global economic policies, fiscal and institutional decentralisation, etc.

The Pearl-Reed Equation

The basic equation representing the spread of an innovation spread from its preservation niche through the whole of the susceptible population, though not penetrating the preservation niche of non-adoption is the well-known Pearl-Reed differential equation (Pearl 1925):

$$\frac{dw}{dt} = \frac{a}{s_+ - s_-}(w - s_-)(s_+ - w)$$
(16.55)

where w is the portion adopters of innovation in an active environment, s_- is the width of the initial adoption niche, i.e. the proportion of adopters captured by the innovation, $1 - s_+$ is the width of the final non-adoption niche, i.e. the proportion of adopters who resist [are immune against] the adoption of the innovation, $s_+ - s_-$ is the share of the susceptible population who adopt the innovation during the innovation spread. The solution of the Pearl-Reed equation has the following form:

$$w(t) = \frac{s_+ + s_- C e^{-at}}{1 + C e^{-at}} = s_- + \frac{s_+ - s_-}{1 + C e^{-at}}$$
(16.56)

The Pearl-Reed equation describes the spread of innovation through the territorial unit. In the beginning of the process, the population is divided into three parts: immediate adopters, immune non-adopters and the susceptible population, who eventually adopt the innovation during the innovation spread. Therefore, at the end of the process, the population of the region will be divided into two parts: adopters and non-adopters.

Introducing the vector W with coordinates $W = (w_1 = w; \ w_2 = 1 - w)$ presenting the distribution of adoption and non-adoption of an innovation in the active environment and putting:

$$s_1 = s_-; \ s_2 = 1 - s_+ \ \text{and} \ s_0 = s_+ - s_-$$
(16.57)

we obtain the solution of the Pearl-Reed Equation (16.56) that can be derived from the logistic growth (16.1) with the help of the Markov matrix:

$$M = \begin{pmatrix} s_o + s_1 & s_1 \\ s_2 & s_o + s_2 \end{pmatrix}$$
(16.58)

So $W = M Y$ which gives the following symbolic matrix form for the Pearl-Reed equation:

$$\frac{d}{dt} \ln (M^{-1} W) = A M^{-1} W$$
(16.59)

presenting the innovation diffusion within an active environment with the help of the antisymmetric interaction matrix A and the Markov matrix M.

Analytical description of the influence of an active environment in the choice process. Since the influence of the active environment on innovation spread results in a redistribution of adopters between the innovation alternatives, the analytical formalism of this redistribution process should express in mathematical form the transfer from one frequency vector to another, i.e. the frequency of choice vector after the intervention (see Sonis 1992c). As the set of all frequency vectors is a multidimensional simplex, the analytical formalism of intervention of active environment in innovation spread will be based on following statement: Let W be a $(n-1)$-dimensional simplex, including all frequency vectors:

$$W = \begin{pmatrix} w_1 \\ w_2 \\ \vdots \\ w_n \end{pmatrix} \quad \text{for} \quad 0 \le w_i \le 1 \quad \text{and} \quad \sum_{i=1}^{n} w_i = 1 \tag{16.60}$$

where **Y** is an arbitrary set in $(m-1)$-dimensional space including vectors:

$$Y = \begin{pmatrix} y_1 \\ y_2 \\ \vdots \\ y_m \end{pmatrix} \quad \text{for } 0 \le y_i \le 1 \quad \text{and} \quad \sum_{i=1}^{m} y_i = 1 \tag{16.61}$$

Therefore each mapping of *Y* into the simplex *W* has the form:

$$w_i = \frac{F_i(Y)}{\sum_{j=1}^{m} F_j(Y)} \quad \text{for } i = 1, 2, ..., n \tag{16.62}$$

where the functions $F_i(Y)$ are positive on the set Y: $F_i Y > 0$, $i = 1, 2, ..., n$. Indeed for each choice of positive on Y functions, the mapping (16.62) transforms Y into W. Further, each mapping of Y into W with the form $w_i = F_i(Y)$, $i = 1, 2, ..., n$ has the properties:

$$F_i(Y) > 0 \quad \text{for} \quad i = 1, 2, ..., n \quad \text{and} \quad \sum_{i=1}^{n} F_i(Y) = 1 \tag{16.63}$$

Thus, each mapping of Y into W has the form (16.62). This statement presents an universal simple analytical procedure for the construction of a multitude of innovation diffusion models that take into consideration frequency redistribution processes, where any initial innovation diffusion model plays the role of a building block. Next, we construct the multinomial generalisation of the Pearl-Reed system, starting from the building block of Verhulst-Volterra system, with the intervention of an active environment, presented with the help of Markov matrix which generalises (16.58).

Multinomial generalisation of the Pearl-Reed differential equation. Let us consider the case where the action of an active environment is represented by the following Markov matrix:

$$M = (m_{ij}) = \begin{pmatrix} s_o + s_1 & s_1 & \cdots & s_1 \\ s_2 & s_o + s_2 & \cdots & s_2 \\ \vdots & \vdots & & \vdots \\ s_n & s_n & \cdots & s_o + s_n \end{pmatrix} \tag{16.64}$$

where $s_1, s_2, ..., s_n$ with $0 \le s_i \le 1$ are the widths of the adoption niches supporting the set of competitive innovations, s_o is the width of the susceptible population, such that:

$$0 \le s_0 \le 1 \quad \text{and} \quad \sum_{i=0}^{n} s_i = 1 \tag{16.65}$$

The rows of this matrix generate the following positive functions:

$$F_i(Y) = \sum_{j=1}^{n} m_{ij} y_j = s_i + s_o y_i \quad \text{for} \quad i = 1, 2, ..., n \tag{16.66}$$

so the transformation (16.62) gives:

$$w_i = s_i + s_o y_i \quad \text{for} \quad i = 1, 2, ..., n \tag{16.67}$$

If the initial building block is the multinomial Verhulst-Volterra system, then the multinomial generalisation of the Pearl-Reed differential equation is:

$$\begin{cases} \left\{ \dfrac{d\ln(w_i - s_i)}{dt} = \dfrac{1}{s_o}\sum_{j=1}^{n} a_{ij}(w_j - s_j) \right. & \text{for } i = 1, 2, ..., n \quad (16.68a) \\[4ex] \sum_{i=1}^{n} w_i = 1 & (16.68b) \end{cases}$$

In the case of totally antagonistic interaction, matrix $A = (a_{ij}) = (a_i - a_j)$, the explicit solution of (16.68) is:

$$w_i(t) = s_i + s_0 / \left(1 + \sum_{\substack{j \neq i \\ j=1}}^{n} \frac{w_j(0) - s_j}{w_i(0) - s_i} \right) \exp((a_j - a_i)t) \qquad \text{for } i = 1, 2, ..., n \quad (16.69)$$

In the general case the matrix M gives the transfer from the multinomial Verhulst-Volterra system (16.5) to the multinomial Pearl-Reed system with the help of direct substitution:

$$W = MY \qquad (16.70)$$

Therefore, the solution of the multinomial Pearl-Reed system describes the following innovation diffusion process: a cluster of new innovations spreading from their territorial 'incubator milieu' [socio-economic niches] on the basis of the efforts of innovating elites within the support system, increases its relative share within the susceptible population in an S-shaped fashion through the gradual exclusion of old alternatives and weaker satellite innovations. When the expansion abilities of this cluster are exhausted, the structural stabilisation of the supply push system prevails, and the innovation cluster became a routine. In the long run, the supply push structure is vulnerable because there is a new cluster of innovations waiting in its incubator niches (Nelson and Winter 1982). When this cluster comes to full maturity, it overcomes the threshold barriers and starts its S-shaped relative growth, accompanied by the S-shaped decline of the old cluster and the growth-decline of the satellite cluster to the level of their preservation niches, or disappearing completely. In this way, short, medium and long-term cycles are generated, and the succesion of two or more cycles results in long term growth-decline waves. Thus, the socio-ecological, socio-spatial aspect of innovation diffusion theory incorporates two basic entities: the internal socio-ecological competition between different clusters of innovating elites, and the external interventions of an active territorial environment on the patterns of innovation spread.

In the case where the bulding block of the analytical redistribution procedure is the space/time multinomial totally antagonistic innovation diffusion, we obtain a dynamic counterpart of the well-known standard Dodit model (Gaudry and Dagenais 1979; Sonis 1984).

The formula (16.69) explains the fact that external territorial interventions generate a process of additional redistribution of adopters, due to the competition between innovating elites in the susceptible population. It is important to underline that the quintessential role of niches has until now been obscured by sociologists and economists.

16.6 Conclusions

The purpose of this contribution was to present a mathematical basis for the description of interconnections between the various forces involved in the innovation diffusion process. These include the connections between the individual and the collective, resulting in competition based on the diffusion abilities of innovations, and between the collective choice of alternative innovations, based on imitation and learning through social interaction. This approach therefore represents a paradigmatic shift from modelling the micro-economic choice behaviour of 'Homo Oeconomicus' to modelling the individual choice behaviour of 'Homo Socialis' within the collective. This means that we have examined the transition from the meso-level to macro-level dynamic choice behaviour regularities. This stands in opposition to the conventional static micro-economic individual utility maximisation viewpoint and the dynamic social statistical dynamics viewpoint, which describes the transition from the micro-level to macro-level choice behaviour and ignores important evolutionary features of individual behaviour within the collective.

The contributions of Schumpeter, Perroux and Volterra are reflected in three main themes of this contribution:

- the presentation of Schumpeter's 'creative gales of destruction' as results of an ecological competition hypergame between innovation alternatives;
- the elaboration of the law of collective choice behaviour as a balance between cumulative adopter interactions and cumulative entropy equalisation of the power of influence abilities of innovation alternatives;
- the analytical presentation of the intervention of an active environment in the form of the redistribution processes superimposed on the collective choice process.

These interventions generate the socio-economic-cultural frameworks [niches] in which the inventions are converted into mature innovation alternatives, and where previous alternatives can be preserved. In this way, the external territorial interventions generate a process of further redistribution of adopters, compressing the competition between innovations into the susceptible population occupying the niche.

Two entities – competition and environment – have been expressed analytically in the form of: i) a dynamic competition hypergame based on an interaction matrix, representing the marginal utilities of transfer from one innovation to other, ii) a Markovian redistribution matrix defining the relative importance of the incubator and preservation niches, and iii) the probabilistic width of the susceptible population. These analytical models which incorporate the competition and redistribution matrices, generate a Schumpeterian wave-like process of territorial innovation diffusion. The dynamic competition hypergame describes Schumpeter's innovation cycle: the cluster of most efficient innovations diffuses within the susceptible population, winning against the cluster of previously stabilised alternatives and new satellite innovation clusters. As a result, short and medium term growth/decline cycles are generated and, furthermore, the succesion of two or more Schumpeterian cycles generates long term innovation waves.

This contribution has focused on the substantive understanding of the demand side of innovation diffusion, choice processes and on a powerful analytical generalisation of the classical Volterra formalism. This latter has been used for the analysis of Schumpeterian innovation diffusion competition. The paper reflects the author's belief that the phenomenon of innovation diffusion is quintessential for further elaboration of the paragmatic transfer from modern complexity theory to its dynamic counterpart – the theory of complication of developing systems.

PART E: Policy Issues

17 Options, Innovation and Metropolitan Development: Novel Insights from Non-Linear Dynamics

Dimitrios Dendrinos
School of Architecture and Urban Design, The University of Kansas

17.1 Background

At the outset, two kinds of comment are needed in order to place metropolitan development, and the public policy-making process tied to it – the subjects proper of this contribution – into their proper foundation and frame of reference, from the viewpoint chosen here to address them, i.e. non-linear dynamics. One comment deals with a basic *paradox* involved in the treatment of metropolitan development and non-linear dynamics; the other is associated with the role of *speculative behaviour* in the course of socio-spatial evolution.

The Paradox

Those involved with the non-linear dynamics type of mathematical modelling, the backbone of complexity theory, seem to operate within an unavoidable paradox: they seek 'truth or truths' through the use of advanced and rigorous analytical methods; at the same time they point to the fact that this quest for 'truth or truths' is elusive, and that a truth might be multifaceted and is quite often extremely difficult to detect and pin down. One might call this exercise a search for 'analytical truth(s).' Non-linear dynamics tend to recognise that there may be not only one but multiple coexisting and coevolving analytical paths present in any context amenable to such analysis. This is, among other factors, because of the simultaneous presence of multiple dynamic equilibria in the relevant phase space, and critical dependence on [possible multiple] initial conditions.

These factors render the analytical search for truth(s) at the least cumbersome and at most intractable. Often, highly inaccurate computer simulations are substituted for this search, which can lead to honest but erroneous conclusions. The realisation that there might be multiple truths which are difficult to analytically pin down is not a message kindly received by many policy-makers who seek fast, simple and easy answers to complex socio-spatial problems.

Public policy-makers may also be faced with conditions where multiple truths may coexist in any metropolitan development context. One might call these 'reality truths' as opposed to the 'analytical truths' of non-linear mathematical models. Nothing guarantees, however, that the possible multiple analytical truths of non-linear dynamics and the multiple reality truths of policy-makers are identical. Seldom do they match. A multiplicity of either model [analytical] or reality truths is of no use to any specific reality truth, or its advocates. All of these truths are vying for the limited resources; competition for dominance in the highly complex ecology of truths is indeed fierce.

And, well beyond the question of multiple analytical or reality truths, public policy-makers are in reality rarely interested in 'truths', given the ambivalence in distinguishing between 'true' or 'false' statements. The need for a vote, a judge or a jury speaks for this basic arbitration. Public policy-makers may have little use, in fact, for those who speak or seek 'the' or 'a' truth. One might say that usually this quest is relegated to or at most sought in religious dogmas.

Thus, what analysts involved in non-linear dynamics actually do might be of little inherent use and value to policy-makers, in the current mode of operation of policy and decision-makers. To seekers of simplistic, linear, unique, robust 'truths' [no matter how ill-conceived], the menu of policy options and their possible effects in the complex dynamics restaurant non-linear analysts operate will have little to offer. There is a clear need, indeed a hope, for a more enlightened and open minded non-linear culture of public policy-making, a culture as yet not present in urban development policy-making circles. The jury is still out as to the feasibility or desirability of such a culture in numerous spatio-temporal contexts. A positive attitude by the public might break the current impasse and paradox experienced in policy circles.

Evolution and Speculation

To those involved in non-linear dynamics and with an interest in both socio-spatial economics and ecological evolution, it is astonishing that the force of speculative behaviour [at both the individual and collective levels] has been largely absent in formulating spatio-temporal models of evolution. Whether these are biological, human or spatial [i.e. urban, regional] models of evolution, in the underlying dynamical foundation usually any reference to speculative behaviour is at best limited, if existent at all.

Economists have extensively considered speculative behaviour in economic growth dynamics, but have delegated this 'speculative' component to a simplistic stochastic consideration – what they have labelled 'rational expectations'. By doing so, they have in effect stripped this powerful force from its basic component, namely the presence of the unknown, and the true uncertainty and risk inherent in any anticipatory decision-making context. It certainly is analytically convenient to define risk probabilistically, as it enables one to treat uncertainty through so-called 'rationality based' assumptions. But it is not satisfactory from a substantive viewpoint to assume that risk is so defined. And, as it turned out, it is also quite costly from a practical viewpoint to make such simplistic assumptions about risk and uncertainty. In this paper some of the significant economic losses that have been incurred partly as a direct result of this simplification [one might characterise it as arrogance in attempting to tame the unknown] will be mentioned, in reference to the use of options by investment firms in the US over the past decade or so.

For this book, which sets out to address the subject of knowledge, to examine how the unknown, expectations, and the presence and treatment of risk and uncertainty is incorporated in an analytical framework seems to be of importance. One forum where the economic treatment of risk occupies a prominent position is that of the pricing of options in options exchanges. This subject is a central topic of this contribution, and we also address here how the pricing of options might be approached in an innovative way. The suggested innovation in the pricing of options can and must be accompanied by concomitant innovations [not addressed here, but only hinted at] in the reforming of options markets and the execution of trades in options within the context of formal markets for metropolitan development options and of the public policy-making processes necessary to regulate and oversee the functioning of these markets.

Our objective here is naturally to seek ways of achieving more efficient and effective development. However, in the complex ecology of policy-making, the factor which determines the prevailing policy at any point in space-time may have less to do with seeking maximum social merit and more with pure chance. It might also have a lot to do with a hybrid between the two extremes, namely an addiction to the past [i.e. historical precedent or what might be called 'path determined' evolution] and imitation of peers or close competitors [i.e. abiding by the principle governing close neighbour behaviour, or what might be called 'space determined evolution']. Any of these forces, whether based on historical determinism or spatial determinism, work counter to the arguments for innovative public policy along the lines suggested in this chapter. Acknowledging them simply points to the high hurdles which have to be overcome in setting up an enlightened public sector. They also point to the reasons why such enlightment has not occurred yet.

17.2 Options and Metropolitan Development

Economic units, be they individual consumers [or aggregations of individuals, such as households], individual producers [firms, or aggregates of firms represented by their respective lobbies, trade or labour unions, etc], developers, or governments always act *speculatively*. For the purposes of this contribution [and without the need to enter here into a full discourse into the various definitions of speculative behaviour, except to note that these definitions vary, at times considerably] speculative behaviour implies that actors act at present in anticipation of [or under expectations about] future conditions. They do so no matter the nature of their action, and irrespective of the source for speculative actions, whether these actions are taken as a consequence of decisions made consciously or unconsciously [i.e. after or without any planning]. Social [individual or collective] agents are speculative beings by their very nature. For observers of and participants in social action this simple observation is taken to be a truism.

By employing the term 'speculation' in a more technical [i.e. in an investment] context, one can argue that social actors, individuals or collectives, act speculatively by, among other things, acquiring and/or *exercising options*. Options are financial [investment] instruments, in the form of legally binding contracts, granting their holder the right [but not the obligation] to acquire a specific amount of a commodity [or equity stake], at a prespecified point in time [referred to as the expiration date of the option], and at a preset price, usually referred to as the *strike* [or the exercise] price. There are many types of options, for example put or call, European or American, etc, but a full discussion of these types of options will not be presented here.

Options contracts themselves command a market price. The manner in which the prices of options contracts is determined at any point in time is a hotly contested issue, to be addressed in this contribution. Contrary to standard conditions, where options prices are set and pegged to the underlying instrument's market price in a trivial manner, here it is argued that options contract prices are also [and should explicitly be] determined by prevailing demand and supply conditions in options contracts markets. Buyers [and thus holders] of such options can acquire them from a variety of suppliers [or issuers of such options] in a formal or informal [i.e. private or public] options market context.

How these options are supplied, and precisely by whom, when and where, are significant issues, particularly when collective actions are involved. These issues, however, will not preoccupy the present analysis or be dealt with at any length; some remarks will be made nonetheless in reference to the factors leading the options markets to fail in obtaining economically efficient outcomes. The analysis will mostly focus on the *demand* side of speculative behaviour, and those demand related factors affecting the price of options contracts in options markets.

A Short Note on Speculation

In this contribution, individuals, as buyers, always act speculatively at the moment of purchase. When they commit a particular quantity of resources in the *acquisition* of a commodity product or service at a specific point in time, they expect or speculate [that is they reason] as follows: over the expected lifespan of the commodity, product or service [the perceived time horizon of consumption at the exact point in time of purchasing the commodity, product or service], the acquisition will result in a currently discounted utility level at least as high as any other possible purchase. Putting it differently, the currently discounted expected utility of committing a specific amount of currently available resources to any currently perceived available alternative - its opportunity cost - is lower than the utility of what is acquired. Exactly the same rationale applies in *consuming* this particular commodity, product or service: the time spent in such consumption is optimally used, i.e. it results in a higher utility level than spending it or consuming anything else.

Producers, as sellers, similarly act speculatively during the *production* process and at the point of *sale*. Time and other resources spent in producing a particular commodity, product or service by a specific producer is perceived as resulting in the highest current profit for the producer, and so is the price at which the commodity, product or service is sold. All these perceptions of course, based on partial and at times erroneous information, give rise to mere *expectations* on the part of a producer that may or may not materialise, or even prove to be wrong. The rationale is similar to the extent that options on commodities, products or services are concerned.

This first note on speculation is offered in order to clearly outline the manner in which options and their markets are treated here, as opposed to the view found in the dominant economic literature, associated with the pricing of options according to the Black-Scholes-Merton scheme, which will be presented later.

A Short Note on Development

Property developers [public or private] are special cases of producers, as they give rise to and supply or at times fill whole metropolitan [spatial] 'patches', where a variety of urban activities occur. Their actions are prima facia examples of speculative behaviour. Metropolitan development markets are excellent forums for studying, experimenting with and analysing development as a speculative activity; at the same time, metropolitan development markets are glaring cases for observing and recording severe market imperfections, imperfections in development and in their markets. Indeed, few other markets are more apt for exploring the use of innovative investment instruments [such as options], as the urban development markets are real market hotbeds. Similar to traditional stock or commodities markets, metropolitan development markets [which include, but are not limited to, real estate markets] are very sophisticated, complex and path breaking for the knowledge, methods and instruments they use. Although extensive, even global, real estate markets are well known for the multiple failures they are beset with, just as urban policy and plan markets are characterised by serious and pervasive imperfections and shortcomings.

Urban plans and policies are a few of the activities usually delegated to local governments, and increasingly to quasi private/public non-governmental organisations. Governments, along with individuals, are generally thought of as agents of speculative behaviour, although numerous government analysts insist that governments do not and should not act speculatively, no matter the definition of speculation they employ. An increasing number of government analysts believe that governments have not acted speculatively enough, and that they often shy away from necessary and desirable risk-taking behaviour. No matter what the ideological position one may wish to adopt in this debate, one thing is clear: governments have escaped close scrutiny for their speculative actions largely because of [but not exclusively due to] the monopoly or monopsony powers they enjoy.

Governments are involved in a variety of activities including, but not limited to, administration, management, plan-making, policy-making, regulation, legislation, arbitration, budgeting, public investment and the implementation of laws, rules, regulations, government directives and orders, etc. In so doing, it is assumed here, they act speculatively to the extent that they expect their actions to be to the public's [and/or their own] net current and future benefit.

Consumption, production, development, and governing activities are the end result of a process which commences well before the actual consumption, production, development or government. Agents act only when they decide to exercise a particular option they hold for action, among possibly many options they perceive as possessing or having acquired. In this simple fact rests a very important element of social, individual and collective, dynamic behaviour that has gone largely unnoticed in the broader social sciences and especially in the metropolitan development literature. The present contribution is devoted to filling this gap.

Thus, to sum up, all acting agents act speculatively in either private or public or quasi-public options markets, i.e. options markets which are either formal and explicit or informal and implicit. There are shades in between these two extremes, possibly encompassing all real markets, containing different proportions of formality/informality.

To the extent that metropolitan development markets are concerned, and as a result of the existence of multiple options markets, a metropolitan areas resources [land, labour, natural resources - air and water included, and amenities in the form of topographical features, etc] acquire value and associated economic rents. Within a relative dynamic inter and/or intra-metropolitan context, the options' prices of all these urban components guide [and signal] urban development. Such is the premise of this contribution.

17.3 Options and Their Prices

The history of options, and their prices, is a rather recent and colourful story. As explained in the book by Bodie, Kane and Marcus (1996), it started in the early 1970s, with the introduction of options contracts on commodities, stocks and bonds, as well as on market indices in the US. Since then, derivatives markets have proliferated the world over, covering an immense number of financial instruments. The classical Black-Scholes formula was used as the basis for the pricing of options, based on work by Fisher Black and Myron Scholes done in 1973 (see Bodie and Merton, 2000).

As options prices were directly linked to the prices of the underlying instrument, issues of 'linkage' began to appear. Concern was raised over the years regarding the role of options markets upon the prices of the underlying primary instruments in regular exchanges. Specifically, the extent to which the behaviour of the derivatives markets was occasionally having severe effects on the stability of the regular markets was becoming a serious concern. The Stock Market Crash of 1987 rekindled this controversy.

In the late 1980s and particularly during the 1990s, the derivatives markets developed a more eventful and informative life, when the so-called Hedge Funds were created. These funds attempted to take advantage of the manner in which risk, uncertainty, hedging and market equilibrium were treated within the derivatives markets with reference to the markets of the underlying instrument. Large sums of money were devoted to the application of the Black-Scholes formula in investment decisions.

With the work by R.C. Merton [who used the so called Ito calculus to modify the original Black-Scholes method for pricing options], and under the new slogan of 'dynamic hedging', novel and ambitious efforts were placed in redefining risk within derivatives and accordingly in refining the original Black-Scholes formula. Market efficiency conditions under efficient and rational speculation and associated market equilibrium notions were defined by employing the work by Lucas (1981), among others, which gave financiers new hope in actually being able to apply these notions in the real world's market places.

What ensued was financial ruin for the largest hedge fund of all! The US federal government was called in to bail them out in the midst of the Asian financial crisis of 1998. The story was even chronicled in a popular television series in the US: one of these was the programme NOVA: The Trillion Dollar Bet, broadcast on February 8, 2000, by the Corporation for Public Broadcasting in the US [web site: www.pbs.org]. The programme told the brief history of Long Term Capital Management, a Wall Street investment firm involved in hedge funds, headed by John Meryweather, and in which Myron Scholes and Robert Merton [co-winners of the 1997 Nobel Prize for Economics] held a financial interest and actively participated in the management. The hedge fund defaulted after extensively speculating on a host of financial instruments spread over the globe.

Fundamentally, the theory of dynamic hedging, as suggested and applied in real markets by Scholes and Merton, calls for the risk factor to be totally removed from the price of options. Risk only enters the equation indirectly, through the price of the underlying instrument, and by the constancy of the interest rate used over the time horizon [all of these variables being explicitly present in the equation giving the options' price]. Thus, the equation for the price of an option deterministically and trivially pegs the behaviour of the option's price to the current price of the underlying instrument. And, of course, in the old economic tradition, it is assumed that all markets are always in [a state of unique] equilibrium and, most importantly, that the conditions for efficient speculation characterise all actors who are individually and collectively engaged in most of the financial markets in the long haul.

Specifically, the generalised form of the Black-Scholes-Merton formula for options prices is as follows:

$$C = F_1\{S\} - F_2\{L, e^{-rT}\} \tag{17.1}$$

where the price of a [call] option, C, is a function of the underlying instruments [stock] current market price, S, and a function of the stock's exercise [strike] price, L, the [constant] interest rate, r, and the time horizon [maturity, or expiration date for the option], T. The formula is expressed in dynamic terms so that if at any time period $S>>L$, then the formula reduces to a simple difference, where $C = S - L \ exp(-rT)$; and when $S<<L$, the formula is reduced to $C = 0$ [i.e. the collapse of the option's price]. For a fuller discussion, see the text by Bodie and Merton (2000).

The key issue is that risk is not explicitly present in the above formula, and by acquiring call options one is capable of simply incurring the benefit of the unlimited upside potential of movement in a stock's price, without taking the risk of a counter motion. What is astonishing in the above formula is the absence of any market conditions [i.e. demand and supply conditions for the particular option] in the shaping of the price of a stock's option. In effect, although market [demand and supply] conditions do shape the current price of a stock, such elementary economic consideration is totally absent from the formula setting the price of an option on that stock. Further, although it is now beyond any doubt [given the trillions of dollars worth of options traded in the various world options exchanges] that the behaviour of options markets have in turn a considerable effect upon the behaviour of the underlying instrument [stock] price, this interdependency too is totally absent from the above formula, as the price of the underlying instrument is supposed to operate under dynamic conditions independent of the formula C.

It could be argued that the Black-Scholes-Merton formula may work well very near to or at market equilibrium in the price of the underlying instrument, and for short time periods. In other words, when there is a relatively low degree of volatility in the price of the stock and when the time horizon is rather short.

One must acknowledge that in the price of a stock, commodity, product, service, etc., the impact of factors associated with their expected life time may be discounted, whereas in the price of a specific option, all factors associated with the specific time horizon and strike price of the option must be accounted for, nothing more and nothing less.

Suggested Options Prices

The main topic of this section is the factors involved in the formation of market prices for metropolitan development related options. First, a description of the various components entering an unspecified equation of options' prices is provided, followed by a discussion of the suggested equation. Then, a brief outline of certain dynamic aspects of this equation is given.

Notation

a. *Time periods.* Three time periods of different lengths are explicitly recognised: T, the maturity date [i.e. the expiration date of the options contract], which we use to designate a [relatively long] time horizon, for instance one year. This identifies the *expiration date* of an option, a date which is found in all options guiding human action. A second time period, shorter than T, is designated as $t = 1, 2, ..., T$, and picks out time periods of the length of one day. Within each t there are even shorter time periods which are designated as τ [for example the length of one hour] such that $\tau = 1, 2, ..., t$. Thus, in the determination of options prices, one can consider time periods ranging from macro time T, to meso time t, and micro time τ.

b. *Variables.* Three variables, all functions of either or both t and τ, are considered. $P(\tau)$ designates the current, or *spot price*, of the base instrument [for example, land. The contract price per unit of the derivative instrument [i.e. the price of one options' contract on land] with strike price π, and expiration date T is designated as $p_{\pi T}(\tau, t)$. This varies during each micro time period, as does P. Finally, the excess supply of contracts for the derivative instrument [option] is designated by $E_{\pi T}(\tau\text{-}1, t)$, with strike price π, and expiration date T, at micro time period τ-1 of meso time period t. This implies that there is a time delay linking the price of the options contract at τ, which is a reaction to the previous micro time period's excess supply conditions. Although one micro time period delay is suggested, nothing precludes a different reaction time. This relaxation does not significantly affect the analysis here, but it might imply considerable computational and dynamical complexities.

c. *Parameters.* Four parameters are involved in the options' price equation: first, the strike price of the base instrument, π, at the expiration date T; in effect, the strike price π is a vector of prices π, although this is not of much interest here. Second, the prevailing interest rate i [assumed to be constant throughout the macro time horizon T]. This constancy during the long term time horizon is indeed a very strong assumption. Third, a parameter depicting the sensitivity of the market to the current 'spread' between the spot price of the base instrument and its strike price in the option, to be designated as λ. And fourth, a constant associated with initial conditions of the options' market in question, designated as c.

The Options Price Function

It is proposed here that any call option, whether on a basic instrument related to metropolitan development or on a commodity, product or service commonly traded in a stock, commodity or service market, have a market price given by the following general dynamic equation [or any variation of it]:

$$p_{\pi T}(\tau, t) = c \; exp\{i[T-t] + \lambda[P(\tau) - \pi]\} - E_{\pi T}(\tau - 1, t) \qquad (17.2)$$

which is an expression containing two fundamental components. The first component captures the 'opportunity cost of time', while the second contains the 'excess supply' conditions for the specific option.

In more detail, the *price* component related to the opportunity cost of time [or any variation of it]:

$$c \, exp\{i[T-t] + \lambda[P(\tau) - \pi]\} \tag{17.3}$$

contains the 'price of time' term, $i(T-t)$, and the 'spread' term, $\lambda[P(\tau) - \pi]$. Being a call option, it is implied that $[P(\tau) - \pi]$ is greater or equal to zero near the expiration date, T, for $p>0$ and collapses to zero when $P<\pi$, near or at equilibrium. The price of time simply discounts the remaining time period at the price of the prevailing market rate of interest, i; whereas the spread between the current market price of the basic instrument, $P(\tau)$, and the strike price, π, represents the current opportunity for temporal arbitrage. These terms are quite similar [although not identical] to the two terms of the Black-Scholes-Merton formula for options prices.

Such a formula captures the danger deriving from either over- or under-reaction, with a speed picked up by the parameter λ, the collective market sensitivity to the spread. Over-reaction is depicted by a relatively high value for the parameter λ, whereas under-reaction implies a relatively low value for the parameter. There is nothing in the Black-Scholes-Merton formula relating to these conditions.

The second term, is the *volume* component [or any variation of it]:

$$E_{\pi T}(\tau - 1, t) \tag{17.4}$$

and identifies the specific excess supply conditions associated with the particular options contract in hand at the *immediately prior* micro time period τ-1. It is a *volume* component, in the sense that its specifications are tied to the open interest of the specific derivative instrument and its market.

Of special interest is the condition under which an option's price collapses even under conditions far from equilibrium. This event occurs analytically, when:

$$p_{\pi T}(\tau, t) = 0 \tag{17.5}$$

which occurs when the opportunity cost of time equals the excess supply conditions for the specific option in hand:

$$c \, exp\{i[T-t] + \lambda[P(\tau) - \pi]\} = E_{\pi T}(\tau - 1, t) \tag{17.6}$$

or

$$i\,[T - t'] + \lambda\,[P(\tau') - \pi] = k\,ln\,E_{\pi T}(\tau'{-}1, t') \tag{17.7}$$

where, $k{=}1/c$. Further, one has that the spread during the 'collapse' meso [and thus also micro] time period, t', could be positive:

$$P(\tau') - \pi = [\{k\,ln\,E_{\pi T}(\tau' - 1, t') - i(T - t')\}/\lambda] \geq 0. \tag{17.8}$$

If excess supply, E, is also a function of $P(\tau)$, as is quite likely in real markets, then condition (17.8) could have multiple [admissible] solutions. In other words, for a number of the underlying primary instrument current prices, the options' contract price could collapse during any micro, meso or macro time periods. This condition points to inherent dynamic instability issues found in options markets and their linkages to spot markets of the underlying primary instrument, a subject which we turn to in the analysis below. It should be noted that when excess supply E is zero, then the price of the option P is directly equivalent to the Black-Scholes-Merton formula.

Dynamics

A closer look at the specifications of the options price function reveals that the possibility for severe dynamical instability is quite strong within these derivative markets. We list below four conditions under which dynamic instability can endogenously appear within these options markets. It should be noted that besides the oscillations which could result from endogenous dynamics, are those which could be attributed to exogenous factors [such as, for instance, the synchronous or asynchronous cycles due to linkages with other markets].

- Dynamical instability could ensue when relatively long micro time and/or meso time lags, and/or memory are built into the options price equation. Time lags in either difference or differential equations have always been thought of as inherently destabilising components in non-linear dynamics [and even in certain linear dynamics]. Chaotic events may be hints of time lags being present, among other possible suspects.
- The existence of strong temporal interdependencies between the price of an option, $p_{\pi T}(\tau, t)$, and its excess supply, $E_{\pi T}(\tau{-}1, t)$, in the form of *expectations* at micro time period $\tau{-}1$ about the [average expected] price, $p_{\pi T}$, at τ comprised of individual perceptions by the suppliers of these options. Such an assumption would not however be a strong one.

- If the 'sensitivity to spread' coefficient, λ, is connected to an average [over t or T] excess supply, $E_{\pi T}$, or *vice versa.*
- Finally, if there is an interdependence, in the form of memory, between the current price at micro time period τ of the underlying primary instrument, $P(\tau)$, and an average past excess supply condition, E :

$$E = \Sigma_\pi \Sigma_\tau E_{\pi \tau}(\tau - 1, t')$$
(17.9)

over preceding meso time periods, t. This key interdependence between the primary instrument and its derivative [option] provides the second leg to the one-legged monster specified under the Black-Scholes-Merton formulation.

Any or all of the above four conditions could result in turbulent endogenous dynamics within the options price of a metropolitan development market. The reasons for such turbulence are found in the basic properties of discrete iterative dynamics first given by May (1973) and elaborated throughout the past 25-year period in the literature on non-linear dynamics and chaos. A whole new industry has been generated around this topic, see for instance Trippi (1995) as an example of the plethora of books on the subject of chaos in financial markets! Among them is also the latest [electronic] journal, Studies in Non-linear Dynamics and Econometrics by MIT Press, which focuses on the subject of chaos in economics, business and finance.

17.4 Derivatives, Speculation and Development

Derivative instruments are innovative investment tools, still in their formative stages in North America. They have been fairly extensively used in metropolitan and regional development processes, and their applications are currently on the increase in many areas of federal, state and local public policy, including the environmental, pollution, transportation and land use sectors of cities and regions. Recognised as enterprises which are inherently speculative, carrying considerable risk, private agents in markets for metropolitan resources, particularly land, have long sought various hedging strategies in order to effectively and efficiently manage uncertainty and exposure to risk through the use of derivative instruments, like options. Metropolitan governments, on the other hand, through the use of derivative instruments have sought to achieve social objectives regarding land conservation, environmental quality control, and regulation of congestion externalities among other policy issues.

Even exposure to risks affecting the metropolitan infrastructure due to natural disasters [hurricanes, earthquakes, flooding, etc.] was covered in the US by the establishment of highly innovative formal options markets on natural catastrophes in 1992 at the Chicago Board of Trade. Use of these sophisticated instruments enables insurance companies and large metropolitan entities, ranging from a host of governmental units to large land holders and developers, to acquire some form of primary [or secondary in the case of insurance companies] insurance against risks.

Next, a brief discussion of derivatives, speculation and their role in metropolitan development is presented. In addition, we provide a limited array of sectors where these instruments have been applied in North America.

Markets of Metropolitan Resources

Metropolitan resources, including land, air, water, right-of-ways, views, open space, etc., and rights to their use, rent, lease, or purchase have long been the objects of elaborate as well as complex market mechanisms which are subject to legal rules and regulations governing their exchange, particularly in developed market economies. Control of knowledge and information, correct or erroneous, honest or deceitful, their timely possession and fast dissemination are key ingredients of these regulations.

In North America, the allocation of scarce metropolitan resources, particularly land and air, have been the subject of speculative as well as highly advanced market allocation schemes ranging from sophisticated real estate markets to a host of derivative instruments for trading land and land development. Such instruments include, but are not limited to land development rights, and their transfer, as well as land use options. Three areas where the use of derivative instruments has made considerable inroads are: land development rights and the transfer of such rights; air pollution rights; and, traffic congestion rights. They are addressed in some detail below.

Land Development Rights and their Transfer

The owning, leasing, buying or otherwise using land for a specific purpose or purposes is traditionally a right granted to a lond owner by a national and/or local unit of government. These rights are to a certain extent limited by the Constitution and/or through the power of zoning and eminent domain exercised by the appropriate governing bodies. The land owner can in turn sell one [or all] of these right(s) to any willing buyer in municipal land rights exchange markets, the most common of them being the buying and selling of ownership rights in municipal real estate markets.

Transactions concerning land and land development can take place directly, or indirectly, by the use of derivative instruments, namely options to develop. Individuals, or developers, can purchase or sell in a legally binding contract the right [but not the obligation] to develop [i.e. own, lease, rent or otherwise use and exploit] a prespecified quantity of land at a particular point in space [location], by a specific date in the future [the maturity], at a preset price [referred to as the

strike or exercise price], which is an array of currently expected possible profit
levels from the development wactivity as envisioned in the development scheme.

These individuals [investors, bankers, insurance companies, land owners,
builders or developers] can in turn sell this contract to any willing buyer in land
[and land development] options markets, at any point in time within the period
prior to and up till the expiration date of the option, at an option contract price set
by these options markets. Further, the individual holder of the option contract may
choose not to exercise the option at the expiration date of the contract, and instead
enter into a monetary exchange with the options seller.

Transfer of development rights [TDRs] enables the transfer of these
development options from one location to another. Such transfers can be carried
out by exchanging these rights not only among individuals [and/or developers] but
also among locations, in land use and development options markets. Employment
of these options markets on land uses and TDRs has been increasing in the US, as
well as in a number of European and Asian countries. In the US, this has been
happening in effect for the good part of this half century. The reader is directed to
a special issue of the Journal of the American Planning Association, and the article
by Pizor (1980), for a more comprehensive coverage of the subject of TDRs.
Initially lightly traded and in very inefficient markets, the instrument is becoming
more effective and the markets more sophisticated in a number of US settings
(Daniels 1981; Wright 1993).

Pollution Rights

Pollution rights are relatively new instruments used for controlling environmental
pollution by fixed source polluters, usually large manufacturing plants. These
rights are supplied by local Environmental Management Districts in metropolitan
areas of the US, which are special purpose governmental units set up by the Clean
Air Act [CAA], enacted in the early 1980s and since amended over the years.

The rather complicated process of allocating pollution rights and establishing
the relative market, a programme referred to as the Regional Clean Air Incentives
Market – or RECLAIM, in the case of the California South Coast Air Quality
District [a pioneer in setting up such programmes in the US] (see Robinson 1993)
– works in the following simple way: a fixed amount of such pollution rights [set
by the Agency in accordance with the desired total amount of pollution to be
emitted] is distributed to fixed source polluting agents participating in the
programme at some initial price [computed as the marginal and real social cost of
pollution], allowing them to emit a prespecified amount of pollutant(s) over a
prespecified time period. If and when these sources of pollution do not meet their
allotted [purchased] total pollution level, they are allowed to sell the remaining
portion of their rights to other fixed sources pollution agents of the region in an
open market context, and at prices set by pollution option markets.

Implementing such programmes in any region or metropolitan setting, carries
potentially considerable implications for the overall development of the region.
An attempt, on a preliminary basis, to provide an initial outline of such effects for
the Southern California region was attempted by Polenske et al. (1992).

Congestion Rights

Congestion rights are even more recent than pollution rights, and are based on the trading of licenses to use congested segments of roads. Such licenses were to a certain extent envisioned by the Intermodal Surface Transportation and Efficiency Act [ISTEA] of 1990 as amended through the years following its enactment. The programme is an extension of the effective congestion pricing scheme called for in the Act, and is currently being tested in a number of demonstration sites in the US.

Under this programme, individual trip makers acquire permits [usually, on a monthly basis] for the 'right to use' particular segments of a certain set of roads, usually through a government rationing scheme. These permits [decals], which are possible to read electronically, allow the holder – during the month for which the permit was purchased – to use particular sections of road fitted with electronic devices. These devices automatically charge the user [trip maker] in real time an amount which varies according to the prevailing congestion conditions.

These 'congestion rights' are currently open-ended, in terms of a [monthly] total cost, to the trip makers. A far more interesting extension of this programme, under consideration by a number of municipalities in the US, is to substitute this rationing scheme with an open market, according to which the government would fix a total allowed cost on each of these permits. This amount would depend on the total social cost of pollution an agency would be prepared to tolerate over a specific time period [say, a month] in the area of its jurisdiction. The municipality would then sell a fixed number of such 'options to congest' at an initial monthly [or any other period] auction.

A trip maker not planning to use all of these 'congestion options' could, in a formal market, exchange for cash the portion of these options remaining unused [due to an excess supply created during the month at the individual level] with any willing buyer [having an excess demand for trips on these congested segments of the designated roads created during the same month]. Alternatively, a speculator could purchase a number of these options, hoping to fetch a higher price at some time later that month.

Derivative instruments, such as options, could substitute the need of a government [or in this case a transportation or environmental pollution agency] to devise a more or less arbitrarily derived rationing scheme to allocate congestion [or pollution] rights. Any options market for congestion [or pollution] rights, no matter how imperfect, would be preferable to an arbitrary rationing scheme. Arguments for this proposition will be provided in Section 17.5.

Speculative Behaviour, Risk and Development

Considering the development of land; deciding to pollute metropolitan air; opting to use a particular congested segment of a road: all of these decisions [and a myriad of others], taken at any point in time by a host of individual [and interacting] agents in a metropolitan setting, carry with them a degree of risk, which may at times be considerable. All these decisions are made under prevailing conditions beset by uncertainty, limited information and at times with little foresight; they are thus inherently speculative actions.

Land development takes place, in a market economy, as a result of private entities being willing to take a bet that the supply of certain land uses at particular points in space and time would satisfy a perceived [and to them 'real'] demand. What makes such a decision to invest speculative is, first, that it is a 'future' demand condition and, second, that it is 'perceived' by a limited number of development agents. These entities are willing to undertake the risk of being wrong, and thus lose all or part of the amount of capital invested. To them, such an investment must have an 'expected' profit which is higher than the opportunity cost of capital at the point in time when the decision is made.

In environmental pollution markets, the decision to pollute the air by a certain amount is an investment decision by fixed source polluters [firms]. This is because such a decision affects the firm's expected profit margins. In turn, any decision affecting expected profits is a decision which affects not only future production schedules, but [and this is of particular interest to spatial analysts] also the present and future location choices of firms. Thus, environmental pollution markets bear indirectly on the ever-changing development prospects [present and future land prices and uses] of metropolitan areas and regions, in that it affects decisions by firms to locate at particular points in space and time.

As far as congestion prices and congestion rights markets go, they too affect [albeit indirectly] the development patterns of cities. The expected 'real social' as well as 'individually perceived' transportation cost incurred by trip makers [particularly for the trip to/from work and from/to place of residence] is a major determinant of residential choice behaviour. In turn, present and expected future access costs are to a large degree capitalised [or captured] by land prices, which in turn determine current and expected future land uses and their densities.

These determinants feed on the current level of transport costs through a strong feedback loop. Thus, a highly interactive and dynamic relationship is formed between the two [transport costs and land prices/densities/uses], which under the presence of the congestion externality and its internalisation conditions drives the metropolitan spatial land allocation pattern towards or away from a short/long run dynamic equilibrium. This will certainly affect the daily travel pattern of metropolitan residents when the is a real time pricing policy [or options to use are exercised] for congested roads. This is a central component of the project tested in a number of metropolitan areas in the US during the early 1990s under the Intelligent Automobile and Highway Systems Program envisioned by ISTEA.

In all three cases – land development as well as pollution and congestion rights – it is the presence of risk and uncertainty that characterises individual choices by land developers, firms and individual trip makers. To incur risk allows for the process of development to go forward. Development and risk are necessary and sufficient for each other to exist. Planning ought not regulate development through direct control by zoning. It should not act in a paternalistic way as an insurance agent, attempting to eliminate risk, because this could stifle development and diminish the willingness of speculative agents to invest. Instead, planning and regulatory functions of government, ought to be focusing their attention on efforts to make the development markets as efficient as possible. Instituting formal and well regulated metropolitan wide derivative markets, such as options markets for metropolitan development, pollution options markets and congestion options markets, should go a long way towards making the direct metropolitan development markets much more efficient and effective than they are currently.

There are numerous reasons why the land development process as well as the pollution and congestion rights/options markets are themselves imperfect. Governments ought to correct such imperfections, to the extent possible, by regulation, a subject to which we now turn the analysis.

17.5 Derivatives Market Failures and Regulation

Land development markets are mired in failure. Instituting a market for derivative instruments [in this case, options] on metropolitan development would improve [and indeed, to the extent that it has been applied, has improved] on the situation which direct development markets experience. In spite of the failures of the derivative instrument markets themselves, their existence is likely to result in an improvement, which will be discussed further below. Similarly, the markets for the exchange of pollution and/or congestion rights/options encounter numerous failures for a host of reasons. However, their presence is expected to facilitate the more rational use of the metropolitan air and transportation capacity.

The main reasons leading to failure in direct development markets providing efficient allocation of the scarce urban resources [land, air, transport capacity] are discussed below. Emphasis is placed on the direct land development markets, as the discussion is largely equivalent for the other two sectors [environmental pollution and transportation congestion]. A set of similar causes characteristic of derivatives markets [again with an emphasis on options on land development markets] are then addressed, as well as the [positive and negative] external effects that the derivative markets might have on direct markets. Furthermore, the way that such externalities and failures can be addressed by metropolitan governments instituting derivative markets in these sectors is outlined.

Primary Metropolitan Development Markets: Their Failures

A number of causes can be cited for the diverse metropolitan development market failures, ranging from the traditional ones found in most markets [i.e. the presence of monopolies or monopsonies, oligopolies or oligopsonies, imperfect information, externalities, lack of foresight, etc.] to some special causes which are unique to the primary metropolitan development markets already mentioned.

(i) *Spatial monopoly conditions*: the presence of space is, in and by itself, the single most important cause for creating monopolistic competition. Every location is unique, and as such it cannot be easily substituted, no matter how similar neighbouring or lots or spots might be in urban settings. Supply of land, thus, occurs under conditions of spatial monopoly.

(ii) *Thinness of markets*: suppliers of development schemes [and thus potential buyers of metropolitan land parcels] are very few. This makes impossible the standard economics requirement of asking for 'a large number of' developers for specific chunks of urban space so that the markets can attain economic efficient outcomes. In fact, the exact opposite may occur, whereby a very few developers correctly perceive the development potential of urban spaces and are willing and able to take a risk.

In cases where there are a number of potential developers [no matter how small that number might be], not all have the same quantity of capital available for development. As increasing returns to scale are present, a small number of very large development firms tend to form in any metropolitan setting. These have comfortable profit margins and generally coexist with a few development companies with very low capitalisation and rather poor profit margins. New York and Los Angeles are examples of very large metropolitan areas in the US where the about two or three large developers dominate the metropolitan landscape. Only these very large developers can take enough risks at enough locations within a metropolitan area to effectively manage uncertainty through a satisfactorily diversified portfolio of land development schemes.

(iii) *Imperfect information*: not only does the impossibility of knowing all relevant information about all locations in an error-free environment hinder the accomplishment of the requirement for participating in perfect [or close to perfect] development markets; but the possibility of insider information gives a much more insidious cause for metropolitan development market failures to occur. In highly developed and formal stock markets [as for instance those of the US], insider trading is legally prohibited [although not possible to totally prevent or eliminate]. Insider trading is in fact pervasive in metropolitan development markets in general. Successful developers usually have access to confidential government decisions, and decision-makers [often the politicians themselves] have a stake in development outcomes which they have helped to bring about.

With so few development players, collusion [and club formation] is possible, thus effectively eliminating risk and setting prices for the development product to consumers. Both of these corrupt practices, insider trading and collusion, have been widespread in metropolitan development of market [and non-market] economies, and have even been the subject of the popular folklore [in the form of novels and movies].

(iv) *Lack of or differences in foresight:* although all economic agents, individual consumers and producers as well as governments have some form of foresight, not everyone's time horizon is the same. Some developers tend to take a longer view of the development game, and can afford to do so since these developers with a longer time horizon tend to be the biggest players as well [i.e. those with the highest level of capitalisation]. Smaller scale developers cannot afford to take bets with a long term payoff period, since they do not have the investment capital necessary to do so. Longer term investments tend to be associated with larger scale capital expenditure, and shorter term investments with smaller capital outlays.

Although no individual operates with no foresight whatsoever [it is naturally impossible], some individuals tend to take a rather short term view of their actions, including their locational decisions. Changes in income, job, the place of work, and other factors, such as market interest rates [and thus mortgage rates] can require certain households to frequently updating their location decisions. Similarly, changes in market conditions require frequent updating by firms of their location choices under a variety of conditions, depending on the firm and its size. Both households and firms are, in turn, the end users of metropolitan development schemes.

Agents with shorter term horizons distort the efficiency of markets more than those with longer term horizons. The reason is quite simple: agents with shorter horizons give signals to developers which become outdated faster than signals given by agents with longer term horizons. Since large capital investments are durable [their economic lifetime extends into decades], errors made as a result of short term signals [say, by decisions of households with a timespan of a year or two] could be very costly. Negative effects from such decisions do not confine themselves to the developer(s) who responded to an erroneous short term signal; but they also extend their costs to the metropolitan areas at large: empty space [excess supply of office space for instance] as a result of overreaction by developers is an element affecting the vitality of neighbourhoods, as well as the fiscal vitality and the image of metropolitan governments and their cities.

(v) *Externalities*: spillover effects of developmental actions [or inaction] by developers are abundant in metropolitan settings. Among the most obvious are neighbourhood externalities: development at particular points in space will have [negative or positive] impacts upon adjacent land uses and prices, in the short and long run. The designation by metropolitan governments of direct 'benefit districts' for metropolitan development actions [by either the public or private sectors] and the effort to appropriately capture these effects by the imposition of a tax or subsidy scheme within such districts is an effective means of internalising developmental externalities.

The insurance and trade of metropolitan-wide developmental options is by far the most efficient way to capture the effects of such development schemes. These options would substitute the spatial rationing schemes operated by governments in establishing 'direct benefit districts', avoiding the possible errors caused by either over- or under-extending the boundaries of such districts, or exaggerating or underestimating the size of the benefit/cost, since markets would precisely peg the spatial and monetary extent of the benefit over time.

Derivative Metropolitan Development Markets: Their Failures

The causes of failures in direct markets for metropolitan development, discussed above, are similar to those in the derivative [options] markets for metropolitan development. On the surface, it might be thought that such failures risk accentuating actual developmental failures, but this is not always the case, as will be explained in the following.

As already noted, developmental monopolies or oligopolies are evident in even the largest of metropolitan settings. But whereas the thinness on the ground of potential developers in these primary [for example, land] markets is the rule rather than the exception, this is not necessarily the case with the derivatives [options on development] markets. As many diverse [and not exclusively developer type] agents could participate in these options markets [for example, governments, land owners, builders, insurance companies, large firms, and individual investors], there is a lesser likelihood of these derivative markets being thin than for the primary markets.

Lack of foresight or varying time horizons could be a cause for market failure in the derivative markets, as it is for the direct ones. However, there is less likelihood for this in the options market, as options are instruments which address head on the existence of specific time horizons among investors: an array of strike prices can be set up for trade on an array of expiration dates. Obviously, the longer the array for strike prices, the more likely they are to result in thinner markets.

Here lies a fundamental trade-off implicit in development theory, which simply becomes explicit by recognising the existence of markets for derivative instruments [options on development]. A more extensive coverage of options by the metropolitan investment public [which would result from their becoming more knowledgeable about their existence, analyzing them more fully and, more importantly, taking action on them], can only take place at the expense of the number of options outstanding.

How many developmental schemes can the diverse investment constituency of a metropolitan community really [effectively and efficiently] absorb at any particular point in time? How is the development process itself affected by this merely logistical factor? It is necessary to accept certain imperfections [like partial information] during the development process? And, in effect, is it necessary [in certain instances of information congestion and overflow] to manage information dissemination by nurturing partial information sets [through filtering out by governments of redundant information, or by pricing information]?

Is there an optimum phasing-in of all developmental schemes? And how is this phasing-in to be effectively accomplished without resorting to a spatio-temporal rationing scheme or to a phasing-in of developmental options markets by government? Is it possible to set up an even broader derivatives market on a phasing-in derivative instruments [options] market? All these questions obviously need to be addressed in order to further the research ideas presented here.

Conditions where imperfect information and insider trading situations might arise seem far more likely in derivatives markets than in the direct [primary] development markets, mostly due to the thinness in trading conditions. This is where effective regulation and vigilance on the part of appropriate governmental institutions is needed. Before addressing this issue, however, we shall take a look at the linkages between the primary and derivative markets.

Linkages Between Primary and Derivative Markets

The central reason for the existence of a derivative market [options or futures contract markets, for example] on any direct [primary] investment [for example, on commodities such as gold or land, or on financial instruments like stocks, bonds or currencies] is to enable the more effective management of risk. Hedging strategies in conditions of exposure to risk are an effective way of managing uncertainty. Losses from a collapsed option's price are usually far less when compared with the potential losses from the direct investment going wrong [for whatever reason, ranging from bad planning to drastic and sudden changes in environmental conditions].

Negative signals by investors from options markets on metropolitan development schemes might prevent a governing unit from going ahead with a project which is bound to disappoint its proponents. This, among other things, might save the public from a bad public investment idea, and thus prevent the local municipality from a drain on its financial resources and fiscal vitality. Such a linkage between the direct development market and its derivative market is a positive externality of one market upon the other. However, this is not the only positive externality present between the two markets. It has been recognised, from extensive experience in stock market behaviour in the US capital markets where derivatives markets have been present for some time, that a significant negative externality also exists between the two markets [primary and derivatives]. This emanates from the derivative market and affects the primary market.

Excessive covering in the options markets [through the presence of excess demand or supply conditions] at time results in severe and excessive swings [cycles] in the price of the primary instrument. Wide swings in stock markets [as well as commodities, currencies, and bond prices] have been often recorded after the institution of derivatives markets on direct investment instruments. Such swings are especially pronounced at the expiration dates of options, futures and options on futures contracts. Thus, swings exclusively due to the presence of derivatives markets could potentially work against the efficient performance of the primary markets, by giving erroneous investment signals to investors particularly in the shorter term.

Governmental Regulations

In view of the above discussion, it has become clear that the government's role should be confined to effectively setting up, formalising, regulating, monitoring and policing the derivative markets for metropolitan development. These roles seem to be far more appropriate for metropolitan governments than their traditional role of rationing [more or less arbitrarily] various development schemes over the metropolitan landscape. In other words, the metropolitan government ought to stop planning the development of cities, leaving development markets to take care of this task, and instead supply an efficient institutional framework to facilitate, as far as possible, the uninterrupted and effective functioning of the primary and derivative development markets.

A number of specific areas were identified earlier as possible arenas where effective action by metropolitan governments is needed. One lies in the regulatory function, having to do with eliminating insider trading in both primary and, by and large, derivative [options] markets. Thin trading in these markets naturally lends itself to insider trading and corruption opportunities by both developers and agents of government. Conditions should also be avoided where potential conflict of interest cases might arise from governmental decision-making itself. Alongside this concern, governments ought to facilitate, to the full extent possible, the dissemination of information on developmental schemes and the options on such schemes, thus supplying all potential participants [those with a stake in the outcome] with a complete, non-redundant and, as far as possible, error-free information base. This is a way of preventing information underflow, or turbulent [possibly noisy] information flow.

A second area, where metropolitan governments could intervene, again to the extent feasible, in securing the efficient functioning of derivative markets, is in alerting the investment public to the formation of developmental oligopolies. Timely announcements of entry into the developmental markets of various developmental players would be one way of dealing with such issues. Furthermore, in view of the risk of oligopolies, a public assistance programme to encourage entry in the primary [and, even more importantly, the derivatives] development markets might help alleviate the problems of cartels being created in developmental markets.

Subsidising entry into derivatives markets could prove an effective means of indirectly policing such oligopolies. In the case of failure to entice firms to enter developmental markets – short of explicitly trying to break down and regulate the size of developmental firms [or firms specialising in the trade of developmental options] – another approach would be to tax developmental oligopolies, i.e. the larger and most dominant development firms within a metropolitan setting. The aim of taxation would be to oblige developmental firms to operate at the point of a Pareto optimum, so that the price of the development option [in case of the derivatives instrument markets] is equal to the marginal social cost of this option [as computed by the metropolitan government].

The third area where metropolitan governments could be active is in the regulation of knowledge and information overflows in the primary and derivative developmental markets. Congestion and bottleneck conditions in developmental market-related information is usually to the benefit of a small number of players who even seek such congestion conditions. Usually, the correct tool to use to prevent this would be a congestion pricing scheme. It would be incumbent upon the metropolitan government to devise a scheme in which an appropriate toll captures the true social cost of the information flow at the margin.

Finally, another role of metropolitan governments in regulating developmental markets has to do with smoothing the impact and suppressing the magnitude of oscillations in the behaviour of the primary market for metropolitan development, resulting from the very presence of the derivatives instrument [options] markets. Oscillations in the behaviour of primary development markets may manifest themselves either through drastic changes [wide amplitude cycles] in the currently expected future [social and individual] costs and benefits; or in up- and down swings in the number of developmental players willing to enter or exit the [primary and derivatives] metropolitan development markets.

Governments could be participants [albeit not principals] in deriving the expected costs of any developmental action. Supplying such estimates, governmental agencies could provide a stabilising force in an environment of sharp cycles in developmental perceptions. Sharp cyclical swings in expected benefits and/or costs from developmental actions could potentially supply erroneous signals to the metropolitan development markets. And, as already stated, metropolitan governments could encourage or discourage entry to or exit from metropolitan development markets, as appropriate. Both of these actions could help to accentuate the role of a metropolitan government as a stabilising force in development dynamics, instead of their traditional role operating monopolies or monopsonies over metropolitan development.

The numerous technological innovations, currently able to monitor stock markets and their derivatives markets in North America, would provide a good basis upon which to structure the institutional and regulatory infrastructure of formal metropolitan development markets.

17.6 Conclusions

Knowledge, information, complexity and innovation are central components of metropolitan development markets. The use of options as derivative instruments for any primary urban asset market would be an effective way of capturing and accounting for all of these attributes.

Preliminary evidence on the application of derivatives [options] in development markets in North America gives encouraging indications as to their effectiveness and desirability in relation to land development. Initiation of similar derivatives markets for options [rights] on pollution and congestion is further evidence of the potentially widespread use of such markets. In spite of certain weaknesses, mostly due to the conditions of 'thinness' encountered in them, there is no doubt that these secondary markets can meet their objectives. It is up to the specific metropolitan governments to pursue a policy of effectiveness and transparency in their performance to the maximum extent possible.

It was argued in this chapter that a drastic revision in the role and function of metropolitan governments would have them abandon their traditional direct intervention in metropolitan development activities. Their monopoly and/or monopsony power theoretically and practically disqualifies them as effective and efficient agents for development. Instead, formal primary and derivatives markets on development ought to be instituted, so that the very large number of potential private [or quasi-public] agents for metropolitan development could participate and guide the development of cities. Governments should concentrate their limited resources in setting up the required formal markets, infrastructure, and seek to monitor and regulate the performance of these markets, so that failures can be avoided or at least mitigated.

In addition, by regulating the functions of the derivative markets, governments would possess a means through which social aims could be effectively attained. As in the case of TDRs, whereby land conservation and renewal as well as socially desired local density conditions could be met through decentralised [market] decision-making, metropolitan governments could use these derivative instruments to achieve a variety of objectives. Pollution control, congestion pricing, and the effective use of public spaces are just some of the domains where these derivative instruments could be of considerable assistance.

During the Workshop on 'Knowledge, Innovation and Complexity' held in Vienna, it was suggested by Professor Michael Sonis that [as an analogy] the concept of options and associated prices suggested here, could be used to identify the differences between inventions and innovations. An invention, according to Sonis translates new information into an applied innovation. The difference at any point in space-time between the new information [invention] and its potential application [adopted innovation] can be considered to be the price of the option in the action of adopting the innovation. The prices of options express the expectations of future gains through the adoption of an innovation, thus reflecting the speculative behaviour of potential adopters. This is a provocative idea which merits further development and research.

Finally, metropolitan plans and policies themselves can be considered as commodities being traded [among agents with a stake in their outcome] in highly imperfect and informal policy options markets. This idea was one of the key suggestions presented in Dendrinos (1992), which provides further details on the theme. The discussion, which dealt with primary [direct] and derivatives [secondary] markets for metropolitan development, applies by extension to the urban plans and policies markets as well. It is also a source of ideas about how to introduce innovative ways of treating pricing and exchange of policies and plans. Whether such innovative policy-making processes and instruments are feasible at a wholesale level in metropolitan development, however, remains an open question.

Acknowledgements. This contribution was presented at the International Workshop on Knowledge, Innovation, and Complexity in Vienna, Austria, July 1-3, 2000. The author wishes to thank Professors Michael Sonis and Manfred Fischer, as well as other members of the Workshop, for helpful comments and suggestions. The main ideas and a first draft of the contribution were presented at the Department of Economic Sciences at the University of Venice on April 22, 1996. The author is grateful to Professor Dino Martellato for organising that seminar and for comments on this paper. Further comments from Professors Sergio Bertuglia and David Batten, as well as Sylvie Occelli and Angela Spence supplied in April 1996 are also acknowledged. The usual caveats nonetheless apply.

18 Spatial Dynamics and Government Policy: An Artificial Intelligence Approach to Comparing Complex Systems

Peter Nijkamp[*], Jacques Poot[**], and Gabriella Vindigni[***]
[*] Department of Spatial Economics, Free University Amsterdam
[**] Department of Economics, Victoria University of Wellington
[***] Department of Economics, University of Catania

18.1 Complexity in a Dynamic Spatial-Economic Context

Complexity is concerned with the unpredictable nature of non-linear and dynamic systems. Complexity can relate to a dynamic causal sequence of events at an object-specific micro-level [such as in the case of the weather, business performance, market impact of innovation, individual well-being, etc.], but it may also refer to the outcomes of repeated experiments in a semi-controlled setting. A comparison of the results of case studies is a good illustration of the latter interpretation of complexity. In this case, it is useful to see to what extent the outcome is shaped by the systemic background and the specific research methodologies used.

The rapidly changing and often unpredictable economic events of the past decades call for a thorough analysis of the behavioural and methodological foundations of contemporary economic systems. This applies not only to national economies, but also to industrial systems at both a regional and a global level. We are witnessing in modern economic research an increase of interest in the behaviour of actors and agents from a micro-economic perspective, with a view to deriving behavioural implications at a meso or macro-level [e.g. supranational economic policy, group decision-making, industrial concentration tendencies, institutional network developments, international mergers, functioning of spatial labour markets, etc.]. This multi-layer interlinked and inherently complex behaviour lies at the heart of modern theories on economic dynamics, such as endogenous growth theory, evolutionary theory, externality theory, game theory, models of search behaviour, network theory, location theory under monopolistic competition, or option theory. Complex economic behaviour appears in a wide range of forms in the space-time context, from slow dynamics [e.g. changes in trade and location patterns, the process of regulatory reforms or evolving technological trajectories] to fast dynamics [e.g. changes in financial markets, in transport patterns or in industrial growth]. Intermediate and hybrid forms of dynamics may emerge as well [e.g. in urban evolution or in the transformation of social security systems]. In this context, institutional configurations may play a mitigating role, but also have a retarding effect, as has been shown in recent studies on transaction costs in the private sector and in the public sector.

The development of economic systems in the industrialised world has in recent years exhibited various dynamic patterns, partly caused by drastic changes in international markets and relations [such as in Central and Eastern Europe, the Asian crisis, etc.], and partly by intrinsic changes in the way that economic systems themselves function [e.g. greater deregulation, outsourcing]. There have been drastic changes in market configurations not only at the local but also at the global level. The internationalisation of business life has led to a situation where local and global markets are, more than ever before, spatially-linked platforms of economic activity. The parallel development in institutional reform [e.g. privatisation of networks] has created enormous economic flexibility and adjustment potential which has had far-reaching implications for the behaviour of actors and agents.

Given these complex economic changes, it is not surprising that during the last decade there has been considerable scientific interest in the causes of economic growth, both local and global. The seminal contributions by Lucas (1988) and Romer (1986) have generated a wave of new research. Initially, the aim was to formulate dynamic general equilibrium models with precisely formulated microeconomic foundations that would enable a clearer understanding of processes such as physical and human capital accumulation, innovation, or knowledge and product differentiation in terms of their impact on long-run economic growth. The extension of the traditional Solow-Swan growth framework with various endogenous learning and technology mechanisms has led to a wealth of literature since the 1980s on endogenous growth (see Nijkamp and Poot 1998 for a survey which also addresses the spatial implications). These growth theories focus on the role of various externalities related to technological change: specialisation and trade, monopoly rents from innovation and 'creative destruction', human capital and government policy. There have also already been several attempts to endogenise technical progress within models designed to address environmental issues and sustainability (e.g. Gradus and Smulders 1993; den Butter, Delling and Hofkes 1995; van den Bergh and Nijkamp 1994; Bovenberg and Smulders 1995). Other treatments of the relationship between technology, growth and externalities have stressed its disequilibrium, uncertainty and evolutionary [or Schumpeterian] character (e.g. Dosi et al. 1988; Aghion and Howitt 1998; and in an environmental context; Faber and Proops 1990; Clark, Perez-Trejo and Allen 1995).

At the same time, an empirical research programme has emerged, setting out to explain the variation in long-run growth rates across regions or countries by means of cross section or panel data. Both the theoretical and the empirical research programmes have reached a stage of maturity at which graduate level text books have been published (Barro and Sala-i-Martin 1995; Aghion and Howitt 1998), providing an excellent overview of what has been accomplished to date. Yet, as the authors of recent books and other surveys of the recent literature have concluded, much research remains to be done. Firstly, the empirical work on growth accounting and convergence at the macro-level has rarely provided a direct verification of the micro-level behavioural mechanisms that drive the growth process in the theoretical models. Secondly, much of the theoretical literature has been concerned with models of single, closed economies. Thirdly, while the theoretical research points to why and how governments can influence the long-run growth rate, the empirical literature on this aspect appears rather fragmented and inconclusive.

Endogenous growth phenomena, as described above in the context of complexity, are also particularly relevant for the study of the interactive behaviour of actors in modern *evolutionary* economics. Recently much attention has been given to evolutionary and learning processes in the game-theoretic literature, see for example Maynard Smith and Price (1973), Kandori, Mailath and Rob (1993), Young (1993), Weibull (1995), Samuelson (1997) and the survey paper of Van der Laan and Tieman (1998). The main stance taken in this literature is that players react to the circumstances with a bounded rationality. They learn 'how to play the game' through time. In most studies of this kind, we begin with a population of players who, together, form a social network. It is often assumed that the players are randomly matched in pairs and play one stage of the game at each round of play. At any time, the players use learning rules to decide how to play. Such learning rules map the past history of play into what to do next. A closely related approach is to assume that the fraction of players using a certain strategy evolves over time through some replicator dynamics, see e.g. Crawford (1989), Björnerstedt and Weibull (1993), Ellison (1993) and Karandikar et al. (1998). The central question addressed in this literature is whether the resulting dynamic system converges and, if so, what is the outcome?

The strategies of researchers who build on the successes and failures of earlier endeavours may also be interpreted in an evolutionary game-theoretic context! Consequently, there is much scope for international comparative research [e.g. through meta-analysis, or value transfer] and it is not suprising, given our earlier observations, that this strand of research is gaining in importance in economics. *Meta-analysis* is a promising approach which has been employed in the present study. In Section 18.2 we offer a concise introduction to meta-analysis as a modern tool for research synthesis. Next, in Section 18.3, we introduce *rough set analysis* as a recently developed artificial intelligence tool, which is able to draw generally valid inferences from a series of classified studies. Then in Section 18.4 we examine some studies on government expenditure and economic growth in order to create a sample of studies on the role of governments in complex spatial systems which can be empirically tested. Rough set analysis will then be used in Section 18.5 as a method for extracting valid results. Section 18.concludes the chapter with some reflections on the findings and implications for public policy.

18.2   Research Synthesis and Meta-Analysis

In the field of economics research, we are seeing the emergence of a veritable 'knowledge economy', and an exponential growth in the number of research findings. In certain areas, such as studies of innovation or of government policy impact, an enormous number of publications exist. This suggests that there is an urgent need for a synthesis of such results. The question is whether a common knowledge pool can be identified from so many individual studies. Meta-analysis is one of the more promising quantitative research techniques for creating a synthesis from different studies on similar issues. While the approach has a relatively long pedigree in certain areas of research [e.g. medical science, physics, psychology], its use in economics is rather more recent (see e.g. Van den Bergh et al. 1997; Florax, Nijkamp and Willis 2000).

The introduction of meta-analysis as a formal analytical procedure has emerged from the need to be able to summarise and infer general results from scientific studies. Glass (1976, pp.3-8), who coined the term meta-analysis, provides a simple definition of this approach: 'Meta-analysis refers to the statistical analysis of a large collection of results from individual studies for the purpose of integrating the findings. It connotes a rigorous alternative to the casual, narrative discussions of research studies which typify our attempt to make sense of the rapidly expanding research literature'.

Meta-analysis is not a single technique, but a collection of quantitative methods that serve to derive additional [cumulated] knowledge from the *ex post* analysis of a well-defined set of independent [case] studies on similar issues. Clearly, case study research focuses essentially on the inference of general or transferable findings and thus also involves optimal experimentation (for further details see Yin 1994). However, a major problem inherent in social science research is the lack of controlled experimentation in empirical investigations. Case studies carried out in different countries rarely have a common design. This is clearly exemplified by the *ceteris paribus* conditions often assumed in such studies. At best, case studies addressing the same phenomenon, may use more or less the same research methodology or employ similar data. This of course makes a rigorous synthesis difficult, and also the degree of transferability may sometimes be questionable. It is a particular problem when case studies which were never meant to be integrated are pooled in a meta-analytical experiment.

The conditions for a correct application of conventional statistical meta-analysis are fairly stringent. After its genesis in medicine and the natural sciences, meta-analysis was introduced in social science research in the 1970s to overcome problems of common application, the lack of large data sets needed to derive general results, and the problem of uncertainty of information. The advantage of meta-analysis is that it provides a systematic framework for the synthesis and comparison of studies, and can also extend and re-examine the results in order to produce more general results than could previously be obtained, by focusing on a common kernel of research.

The meta-analysis approach thus offers a series of techniques permitting a quantitative aggregation of results across different studies. In so doing, it can serve to calculate more accurate numerical values from available data, for instance in studies of economic costs and benefits. When reviewing the usefulness of parameters derived from prior studies, it can also act as a supplement to more common literary-type approaches, and help direct new research to related areas. Finally, it may also be of help in establishing the robustness of certain findings by using the research synthesis as a kind of sensitivity analysis.

A related and commonly-used method in social science research is *meta-regression analysis*. This statistical technique has been widely and successfully applied in biometrics and sociometrics. The primary characteristics of a meta-regression analysis are similar to those used in standard regression analysis, i.e. the search for a statistical linkage between one of the variables [the dependent variable] and other variables [the independent variables]. Since a statistical tool is used, the input data must of course be quantitative. The main problem is of course the existence of variance in the original case study data.

The application of a meta-regression analysis, which identifies average parameter values, helps to generate meaningful comparative results from a survey of literature or a research synthesis. Having obtained the regression results, tests must be carried out to verify their correctness. Such tests will generally try to assess the magnitude of any effects being explored by the study. For instance, we can test the extent to which an elasticity, chosen to investigate certain effects, depends upon the design of the research study, or how different estimates can be combined into one.

A general guideline for deciding whether or not a particular study should be considered in a meta-analytical formulation is the existence of a basic similarity between the studies. A meta-regression analysis rests upon the following general rules: all studies included must focus on the same phenomenon, they must use similar measures of any given outcome and have the same population characteristics. Finally, they must have a similar underlying research objective. In order to identify relevant studies to be included in the analysis, it is essential to establish selection criteria. As already mentioned above, the available data needs to be of a quantitative nature. Moreover, it is necessary to verify uniformity in order to minimise possible errors in the calculation. It may be necessary to conduct further experiments or carry out new analyses of the data presented in the individual studies (see Van den Bergh et al. 1997).

In addition to meta-regression analysis, there is also a wide variety of other techniques for research synthesis of a sample of individual studies. These include content analysis, fuzzy set analysis, rough set analysis, multicriteria analysis, discriminant analysis and the like. Especially in case of categorical data [such as nominal or binary data], rough set analysis is a very promising tool for research synthesis. This method will be discussed in the next section.

In conclusion, meta-analysis has several interesting features as a method of research synthesis. First, while it deals predominantly with quantitative knowledge, it can also be used with qualitative knowledge under certain conditions. Second, meta-analytical techniques can isolate relevant knowledge from a well-defined collection of previous studies. Compared to other techniques, more knowledge is acquired and made available for value transfers. Third, in certain circumstances, key parameters can be quantified and corrected for bias (Hunter and Schmidt 1990). Fourth, meta-analysis reduces the context-dependency of research findings. In a value transfer context, there is greater transparency. Hence, meta-analysis encourages the building of a more objective body of knowledge. Not only are moderator variables present, they are also quantified to some degree. In contrast with the information contained in individual case studies, a more objective, relational structure becomes available for value transfer. Quantifying allows relationships to be ordered. This ordering process may have interesting features that facilitate the application of a value transfer. Moderator variables can be determined within a subset. A classification of constants and variables is also possible. In a practical context, such as the valuation of a recreational site, common variables can be distinguished from specific values.

18.3 Rough Set Analysis as an Artificial Intelligence Tool for Research Synthesis

Many empirical studies focus on 'what...if' questions, in particular in impact assessment and effect analysis. The empirical findings often show considerable variation, be it quantitative [e.g. in terms of elasticities] or qualitative. Rough set analysis provides a powerful tool for research synthesis based on multiple classification analysis. Rough set analysis belongs to the family of artificial intelligence approaches, and aims to identify which statements on cause-effect relationships [if...then] are consistent with a collection of empirical, cross-classified data. It seeks to compose 'decision rules' that – at least across the sample of individual study results – lead to an unambiguous mapping of empirically justified statements.

The aim of rough set analysis is to recognise possible cause-effect relationships in the available data, as well as to underline the importance and strategic role of certain data and irrelevance of other data (Pawlak 1986, 1991). The approach focuses on regularities in the data in order to draw aspects and relationships from them which are less evident immediately, but which amy be useful for further research and policy making.

Rough set analysis is one of the new mathematical tools designed to investigate the meaning of knowledge and its representation, i.e. to organise and classify data. It is evident that such a method is potentially very useful for the analysis of assessment problems. The data from which a decision-maker determines an evaluation are often disorganised, they contain useless details, or are incomplete and vague. This type of data does not represent structured and systematic knowledge.

Knowledge, according to the rough set philosophy, is generated when we are able to define a classification of relevant objects, e.g., states, processes, events. By doing this we divide and cluster objects within the same pattern classes. These classes are the building blocks [granules, atoms] of the knowledge we employ to define the basic concepts used in rough set analysis. But how can we tackle the problem of imprecision which occurs when the granules of knowledge can be expressed only vaguely?

'In the rough set theory each imprecise concept is replaced by a pair of precise concepts called its lower and upper approximation; the lower approximation of a concept consists of all objects which surely belong to the concept whereas the upper approximation of the concept consists of all objects which possibly belong to the concept in question' (Pawlak 1992, p.1).

By using the lower and upper approximation, we address the problem of vague information, and focus in particular on the problem of dependency and the relationships among attributes. A crucial aspect in the assessment process is the need to distinguish between the conditions through which we make a decision and the attributes that describe the various options. Rough set analysis can examine, on the one hand, the dependencies among attributes but, on the other hand, can also describe these objects in terms of available attributes in order to find essential differences between objects. This latter analysis, which represents the knowledge representation system [or dissimilarity analysis], assumes an important role in many decision-making processes in which it is necessary to indicate the differences among possible options, in order to eliminate superfluous information for a proper decision choice.

Let us consider a finite universe of *objects* which we would like to examine and classify. For each object we can define a number of *attributes* in order to create a sufficient basis for the required characterisation of the object. If the attribute is qualitative, the domain will consist of a discrete set of levels. If the attribute is quantitative, we divide its continous domain into discrete sub-intervals to obtain an categorical description of the object. We have classified our objects by means of the attributes, and thus we can assign to each object a vector of attributes. The table containing all this organised information is called the *information table*. From this table, we can immediately observe which objects share the same levels of each of the attributes. Two separate objects have an *indiscernibility* relation when they have the same descriptive attributes. Such a binary relation is reflexive, symmetric and transitive.

We can now introduce a fundamental concept in the rough set analysis procedure. Let us imagine that Q is the set of attributes that describe the set of objects U. Let P represent a sub-set of the set of attributes Q, and X represent a sub-set of the set of objects U. We define as a sub-set of X those objects which all have the attributes belonging to set P. Such a set is called the *P-lower approximation* of set X, and is denoted as $P_L X$. We then define as *P-upper approximation* of X, denoted as $P_U X$, the sub-set of U having as its elements all objects belonging to the P set of attributes and which has at least one element in common with set X.

The definition of the upper and lower approximation sets has an important role in the rough set methodology. Through these sets we can classify and examine the importance of any uncertain information that we have collected. Consequently, this approach could lead to an imprecise representation of reality by reducing the information-specific sets. This objection may be better understood if we recall that the capacity to manipulate uncertain information and consequent ability to reach conclusions is one of the most essential assets of the human mind in obtaining knowledge. Therefore, the representation of reality by means of rough set analysis is indeed a reduction of the perceived real phenomenon, but it is done in such a way as to enable us to classify, distinguish and express judgements about it.

So far we have focused on the classification of uncertain data. Let us now examine the case where we wish to study the conditions for making a choice among different alternatives; i.e. when we confront an assessment problem. In this case, we can distinguish two classes among the set of attributes in the information table: a class of *condition* attributes and a class of *decision* attributes. The class of condition attributes describe the object following the procedure described above. The class of decision attributes is defined by all the attributes the object must have in order to be selected as an acceptable alternative. For instance, a set of objects can be described by values of condition attributes, while expert judgements may be represented by values of decision attributes.

At this point we must define a *decision rule* as an implication relation between the description of a condition class and the description of a decision class. The decision rule can be *exact* or deterministic when the class of decision is contained in the set of conditions, i.e. all the decision attributes belong to the class of the condition attributes. We have an *approximate* rule when more than one value of the decision attributes corresponds to the same combination of values of the condition attributes. Therefore, an exact rule offers a sufficient condition for belonging to a decision class; an approximate rule only admits the possibility of this.

The decision rules and the information table are the basic elements needed to solve multi-attribute choice and ranking problems. The binary preference relations between the decision rules and the description of the objects by means of the condition attributes determine a set of potentially acceptable actions. In order to rank such alternatives, we need to conduct a final binary comparison among the potential actions. This procedure will define the most acceptable action or alternative.

The technique of rough set analysis is rather cumbersome, although at present some interesting and user-friendly software algorithms do exist. A basic question in all investigations is whether a *core* does exist. A core is a common explanatory factor that shows up in all decision rules. In case of a core, it is clear that we have found a situation where the core variables are the most critical explanatory factors. Several applications of rough set analysis to research synthesis can be found in Van den Bergh et al. (1998), Capello, Nijkamp and Pepping (1999) and Florax, Nijkamp and Willis (2000).

18.4 A Sample of Studies on Government and Growth

After the theoretical and methodological expositions above, we will now present the empirical basis for our applied meta-analysis. The basic question is whether conclusive regularities or explanatory factors can be identified from the great many historical studies on the relationship between the nature and intensity of government policy and coincident economic growth. Since the mid 1980s there have been many empirical analyses of the relationship between government and growth, either as a by-product of tests of conditional convergence between countries or regions, or to address the issue explicitly. Not all studies have a solid theoretical framework or an appropriate econometric methodology. From this vast literature of several hundred published and unpublished papers, a selection was made of 93 published articles from the period 1983-98. Features of each of the articles are reported in Appendix A-18.1 and further summaries of each article can be found in Poot (2000).

All the selected articles were published in refereed international journals in the English language. Being cited in later research was also a criterion for inclusion. However, by focusing specifically on relatively high-quality commonly cited papers, it was felt that useful generalisations could be made from this body of knowledge. The synthesis attempted here undoubtedly suffers from the so-called *publication bias* in that significant findings are likely to be more prominent in the papers summarised here than in the excluded papers (Begg 1994). However, the ultimate objective of the exercise is to assess the difference in robustness of the findings across different areas of government behaviour and it is not clear that publication bias would systematically differ across these different areas.

A coded summary of the information contained in each of the studies is provided in Appendix A-18.1. Following the terminology of rough set analysis described in the previous section, each individual study is referred to as an object, the features of the studies that are reported in Appendix A-18.1 are referred to as attributes [A1 to A9] and the conclusion is referred to as the decision variable [D]. Because several growth studies considered more than one policy area, the 93 articles yielded 123 objects for the information table in Appendix A-18.1.

With respect to fiscal policy, five policy areas were considered: general government consumption in relation to overall GDP [also referred to as government size], taxation policy, education levels, defence spending and infrastructure. These are coded as values one to five respectively of study attribute A1. Quantitative research characteristics that are included in Appendix A-18.1 are the variables: year of publication [A2], number of observations [A4], year of earliest observation [A5] and the year of the most recent observation [A6]. The qualitative [categorical] variables included in the table are the spatial level of the data [A3]; the level of development [A7], the method of research [A8] and the ranking of the journal in which the results were published [A9].

Before a detailed analysis of the data summarised in Appendix A-18.1 is undertaken in the next section, it is useful to point out some general features of this body of research. Firstly, the vast majority of studies have used standard regression techniques. Ordinary Least Squares [OLS] was the most commonly adopted technique [attribute A8 at level 1]. The suitability of this method was rarely tested by means of diagnostic statistics which could have signalled that the use of OLS was problematic for the data in hand.

Thirty five of the 123 observations relied on cross-section [CS] data, but there has been an increasing use of pooled cross-section time-series data [A8 at level 3], as the availability of such data has improved. Forty seven studies used pooled data. This is a welcome trend, as the pooled data studies show that region and period effects are important. In time series studies [A8 at level 2], we see in recent years a growing use of vector autoregressions, Granger causality tests and the co-integration framework (see also Anwar, Davies and Sampath 1996).

The dearth of studies adopting other methods [A8 at level 4], such as dynamic simulation modelling approaches is rather surprising. The sample of 93 articles included only two studies that adopted a calibration/simulation approach. There also appear to be few studies that have adopted a computable general equilibrium [CGE] model approach. A recent example of a dynamic CGE model of the impact of infrastructure on growth is Kim (1998). An additional weakness of many past regression studies is that these purport to provide information on long-run growth, but use observations over only a relative short time span of 5 to 30 years [attributes A4 and A5]. For example, it is possible that public infrastructure does raise the [local] long-run growth rate, *ceteris paribus*, but that the effect emerges only very gradually over time, for example because of a complementarity with certain types of private capital that may, for various reasons, only be undertaken at a slow rate. It this case, it may be very hard to detect the effect of an additional amount of public investment compared with the [unobservable] counterfactual.

In an influential paper, Levine and Renelt (1992) use Extreme Bounds Analysis [EBA] to show that many of the results from CS regression analyses of the determinants of long-run growth are not robust. However, their conclusion does not appear to have discouraged others from continuing to carry out CS regression analyses, although time series and pooled data analyses have become far more prominent in recent years. Indeed, Sala-i-Martin (1997) argued that the EBA criterion of fragility is too strict to be of any use. Assessing instead the robustness of a variable by the probability that the coefficient is on one side of zero in the cumulative distribution function of the regressions which include this variable, Sala-i-Martin finds that 22 out of 59 possible determinants of growth are 'significant'. Interestingly, no measure of government spending [including investment] is among these 22 variables. Moreover, Evans (1996) finds by means of long-run data (1870-1989) for thirteen countries, that there is much evidence that these countries converge to a common trend, i.e. that policies and other shocks influence the growth rate only temporarily.

With respect to the publication outlet, four categories are considered. These are based on the Towe and Wright (1995) classification. These authors distinguish four groups: the top 12 journals in terms of citations, a second group of 23 journals, a third-ranked group of 36 journals and a fourth group of all other journals. The relative frequencies of these journals in our sample of articles are 17 percent, 16 percent, 24 percent and 43 percent respectively.

Virtually all studies of government and growth are *primary* analyses (Glass 1976). Each study has rather unique features in terms of the specification of the model, the sample of countries or regions considered, the time period of observation and the range and definitions of the variables used. Few authors have carried out replications or extensions of earlier research [so-called *secondary* analysis], although *tertiary analysis* in the form of a survey is more common. Among the articles included in our sample, there is only one example of *meta-regression* analysis – a study by Button (1998) on infrastructure and growth.

A final general finding from our sample of publications is that most of the studies on the relationship between government and growth have focused on government at the national level [attribute A3 at level 0] and have consequently used country data. Only about one fifth of the studies use regional data. This sample will now be exposed to the 'meta-scope' in the form of rough set analysis.

18.5 The Generalised Rough Set Method: A Multilevel Strategy Learning Approach

This section proposes and evaluates a multilevel strategy learning approach, based on rough set analysis, that seeks to retain most of the knowledge gains of multiple data set classification. This is done by dividing the full database into different subsets composed of the number of growth studies grouped according to the various levels of each attribute, while the original sample of studies is of course still also considered in its entirety. Given the discretisation which we apply to continuous attributes [see below], this yields a total of 31 benchmark data sets for rough set analysis [30 subsets plus the original dataset with 123 objects]. For the sake of brevity, we will present here only the results related to the type of government policy and the level of the development in different countries.

The sample database of studies is first treated as a general information system (Appendix A-18.1). The articles from which the information was derived are listed in Appendix A-18.2. Each pattern [object, study] consists of the same set of multiple attributes [study features] described in the previous section, and each pattern has a known *class identity* associated with it. The class identity is equal to the conclusion of the study, i.e. a negative, none/inconclusive or positive impact of government policy on growth.

The first phase of our analysis deals with the application of rough set theory to the full sample of studies concerning economic growth with the aim of finding robust relationships between variables incorporated in the modern endogenous growth literature and the impacts found in the studies. For example, we distinguish between cross section, time series and pooled analyses. This will enable us to identify the importance of path dependencies for persistent differences in growth performance. Meta-variables such as the year of the publication and level of the journal in which the articles have been published are also included to understand better the evolution of scientific inquiry in this literature during the last twenty years.

The collected data set is characterised by a certain number of continuous variables, namely the attributes A2, A4, A5, A6. Due to the fact that the learning algorithms used in this paper accept exclusively discrete data as input, those attributes in the decision table which are continuous variables in the real domain have been transformed into discrete attributes [for an overview on such data pre-processing, see Famili et al. 1997). This transformation requires the specification of so-called cutting points [cuts]. The cutting points, defined over the domains of the continuous attributes, divide these into consecutive intervals. The real values of the attributes are converted into discrete ones by assigning each value to the number of the interval to which it belongs (Susmaga 1997).

Numerous algorithms have been developed for this discretisation. These utilise different methods of searching for appropriate cutting points. Discretisations featuring intervals defined by cutting points are called *hard discretisations*. Because such a 'knife-edge' approach may be viewed as too categorical in some situations, new ideas have emerged in which some additional 'softening' thresholds were introduced. More recent approaches employ other types of discretisations, the so-called *fuzzy discretisations*, in which the hard intervals defined by the cutting points are replaced with fuzzy intervals defined by fuzzy numbers with overlapping bounds. However, there is still some ambiguity in methodology and terminology (see Dougherty, Kohavi and Sahami 1995; Fayyad and Irany 1993, Kerber 1992). For example, besides hard versus fuzzy discretisation, a distinction is also made between supervised versus unsupervised, local versus global, and parametrized versus non-parametrised discretisation.

Supervised methods take into consideration the reported values of the decision attribute [or attributes] when producing the discretisation, the unsupervised methods are, on the other hand, information blind [or class-blind]. Local methods discretise only one condition attribute at a time, while the global ones attempt simulaneous discretisation of all attributes. Parametrised methods are those for which the maximal number of intervals generated for a given feature is specified in advance, while the non-parametrised methods determine this value algorithmically.

In spite of all these extensions, hard discretisation still remains the most commonly used form. It is also the form most frequently adopted in existing discretisation algorithms. Consequently, it has been adopted here also and the discretisation has been implemented in the following way:
- A2 : [before 1985 = 0; from 1985-1996 = 1; after 1996 = 2],
- A4 : [up to 7 = 0 ; from 7- 49 = 1 ; more than 49 =2],
- A5 : [before 1959 = 0; from 1959-1969 = 1; after 1969 = 2],
- A6 : [before 1984 = 0; from 1984-1993 = 1; after 1993 = 2].

As noted in the previous section, Appendix A-18.1 constitutes an information system in which the rows represent the case studies and the columns the attributes considered. Thus, two central questions are important in the process of knowledge representation and acquisition: how the knowledge is organised and how it can be increased. The first relates to the classification ability of the constructed system, i.e. how well it is suited to a meta-analytical perspective, also in the sense that it can be used to classify newly published case studies on economic growth. Classification, which involves the process of finding rules to partition a given data set into disjoint groups, is one class of data-mining problems. It is the process that finds the common properties among a set of objects in a database and divides them into different classes, according to a classification model. The performance of the system as a classifier is measured by the classification accuracy, which can be viewed as the main evaluation criterion.

The second aspect regarding the application of rough set analysis concerns the accumulation of knowledge acquired through learning from case studies or examples. Knowledge is usually represented here in form of *rules*. The generation of rules aims to create a symbolic representation of the knowledge contained in the data; it consists of the process of creating valid generalisations or detecting data patterns.

Hence three approaches were used to evaluate the system. These are considered in the following order: (i) classification accuracy and quality, (ii) core of the attributes, and (iii) rules. The results from carrying out rough set analysis applied to the full database are presented in Table 18.1, which gives the lower and upper approximations, the accuracy and the quality of the classification, and the core of the attributes. Table 18.1 shows that 32 case studies concluded that the impact of government policy on growth was negative, 44 studies were inconclusive and 47 suggested that the impact was positive.

The accuracy of the classification is 0.84. All attributes, except the geographic unit of measurement [countries or regions] and the year of the earliest observation, form the core of the attributes; these are the attributes that cannot be eliminated without disturbing the quality of classification.

Table 18.1   Classification and core of the attributes

Class	Lower Approximation	Upper Approximation	Accuracy
Class 1 – 32 objects	31	34	0.9118
Class 2 – 44 objects	38	49	0.7755
Class 3 – 47 objects	43	51	0.8431
Quality of the Classification:	0.9106		
Accuracy of the Classification:	0.8358		
Core A1, A2, A4, A5, A7, A8, A9			

In the next step, the induction of decision rules was performed. Rules are generated by repeated application of an algorithm to the data. The quality of the rules, and hence the knowledge discovered, is heavily dependent on the algorithm used to analyse the data. Thus, the central problem in knowledge extraction is the choice of techniques to generate such rules. In our experiment we adopted an approach that uses two learning systems for knowledge representation: first in the form of decision rules corresponding to the so-called minimal covering, secondly in the form of generating general knowledge. The rule induction algorithm called 'learning from examples' [LEM2] was used first. This search technique has the technical advantage that it is a 'depth-first' simpler equivalent of the search technique used in the algorithm *Explore* (for details see Stefanoski and Wilk 1999).

The output of LEM2 is a set of rules that is minimal and provides a description of all classes defined only by the examples supporting it [positive examples]. The large number of rules generated [56 exact rules and four approximate ones] and their low quality, expressed in terms of support found by the numerical computation, showed a high level of fragmentation of the data system. The average strength is only three observations per rule (see Table 18.2). It can be shown that for studies reporting a negative or inconclusive impact on growth the highest level of support is represented by only ten percent of these studies.

Table 18.2   Results of applying the LEM 2 algorithm to induce rules

Number of Exact Rules	Number of Approximate Rules	Average Strength	Discrimination Level
56	4	3	100 %

However, the main aim of the LEM2 algorithm is to derive the so-called minimal set of rules covering – ideally, all learning examples. However, it is also useful to consider an induction algorithm that generates rules which cover the most general examples (Stefanoski 1998). This led us to choose another rule induction algorithm. The algorithm *Explore*, originally proposed by Stefanowski and Vanderpoten (see Mienko *et al.* 1996) was chosen as a tool for inducing general rules. This algorithm generates all short and sufficiently 'strong' general rules for a given data set. Examples uncovered by these rules can be used to identify exceptions and atypical cases. The adjective 'strong' means here that the rules are satisfied by a relatively large number of learning examples.

The *Explore* algorithm induces all decision rules which satisfy predefined requirements with respect to the rule strength [that is, the relative number of learning examples that support the rule], the length of the conditioning part [i.e. 'if <conditioning part>, then...'], the level of confidence in the rule, as well as requirements to the syntax of conditions. As argues by Mienko et al. (1996), focusing mainly on the strength of rules will result in obtaining a limited number of rules which cover only a certain subset of the learning examples that represent general information patterns in the data, while leaving uncovered some more difficult and specific examples.

The search for rules in the algorithm *Explore* is controlled by parameters called *stopping conditions SC* that reflect the user's requirements. As our main selection criterion will be the strength of the rules, the definition of *SC* is connected with determining the threshold value for the minimal strength of the conjunction being candidate for the conditioning part of the rule. If its strength is lower than *SC*, it is discarded, otherwise it can be evaluated further. Additionally, one can define a threshold d that expresses the minimum value of the level of discrimination D (R) of the rules to be generated. The algorithm *Explore* is based on the *breadth-first* strategy, which generates rules of increasing size starting from the shortest ones (Stefanoski and Wilk 1999).

In general, *Explore* is an algorithm that incorporates searching strategies to adapt the discovery processes to the characteristics of the applications. Knowledge discovery is evaluated locally in the form of the filtering of redundant rules, that is by finding something that is useful to the user (Deogun et al. 1996). In fact, *Explore* is specifically designed to work with data that changes regularly. This is helpful in our context, as new studies on economic growth are continually emerging and it will be useful to be able to assess whether these new findings are consistent with the knowledge discovered in the past.

In our experiment, we used the following procedure of parameter tuning. Firstly, initial values of threshold were calculated on the basis of the strength of rules generated by the LEM2 algorithm because, as noted above, it is an efficient algorithm to derive the minimal number of rules with a good classification ability. Thus, the following threshold values were adopted: discrimination level = 100 percent, minimal strength of support within the class = 15 percent, maximum length of the conditioning items = five.

Table 18.3   Set of decision rules generated using *Explore* algorithm

---

rule 1. (A1 = 2) & (A2 = 0) => (D1=1); [5, 5, 15.63%, 100.00%] [5, 0, 0]
    [{ 1, 5, 7, 8, 9 }, {}, {}]
rule 2. (A4 = 2) & (A5 = 0) & (A6 = 0) => (D1=1); [5, 5, 15.63%, 100.00%] [5, 0, 0]
    [{ 8, 12, 13, 20, 27 }, {}, {}]
rule 3. (A5 = 0) & (A6 = 0) & (A8 = 3) => (D1=1); [5, 5, 15.63%, 100.00%] [5, 0, 0]
    [{ 8, 12, 13, 20, 27 }, {}, {}]
rule 4. (A1 = 4) & (A8 = 2) => (D1=3); [12, 12, 25.53%, 100.00%] [0, 0, 12]
    [{}, {}, { 6, 25, 26, 37, 51, 53, 60, 97, 100, 103, 112, 114 }]
rule 5. (A1 = 4) & (A2 = 1) & (A5 = 0) => (D1=3); [10, 10, 21.28%, 100.00%] [0, 0, 10]
    [{}, {}, { 25, 26, 37, 47, 51, 59, 60, 97, 100, 103 }]
rule 6. (A1 = 4) & (A3 = 0) & (A4 = 1) => (D1=3); [11, 11, 23.40%, 100.00%] [0, 0, 11]
    [{}, {}, { 6, 25, 26, 37, 51, 53, 60, 97, 103, 109, 122 }]
rule 7. (A1 = 4) & (A3 = 0) & (A9 = 1) => (D1=3); [8, 8, 17.02%, 100.00%] [0, 0, 8]
    [{}, {}, { 24, 37, 53, 97, 109, 115, 121, 122 }]
rule 8. (A1 = 4) & (A4 = 1) & (A5 = 0) => (D1=3); [9, 9, 19.15%, 100.00%] [0, 0, 9]
    [{}, {}, { 6, 25, 26, 37, 47, 51, 60, 97, 103 }]
rule 9. (A1 = 4) & (A4 = 1) & (A6 = 1) => (D1=3); [12, 12, 25.53%, 100.00%] [0, 0, 12]
    [{}, {}, { 25, 26, 37, 38, 43, 51, 52, 53, 60, 97, 103, 122 }]
rule 10. (A1 = 4) & (A4 = 1) & (A9 = 1) => (D1=3); [8, 8, 17.02%, 100.00%] [0, 0, 8]
    [{}, {}, { 37, 38, 43, 53, 97, 109, 119, 122 }]
rule 11. (A1 = 4) & (A5 = 0) & (A6 = 1) => (D1=3); [11, 11, 23.40%, 100.00%] [0, 0, 11]
    [{}, {}, { 25, 26, 37, 51, 59, 60, 97, 100, 103, 112, 114 }]

Table 18.3 shows that the resulting output consists of eleven rules. Given that LEM2 generated 60 rules, we can conclude that significant progress has been made towards knowledge reduction. Interestingly, no rules were found for the class of inconclusive studies regarding the impact of public policy on economic growth. Regarding a negative impact of policy on growth, only three rules were obtained, all of which are supported by five examples [studies] and hence with a support of 5/32 [15.63 percent]. Specifically, we observe that:

- *If* the study focuses on defence expenditure and the year of publication is before 1985 then negative impact (Table 18.3 rule 1);
- *If* the number of observations is more than 49 and the year of the earliest observation is before 1959 and the year of the most recent observation is before 1984 then negative impact (Table 18.3, rule 2);
- *If* the year of the earliest observation is before 1959 and the year of the most recent *observation* is before 1984 and the method adopted is pooled cross section time series analysis then negative impact (Table 18.3, rule 3).

Furthermore, we obtained a significant of rules with respect to the class of a positive policy impact on economic growth. The support ranged from 8 to 12 examples while the relative support consequently ranged from 17.02 percent to 25.53 percent. All eight rules referred to studies of the impact on growth of public investment in infrastructure. In particular, we have:

- *If* the study focuses on infrastructure expenditure and the method adopted is time series analysis then positive impact (Table 18.3, rule 4);
- *If* the study focuses on infrastructure expenditure and year of the publication is between 1985 and 1996 and year of the earliest observation is before 1959 then positive impact (Table 18.3, rule 5);
- *If* the study focuses on infrastructure expenditure and study data are at country level and number of observations are between 7 and 49 then positive impact (Table 18.3, rule 6);
- *If* the study focuses on infrastructure expenditure and study data are at the country level and published at the unclassified journal level then positive impact (Table 18.3, rule 7);
- *If* the study focuses on infrastructure expenditure and the number of observations is between 7 and 49 and year of the earliest observation is before 1959 then positive impact (Table 18.3, rule 8);
- *If* the study focuses on infrastructure expenditure and number of observations is between 7 and 49 and year of the most recent observation is between 1984 and 1993 then positive impact (Table 18.3, rule 9);
- *If* the study focuses on infrastructure expenditure and number of observations is between 7 and 49 and published in the unclassified journal level then positive impact (Table 18.3, rule 10);
- *If* the study focuses on infrastructure expenditure and year of the earliest observation is before 1959 and year of the most recent observation is between 1984 and 1993 then positive impact (Table 18.3, rule 11).

Such class descriptions have of course great potential and can be used to classify further studies in the field of economic growth research. Nevertheless, if this process is used to classify new additional knowledge, we are interested in verifying the possibility that additional information is likely to improve our knowledge. Because of this concern with knowledge, rather than simply looking for accurate prediction, a more in-depth analysis was sought. There are also practical reasons for doing this. Meta-analysis makes specific demands on rough set algorithms. Even when predictive decision rules are the sole goal, comprehensibility is an important condition because it facilitates the process of interactive refinement that is at the heart of most successful artificial intelligence applications.

Thus, the subset of studies concerning government spending was divided into five data sets: fiscal policy, defence expenditure, taxation policy, education policy and infrastructure policy. As in the previous rule generation experiment on the overall data set, we used a parametric fine-tuning procedure. Here we chose the same *SC* as for the full data set, but with the difference that an absolute strength equal to five was adopted. However, an absolute class strength of three was used in the case of taxation studies due to the limited number of examples [10] present in the data set.

Table 18.4 shows that in the case of studies of pure fiscal policy [i.e. government expenditure as a proportion of GDP], the attributes 'year of the publication', 'year of the earliest observation', 'level of development', 'method' and 'journal level' make up the core of the attributes. The algorithm generated the following rules:

- *If* the year of publication is between 1985 and 1996 *and* the number of observations is more than 49 *and* the year of the first observation is between 1959-1969 and low journal level *then* the impact is inconclusive (Table 18.5, rule 1);
- *If* year of the publication is between 1985 and 1996 *and* the year of the recent observation is between 1959-1969 *and* year of the most recent observation is between 1984 and 1993 *and* low journal level *then* the impact is inconclusive (Table 18.5, rule 2);
- *If* country level data *and* year of the most recent observation is between 1984 and 1993 *and* data on developed countries *and* low journal level *then* the impact is inconclusive (Table 18.5, rule 3).

Table 18.4 Subset of fiscal policy studies (A1=1): Approximation, quality and accuracy of the classification and core of the attributes

Class: 1	Class: 2	Class: 3
Objects: 12	Objects: 22	Objects: 7
Lower: 11 ; Upper: 14	Lower: 18;Upper: 25	Lower: 5;Upper: 9
Accuracy: 0.7857	Accuracy: 0.7200	Accuracy: 0.5556
Quality: 0.8293; Accuracy: 0.7083		
Core: 0.7805; Attributes: A2, A5,A7,A8,A9		

Table 18.5 Subset of fiscal policy studies (A1=1): Rules

Parameters: Class: *General* ->   SC1 (absolute) = 5   SC2 = 5   D = 100
rule 1. (A2 = 1) & (A4 = 2) & (A5 = 1) & (A9 = 1) => (D1=2); [5, 5, 22.73%, 100.00%] [0, 5, 0]
rule 2. (A2 = 1) & (A5 = 1) & (A6 = 1) & (A9 = 1) => (D1=2); [6, 6, 27.27%, 100.00%] [0, 6, 0]
rule 3. (A3 = 0) & (A6 = 1) & (A7 = 3) & (A9 = 1) => (D1=2); [5, 5, 22.73%, 100.00%] [0, 5, 0]

Table 18.6 shows that the subset of 22 defence expenditure studies has the following core attributes: 'year of publication', 'number of observations', 'year of

the earliest observation', 'level of development' and 'method used'. Table 18.7 shows that only one rule was generated:

- *If* data at *country* level *then* negative impact (Table 18.7, rule 1).

Table 18.6   Subset of defence policy studies (A1=2): Approximation, quality and accuracy of the classification and core of the attributes

Class: 1	Class: 2	Class: 3
Objects: 12	Objects: 9	Objects: 1
Lower: 12; Upper: 12	Lower: 9;Upper: 9	Lower: 1; Upper: 1
Accuracy: 1.0000	Accuracy: 1.0000	Accuracy: 1.0000
Quality: 1.0000; Accuracy: 1.0000		
Core: 1.0000; Attributes :A2, A4,A5,A7,A8		

Table 18.7   Subset of defence policy studies (A1=2): Rules

Parameters: Class: *General* ->   SC1 (absolute) = 5   SC2 = 5   D = 100
rule 1. (A2 = 0) => (D1=1); [5, 5, 41.67%, 100.00%] [5, 0, 0]

Table 18.8 shows that the subset of 10 tax policy studies has an empty core. The rules found are given in Table 18.9:

- *If* data at regional level *then* negative impact (Table 18.9, rule 1);
- *If* low journal level *then* negative impact (Table 18.9, rule 2);
- *If* year of the publication between 1985 and 1996 and high journal level *then* inconclusive impact (Table 18.9, rule 3).

Table 18.8   Subset of tax policy studies (A1=3): Approximation, quality and accuracy of the classification and core of the attributes

Class: 1	Class: 2
Objects: 6	Objects: 4
Lower: 6; Upper: 6	Lower: 4; Upper: 4
Accuracy: 1.0000	Accuracy: 1.0000
Quality: 1.0000; Accuracy: 1.0000	
Core: 0.0000	

Table 18.9   Subset of tax policy studies (A1=3): Rules

Parameters: Class: *General* ->   SC1 (absolute) = 3   SC2 = 5   D = 100
rule 1. (A3 = 1) => (D1=1); [3, 3, 50.00%, 100.00%] [3, 0]
rule 2. (A9 = 1) => (D1=1); [3, 3, 50.00%, 100.00%] [3, 0]
rule 3. (A2 = 1) & (A9 = 3) => (D1=2); [3, 3, 75.00%, 100.00%] [0, 3]

In the subset of 38 infrastructure studies, the core is composed of 'number of observations', 'the year of the earliest observation' and 'journal level' (see Table 18.10). The algorithm generated twelve rules for a positive impact. These are fully reported in Table 18.11. For example, rule 10 states that:

- *If* the year of the earliest observation is before 1959 *and* the year of the most recent *observation* is between 1984 and 1993 *then* positive impact (Table 18.11, rule 10).

Table 18.10 Subset of infrastructure policy studies (A1=4): Approximation, quality and accuracy of the classification and core of the attributes

Class: 1	Class: 2	Class: 3
Objects: 2	Objects: 8	Objects: 28
Lower: 2; Upper: 2	Lower: 7; Upper: 9	Lower: 27; Upper: 29
Accuracy: 1.0000	Accuracy: 0.7778	Accuracy: 0.9310
Quality: 0.9474; Accuracy: 0.9000		
Core: 0.7895; Attributes A4; A5,A9		

Table 18.11 Subset of infrastructure policy studies (A1=4): Rules

Parameters: Class: *General* -> SC1 (absolute) = 5   SC2 = 5   D = 100

rule 1. (A2 = 2) => (D1=3); [7, 7, 25.00%, 100.00%] [0, 0, 7]
rule 2. (A8 = 2) => (D1=3); [12, 12, 42.86%, 100.00%] [0, 0, 12]
rule 3. (A2 = 1) & (A5 = 0) => (D1=3); [10, 10, 35.71%, 100.00%] [0, 0, 10]
rule 4. (A3 = 0) & (A4 = 1) => (D1=3); [11, 11, 39.29%, 100.00%] [0, 0, 11]
rule 5. (A3 = 0) & (A9 = 1) => (D1=3); [8, 8, 28.57%, 100.00%] [0, 0, 8]
rule 6. (A4 = 1) & (A5 = 0) => (D1=3); [9, 9, 32.14%, 100.00%] [0, 0, 9]
rule 7. (A4 = 1) & (A5 = 2) => (D1=3); [7, 7, 25.00%, 100.00%] [0, 0, 7]
rule 8. (A4 = 1) & (A6 = 1) => (D1=3); [12, 12, 42.86%, 100.00%] [0, 0, 12]
rule 9. (A4 = 1) & (A9 = 1) => (D1=3); [8, 8, 28.57%, 100.00%] [0, 0, 8]
rule 10. (A5 = 0) & (A6 = 1) => (D1=3); [11, 11, 39.29%, 100.00%] [0, 0, 11]
rule 11. (A5 = 0) & (A9 = 3) => (D1=3); [6, 6, 21.43%, 100.00%] [0, 0, 6]
rule 12. (A7 = 3) & (A9 = 3) => (D1=3); [6, 6, 21.43%, 100.00%] [0, 0, 6]

For the sake of brevity, we will not describe the others in sentence form. Table 18.12 shows that the subset of education expenditure studies has an empty core. Table 18.13 shows that two rules were generated by the algorithm:

- *If* year of publication is after 1996 *then* positive impact (Table 18.13, rule 1);
- *If* the number of observations is between 7 and 49 *then* positive impact (Table 18.13, rule 2).

As noted earlier, we also divided the original data set into three subsets with respect to the level of development of the countries concerned. The three subsets relate to: less developed countries, mixed countries and developed countries. Due to the different sizes of these subsets, a new parameter tuning procedure was adopted. Specifically, the following *SC* have been used: absolute strength = 3 for the less developed and mixed countries subsets; absolute strength = 8 for the developed countries subset, and maximum conditioning length = 5, discrimination level = 100 percent for all three subsets.

Table 18.12   Subset of education policy studies (A1=5): Approximation, quality and accuracy of the classification and core of the attributes

Class: 2	Class: 3
Objects: 1	Objects: 11
Lower: 0; Upper: 2	Lower: 10; Upper: 12
Accuracy: 0.0000	Accuracy: 0.8333
Quality: 0.8333; Accuracy: 0.7143	
Core: 0.0000	

Table 18.13   Subset of education policy studies (A1=5): Rules

Parameters: Class: *General* ->   SC1 (absolute) = 5   SC2 = 5   D = 100
rule 1. (A2 = 2) => (D1=3); [5, 5, 45.45%, 100.00%] [0, 5]
rule 2. (A4 = 1) => (D1=3); [5, 5, 45.45%, 100.00%] [0, 5]

A very significant result is that the type of government policy [government size, taxation policy, etc.] is part of the core for all three subsets. Table 18.14 shows that in the subset of less developed countries the attributes of the core are 'government spending', 'year of the publication', 'number of observations', 'year of the earliest observation' and 'method'. The principal rule in this subset is:

• *If* year of the publication between 1985-1996 *and* year of the earliest observation is between 1959 and 1969 *and* year of the most recent observation is between 1984 and 1993 *then* inconclusive impact (Table 18.15, rule 3).

Table 18.14   Subset of less developed countries (A7=1): Approximation, quality and accuracy of the classification and core of the attributes

Class: 1	Class: 2	Class: 3
Objects: 11	Objects: 10	Objects: 7
Lower: 11;Upper: 11	Lower: 10; Upper: 10	Lower: 7; Upper: 7
Accuracy: 1.0000	Accuracy: 1.0000	Accuracy: 1.0000
Quality: 1.0000; Accuracy: 1.0000		
Core A1, A2, A4, A5, A8		

Table 18.15   Subset of less developed countries (A7=1): Rules

Parameters: Class: *General*   ->   SC1 (absolute) = 3   SC2 = 5   D = 100
rule 1. (A1 = 2) & (A8 = 2) => (D1=2); [3, 3, 30.00%, 100.00%] [0, 3, 0]
rule 2. (A1 = 2) & (A2 = 1) & (A5 = 1) => (D1=2); [3, 3, 30.00%, 100.00%] [0, 3, 0]
rule 3. (A2 = 1) & (A5 = 1) & (A6 = 1) => (D1=2); [4, 4, 40.00%, 100.00%] [0, 4, 0]
rule 4. (A2 = 1) & (A5 = 1) & (A8 = 3) => (D1=2); [3, 3, 30.00%, 100.00%] [0, 3, 0]
rule 5. (A2 = 1) & (A3 = 0) & (A4 = 2) & (A5 = 1) => (D1=2); [3, 3, 30.00%, 100.00%] [0, 3, 0]

In the subset of mixed countries, the core is made up of the attributes 'government spending', 'year of the publication', 'method' and 'journal level' (Table 18.16). Two strong rules were generated:

- *If* year of the publication is after 1996 *and* year of the earliest observation is after 1969 *then* positive impact (Table 18.17, rule 3);
- *If* year of the earliest observation is after 1969 *and* lowest journal level *then* positive impact (Table 18.17, rule 5).

Table 18.16   Subset of mixed countries (A7=2): Approximation, quality and accuracy of the classification and core of the attributes

Class: 1	Class: 2	Class: 3
Objects: 8	Objects: 15	Objects: 12
Lower: 7; Upper: 10	Lower: 10; Upper: 19	Lower: 9;Upper: 15
Accuracy: 0.7000	Accuracy: 0.5263	Accuracy: 0.6000
Quality: 0.7429; Accuracy: 0.5909		
Core attributes A1, A2, A8, A9		

Table 18.17   Subset of mixed countries (A7=2): Rules

Parameters: Class: *General*   ->   SC1 (absolute) = 3   SC2 = 5   D = 100
rule 1. (A1 = 4) & (A5 = 2) => (D1=3); [3, 3, 25.00%, 100.00%] [0, 0, 3]
rule 2. (A2 = 2) & (A4 = 1) => (D1=3); [3, 3, 25.00%, 100.00%] [0, 0, 3]
rule 3. (A2 = 2) & (A5 = 2) => (D1=3); [5, 5, 41.67%, 100.00%] [0, 0, 5]
rule 4. (A2 = 2) & (A8 = 1) => (D1=3); [3, 3, 25.00%, 100.00%] [0, 0, 3]
rule 5. (A5 = 2) & (A9 = 1) => (D1=3); [4, 4, 33.33%, 100.00%] [0, 0, 4]

In the subset of developed countries, the core consists of the attributes 'government spending', 'number of observations', 'year of the earliest observation' and 'journal level' (Table 18.18). Six strong rules were generated which reconfirmed the results obtained in the full set of studies. For example, rule 1 in Table 18.19 is the same as rule 4 in Table 18.3, rule 2 in Table 18.19 is the same as rule 5 in Table 18.3, etc.

Table 18.18 Subset of developed countries (A7=3): Approximation, quality and accuracy of the classification and core of the attributes

Class: 1	Class: 2	Class: 3
Objects: 13	Objects: 19	Objects: 28
Lower: 13; Upper: 13	Lower: 18; Upper: 20	Lower: 27; Upper: 29
Accuracy: 1.0000	Accuracy: 0.9000	Accuracy: 0.9310
Quality: 0.9667; Accuracy: 0.9355		
Core attributes: A1, A4, A5, A9		

Table 18.19 Subset of developed countries (A7=3): Rules

Parameters: Class: *General* ->	SC1 (absolute) = 8 SC2 = 5 D = 100

rule 1. (A1 = 4) & (A8 = 2) => (D1=3); [12, 12, 42.86%, 100.00%] [0, 0, 12]
rule 2. (A1 = 4) & (A2 = 1) & (A5 = 0) => (D1=3); [10, 10, 35.71%, 100.00%] [0, 0, 10]
rule 3. (A1 = 4) & (A3 = 0) & (A4 = 1) => (D1=3); [9, 9, 32.14%, 100.00%] [0, 0, 9]
rule 4. (A1 = 4) & (A4 = 1) & (A5 = 0) => (D1=3); [9, 9, 32.14%, 100.00%] [0, 0, 9]
rule 5. (A1 = 4) & (A4 = 1) & (A6 = 1) => (D1=3); [11, 11, 39.29%, 100.00%] [0, 0, 11]
rule 6. (A1 = 4) & (A5 = 0) & (A6 = 1) => (D1=3); [11, 11, 39.29%, 100.00%] [0, 0, 11]

18.6 Conclusions

After the computations of the previous section, we may conclude that significant progress has been made towards achieving the goals of accuracy, stability and comprehensibility of the meta-analysis of growth studies. It turned out in our empirical analysis that the classification showed a stable result from the point of view of accuracy, but the overarching goal of output comprehensibility remains elusive. The present study has aimed to move closer to this ideal by having proposed a learning method combining some of the accuracy and stability gains of a multiple data set with the comprehensibility of a single overall data set.

Of course, the results by themselves do not explain the uneven distribution of growth among different countries and regions. Output comprehensibility is more difficult to measure than accuracy, since it is ultimately subjective. The flexibility of the approach, while essential to improving our knowledge in the field, has the disadvantage of allowing data mining algorithms to be overly responsive to the full data set, producing models that can change dramatically with small changes in the data. Of course there is an analogy with classical regression analysis that model over-fitting leads to excellent goodness of fit statistics but bad forecasts. In the same way, instability undermines the claim of artificial intelligence systems to be efficient in the production of knowledge. However, the rough set approach has shown a great potential for knowledge reduction, which is the main focus of this line of research (Famili et al. 1997).

In this contribution we have attempted to identify a robust set of predictive statements regarding the impact of government on long-run growth. In conventional econometric analyses, this type of endeavour would have involved the collection of a range of characteristics for a sample of constituencies and the statistical inference of causal relationships from such characteristics to the economic growth rates of the constituencies. However, the impossibility of carrying out such causal analyses in controlled experimental settings have in the past thwarted our ability to derive strong predictive statements. Missing relevant variables and a space-time dependency of the causal relationships have contributed to this failure, and are partly responsible for the 'dismal science' label of our discipline.

However, the prospects for empirical verification of the body of current theories are becoming brighter. There are at least four reasons. Firstly, the information and communication technology [ICT] revolution has contributed to making the findings of empirical research a greater availability far more rapidly and readily. Working papers can be easily downloaded from the Internet and published results that would traditionally have only been available from a few local libraries can now be accessed throughout the world. Secondly, ICT developments have also decreased the cost of data collection in the form of surveys. The growing number of purpose-designed surveys is providing large new micro-level data sets, often of a panel nature. Thirdly, modern ICT generates vast amounts of data as the by-product of financial and administrative transactions. Although privacy issues are of increasing concern, there is no doubt that there has been a great improvement in the availability of rich information on many socio-economic phenomena [although special features of information, such as monopoly supply and near zero marginal costs of distribution create special pricing problems]. Fourthly, new methods have become available for the analysis of such data. These include neural network analysis, non-linear dynamics and chaos theory, logit models, GIS systems, multi criteria analysis, rough set analysis and other methods for the study of soft or qualitative data.

In this contribution we have adopted the method of rough set analysis to identify specific predictive patterns in a sample of 123 empirical studies of the impact of fiscal policies on the long-run growth rate. The vast majority of these studies were conventional cross-section or panel data regression models. Our results reconfirm other recent evidence, provided for instance by Kneller, Bleaney and Gemmell (1999), that productive government expenditure [education and infrastructure] enhance growth while non-productive expenditure [public services, administration, transfer payments] do not. However, the results are sensitive to various characteristics of the studies. For example, early studies were more supportive of a 'peace dividend' than more recent ones.. Also, we find that the impact of infrastructural investment cannot be detected in the short-run: the positive impact of infrastructure on growth can only be detected in time-series analyses, not in cross section analyses. Furthermore, the impact of government expenditure on growth appears to depend on whether the studies are concerned with developed or less developed countries. Recently, Obersteiner and Wilk (1999) also showed by means of rough set analysis of the growth experience of a cross section of countries that a partitioning into industrialised and non-industrialised countries is essential.

While rough set analysis has enabled us to identify a few firm conclusions from a large body of empirical literature, the method has also some limitations. One problem is that there is a trade-off between sample size and the range of attributes. Large samples may be considered more representative of the available literature than small samples, but they are also more heterogeneous and the range of attributes that are common to all studies is correspondingly rather small. A much wider range of attributes may be available for a small range of more narrowly focussed studies, e.g. all studies concerned with the impact on growth of an increase in R&D subsidies. In this case meta-regression analysis may also become feasible, but the value transfer of the findings from a few location, time and method-specific studies may then be in doubt.

A weakness that rough set analysis has in common with some other meta-analytic techniques is the potential sensitivity of the results to publication bias. If it is true that it is easier to get results that report significant findings published than insignificant ones, then the proportion of studies in our sample revealing a conclusive positive or negative fiscal impact, 64 percent, may be overestimated. One indication of publication bias is the absence of a link between the number of observations and the levels of the decision variable. Statistical theory would suggest that conclusive results [positive or negative] are less common among data sets with a large number of observations, but our sample does not appear to support this prediction. Future research will investigate to what extent pre-testing and specification searches have affected published findings in this literature.

While our rough set analysis highlights the importance of education and infrastructure expenditure, the available information is unfortunately not specific enough to translate the findings into policy recommendations. Moreover, much of the literature ignores the budget constraints under which governments must operate. Another issue is the potential endogeneity of the fiscal variables. Wagner's law suggests that the share of government expenditure in total income is positively related to per capita income. This may explain the growth in the public sector in developed economies in the second half of the 20th century, although Olekalns (1999) recently showed that some of this correlation was due the changing age distribution of the population. However, without substituting suitable instruments for the fiscal variables in the growth regression models, the results may again be biased, and this would also affect our rough set analysis. Indeed, there appears to be scope for investigating causality and endogeneity issues in rough set analysis by assessing the implications of switching condition and decision attributes.

Finally, the present analysis is too narrowly focussed to inform policy. Much of government expenditure is concerned with equity considerations rather than efficiency. The issue of whether government faces a trade-off in its pursuit of long-run economic growth and an equitable distribution of income has not yet been resolved (see Aghion, Caroli and Garcia-Penalosa 1999 for a recent survey). Nonetheless, there would be merit in extending the rough set analysis of the impact of fiscal variables to one with multiple decision variables, in which each variable measures one of a range of public policy objectives.

Appendix A-18.1: The Information Table

Objects / Authors [see Appendix A-18.2]	Observation Number	A1 Type of Government Spending C=1, D=2, T=3, I=4, E=5	A2 Year of Publication	A3 Countries (0) or Regions (1)	A4 Number of Observations	A5 Year of Earliest Observation	A6 Year of Most Recent Observation	A7 Level of Development 1=LDC 2=mixed 3=DC	A8 Method CS=1 TS=2 CSTS=3 Other=4	A9 Journal Level 1=low to 4=top journal	D1 Fiscal Impact 1=negative 2=inconclusive 3=positive
Deger and Smith	1	2	1983	0	50	1965	1973	2	3	1	1
Gemmell	2	1	1983	0	54	1960	1970	2	3	2	2
Landau	3	5	1983	0	96	1960	1977	2	1	2	3
Landau	4	1	1983	0	96	1960	1977	2	1	2	1
Lim	5	2	1983	0	54	1965	1973	1	1	2	1
Ratner	6	4	1983	0	24	1949	1973	3	2	3	3
Cappelen, Gleditsch and Bjerkholt	7	2	1984	0	85	1960	1980	3	3	1	1
Faini, Annez and Taylor	8	2	1984	0	1,242	1952	1970	2	3	2	1
Lindgren	9	2	1984	0	41	1968	1984	3	4	1	1
Helms	10	3	1985	1	672	1965	1979	3	3	4	1
Kormendi and Meguire	11	1	1985	0	47	1950	1977	2	1	3	2
Landau	12	1	1985	0	384	1952	1976	3	3	2	1
Landau	13	4	1985	0	384	1952	1976	3	3	2	1
Saunders	14	1	1985	0	46	1960	1981	3	3	1	2
Biswas and Ram	15	2	1986	0	116	1960	1977	1	3	2	2
Landau	16	1	1986	0	1,152	1960	1980	2	3	2	1
Landau	17	4	1986	0	1,152	1960	1980	2	3	2	2
Ram	18	1	1986	0	230	1960	1980	2	3	4	3
Ram	19	1	1986	0	2,300	1960	1980	2	3	4	3
Canto and Webb	20	3	1987	1	960	1957	1977	1	3	2	1
da Silva Costa, Ellson and Martin	21	4	1987	1	48	1972	1972	3	1	2	3
Bairam	22	1	1988	0	20	1960	1980	3	2	1	3
Grossman	23	1	1988	0	34	1949	1984	3	2	2	2
Aschauer - a	24	4	1989	0	133	1966	1985	3	3	1	3
Aschauer - b	25	4	1989	0	36	1949	1985	3	2	3	3
Aschauer - c	26	4	1989	0	33	1953	1986	3	2	3	3
Grier and Tullock	27	1	1989	0	565	1950	1981	2	3	3	1

ctd.

Study											
Gyimah-Brempong	29	2	1989	0	390	1973	1983	1	3	1	1
Grobar and Porter	28	2	1989	0	29	1972	1988	1	4	1	1
Koester and Kormendi	30	3	1989	0	63	1970	1979	2	1	3	2
Rao	31	1	1989	0	230	1960	1980	2	3	4	2
Rao	32	1	1989	0	2,300	1960	1980	2	3	4	2
Scully	33	1	1989	0	115	1960	1980	1	3	2	1
Bairam	34	1	1990	0	300	1960	1985	2	3	1	2
Grossman	35	1	1990	0	48	1970	1983	3	1	1	2
Mullen and Williams	36	4	1990	1	29	1963	1966	2	1	2	2
Munnell - a	37	4	1990	0	38	1949	1987	3	2	2	3
Munnell - b	38	4	1990	1	48	1970	1986	3	1	1	3
Barro	39	5	1991	0	98	1960	1985	2	1	4	3
Barro	40	1	1991	0	98	1960	1985	2	1	4	1
Barro	41	4	1991	0	98	1960	1985	2	1	4	2
Chowdhury	42	2	1991	0	1,430	1961	1987	1	2	1	2
Eisner	43	4	1991	1	48	1970	1986	3	4	1	3
Hulten and Schwab	44	4	1991	1	144	1970	1986	3	3	2	2
Hulten and Schwab	45	4	1991	1	144	1970	1986	3	4	2	2
Moomaw and Williams	46	5	1991	1	48	1954	1976	3	1	2	3
Moomaw and Williams	47	4	1991	1	48	1954	1976	3	3	2	3
Yu, Wallace and Nardinelli	48	3	1991	1	336	1929	1985	3	1	1	1
Levine and Renelt	49	5	1992	0	103	1960	1989	2	1	4	2
Levine and Renelt	50	1	1992	0	103	1960	1989	2	1	4	2
Lynde and Richmond	51	4	1992	0	32	1958	1989	3	2	4	3
Munnell	52	4	1992	1	38	1973	1992	3	4	2	3
Bajo-Rubio and Sosvilla-Rivero	53	4	1993	0	25	1964	1988	1	2	1	3
Binswanger et al.	54	4	1993	1	85	1960	1981	3	1	3	3
Durden and Elledge	55	1	1993	1	48	1982	1993	3	1	1	1
Easterly and Rebelo	56	3	1993	0	100	1970	1988	2	1	3	2
Easterly and Rebelo	57	4	1993	0	100	1970	1988	2	1	3	3
Easterly and Rebelo	58	3	1993	0	3,304	1870	1988	3	3	3	2
Easterly and Rebelo	59	4	1993	0	3,304	1870	1988	3	3	3	3
Lynde and Richmond	60	4	1993	0	32	1958	1989	3	2	4	3

ctd.

61	Mohammed	2	1993	0	390	1973	1983	1	4	1	2
62	Park	2	1993	0	25	1963	1987	1	2	1	2
63	Sattar	1	1993	0	560	1950	1985	1	3	1	3
64	Sattar	1	1993	0	280	1950	1985	3	3	1	2
65	Sheehey	1	1993	0	102	1960	1980	2	1	1	2
66	van Sinderen	3	1993	0	1	1985	1985	3	4	1	
67	Assane and Pourgerami	1	1994	1	46	1970	1990	1	3	1	1
68	Evans and Karras	5	1994	1	768	1970	1986	3	3	4	3
69	Evans and Karras	1	1994	1	768	1970	1986	3	3	4	2
70	Evans and Karras	4	1994	1	768	1970	1986	3	3	4	1
71	Hansen - a	1	1994	0	242	1966	1988	3	3	1	2
72	Hansen - b	1	1994	0	92	1968	1991	3	2	1	2
73	Hansson and Henrekson	5	1994	0	153	1970	1987	3	1	2	3
74	Hansson and Henrekson	1	1994	0	153	1970	1987	3	1	2	1
75	Hansson and Henrekson	4	1994	0	153	1970	1987	3	1	2	2
76	Holtz-Eakin	4	1994	1	816	1969	1986	3	3	4	2
77	Hsieh and Lai	1	1994	0	714	1885	1987	3	2	1	2
78	Kusi	2	1994	0	1,386	1971	1989	1	2	1	2
79	Lee and Lin	1	1994	0	114	1960	1985	2	1	4	2
80	Lin	1	1994	0	20	1960	1985	3	1	1	2
81	Lin	1	1994	0	42	1960	1985	1	1	1	2
82	Sala-i-Martin	1	1994	0	12	1986	1993	2	4	3	2
83	Sala-i-Martin	5	1994	0	12	1986	1993	2	4	3	3
84	Andrews and Swanson	4	1995	1	768	1970	1986	3	3	3	3
85	Berthelemy, Herrera and Sen	2	1995	0	2	1972	1972	1	4	1	1
86	Chletsos and Kollias	2	1995	0	17	1974	1990	3	2	1	2
87	Garrison and Lee	1	1995	0	67	1960	1987	2	1	1	2
88	Garrison and Lee	3	1995	0	67	1960	1987	2	1	1	1
89	Holtz-Eakin and Schwartz	4	1995	1	720	1971	1986	3	3	1	2
90	Karikari	1	1995	0	21	1963	1984	1	2	1	1
91	Macnair et al.	2	1995	0	370	1951	1988	3	3	2	1
92	Macnair et al.	1	1995	0	370	1951	1988	3	3	2	3
93	Andres, Domenech and Molinas	1	1996	0	720	1960	1990	3	3	3	2

396   *P. Nijkamp, J. Poot and G. Vindigni*

Author											ctd
Devarajan, Swaroop and Zou	94	1	1996	0	860	1970	1990	1	3	3	3
Devarajan, Swaroop and Zou	95	4	1996	0	860	1970	1990	1	3	3	1
Dunne	96	2	1996	0	54	1973	1996	1	4	1	2
Harmatuck	97	4	1996	0	36	1949	1985	3	2	1	3
Kocherlakota and Yi	98	3	1996	0	71	1917	1988	3	2	4	2
Kocherlakota and Yi	99	2	1996	0	71	1917	1988	3	2	4	2
Kocherlakota and Yi	100	4	1996	1	71	1917	1988	3	2	4	3
Morrison and Schwartz	101	4	1996	0	816	1970	1987	3	3	4	3
Roux	102	2	1996	0	30	1960	1990	3	2	1	1
Wylie	103	4	1996	0	45	1946	1991	3	2	3	3
Ansari and Singh	104	5	1997	0	36	1951	1987	1	2	1	3
Barro	105	5	1997	0	3,000	1960	1990	2	3	1	3
Barro	106	1	1997	0	3,000	1960	1990	2	3	3	1
Brumm	107	2	1997	0	88	1974	1989	1	1	1	3
Glomm and Ravikumar	108	5	1997	0	31	1983	1994	2	4	1	3
Glomm and Ravikumar	109	4	1997	0	31	1983	1994	2	4	1	3
Guseh	110	1	1997	0	1,475	1960	1985	1	3	1	1
Kocherlakota and Yi	111	3	1997	0	320	1831	1991	3	2	3	1
Kocherlakota and Yi	112	4	1997	0	320	1831	1991	3	2	3	3
Kollias and Makrydakis	113	2	1997	0	39	1954	1993	1	2	2	2
Lau and Sin	114	4	1997	0	64	1925	1989	3	2	2	3
Odedokun	115	4	1997	0	960	1970	1990	1	3	1	3
Singh and Weber	116	5	1997	0	44	1950	1994	3	2	1	3
Baffes and Shah	117	2	1998	0	420	1965	1984	1	3	2	1
Baffes and Shah	118	5	1998	0	420	1965	1984	1	3	2	3
Button	119	4	1998	1	28	1973	1994	3	4	1	3
Cronovich	120	1	1998	0	30	1970	1990	2	1	2	3
Sanchez-Robles	121	4	1998	0	57	1970	1992	2	1	1	3
Sanchez-Robles	122	4	1998	0	19	1970	1985	1	1	1	3
Zhang and Zou	123	1	1998	0	420	1978	1992	1	3	3	2

Appendix A-18.2:
The 93 articles used in compiling Appendix A-18.1

Andrés J., Doménech R. and Molinas C. (1996): Macroeconomic Performance and Convergence in OECD Countries, *European Economic Review* 40, 1683-1704

Andrews K. and Swanson J. (1995): Does Public Infrastructure Affect Regional Performance? *Growth and Change* 26 (2), 204-216

Ansari M.I. and Singh S.K. (1997): Public Spending on Education and Economic Growth in India: Evidence From VAR Modelling, *Indian Journal of Applied Economics* 6 (2), 43-64

Aschauer D.A. (1989a): Public Investment and Productivity Growth in the Group of Seven, *Economic Perspectives* 13 (5), 17-25

Aschauer D.A. (1989b): Is Public Expenditure Productive? *Journal of Monetary Economics* 23 (2), 177-200

Aschauer D.A. (1989c): Does Public Capital Crowd Out Private Capital? *Journal of Monetary Economics* 24, 171-188

Assane D. and Pourgerami A. (1994): Monetary Co-operation and Economic Growth in Africa: Comparative Evidence From the CFA-Zone Countries, *Journal of Development Studies* 30 (2), 423-442

Baffes J. and Shah A. (1998): Productivity of Public Spending, Sectorial Allocation Choices and Economic Growth, *Economic Development and Cultural Change* 46 (2), 291-303

Bairam E. (1988): Government Expenditure and Economic Growth: Some Evidence From New Zealand Time Series Data, *Keio Economic Studies* 25 (1), 59-66

Bairam E. (1990): Government Size and Economic Growth: The African Experience, 1960-1985, *Applied Economics* 22 (10), 1427-1435

Bajo-Rubio O. and Sosvilla-Rivero S. (1993): Does Public Capital Affect Private Sector Performance? An Analysis of the Spanish Case, 1964-88, *Economic Modelling* 10, 179-185

Barro R.J. (1991): Economic Growth in a Cross-Section of Countries, *Quarterly Journal of Economics* (May), 407-443

Barro R.J. (1997): *Determinants of Economic Growth: A Cross-Country Empirical Study*, MIT Press, Cambridge [MA]

Berthélemy J.C., Herrera R. and Sen S. (1995): Military Expenditure and Economic Development: An Endogenous Growth Perspective, *Economics of Planning* 28 (2-3), 205-233

Binswanger H.P., Khandker S.R. and Rosenzweig M.R. (1993): How Infrastructure and Financial Institutions Affect Agricultural Output and Investment in India, *Journal of Development Economics* 41, 337-366

Biswas B. and Ram R. (1986): Military Expenditures and Economic Growth in Less Developed Countries: An Augmented Model and Further Evidence, *Economic Development and Cultural Change* 34, 361-372

Brumm H. (1997): Military Spending, Government Disarray, and Economic Growth: A Cross-Country Empirical Analysis, *Journal of Macroeconomics* 19 (4), 827-838

Button K. (1998): Infrastructure Investment, Endogenous Growth and Economic Convergence, *The Annals of Regional Science* 32 (1), 145-162

Canto V. and Webb R.I. (1987): The Effect of State Fiscal Policy on State Relative Economic Performance, *Southern Economic Journal* 54 (July), 186-202

Cappelen A., Gleditsch N.P. and Bjerkholt O. (1984): Military Spending and Economic Growth in OECD Countries, *Journal of Peace Research* 21(4), 361-373

Chletsos M. and Kollias C. (1995): Defence Spending and Growth in Greece 1974-1990: Some Preliminary Econometric Results, *Applied Economics* 27 (9), 883-890

Chowdhury AR (1991): A Causal Analysis of Defense Spending and Economic Growth, *Journal of Conflict Resolution* 35 (1), 80-97

Cronovich R. (1998): Measuring the Human Capital Intensity of Government Spending and its Impact on Economic Growth in a Cross Section of Countries, *Scottish Journal of Political Economy* 45 (1), 48-77

da Silva Costa J., Ellson R.W. and Martin R.C. (1987): Public Capital, Regional Output, and Development: Some Empirical Evidence, *Journal of Regional Science* 27 (3), 419-437

Deger S. and Smith R. (1983): Military Expenditure and Growth in Less Developed Countries, *Journal of Conflict Resolution* 27 (2), 335-353

Devarajan S., Swaroop V. and Zou H. (1996): The Composition of Public Expenditure and Economic Growth, *Journal of Monetary Economics* 37 (2), 313-344

Dunne J.P. (1996): Economic Effects of Military Expenditure in Developing Countries: A survey. In: Gleditsch N.P., Bjerkholt O., Cappelen A., Smith R.P., Dunne J.P. (eds.), *The Peace Dividend*, Elsevier Science, Amsterdam, pp. 195-211

Durden G. and Elledge B. (1993): The Effect of Government Size on Economic Growth: Evidence From Gross State Product Data, *Review of Regional Studies* 23 (2), 183-190

Easterly W. and Rebelo S. (1993): Fiscal Policy and Economic Growth: An Empirical Investigation, *Journal of Monetary Economics* 32 (3), 417-458

Eisner R. (1991): Infrastructure and Regional Economic Performance: Comment, *New England Economic Review* 74 (Sep/Oct), 47-58

Evans P. and Karras G. (1994): Are Government Activities Productive? Evidence From a Panel of U.S. States, *Review of Economics and Statistics* 76 (1), 1-11

Faini R., Annez P. and Taylor L (1984): Defense Spending, Economic Structure, and Growth: Evidence Among Countries and Over Time, *Economic Development and Cultural Change* 32 (3), 487-498

Garrison C.B. and Lee F.Y. (1995): The Effect of Macroeconomic Variables on Economic Growth Rates: A Cross-Country Study, *Journal of Macroeconomic* 17 (2), 303-317

Gemmell N. (1983): International Comparison of the Effects of Non-Market Sector Growth, *Journal of Comparative Economics* 7, 368-381

Glomm G. and Ravikumar B. (1997): Productive Government Expenditures and Long-Run Growth, *Journal of Economic Dynamics and Control* 21 (1), 183-204

Grier K.B. and Tullock G. (1989): An Empirical Analysis of Cross-National Economic Growth, 1951-80, *Journal of Monetary Economics* 24, 259-276

Grobar L.M. and Porter R.C. (1989): Benoit Revisited: Defense Spending and Economic Growth in LDCs, *Journal of Conflict Resolution* 33 (2), 318-345

Grossman P.J. (1988): Growth in Government and Economic Growth: The Australian Experience, *Australian Economic Papers* 27 (50), 33-43

Grossman P.J. (1990): Government and Growth: Cross-Sectional Evidence, *Public Choice* 65 (3), 217-227

Guseh J.S. (1997): Government Size and Economic Growth in Developing Countries: A Political-Economy Framework, *Journal of Macroeconomic* 19 (1), 175-192

Gyimah-Brempong K. (1989): Defense Spending and Economic Growth in Subsaharan Afria: An Econometric Investigation, *Journal of Peace Research* 26 (1), 79-90

Hansen P. (1994a): Investment Data and the Empirical Relationship Between Exporters, Government and Economic Growth, *Applied Economics Letters* 1994 (1), 107-110

Hansen P. (1994b): The Government, Exporters and Economic Growth in New Zealand, *New Zealand Economic Papers* 28 (2), 133-142

Hansson P. and Henrekson M. (1994): A New Framework for Testing the Effect of Government Spending on Growth and Productivity, *Public Choice* 81 (3-4), 381- 401

Harmatuck D.J. (1996): The Influence of Transportation Infrastructure on Economic Development, *Logistics and Transportation Review* 32 (1), 63-76

Helms L.J. (1985): The Effects of State and Local Taxes on Economic Growth: A Time Series-Cross Section Approach, *Review of Economics and Statistics* 67 (Nov.), 574-582

Holtz-Eakin D. (1994): Public Sector Capital and the Productivity Puzzle, *Review of Economics and Statistics* 76 (1), 12-21

Holtz-Eakin D. and Schwartz A.E. (1995): Infrastructure in a Structural Model of Economic Growth, *Regional Science and Urban Economics* 25 (2), 131-151

Hsieh E. and Lai K.S. (1994): Government Spending and Economic Growth: The G-7 Experience, *Applied Economics* 26 (5), 535-542

Hulten C. and Schwab R. (1991): Public Capital Formation and the Growth of Regional Manufacturing Industries, *National Tax Journal* 44 (4), 121-134

Karikari J.A. (1995): Government and Economic Growth in a Developing Nation: The Case of Ghana, *Journal of Economic Development* 20 (2), 85-97

Kocherlakota N. and Yi K.M. (1996): A Simple Time Series Test of Endogenous versus Exogenous Growth Models: An Application to the United States, *Review of Economics and Statistics* 78, 126-134

Kocherlakota N. and Yi K.M. (1997): Is There Endogenous Long-Run Growth? Evidence from the United States and the United Kingdom, *Journal of Money, Credit and Banking* 29 (2), 235-262

Koester R.B. and Kormendi R.C. (1989): Taxation, Aggregate Activity and Economic Growth: Cross-Country Evidence on Some Supply-Side Hypotheses, *Economic Inquiry* 27, 367-386

Kollias C. and Makrydakis S. (1997): Defence Spending and Growth in Turkey 1954-1993: A Causal Analysis, *Defence and Peace Economics* 8 (2), 189-204

Kormendi R.C. and Meguire P.G. (1985): Macroeconomic Determinants of Growth: Cross-Country Evidence, *Journal of Monetary Economics* 16 (2), 141-163

Kusi N.K. (1994): Economic Growth and Defense Spending in Developing Countries: A Causal Analysis, *Journal of Conflict Resolution* 38 (1), 152-159

Landau D.L. (1983): Government Expenditure and Economic Growth: A Cross-Country Study, *Southern Economic Journal* 49 (3), 783-792

Landau D.L. (1985): Government Expenditure and Economic Growth in the Developed Countries: 1952-76, *Public Choice* 47, 459-477

Landau D.L. (1986): Government and Economic Growth in the Less Developed Countries: An Empirical Study for 1960-1980, *Economic Development and Cultural Change* 35 (1), 34-75

Lau S.H.P. and Sin C.Y. (1997): Public Infrastructure and Economic Growth: Time-Series Properties and Evidence, *Economic Record* 73 (221), 125-135

Lee B.S. and Lin S. (1994): Government Size, Demographic Changes, and Economic growth, *International Economic Journal* 8 (1), 91-108

Levine R. and Renelt D. (1992): A Sensitivity Analysis of Cross-Country Growth Regressions, *American Economic Review* 82 (4), 942-963

Lim D. (1983): Another Look at Growth and Defense in Less Developed Countries, *Economic Development and Cultural Change* 31 (January), 377-384

Lin S.A.Y. (1994): Government Spending and Economic Growth, *Applied Economics* 26 (1), 83-94

Lindgren G. (1984): Review Essay: Armaments and Economic Performance in Industrialized Market Economies, *Journal of Peace Research* 21 (4), 375-387

Lynde C. and Richmond J. (1992): The Role of Public Capital in Production, *Review of Economics and Statistics* 74, 37-44

Lynde C. and Richmond J. (1993): Public Capital and Total Factor Productivity, *International Economic Review,* 34 (2), 401-444

Macnair E.S., Murdoch J.C., Pi C.R. and Sandler T. (1995): Growth and Defense: Pooled Estimates for the NATO Alliance, 1951-1988, *Southern Economic Journal* 61 (3), 846-860

Mohammed N.A.L. (1993): Defense Spending and Economic Growth in Subsaharan Africa: Comment on Gyimah-Brempong, *Journal of Peace Research* 30 (1), 95-99

Moomaw R. and Williams, M. (1991) Total Factor Productivity Growth in Manufacturing: Further Evidence from the States, *Journal of Regional Science* 31 (1), 17-34

Morrison C.J. and Schwartz A.E. (1996): State Infrastructure and Productive Performance. *American Economic Review* 86 (5), 1095-1111

Mullen J. and Williams M. (1990): Explaining Total Factor Productivity Differentials in Urban Manufacturing, *Journal of Urban Economics* 28, 103-123

Munnell A.H. (1990a): Why Has Productivity Growth Declined? Productivity and Public Investment, *New England Economic Review* 30 (January/February), 3-22

Munnell A.H. (1990b): How Does Public Infrastructure Affect Regional Performance? *New England Economic Review* 30 (Sept/Oct), 11-32

Munnell A.H. (1992): Infrastructure Investment and Economic Growth, *Journal of Economic Perspectives* 6 (4), 189-198

Odedokun M.O. (1997): Relative Effects of Public Versus Private Investment Spending on Economic Efficiency and Growth in Developing Countries, *Applied Economics* 29 (10), 1325-1336

Park K.Y. (1993): Pouring New Wine into Fresh Wineskins: Defense Spending and Economic Growth in LDCs with Application to South Korea, *Journal of Peace Research* 30 (1), 79-93

Ram R. (1986): Government Size and Economic Growth: A New Framework and Some Evidence From Cross-Section and Time-Series Data, *American Economic Review* 76 (1), 191-203

Rao V.V.B. (1989): Government Size and Economic Growth: A New Framework and Some Evidence From Cross-Section and Time-Series Data. Comment, *American Economic Review* 79 (1), 272-280

Ratner J.B. (1983): Government Capital and the Production Function for U.S. Private Output, *Economics Letters* 13, 213-217

Roux A. (1996): Defense Expenditure and Economic Growth in South Africa, *Journal for Studies in Economics and Econometrics* 20 (1), 19-34

Sala-i-Martin X. (1994): Economic Growth: Cross-Sectional Regressions and the Empirics of Economic Growth, *European Economic Review* 38 (3-4), 739-747

Sanchez-Robles B. (1998): Infrastructure Investment and Growth: Some Empirical Evidence, *Contemporary Economic Policy* 16 (1), 98-108

Sattar Z. (1993): Government Control and Economic Growth in Asia: Evidence From Time Series Data, *The Pakistan Development Review* 32 (2), 179-197

Saunders P. (1985): Public Expenditure and Economic Performance in OECD Countries, *Journal of Public Policy* 5 (1), 1-21

Scully G.W. (1989): The Size of the State, Economic Growth and the Efficient Utilization of National Resources, *Public Choice* 63 (2), 149-164

Sheehey A.J. (1993): The Effect of Government Size on Economic Growth, *Eastern Economic Journal* 19 (3), 321-328

Singh R.J. and Weber R. (1997): The Composition of Public Expenditure and Economic Growth: Can Anything Be Learned From Swiss Data? *Schweizerische Zeitschrift fur Volkswirtschaft und Statistik/Swiss Journal of Economics and Statistics* 133 (3), 617-634

Van Sinderen J. (1993): Taxation and Economic Growth, *Economic Modelling* 10 (3), 285-300

Wylie P.J. (1996): Infrastructure and Canadian Economic Growth, 1946-1991, *Canadian Journal of Economics* 29 [Special Issue Part 1], 350-355

Yu W., Wallace M.S. and Nardinelli C. (1991): State Growth Rates: Taxes, Spending and Catching Up, *Public Finance Quarterly* 19 (1), 80-93

Zhang T. and Zou H.F. (1998): Fiscal Decentralization, Public Spending and Economic Growth in China, *Journal of Public Economics* 67 (2), 221-240

19 Does R&D Infrastructure Attract High-Tech Start-Ups?

Dirk Engel and Andreas Fier
Center of European Economic Research Department of Industrial Economics and International Management

19.1 Introduction

There is a wide agreement that high-tech start-ups can be regarded as a driving force of economic growth in general. In particular, they have been seen as a crucial element in the attempt to close the productivity gap between Eastern and Western Germany. This chapter makes a modest attempt to identify regional differences in start-up activities in Eastern Germany. Particular attention is paid to the question of whether R&D infrastructure is able to attract high-tech start-ups. The impact of the proximity and size of publicly financed R&D institutions is emphasised, as well as the role of large firms as incubators for start-up activities, and the importance of the concentration of economic activities is analysed.

The chapter starts with some theoretical reflections in Section 19.2, then briefly defines some crucial variables in the study and describes the data available in Section 19.3. Section 19.4 provides a characterisation of the most successful East German high-tech regions, while key variables affecting start-up activities in high-tech industries are identified by means of econometric analysis in Section 19.5. The chapter concludes with some policy implications.

19.2 Theoretical Reflections

According to Markusen, Hall and Glasmeier (1986), there does not exist a comprehensive theoretical model for the siting of high-tech start-ups. It is necessary to fall back on theory-based studies on firm foundations, as well as the theoretical approaches of innovation economics and regional economics, which have to be expanded by the characteristics of technology-intensive [high-tech] start-ups. Using this approach, we have derived hypotheses on the regional concentration of high-tech start-ups and will examine their validity in the empirical part of the study.

Start-ups and Market Entries

In contrast to inter-industrial cross-sectional models that focus on industry characteristics and entry barriers as key determinants of market entry behaviour (see also Cable and Schwalbach 1991), 'Models of Self-employment' derive and include the motives of the individual in starting-up a company. These theoretical approaches go back to studies by Chapman and Marquis (1912) and Knight (1921a) and are based on the individual's decision to choose to be self-employed rather than not self-employed. The individual probability of self-employment $Pr(e)$ is determined by the difference between the expected income gained through market entry π^* and the expected income w^* provided by the alternative, non self-employment:

$$Pr(e) = g\,(\pi^* - w^*) \tag{19.1}$$

The income alternatives depend on factors such as the individual's skills and personality or industry-specific characteristics (see Evans and Leighton 1989; Evans and Jovanovic 1989; Pfeiffer 1994). Bania, Eberts and Fogarty (1993) apply an expansion of the model by extending by Equation (19.1) with a regional dimension and the integration of site selection by individuals. Here, regional deviation in the number of firms is explained by differences in regional resource availability, and industry and infrastructure. Thus, the basic model of self-employment is transformed from the individual level to a regional level i. The new equation applies the empirical implementation of the 'self-employment decision to a regional level' $Pr(e_i)$ (see e.g. Bania, Eberts and Fogarty 1993; Audretsch and Vivarelli 1996), where characteristics of potential founders and industry-specific characteristics are aggregated.

Regional Spillover Effects

Moreover, the relevance of spillover effects resulting from the establishment of R&D specific infrastructure and from the concentration of economic activities have often been discussed in connection with start-up activities (see e.g. Harhoff 1995; Audretsch and Feldmann 1996). These spillover effects increase the probability $Pr(e_i)$ of a high-tech start-up in a specific region i. The strength of the influence depends on the range and size of spillovers.

The innovation process was long seen as a linear model, starting with research on product development and ending with the market entry of the new product.

Today, the process has been expanded to include interaction and feedback among the participating partners (see Freeman 1982, Fischer 1999). Hence, the stimulus for innovation activities is not limited to basic research, but also includes the practical experience that leads to product improvement (see von Hipple 1988). The successful market entry of an innovation depends on the interplay of all partners in the innovation process. While in the early discovery stages, university and other research establishments may add various elements of novel knowledge to the innovation process, especially in science-based industries, R&D activities tend to play a major role in the development phase (see Nelson 1986; Tassey 1991). For the commercialisation stage of the innovation process [market entry], according to Coffey and Polese (1987) company input with respect to undertaking authorisation procedures, tax issues and marketing plays an important role.

On the basis of this approach to innovation, numerous studies indicate that knowledge, in particular tacit forms of knowledge, is only available within certain geographical distances (see Krugman 1991b). Along with Feldman (1994), he claims that the spread and transfer knowledge in a region is strongly influenced by the ability to transform know-how into language and communication. Spatial proximity is therefore expected to provide an advantage in knowledge transfer (see Feldman 1994a; Audretsch and Feldman 1996; Helpman and Trajtenberg 1998), on the one hand because of the increasing complexity of new technologies, the significance of informal means of communication and the personal exchange of experience, and on the other due to increasing transaction costs over distance (see Schrader 1991; Beise, Licht and Spielkamp 1995, and PART C of this volume).

Recent approaches to site and location theories discuss a number of different agglomeration effects (see Romer 1986, 1990; Krugman 1991b). While factors that lead to a concentration of similar economic activities are often referred to as 'economies of localisation', the forces contributing to the agglomeration of different economic activities especially in urban areas are called 'economies of urbanisation' (see Henderson 1988; Stahl 1995). The advantages of an agglomeration of different kinds of economic activity are derived in a similar way to the concentration advantages of similar firms. Spatial proximity to firms producing complementary industrial goods or services, as well as proximity to suppliers and customers can have favourable due to effects on transportation and storage costs.

The complexity of new technologies and the ability to adapt and implement technological know-how is strongly linked to human capital, especially the entrepreneur's technical skills (Storey and Tether 1996). Markusen, Hall and Glasmeier (1986), Oakey, Rothwell and Cooper (1988) and Saxenian (1990) point out that well-known high-tech areas such as Silicon Valley have primarily developed because of a favourable R&D infrastructure and technological know-how in the region. Agglomerations are favourable for creating spillover effects.

19.3 Definitions and Data

The analysis of the start-up situation in Eastern Germany is based on data from the ZEW-Foundation Panel East. Firm-specific data have been collected by the largest German credit rating agency CREDITREFORM since 1991. The panel now consists of about one million firms in Eastern Germany (see Almus, Engel and Prantl 2000 for further explanation). Of special interest to us are the location factors that improve the long run conditions. The analysis is therefore restricted to firms founded between January 1995 and December 1998.

Identification of Start-Ups

The definition of the type of start-up plays a crucial role in the analysis of regional differences in the frequency of firm foundation (see Geroski 1995). We have therefore differentiated with respect to prior structural change in the firm, as well as their independence. In this study, only newly founded firms which are independent have been considered. They must represent a genuinely new enterprise undertaken by one or more natural persons, rather than branches of existing firms (see Nerlinger 1998), so the setting up of affiliated firms is not considered. A great many companies in Eastern Germany evolved from firms of the former GDR that were later privatised, reprivatised or partly privatised (see Felder, Fier and Nerlinger 1997). These converted firms ['derivative foundations'] are characterised by the fact that the site selection has already taken place. Firms that were held partly by the German privatisation agency 'Treuhandanstalt' or its successor will not be considered. In some cases a capital structure is not made available. In order to differentiate these firms from genuine new start-ups, information about the number of employees needs to be taken into account. Therefore, firms with more than fifty employees at the foundation date are often excluded in the statistics and also in our study (see Audretsch and Fritsch 1992).

Identification of High-Tech Start-Ups

Information about the innovation *output* of firms is not available in the CREDITREFORM database. This database applies a definition of high-tech start-ups according to innovation *input*, which is based on the classification of the 'technology-intensive' goods derived by the OECD (see Gehrke et al. 1997). The technology-intensive manufacturing industries are divided, according to their R&D intensity, into 'High-Tech Industries' and 'Other Manufacturing'. High-tech industries are considered to have an average R&D intensity rate more than or equal to 3.5 percent.

Recent empirical studies (see Harhoff et al. 1996; Nerlinger 1998) have shown that many firms in the service sector also carry out a considerable amount of R&D and innovation activities. By analogy, these service industries are considered to belong to the high-tech service sector (see Table 19.1)

Table 19.1   Technology-intensive services

Technology Intensive Services [TIS]	
72	Computer and related activities
731	R&D on natural sciences, engineering and medicine
742	Architectural and engineering activities and related technical consultancy
743	Technical testing and analysis
Non-Technical Consulting [NTC]	
732	R&D on social sciences and humanities
7411, ..., 7414	Legal activities, Accounting, book-keeping and auditing activities; tax consultancy
	Market research and public opinion polling, Business and management consultancy activities
744	Advertising

Source: Almus, Egeln and Engel (1999), Nerlinger (1998), according to NACE 93 Codes

19.4 High-Tech Start-Ups in Eastern Germany

In the following section, we shall analyse the common characteristics of the most successful East German high-tech regions. Such characteristics represent potential factors for explaining the regional differences observed in the number of high-tech start-ups. The existence of R&D infrastructure, the presence of large firms, the nature of traffic conditions and the industrial structure will be evaluated in these regions. Fig. 19.1 shows the most technology-intensive postcode areas, based on the number of start-up activities. In our analysis, we have considered a postcode area to be technology-intensive if it has an absolute and relative number of high-tech start-ups above the 65 and 80 percentile[3] respectively. The relative number was calculated as the absolute number divided by the sum of employees in the postcode area.

Let us now examine five regions where there was clear evidence of a very large number of high-tech start-ups between 1995 and 1998: Dresden, Ilmenau-Erfurt-Jena, Magdeburg-Halle-Leipzig-Dessau, Berlin and its fringe, and the two main cities of Mecklenburg-Westpomerania, Rostock and Schwerin. It should be noted that there was a notable concentration in postcode areas around the big cities, but not within.

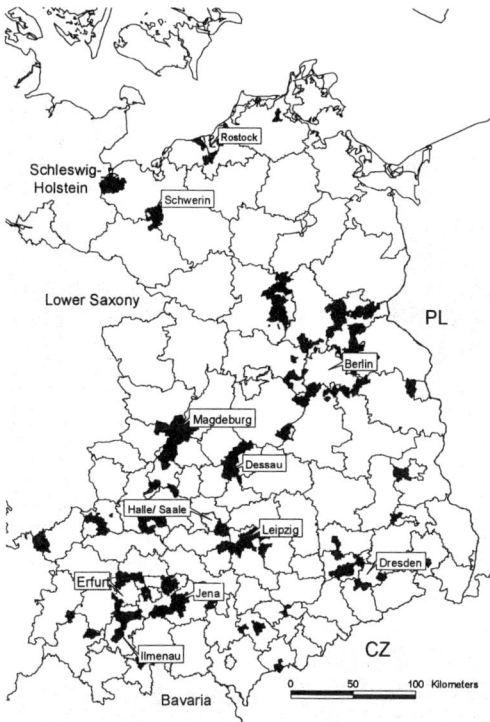

Source: ZEW-Foundation Panel East, own calculation

Fig. 19.1   High-tech start-ups [1995-1998]

a) *The Dresden region:* The economic structure of this region is dominated by micro-electronics and related fields. Mechanical engineering is the main sector, followed by R&D and the manufacture of medical, and precision and optical instruments. Specialised technology areas are electronic engineering, new materials, and software and simulation. Moreover, pharmaceuticals and biotechnology strongly influence the sectoral structure. Large R&D institutions that undertake both pure and applied research have been set up in the region. The most important are the Technological University of Dresden [TU Dresden], institutes associated with the Max-Planck-Society and the Fraunhofer-Society, and the universities of Applied Science [Fachhochschulen]. Regarding industry, numerous large companies with international operations have settled in or near Dresden, such as *Advanced Micro Devices Inc.*, *Infineon Technologies*, *Volkswagen* and *Mannesmann/Vodaphone*. These companies are global suppliers of integrated circuits, semiconductors, telecommunications, computers, automobiles, etc. High-tech start-up companies benefit from opportunities as co-operation partners of these large firms, and gain from their international fame.

b) *Ilmenau–Erfurt–Jena:* In the state of Thuringia, many traditional industries and large companies have collapsed since the political upheaval in 1989/90. New industries can be found today in sectors such as microelectronics, optics and precision technology, vehicle assembly, automotive suppliers as well as the former glass, wood and furniture industry. *Siemens AG, Jenoptik AG, Jenapharm GmbH & Co KG, Carl Zeiss Jena GmbH* are some of the best-known companies in the region. The metropolitan area of Ilmenau–Erfurt–Jena has universities of applied science [Fachhochschulen], laboratories and other businesses which emphasise the wealth of human capital [about 24,000 students] and the large number of potential high-tech entrepreneurs. Most impressive is the fact that this metropolitan area is circled by a ring of intense start-up activity. The vicinity of former start-up companies, universities and other research facilities seems to be crucial for new high-tech firms. While other regions of Eastern Germany have islands of start-ups, the Ilmenau-Erfurt-Jena region concentrated their regional strength until 1998 in a network of students, science and industry.

c) *Halle–Leipzig–Dessau:* This is one of the strongest growing economic regions in Germany in which a variety of industries have located. Favourable conditions have been created for the restructuring of traditional industries and the development of innovations in key future areas. Strong co-operation between industries and research institutions has been emphasised. About eight higher education institutions and almost sixty R&D and technology centres work hand in hand with firms in the region. Today, the firms from the chemical industry, biotechnology, industrial and environmental engineering, mining and power industries as well as the electronical industry have located here. Some of the national and international companies in the Halle-Leipzig-Dessau region are: *Montedison SpA [Italy], Kvaerner plc. [UK], Thyssen Rheinstahl Technik GmbH, Klöckner AG, Bayer Bitterfeld GmbH* or *Dow Chemical Co. [USA].* The region recently gained from major investment projects in the chemical sector carried out by *Buna Sow Leuna Olefinverbund GmbH, Elf Aquitaine, Mitteldeutsche Erdöl-Raffinerie GmbH* or *Infraleuna.* Moreover, with *Leipziger Messe GmbH* and the mail order company *Quelle,* Leipzig has become an important trade, service and media centre for Eastern Germany.

d) *The metropolitan region of Berlin* and its surroundings, especially the north-east fringe of the region has been one of the most attractive regions for high-tech start-ups in the past few years. In Brandenburg, special expertise exists in traffic technology, microsystems technology, energy and construction technologies, electronics, optics, synthetic chemistry and biotechnology. Some major industrial players like *Rolls-Royce Deutschland GmbH* with an air engine lab and production facility, DaimlerChrysler [train technology and truck factory], *Heidelberger Printing Machines* with a production site, *ABB* with its Automation Branch, *VEBA* with an oil-refinery in *Schwedt, MAN heavy-duty Machine, Bosch-Siemens Appliances* or *BASF* are located in Brandenburg. Young high-tech companies were founded in the second half of the nineties near Berlin and along the highway A 11 to Poland. There are also some high-tech start-ups on the border of Berlin and just 30 miles to the north-west in Neuruppin, and north-east in Eberswalde.

Whereas the existence of many small high-technology oriented firms is typical of East Berlin, Brandenburg has some large firms in the vehicle engineering sector. Brandenburg has unrivalled prospects for technology transfer with the three universities at Cottbus, Frankfurt/Oder and Potsdam, five universities of applied science and 18 technology centres in addition to Berlin's four universities, nine universities of applied science and several R&D institutions. Of further significance is the re-emergence of a booming media industry in and around the cradle of the German film industry at Potsdam-Babelsberg.

e) **Mecklenburg-Westpomerenia** is raising its efforts to strengthen the industry of the region. For years, the construction industry has been stagnating, thus the region is trying to emphasise one of is core competences, the shipbuilding industry [e.g. *Kvaerner Warnow Werft, Aker MTW Werft*] and its state-of-the-art shipyards. The region also concentrates on exports and is aiming to strengthen its ties with the Baltic region. Along with its two traditional universities [Rostock and Greifswald] and a number of technology and foundation centres [the private institute of microbiological research, Institute for Applied Biosciences], the area is fostering the development of technology-based firms and start-ups, and is thereby actively supporting structural change. Only the main cities, Rostock and Schwerin, have had high numbers of start-ups since 1990. The other regions of Mecklenburg-Westpomerenia are 'deserts' in terms of high-tech activities. Today, the high-technology industry is increasingly focusing on the new biotech industries like *Bioserv AG, BioTechnikum Greifswald, DNA Diagnostik Nord GmbH* or *PlasmaSelect AG*.

We have identified a tendency on the part of high-tech start-up companies to locate their businesses around the outskirts of the larger cities of East Germany. It seems a likely supposition that industrial estates around the city provide better possibilities for growth than most districts of the inner city. Lower tax rates, cheaper business premises, the availability of parking space, the visibility of large well-known companies or small and medium-sized high-tech enterprises, and faster links to traffic infrastructure are all reasons to set up a business in the periphery of large cities. The newly established large companies in East Germany are often in R&D intensive industries. In all of the regions examined we have found clear evidence that young high-tech firms choose to locate in places where there is already a critical mass of R&D activities [universities, R&D-intensive industries, R&D institutes]. Proximity to agglomeration centres seems to be important for high-tech start-ups. Moreover, the region's history regarding industrial structure seems to have influenced the siting of firms in recent years.

19.5 Determinants for Regional Differences in the Number of High-Tech Start-Ups

In the econometric analysis that follows, we identify some key variables that appear to affect start-up activities in high-tech industries. The variables taken into consideration include:
- R&D specific human capital
- The presence of large firms
- Urbanisation effects and specific infrastructure
- Other regional influences on start-up activities.

In order to achieve a better understanding and interpretation of observed effects, the results for high-tech start-ups are compared with those for other industries. We therefore distinguish six categories of industry:
- STI and HTI are respectively the superior and high technology-intensive industries in the manufacturing sector, according the NIW/ISI list of technology-intensive industries (see Gehrke et al. 1997),
- TIS refers to the high-tech service sector,
- NTCS contains the remaining knowledge intensive services (see Table 19.1),
- NTI represents the non high-tech industries in the manufacturing sector,
- OS includes the non high-tech industries in the service sector.

The average R&D intensity rate of STI is above or equal to 8.5 percent; those of HTI between 3.5 and below 8.5 percent. The average R&D intensity rate of industries in the TIS is unknown. Table 19.2 shows the average innovation intensity rate for TIS and the average R&D intensity rate for the manufacturing sector based on 'Mannheimer Innovation Panel'. In general, the results confirm the differentiation of high-tech industries.

Table 19.2 R&D intensity and innovation intensity of technology-intensive industries

Industry	R&D-Intensity [%]	Innovation-Intensity [%]
STI	6.4	11.0
HTI	2.8	4.6
NTI	0.9	3.7
TIS	-	7.1
Other	-	2.0

Source: Mannheim Innovation Panel, own calculation

Notes: Results for Western Germany, average R&D-intensity (R&D expenditures in relation to turnover) for 1993-1997; average innovation-intensity (innovation expenditures in relation to turnover) for 1993-1998

The number of newly founded firms in a postcode area within a certain period of time can be described with a positive numbered random variable Y. A count data model, e.g. a Poisson model or a negative binomial model, represents a suitable econometric approach to identify key variables effecting start-up activities. The model is parameterised in such a manner that the logarithmic expected value of the number of foundations is a linear function of the explanatory variables:

$$\ln E(Y) = X\beta \tag{19.2}$$

The matrix X contains the k variables through which the relevance of the above variables is examined. Our specification comprises different types of exogenous variables. The coefficients of logarithmic continuous variables [e.g. *employees*] present elasticities, i.e. an increase of about one percent of the explanatory variable leads to a variation in the anticipated number of start-ups of about β_k percent. All coefficients of shares [e.g. *share of employees*] with values between 0 and 1 almost equal the percentage variation of the number of the new firms if the share goes up by one percent.

Table 19.3   The number of start-ups in East German postcode areas

Number of Start-Ups in Postcode Area	Frequency Distribution of Start-Ups in Postcode Areas					
	STI	HTI	NTI	TIS	NTCS	OS
0	81.07	55.84	14.91	21.92	36.83	3.86
1	12.70	20.98	11.99	16.09	12.54	1.34
2	3.71	9.38	9.86	9.07	9.07	2.21
3 to 5	2.21	9.70	23.26	20.11	17.27	5.28
6 to 10	0.24	3.47	20.82	12.46	11.28	11.99
11 to 20	0.08	0.63	15.30	11.04	8.60	20.27
21 to 50			3.86	8.68	4.10	28.08
50 and more				0.63	0.32	26.97
Cummulative	100	100	100	100	100	100
Number of Observations	1,268	1,268	1,268	1,268	1,268	1,268
Mean	0.31	1.08	6.06	6.55	4.46	42.82
Standard Deviation	0.87	1.89	6.44	9.52	7.68	52.07

Source: ZEW-Foundation Panel East, own calculation

Notes: STI denotes Superior Technology Industries, HTI High Technology Industries, NTI Non-Technology Industries, TIS Technology-Intensive Services, NTCS Non-Technical Consulting Services, OS Other Services

The Poisson assumption that the mean equals the variance is a shortcoming of the Poisson model (Greene 1997, p. 939). The negative binomial model seems to be the correct econometric approach for the current research question because the coefficient of the heterogeneity component ε differs significantly from zero. Therefore, only the estimation results based on the negative binomial model are presented in Table 19.3. Of course, the use of this model suffers from some restrictive assumptions. Count data models have the disadvantage that at present there is no model that considers spatial auto-correlation between regions (see Nerlinger 1998; Steil 1999). However, since we are modelling distance directly in some exogenous variables, we believe that the reasons for spatial correlation in other studies are already considered in our exogenous variables. Data constraints hinder a panel data approach which would allow us to consider unobserved heterogeneity. The results obtained are summarised in Table 19.4.

Table 19.4 Determinants for regional differences in the number of start-ups in technology-intensive industries founded between 1995-1998

Independent Variables	STI	HTI	NTI	TIS	NTCS	OS
	\multicolumn{6}{c}{**Parameter Estimates**}					
R&D Specific Human Capital						
• R&D Employees in Industry [ln]	0.00 (0.08)	-0.02 (0.04)	0.00 (0.03)	0.02 (0.03)	-0.01 (0.04)	0.00 (0.02)
• Sum of R&D Employees in Industry 50 km around [ln]	0.27 (0.18)	-0.17 (0.13)	0.03 (0.05)	0.11 (0.06)	0.07 (0.09)	0.06 (0.04)
• R&D Employees in Public Institutions [ln]	-0.12* (0.05)	0.00 (0.03)	0.01 (0.02)	-0.03 (0.02)	0.01 (0.02)	0.00 (0.01)
• Sum of R&D Employees in Public Institutions 50 km around [ln]		0.05 (0.10)	0.02 (0.03)	0.08* (0.04)	-0.01 (0.05)	0.04 (0.02)
• Share of Highly Qualified Employees	0.07 (0.05)	-0.06* (0.03)	-0.05* (0.02)	-0.02 (0.02)	-0.03 (0.03)	-0.02 (0.02)
• Direct Distance in km [ln] to the Next Fraunhofer-Institute	-0.09 (0.11)	-0.12 (0.08)	0.04 (0.05)	-0.06 (0.05)	-0.05 (0.06)	-0.02 (0.05)
• Direct Distance in km [ln]to the Next Max-Planck-Institute	-0.10 (0.12)	0.07 (0.10)	0.00 (0.05)	-0.09* (0.05)	-0.13 (0.07)	-0.04 (0.04)
• Min [ln (Direct Distance in km/ Scientific Staff in Universities)]						
– Engineering and Computer Sciences	-17.71** (6.66)	-8.31* (4.16)	-3.03 (2.28)	-8.80** (2.49)	-3.49 (3.15)	-3.13 (2.01)
– Engineering and Computer Sciences (squared)	99.35* (41.73)	50.15 (27.81)	18.55 (14.09)	49.15** (15.23)	19.61 (19.08)	20.54 (11.52)
– Natural Sciences	5.55 (4.21)	-4.74* (2.35)	-0.31 (1.34)	-2.46 (1.72)	-4.34* (1.99)	-1.85 (1.25)
– Medicine	-1.75 (3.63)	2.29 (1.98)	0.36 (0.99)	-0.33 (1.16)	1.09 (1.34)	0.08 (0.79)
– Law, Administation and Economic Sciences	1.82 (1.28)	0.66 (0.68)	0.02 (0.43)	0.76 (0.59)	0.64 (0.64)	0.41 (0.42)
Large Firms						
• Direct Distance in km [ln] to the next Firm						
– Manufacturing Sector [founded till 1995]	-0.12 (0.07)	-0.06 (0.05)	-0.07** (0.02)	-0.06* (0.03)	-0.09** (0.03)	-0.08** (0.02)
– Manufacturing Sector [founded after 1995]	-0.16* (0.07)	-0.09* (0.04)	-0.05* (0.02)	-0.14** (0.03)	-0.12** (0.03)	-0.09** (0.02)
– Other Sectors [founded till 1995]	-0.18* (0.07)	-0.11** (0.04)	-0.02 (0.02)	-0.10** (0.03)	-0.13** (0.03)	-0.08** (0.02)
– Other Sectors [founded after 1995]	-0.04 (0.08) (0.09)	-0.02 (0.06) (0.05)	0.01 (0.03) (0.03)	-0.01 (0.03) (0.03)	-0.03 (0.04) (0.04)	0.01 (0.03) (0.02)

ctd.

Urbanisation Effects and Specific Infrastructure

- Direct Distance in km [ln] to the next

– Motorway Entrance	-0.25	0.05	0.31*	0.10	-0.32	-0.10
	(0.39)	(0.25)	(0.13)	(0.15)	(0.19)	(0.11)
– Motorway Entrance [squared]	0.03	-0.01	-0.07*	-0.04	0.05	0.01
– Technology and Foundation Centre	-0.37**	-0.17*	-0.06*	-0.20**	-0.14**	-0.09**
	(0.09)	(0.06)	(0.03)	(0.03)	(0.04)	(0.03)
– Town	-0.03	0.10	-0.05	-0.21**	-0.26**	-0.18**
	(0.14)	(0.09)	(0.05)	(0.05)	(0.07)	(0.05)

Other Regional Influences on Start-up Activities

• Employees 1996 [ln]	0.53**	0.73**	0.69**	0.64**	0.70**	0.61**
	(0.11)	(0.07)	(0.03)	(0.05)	(0.05)	(0.04)
• Employees 1996, 50 km around the Postcode Area [ln]	-0.57	0.07	-0.21	-0.59**	-0.35	-0.38**
	(0.41)	(0.27)	(0.13)	(0.15)	(0.19)	(0.10)
• Unemployment Rate 07/97	-3.71	-3.45	-1.52	-1.39	-1.28	-1.25
	(3.68)	(2.17)	(1.12)	(1.25)	(1.65)	(0.99)
• Share of Employees in						
– Manufacturing	3.41	6.34**	4.29**	1.21	3.33**	1.33*
	(2.76)	(1.48)	(0.78)	(0.92)	(1.09)	(0.60)
– Manufacturing [squared]	-5.30	-7.77**	-6.72**	-3.12*	-6.71**	-3.38**
	(4.29)	(2.33)	(1.29)	(1.32)	(1.73)	(0.86)
– Construction	2.62*	0.07	0.02	0.28	0.02	-0.19
	(1.08)	(0.71)	(0.39)	(0.47)	(0.58)	(0.36)
– Other Private Service Sectors	0.05	0.42	-0.37	0.68	0.70	1.18**
	(0.91)	(0.57)	(0.27)	(0.36)	(0.39)	(0.28)

Constant	0.85	-4.09	-1.19	5.25**	2.70	4.31**
	(4.81)	(3.06)	(1.40)	(1.74)	(2.12)	(1.20)
Alpha	1.18**	0.73**	0.33**	0.50**	0.74**	0.41**
	(0.25)	(0.09)	(0.03)	(0.04)	(0.06)	(0.03)
Number of Observations	1,268	1,268	1,268	1,268	1,268	1,268
Log-Likelihood	-749.18	-1558.38	-3188.58	-3080.99	-2643.43	-5351.84
Pseudo R²	0.14	0.13	0.13	0.16	0.15	0.11
LR-Tests: $\chi^2(df)$						
min (Direct Distance/Scientific Staff at Universities) (2)	7.08*	3.99	1.83	12.54**	1.24	3.19
Motorway Entrance (2)	1.24	0.04	6.48*	8.88*	5.57*	2.52
Dummies: Provinces	28.15**	9.14	13.65**	20.37**	7.10	39.49**
Type of District	12.21	15.18	14.51	25.68**	13.97*	25.98**

Source: ZEW-Foundation Panel East, BBR, own estimation

Notes: STI denotes Superior Technology Industries, HTI High Technology Industries, NTI Non-Technology Industries, TIS Technology-Intensive Services, NTCS Non-Technical Consulting Services, OS Other Services; Reference Group: Share of Employees in Trade, Energy Supply and Mining Industries, Banking and Insurances, Stata Institutions and Organisations/Social Security.
** significant at the 1%-level, * significant at the 5%-level.

Linear models are sometimes used as an alternative if the endogenous variable is continuous. The OLS-model is not suitable in our case because we observe only a small number of different values in high-tech industries on the level of postcode areas. Moreover, no high-tech start ups have been founded in a large number of postcodes (see Table 19.3). The logarithm of this value is not defined.

Table 19.4 summarises the main estimation results for explaining the number of start-ups in these industries, founded between 1995 and 1998, in Eastern German postcode areas.

R&D Specific Human Capital

Marshall (1890) formulated three core hypotheses to explain the local concentration of industries. In the present study we have considered the information relations between firms to be relevant and also a region's externally available knowledge, for instance through public research institutions. Highly qualified human capital is concentrated at the locations of R&D intensive firms and near public research institutions. A large amount of entrepreneurial potential and knowledge spillovers is evident in such areas.

It is assumed that such spillovers increase product and process innovations through the continuous information flows between employees and firms. Czarnitzki et al. (2000) obtained the result that 12 percent of new products introduced to the market relate directly to research activities carried out in publicly financed institutions. Universities were found to be the most important sources of innovation for firms, which emphasises that institutions with public finance are essential for achieving innovation. The Fraunhofer-Institutes and Max-Planck-Institutes[4] were found to be less important. In the study by Picot, Laub and Schneider (1989), proximity to universities and an 'industrial milieu' were declared by the founders of new firms to be the second and third most crucial factors. The most important factor was the distance to founder's place of residence. This is in accordance with the fact of the positive relationship between founder potential and start-up activities in the same region (see Schmude 1994). A recent paper of Sternberg at al. (2000) confirms this result. 65 percent of founders of businesses in knowledge-intensive industries investigated in Cologne studied at one of Cologne's universities or colleges. The results give one reason for analysing the impact of founder potential and knowledge spillovers at R&D-intensive institutions or firms for start-up activities.

> *First Hypothesis: The number of high-tech start ups is positively related to specific human capital. The existence of universities has a greater impact for explaining the regional differences in the start-up activities compared with other publicly financed institutions.*

The probability of a high-tech start-up rises with proximity to a university/ university of applied science and with the size of such institutions. We calculated for each postcode area the ratios of distance to each higher-education institution divided by the size of the scientific staff. We took the distance as a nominator and scientific staff as denominator in order to interpret distance effects as in the economic literature (see e.g. McCallum 1995). Our specific interest was linked to effects of the nearest and biggest institution on the probability of a high-tech start-up, so we took only the minimum of all ratios for each postcode area.

Potential founders of high-tech start-ups often need a qualification in one of the technical sciences (see Bruederl, Preisendoerfer and Ziegler 1996). Such a qualification frequently requires a degree in natural or engineering sciences. We differentiated between some specific faculties in order to measure these effects. The specific human capital and knowledge spillovers at the *Institutes of Fraunhofer-Society* and the *Max-Planck-Society* were measured by taking the direct distance between each of the postcode areas to the nearest institute. Moreover, the effects of the concentration and the accumulation of specific human capital were measured in terms of the number of R&D employees in *private firms* in the manufacturing sector and in *non-university R&D institutions* and the *share of highly qualified employees.* These variables are only available at the county level. We considered the effects of both variables in a radius of 50 kilometres around the postcode area in addition to effects within the county.

The empirical results (summarised in Table 19.4) show that R&D employees in *private firms* of the manufacturing sector within the county and around the postcode area are insignificant in all specifications. Both results are consistent with the results of Nerlinger (1998) for Western German counties. Felder, Fier and Nerlinger (1997) get a positive impact of R&D employees in private firms on the number of start-ups in technology-intensive industries using data for East Germany. The share of highly qualified employees at the county level is negatively correlated with start-ups in NTI, however we obtained an unexpected negative impact for start-ups in HTI too. We get a significant impact of the total sum of R&D employees in public institutes 50 km around the postcode area for TIS, but a negative coefficent of R&D employees in the county for STI. These contradictory results led us to assume that these variables had low suitability at the county level for measuring spillover effects through the accumulation and concentration of R&D specific human capital. East German counties are very large, which may be give one reason for contradictory results in the observance of proximity effects. This emphasises the importance of using suitable variables.

Proximity to *Fraunhofer-Institute* is shown to have no significant impact on the number of start-ups. However, the closer we come to a *Max-Planck-Institut* the higher the number of start-ups, but only for TIS. Some start-ups in this sector are in 'technical testing and analysis'. Estimation results show that specialised human capital and knowledge spillovers at non-university public institutions have no significant influence on regional differences in high-tech start-up activities. One possible reason is that the various specialisations of such institutes are important for only a specific subgroup of high-tech start-ups. Moreover, the relationships between these institutes and SMEs [small-medium sized enterprises, start-ups included] are weaker than with established firms (see Czarnitzki et al. 2000).

The proximity and size of *universities* and *universities of applied science* have a greater relevance for explaining regional differences in high-tech start-up activities. The estimation results support the findings of Czarnitzki et al. 2000 concerning the relative importance of universities for innovation processes in firms in comparison with the Fraunhofer and Max-Planck-Institutes. The human resources at higher-education institutions have a positive impact on the number of start-ups in the high-tech industries of the types STI, HTI, TIS and on start-up activities in the remaining knowledge-intensive sector, NTCS.

Universities with faculties of engineering and computer science seem to be very attractive for start-ups in STI, HTI and TIS. Increasing the distance to higher-education institutions as well as reducing the scientific staff in engineering and computer science leads to a decrease in the number of start-ups in those industries. Specification tests show that we only get a convex connection for the variable 'MIN[ln(Direct Distance in km/scientific staff in Engineering and Computer Science)]'. However, for HTI only the negative impact is significant. We interpret this in the following way: the positive impact of 'engineering and computer sciences' is stronger for HTI compared with STI and TIS. Start-ups in STI and TIS are only concentrated close to larger universities or universities of applied science. Smaller higher-education institutions or those further away have no influence on regional differences in the start-up activities in these types of high-tech industry. In comparison, new firms in HTI are concentrated also near smaller universities, and the number of start-ups decreases only at a great distance. Nerlinger (1998) noticed that in West German counties universities with an engineering faculty located a maximum distance of 50 kilometres away had a positive effect on start-up activities of only STI.

Proximity to universities with a faculty of natural sciences is also important for strat-ups in HTI and NTCS. The correlation between these variables is strongly negative. So we observe fewer start-ups if the direct distance increases or the number of staff decreases. One reason for the significant impact for NTCS may be that a large number of graduates in mathematics work in non-technical consulting services. The insignificance of the variables just described for non-technology industries confirm our expectations about the relevance of human capital and knowledge spillovers at universities and universities of applied science only for high-tech start-ups.

Large Firms

Aside from the Marshall theory, it can be observed that a local concentration of industry facilitates the establishment of a supply industry in the respective area. This can be explained by the effect of economies of scale in permitting lower prices for intermediate goods. Proximity to market is an important factor, especially in the early stage of a firm's existence (see, for example, Nerlinger 1998; Steil 1999). Moreover, the market and customer-supplier relationships are among the most important sources of innovation for firms (see Czarnitzki et al. 2000). It emerges that high-tech start-ups are more likely to have firms as customers than other start-ups, so we can assume that the existence of large firms has a greater impact for them.

Second Hypothesis: The existence of large firms in the manufacturing sector increases the attractiveness of a region for new start-ups especially in high-tech industries.

The hypothesis was tested by measuring the direct distance to the nearest firm in the *manufacturing sector* with more than 250 employees and also the distance to the nearest large firm in *other sectors*. The number of start-ups in almost all sectors was found to decrease in the former case, i.e. with distance to the nearest firm in the manufacturing sector with more than 250 employees (see Table 19.3). However, there was no evidence of the expected differences between industries. It is possible that start ups in the six types of industry have different relations to each of industry in the manufacturing sector which was not taken into account in the variable used. Large firms in other sectors which had been established recently did not attract start-ups. The results confirm the hypothesis in part, in the sense that that the proximity of large firms in the manufacturing sector is of considerable importance for start-up activities in general, but not for high-tech start-up activities in particular.

Urbanisation Effects and Specific Infrastructure

According to Ohlin (1933) and Hoover (1937), economies of urbanisation entail both the positive and negative effects of the concentration of socio-economic activities. These effects derive from the existence of major local markets and a highly developed technical, social and cultural infrastructure. The siting of new firms close to already established large firms helps to create a heterogenous industrial structure. Jacobs (1969) emphasises that a heterogeneous industrial structure and localised knowledge externalities offer advantageous conditions for the growth of agglomerations [including the location of new start-ups]. Following this reasoning, we can assume that a concentration of different industries in a region could decrease the risk of large-scale structural, technological or economic change occurring. Apart from this industry-specific argument, large urban areas find it easier to attract potential start-ups due to specifialised infrastructure which decreases the costs of firms. Several studies have emphasised, for example, the importance of well established traffic links for firms, especially those in transportation cost-intensive industries. A large number of Technology and Foundation Centres[5] have been established in East Germany to stimulate start-up and innovation activities and to increase their success. Therefore, we formulate the following hypothesis:

Third Hypothesis: Economies of urbanisation stimulate creativity, technological efficiency and success through diversity. This explains the attractiveness of setting up innovative businesses in large cities. Infrastructure influences the regional differences in start-ups.

Urbanisation effects were measured with the direct distance to the next *town*. The demographic density of counties was considered a suitable proxy for the agglomeration effects. We took nine dummy variables to take into account the different size and density of agglomerations according the classification of the Federal Office for Regional Planning [BBR]. The consideration of so may variables was necessary to avoid mismeasuring the effects of R&D infrastructure. Such infrastructure [universities etc.] is often located within or around big cities. The distance to the nearest *motorway entrance* and the next *technology and foundation centre* were used to discover the relevance of specific infrastructure for start-ups.

The proximity to the next town administered as district in its own right was found to have a positive influence for start-ups in all service sectors (see Table

19.4). The number of start-ups decreased with an increase in direct distance. The test of dummy variables for agglomeration size and density showed up significant differences in start-up activities between densely populated counties and others. Market size and proximity to potential customers was shown to be more important for firms in the service sector than manufacturing firms. We therefore could not confirm the hypothesis of the greater importance of urbanisation effects for new firms in high-tech industries. As expected, the proximity to the nearest motorway entrance was important only for start-ups in NTI, which often sell traditional products. Transportation costs and also traffic links were more important for firms in this sector than STI or HTI. In fact, many industrial parks for manufacturing firms can be found near important traffic links. Our survey showed a positive impact for a distance of up to about 10 kilometres; beyond this threshold the number of start-ups in NTI decreased.

We now need to see whether the establishment of Technology and Foundation Centres [TFC], as one possible policy of state intervention, does in fact lead to an increase in the number of start-ups in technology-oriented industries. The results estimated showed *ceteris paribus* a strong negative connection between the direct distance to a TFC and the number of start-ups. Furthermore, the relationship was more negative for technology-intensive industries than the others. This meant that high-tech start-ups were more concentrated within or around TFC's, implying that the establishment of TFCs is successful in stimulating the start-ups in high-tech industries.

Other Regional Influences on Start-Up Activities

As already stated, it has been observed that firms are often set up in close to the place of residence of the founder (see e.g. Schmude 1994). The size of regional entrepreneurial potential could consequently influence the number of new firms. There were however no statistics available about the labour force at the level of postcode areas. The *workforce in a postcode area* therefore had to be calculated as an aggregation of the workforce in various districts. Some of the districts are administered as towns in their own right and include more than one postcode area. In this case the total workforce was divided by the number of postcode areas. In general, we may assume a positive relationship between the workforce and the frequency of start-ups. The coefficient for postcode area workforce [ln] can be interpreted as an elasticity. It was positively significant and varied between 0.53 and 0.73 for sectors. Other studies at the level of counties have obtained a coefficient around one (see e.g. Almus, Egeln and Engel 1999; Nerlinger 1998). The number of start-ups increased by less than one percent if the workforce increased by one percent. The variable 'Sum of Employees' was used to measure the effects of neighbourhood size on the number of start-ups. We observed urbanisation tendencies for TIS and OS. The negative impact of the size of the neighbourhood on start-up activities was in accordance with the importance of market size for services. This means that new firms in TIS and OS are more likely to concentrate near large agglomerations.

An important motive for starting up a new firm is represented by existing or anticipated unemployment (see Evans and Leighton 1989; Reize 1999). The probability of becoming unemployed should be positively correlated with the unemployment rate in a region. However, an opposite effect may arise due to the fact that a high rate of unemployment tends to indicate the existence of economic problems and represents an unfavourable condition for demand in a region. Cross-section analyses generally reveal a negative coefficient for the rate of unemployment (see e.g. Felder, Fier and Nerlinger 1997; Almus, Egeln and Engel 1999). According to our estimation, a county's unemployment rate has no significant impact on the number of start-ups. The lack of data on differentiation of the unemployed according to industry or last job, and the small number of counties after the new classification made in 1993, may be some reasons for this result.

The economic structure of a region can also explain regional differences in start-up activities. According the studies by Bruederl, Preisendoerfer and Ziegler (1996) and Pfeiffer (1994) for West Germany, 60 percent and 75 percent of founders were respectively either previously employed in the same industry or had industry-specific skills. The regional concentration of a industry should support the supply of qualified workers and induce knowledge spill-over. Several studies (Steil 1999; Almus, Egeln and Engel 1999) show significantly positive correlations between the *share of employees* and the number of start-ups in the respective sector. Furthermore, we can assume that interactions exist between the manufacturing sector and number of start-ups in services (Klodt, Maurer and Schimmelpfennig 1997; Almus, Egeln and Engel 1999) and that these contribute to the explanation of start-up dynamics. A convex relationship was obtained between the share of employees in the manufacturing sector and the number of start-ups in HTI, NTI, NTCS and OS. The positive correlation confirms the hypothesis of interactions between firms in the service and the manufacturing sectors. The negative coefficient can be interpreted as a negative impact of a high-concentration of employees in the manufacturing sector for start-up activities. Regions with a well-established manufacturing sector are not attractive for start-ups in most technology-intensive industries. The share of employees in services has impact on start-up activities only in OS. Finally, the share of employees in construction has a positive significant effect only on the number of start-ups in STI. This is a surprising result because there is no apparent explanation.

19.6 Conclusions and Policy Implications

This chapter has examined possible determinants of regional differences in the number of high-tech start-ups founded between 1995-1998 calculated at the level of postcode areas in East Germany. The results were compared with those for non-technology-intensive industries in order to achieve a better understanding of the effects observed. Variables measuring the effects of specialised human capital and knowledge spillovers at the level of counties led to contradictory results. It was concluded that using data at the county level increases the risk of misspecification. The application of distance variables allowed us to evaluate the impact of the proximity of specific infrastructure, such as higher education or research institutes, in more detail and to directly take into consideration neighbourhood effects.

The availability of business premises near incubator organisations is particularly beneficial for high-tech start-up activities. High-tech start-ups were found to be rather more concentrated around or within technology and foundation centres. The establishment of science parks in well-suited districts can be helpful to increase the number of start-ups. Moreover, the existence of 'industrial centres' seems to be important for start-up activities in all sectors. The lack of large firms in East Germany reduces regional demand and has had a negative impact on start-up activities.

The estimation results revealed that R&D-oriented infrastructure had positive externalities for start-up activities in high-tech industries. Districts possessing universities or applied sceince institutes with faculties in engineering or computer science seem to be of particular interest for start-ups in superior/high-tech industries and technology-intensive service sectors. Districts with natural science faculties were preferred by start-ups in high-tech industries and non-technical consulting services. These effects were tested, but no evidence found for the non-technology-intensive industries. Non-university publicly financed institutions like the Fraunhofer-Society and the Max-Planck-Society were shown to be less important for stimulating high-tech start-up activities. Only new firms in the technology-intensive service sector tend to concentrate close to institutes of the Max-Planck-Society.

Higher education institutions are the most important incubators of the R&D-oriented infrastructure for high-tech start-up activities. Graduates at higher education institutions are a major source of 'founder' potential in the region where those institutes are located. The positive impact of both ordinary universities and universities of applied science confirms the stimulus of specific programmes, such as EXIST, on university-based start-ups. EXIST aims to teach entrepreneurial knowledge and supports relations between potential founders and those involved in regional networks, increasing the probability of people deciding to take up self-employment. These programmes are of particular interest because the fall in the number of graduates in natural and engineering sciences in recent years could lead to a decrease in high-tech start-up activities in the future. The creation of a 'culture of entrepreneurship' in teaching, research and administration in higher education systems is an important way to increase the probability of graduates or researchers in high-tech industries setting up their own businesses. Such promotion should be an original line of duty of each higher education institution.

Endnotes

1   We gratefully acknowledge critical comments on this chapter by Georg Licht. We are also indebted to the valuable contribution of Sandra Gottschalk, Jürgen Moka, Juliane Lauer, Heni Haász, Paul Jurkiewicz and Emil Marinov in preparing the data.

2   Berlin is excluded, because of the strong differences in the development in this city and other East German counties.

3   Some postcode areas are very small. A definition of 'most technology-intensive regions' is also possible by taking the relative number, even when only a small number of high-tech start ups in those areas is observed. For this reason we accepted a postcode area as most technology-intensive only when the absolute number was above a lower critical value [65 percentile]:

4   The Fraunhofer-Society is the leading institute of applied research in Germany. A staff of around 9,000 is employed at 47 research establishments throughout Germany, most of them scientists and engineers. The work of the Max-Planck-Society concentrates on basic research, especially in key areas not established at universities. The emphasis of research is on physics, chemistry, biology and medical science.

5   Technology and Foundation Centres [TFC] have existed since 1983 in West Germany and since 1990 in East Germany. In East Germany there are about 50 such centres.

References

Abbott J. (2000): Virgin Voyagers: 200km/hr DEMUs for the CrossCountry Network, *Modern Railways*, April, 35-44

Abernathy W.J. and Utterback J.M. (1975): A Dynamic Model of Product and Process Innovation, *Omega* 3, 3-22

Abernathy W.J. and Utterback J.M. (1978): Patterns of Industrial Innovation, *Technology Review* (June/July 1978) 41-47

Ács Z. (1993): *U.S. High Technology Clusters*, Discussion Paper No. 9315, Centre for Research Into Industry Enterprise, Finance and the Firm, Department of Economics, St. Salvator's College, Fife, Scotland

Ács Z. (2000) (ed.): *Regional Innovation, Knowledge and Global Change*, Pinter, London

Ács Z. and Morck R. (1999) (eds.): *Small and Medium-Sized Enterprises in the Global Economy*, The University of Michigan Press, Ann Arbor

Ács Z. and Varga A. (1999): *Geography, Endogenous Growth and Innovation*, Vienna University of Economics and Business Administration, Vienna [draft, available from the authors]

Ács Z., Audretsch D. and Feldman M. (1991): Real Effects of Academic Research: Comment, *American Economic Review* 81, 363-367

Ács Z., Audretsch D. and Feldman M (1994): R&D Spillovers and Recipient Firm Size, *The Review of Economics and Statistics* 76, 336-340

Aghion, P and Howitt P. (1998): *Endogenous Growth Theory*, MIT Press, Cambridge [MA]

Aghion P., Caroli E. and Garcia-Penalosa C. (1999): Inequality and Growth: The Perspective of the New Growth Theories, *Journal of Economic Literature* 37 (4), 1615-1660

Albin P. (1975): *The Analysis of Complex Socio-Economic Systems*, Lexington Books, Lexington [MA]

Alderman N. (1999): Local Product Development Trajectories: Engineering Establishments in Three Contrasting Regions. In: Malecki E.J., Oinas P. (eds.) *Making Connections: Technological Learning and Regional Economic Change*, Ashgate, Aldershot, pp. 79-107

Alderman N., Thwaites A.T. and Maffin D. (2000): *On the Geography of Innovation Networks in Capital Goods Projects*, Paper Presented at the RGS/IBG Conference, Brighton, 4-7 January 2000

Alderman N., Maffin D., Thwaites A. and Vaughan R. (1998a): Supply Chain Management in the Low Volume Capital Goods Context. In: Procter S., Palwar K. (eds.) *Proceedings of the 3rd International Conference on Managing Innovative Manufacturing (MIM'98)*, University of Nottingham, Nottingham, pp. 121-126

Alderman N., Maffin D., Thwaites A.T., Vaughan R., Braiden P. and Hills W. (1997): Providing Customer Value: A Business Process Analysis Approach. In: Wright D.T., Rudolph M.M., Hanna V., Gillingwater D., Burns N.D. (eds.) *Managing Enterprises - Stakeholders, Engineering, Logistics and Achievement*, Mechanical Engineering Publications Ltd, London, pp. 203- 209

Alderman N., Braiden P., Hills W., Maffin D., Thwaites A.T. and Vaughan R. (1998b): Business Process Analysis and Technological Change in the Capital Goods Industry, *International Journal of Computer Applications in Technology* 11 (6), 418-427

Allen P.M. (1998): Modelling Complex Economic Evolution. In: Schweitzer F., Silverberg G. (eds.) *Evolution and Self-Organization in Economics,* Duncker & Humblot, Berlin, pp. 47-75

Allen P.M. (2000): Knowledge, Ignorance and the Evolution of Complex Systems, *World Futures* 55, 37-70

Almus M., Egeln J. and Engel D. (1999): Determinanten regionaler Unterschiede in der Gründungshäufigkeit wissensintensiver Dienstleister, ZEW Discussion Paper 99-22, Mannheim, Germany

Almus M., Engel D. and Prantl S. (2000): The 'Mannheim Foundation Panels' of the Centre for European Economic Research (ZEW), ZEW-Documentation 00-02, Mannheim, Germany

Alonso W. (1971): The Economics of Urban Size, *Proceedings of the Regional Science Association* 26, 67-83

American Public Works Association (1976): *History of Public Works in the United States, 1776-1976,* U.S. Government Printing Office, Washington

Amin A. (1999): An Institutionalist Perspective on Regional Development, *International Journal of Urban and Regional Research* 22, 365-378

Andersen P.W.; Arrow K.J. and Pines D. (eds.) (1998) *The Economy as an Evolving Complex System,* Addison Wesley, Reading [MA]

Andersson Å.E. (1981): Structural Change and Technological Development, *Regional Science and Urban Economics* 11, 351-361

Andersson Å.E. and Mantsinen J. (1980): Mobility of Resources; Accessibility of Knowledge and Economic Growth, *Behavioural Science* 25, 353-366

Andersson Å.E., Batten D.F. and Karlsson C. (1989) (eds.): *Knowledge and Industrial Organisation,* Springer, Berlin, Heidelberg, New York

Andersson Å.E., Batten D.F., Kobayashi K. and Yoshikawa K. (1993) (eds.): *The Cosmo-Creative Society. Logistical Networks in a Dynamic Economy,* Springer, Berlin, Heidelberg, New York

Angel D.P. (1995): *Interfirm Collaboration in Technology Development,* Economic Development Administration, U.S. Department of Commerce, Washington, D.C.

Anselin L. (1995): Local Indicators of Spatial Association – LISA, *Geographical Analysis* 27, 93-115

Anselin L. (1997): The Moran Scatterplot as an ESDA Tool to Assess Local Instability in Spatial Association. In: Fischer M, Scholten H., Unwin D. (eds.): *Spatial Analytical Perspectives on GIS,* Taylor and Francis, London, pp. 111-125

Anselin L., Varga A. and Ács Z. (1997): Local Geographic Spillovers between University Research and High Technology Innovations, *Journal of Urban Economics* 42, 422-448

Anselin L., Varga A., and Ács Z. (1999): Geographic and Sectoral Characteristics of Academic Knowledge Externalities, REAL Research Paper, University of Illinois at Urbana Champaign [forthcoming in Papers *in Regional Science*]

Anselin L., Varga A. and Ács Z. (2000): Geographical Spillovers and University Research: A Spatial Econometric Perspective, *Growth and Change* 31 (4), 501-515

Anwar M.S., Davies S. and Sampath R.K. (1996): Causality between Government Expenditures and Economic Growth: An Examination Using Cointegration Techniques, *Public Finance* 5 (2), 166-184

Appold S.J. (1995): Agglomeration, Interorganizational Networks and Competitive Performance in the U.S. Metalworking Sector, *Economic Geography* 71, 27-54

Archibugi D. (1992): Patenting as an Indicator of Technological Innovation: A Review, *Science and Public Policy* 19, 357-368

Archibugi D. and Iammarino S. (1999): The Policy Implications of the Globalisation of Innovation. In: Archibugi D.,Howells J., Michie J. (eds.) *Innovation Policy in a Global Economy*, Cambridge University Press, Cambridge, pp. 242-271

Archibugi D. and Iammarino S. (2000): Innovation and Globalisation: Evidence and Implications. In: Chesnais F., Ietto-Gillies G., Simonetti R. (eds.) *European Integration and Global Corporate Strategies*, Routledge, London, New York, pp. 95-120

Archibugi D. and Lundvall B.-Å. (eds.) (2000): *The Globalising Learning Economy*, Oxford University Press, Oxford

Archibugi D. and Michie J. (1995): The Globalisation of Technology: A New Taxonomy, *Cambridge Journal of Economics* 19, 121-140

Archibugi D. and Pianta M. (1992): *The Technological Specialisation of Advanced Countries. A Report to the EEC on International Science and Technology Activities*, Kluwer, Dordrecht, Boston

Archibugi D. and Pianta, M. (1996): Measuring Technological Change through Patents and Innovation Surveys, *Technovation* 16, 451-468

Archibugi D., Howells J. and Michie J. (eds.) (1999): *Innovation Policy in the Global Economy*, Cambridge University Press, Cambridge

Armington C. (1998): *Statistics of U. S. Business – Microdata and Tables of SBA/Census Data on Establishment Size*, Office of Advocacy, U. S. Small Business Administration, Washington D. C.

Armington C. and Ács Z. (2000): The Determinants of New Firm Formation, University of Baltimore, October 2000

Arrow K.J. (1962): The Economic Implications of Learning by Doing, *Review of Economic Studies* 29, 155-173

Arthur W.B. (1989): Competing Technologies, Increasing Returns, and Lock-In by Historical Events, *Economic Journal* 99, 116-131

Arthur W.B. (1993): On Designing Economic Agents that Behave Like Human Agents, *Journal of Evolutionary Economics* 3, 1-22

Arthur W.B. (1994): Increasing Returns and Path Dependence in the Economy, The University of Michigan Press, Ann Arbor

Arthur W.B. (1995): Complexity in Economic and Financial Markets, *Complexity* 1, 20-25

Arthur W.B., Durlauf S.N. and Lane D.A. (eds.) (1997): *The Economy as an Evolving Complex System* II, Addison-Wesley, Reading [MA]

Arthur W.B., Holland J.H., LeBaron B., Palmer R. and Tayler P. (1997): Asset Pricing under Endogenous Expectations in an Artificial Stock Market. In: Arthur, W.B., Durlauf, S.N., D.A. Lane (eds.) *The Economy as an Evolving Complex System II*, Addison-Wesley, Reading [MA], pp. 15-44

Asheim B.T. and Cooke P. (1999): Innovation and Local Development: The Neglected Role of Large Firms. In: Malecki E.J., Oinas P. (eds.) *Local Learning and Interactive Innovation Networks in a Global Economy*, Ashgate, Aldershot, pp. 145-178

Audretsch D.B. (1996), Industrieökonomik. In: Hagen von, J., Börsch-Supan, A., Welfens, P. (eds.) *Handbuch der Volkswirtschaftslehre I*, Springer, Berlin, Heidelberg, New York, pp. 178-218

Audretsch D.B. and Feldman M. (1994): Knowledge Spillovers and the Geography of Innovation and Production, Discussion Paper 953, Centre for Economic Policy Research, London

Audretsch D.B. and Feldman M. (1995): Innovative Clusters and the Industry Life Cycle, Discussion Paper 1161, Centre for Economic Policy Research, London

Audretsch D.B. and Feldman M. (1996): R&D Spillovers and the Geography of Innovation and Production, *American Economic Review* 86 (3), 630-640

Audretsch D.B. and Feldman M. (2000): The Telecommunications Revolution and the Geography of Innovation. In: Wheeler J.O., Aoyama Y., Warf B. (2000) (eds.) *Cities in the Telecommunications Age. The Fracturing of Geographies*, Routledge, London, New York, pp. 181-199

Audretsch D.B. and Fritsch M. (1992): Interregional Differences in New-Firm Formation: Evidence from West Germany, *Regional Studies* 25, 233-241

Audretsch D.B. and Fritsch M. (1994): The Geography of Firm Births in Germany, *Regional Studies* 28, 359-365

Audretsch D.B. and Stephan P.E. (1996): Company-Scientist Locational Links: The Case of Biotechnology, *The American Economic Review* 86, 641-652

Audretsch D.B. and Vivarelli M. (1996): Determinants of New-Firm Startups in Italy, *Empirica* 23, 91-103

Aydalot P. (ed.) (1986): *Milieux Innovateurs en Europe*, GREMI, Paris

Aydalot P. and Keeble D. (eds.) (1988): *High Technology Industry and Innovative Environment*, Routledge, London, New York

Bagnasco A. and Trigilia C. (1984): *Società e Politica nelle Aree di Piccola Impresa: Il Caso di Bassano*, Arsenale, Venice

Bahr D.B. and Passerini E. (1998): Statistical Mechanics of Opinion Formation and Collective Behaviour, *Journal of Mathematical Sociology* 23, 1-27.

Bania N., Eberts R.W. and Fogarty M.S. (1993): Universities and the Startup of New Companies: Can we Generalize from Route 128 and Silicon Valley? *The Review of Economics and Statistics* 65, 761-765

Barley S.R. and Freeman J. (1991): Niches as Networks: The Evolution of Organisational Fields in the Biotechnology Industry, Cornell University.

Barney J.B. and Hesterly W. (1999): Organizational Economics: Understanding the Relationship Between Organzations and Economic Analysis. In: Clegg, S., Hardy, C. (eds.) *Studying Organization. Theory & Method*, Sage, London, pp. 109-141

Barrett C.L., Thord R. and Reidys C. (1998): Simulations in Decision Making for Socio-Technical Systems. In: M.J. Beckmann, B. Johansson, F. Snickars, R. Thord (eds.) *Knowledge and Networks in a Dynamic Economy*, Springer, Berlin, Heidelberg, New York, pp. 59-82

Barro R.J. and Sala-i-Martin X. (1995): *Economic Growth,* McGraw-Hill, New York

Basberg B. (1987): Patents and the Measurement of Technological Change: A Survey of the Literature, *Research Policy* 16, 131-141

Batten D.F. (2000): *Discovering Artificial Economics: How Agents Learn and Economies Evolve,* Westview Press, New York

Batten D.F., Casti J. and Thord R. (1995) (eds.): *Networks in Action. Communication, Economics and Human Knowledge*, Springer, Berlin, Heidelberg, New York

Batten D.F., Kobayashi K. and Andersson Å.E. (1989): Knowledge, Nodes and Networks: An Analytical Perspective. In: Andersson, Å.E., Batten, D.F., Karlsson, C. (1989) (eds.) *Knowledge and Industrial Organisation*, Springer, Berlin, Heidelberg, New York, pp. 31-46

Batty M., Couclelis H. and Eichen M. (1997): Urban Systems as Cellular Automata, *Environment and Planning B* 24, 159-164

Beardsell M. and Henderson V. (1999): Spatial Evolution of the Computer Industry in the United States, *European Economic Review* 43, 431-456

Becattini G. (1979): Dal Settore Industriale al Distretto Industriale: Alcune Considerazioni sull'Unità di Indagine della Politica Industriale, *Economia e Politica Industriale* 1, pp. 1-79

Becattini G. (1990): The Marshallian Industrial District as a Socio-Economic Notion. In: Pyke F., Becattini G., Sengenberger W. (eds.) *Industrial Districts and Inter-firm Cooperation in Italy*, ILO, Geneva, pp. 37-51

Beckmann M.J., Johansson B., Snickars F. and Thord R. (1998) (eds.): *Knowledge and Networks in a Dynamic Economy*, Springer, Berlin, Heidelberg, New York

Beeson P. (1992): Agglomeration Economies and Productivity Growth. In: Mills E., McDonald F. (eds.) *Sources of Metropolitan Growth*, Center for Urban Policy Research, New Brunswick, New Jersey, pp. 19-35

Begg, C.B. (1994): Publication Bias. In: Cooper H., Hedges L.V. (eds.) *The Handbook of Research Synthesis*, Russell Sage Foundation, New York

Beise M., Licht G. and Spielkamp A. (1995): *Technologietransfer an kleine und mittlere Unternehmen: Analysen und Perspektiven für Baden-Württemberg*, Schriftenreihe des ZEW 3, Nomos, Baden-Baden

Bellet M., Colletis G. and Lung Y. (eds.) (1993): Economies de Proximités, Special Issue of the *Revue d'Economie Régionale et Urbaine* 3

Bergh J.C.J.M. van den and Nijkamp P. (1994): Dynamic Macro Modelling and Materials Balance: Economic-Environmental Integration for Sustainable Development, *Economic Modelling* 11, 83-307

Bergh J.C.J.M. van den, Button K., Nijkamp P. and Pepping G. (1997): *Meta-Analysis in Environmental Economics*, Kluwer, Dordrecht, Boston

Bertalanffy L. von. (1950): The Theory of Open Systems in Physics and Biology, *Science* 111, 23-29 [reprinted in Emery F.E. (ed.) *Systems Thinking*, Penguin Books, Harmondsworth (1969), pp. 83-99]

Biebricher C.K.; Nicolis G. and Schuster P. (1995): Self-Organization in the Physico-Chemical and Life Sciences, EU Report 16546

Björnerstedt B.J. and Weibull J. (1993): Nash Equilibrium and Evolution by Imitation. In: Arrow K.J., Colombatto E. (eds.) *Rationality in Economics,* Macmillan, New York, pp. 81-98

Black D. and Henderson V. (1999): Spatial Evolution of Population and Industry in the United States, *American Economic Review* 89, 321-327

Bluestone B. and Harrison B. (2000): *Growing Prosperity*, Houghton, Mifflin, New York

Bodie Z. and Merton R. (2000): *Finance*, Prentice Hall, Upper Saddle River [NJ]

Bodie Z., Kane A. and Marcus A. (1996): *Investments,* Irwin Press, Chicago [4th edition]

Bona J.L. and Santos M.S. (1997): On the Role of Computation in Economic Theory, *Journal of Economic Theory* 72, 241-281

Boon J.P. (ed.) (1992): Lattice Gas Automata: Theory, Simulation, Implementation, *Journal of Statistical Physics* 68 (3-4) [Special Issue]

Bos H. (1965): *Spatial Dispertion of Economic Activities*, Rotterdam University Press, Rotterdam

Bovenberg A.L. and Smulders S.A. (1995): Environmental Quality and Pollution Augmenting Technological Change in a Two-Sector Endogenous Growth Model, *Journal of Public Economics* 57, 369-391

Braczyk H. J., Cooke P. and Heidenreich M. (eds.) (1998): *Regional Innovation Systems: The Role of Governance in a Globalized World*, UCL Press, London

Bramanti A. and Maggioni M. (eds.) (1997): *La Dinamica dei Sistemi Produttivi Territoriali: Teorie, Tecniche, Politiche*, Franco Angeli Editore, Milan

Branstetter L. (1996): Are Knowledge Spillovers International or Intranational in Scope? Microeconometric Evidence from the U.S. and Japan, National Bureau of Economic Research, Working Paper 5800

Breschi S. and Malerba F. (1997): Sectoral Innovation Systems: Technological Regimes, Schumpeterian Dynamics, and Spatial Boundaries. In: Edquist C. (ed.) *Systems of Innovation: Technologies, Institutions and Organisations*, Pinter, London, pp. 130-156

Bretschger L (1999): Knowledge Diffusion and the Development of Regions, *The Annals of Regional Science 33*, 251-268

Brown L.E. (1981): *Innovation Diffusion: A New Perspective,* Methuen, London

Bruckner E., Ebeling W., Jimenez-Montano M.A. and Scharnhorst A. (1994): Hyperselection and Innovation Described by a Stochastic Model of Technological

Change. In: Leydesdorff L., Besselaar P. van den (eds.) *Evolutionary Economics and Chaos Theory: New Directions in Technology Studies,* Pinter, London, pp. 89-90

Bruederl J., Preisendoerfer P. and Ziegler R. (1996): *Der Erfolg neugegründeter Betriebe: Eine empirische Studie zu den Chancen und Risiken von Unternehmensgründungen,* Duncker & Humblot, Berlin

Buchanan J. (1965): An Economic Theory of Clubs, *Economica* 32 [February], 1-14

Bundesministerium für Wissenschaft und Verkehr (1993): Arbeitsberichte der Institutsvorstände gemäß § 95 UOG '75 über das Studienjahr (1991/92), Vienna

Butter F.A.G. den, Delling R.B. and Hofkes M.W. (1995): Energy Levies and Endogenous Technology in an Empirical Simulation Model for The Netherlands. In: Bovenberg A.L., Cnossen S. (eds.) *Public Economics and the Environment in an Imperfect World,* Kluwer, Dordrecht, Boston, pp. 315-335

Button K. (1998): Infrastructure Investment, Endogenous Growth and Economic Convergence, *The Annals of Regional Science* 32 (1) 145-162

Bygrave W. and Timmons J. (1992): *Venture Capital at the Crossroads,* Harvard Business School Press, Cambridge [MA]

Caballero R. and Jaffe A. (1994): How High Are the Giants' Shoulders? 1994 Macroeconomics Annual, National Bureau of Economic Research

Cable J. and Schwalbach J. (1991): International Comparisons of Entry and Exit, In: Geroski P., Schwalbach J. (eds.) *Entry and Market Contestability,* Blackwell, Oxford, pp. 257-282

Callon M. (1992): The Dynamics of Techno-Economic Networks. In: Coombs R., Saviotti P.P., Walsh V. (eds.) *Technical Change and Company Strategies,* Academic Press, London, pp. 349-369

Callon M. (1993): Diversity and Irreversibility in Networks of Technique Conception and Adoption. In: Foray D., Freeman C. (eds.) *Technology and the Wealth of Nations,* Pinter, London, pp. 232-268

Camagni R. (ed.) (1991): *Innovation Networks: Spatial Perspectives,* Belhaven-Pinter, London

Camagni R. (1995): Global Network and Local Milieux: Towards a Theory of Economic Space. In: Conti S., Malecki E., Oinas P. (eds.) *The Industrial Enterprise and its Environment: Spatial Perspective,* Avebury, Aldershot, pp. 195-216

Camagni R. (1999): The City as a Milieu: Applying the Gremi Approach to Urban Evolution, *Révue d'Economie Régionale et Urbaine* 3, 591-606

Campbell J.P. and Goodman S. (1985): High-Technology Employment in Texas–A Labour Market Analysis, Bureau of Business Research, Graduate School of Business, The University of Texas at Austin, Austin [TX]

Cantner U. and Pyka A. (1998): Technological Evolution - An Analysis Within the Knowledge-Based Approach, *Structural Change and Economic Dynamics* 9, 85-107

Cantwell J.A. (1995): The Globalisation of Technology: What Remains of the Product Cycle Model? *Cambridge Journal of Economics* 19, 155-174

Capello R. (1999a): Spatial Transfer of Knowledge in High-technology Milieux: Learning versus Collective Learning Processes, *Regional Studies* 33 (4), 353-365

Capello R. (1999b): A Measurement of Collective Learning Effects in Italian High-tech Milieux, *Révue d'Economie Régionale et Urbaine* 3, 449-468

Capello R. (2001): The Determinants of Innovation in Cities: Dynamic Urbanisation Economies versus Milieu Economies in the Metropolitan Area of Milan. In: Simmie J. (ed.) *Innovative Cities,* Spon, London [forthcoming]

Capello, R., Nijkamp, P. and Pepping, G. (1999): *Sustainable Cities and Energy Policy,* Springer, Berlin, Heidelberg, New York

Carlino G. (1980): Constrast in Agglomeration: New York and Pittsburgh Reconsidered, *Urban Studies* 17, 343-351

Carlsson B. (1992): *Technological Systems and Economic Development Potential: Four Swedish Case Studies,* Presented at the International Joseph A. Schumpeter Conference, Kyoto, August 1992

Carlsson B. (1994): Technological Systems and Economic Development Potential: Four Swedish Case Studies. In: Shionoya Y., Perlman M. (eds.) *Technology, Industries and Institutions: Studies in Schumpeterian Perspectives,* The University of Michigan Press, Ann Arbor, pp. 49-69

Carlsson B. and Stankiewicz R. (1991): On the Nature, Function and Composition of Technological Systems, *Journal of Evolutionary Economics* 1, 93-118

Casetti E. (1969): Why Do Diffusion Processes Conform to Logistic Trends, *Geographical Analysis* 1, 101-105

Casti J.L. (1997): *Would-Be Worlds,* Wiley, New York

Chandler A.D. (1962): *Strategy and Structure,* MIT Press, Cambridge [MA]

Chandler A.D. (1977): *The Visible Hand,* Harvard University Press, Cambridge [MA]

Chandler A.D. (1990): *Scale and Scope: The Dynamics of Industrial Capitalism,* Harvard University Press, Cambridge [MA]

Chapman S.J. and Marquis F.J. (1912) : The Recruiting of the Employing Class from the Ranks of the Wage-Earners in the Cotton Industry, *Journal of Royal Statistical Society* 75, 293-313

Chesnais F., Ietto-Gillies G. and Simonetti R. (eds.) (2000): *European Integration and Global Corporate Strategies,* Routledge, London, New York

Chinitz B. (1961): Contrast in Agglomeration: New York and Pittsburgh, *American Economic Review* 51, 279-289

Christaller W. (1933): *Die zentralen Orte in Süddeuschland,* Gustav Fischer Verlag, Jena

Cladis P.E. and Palffy-Muhoray P. (eds.) (1995): *Spatio-Temporal Patterns in Non-Equilibrium Complex Systems,* Addison-Wesley, Reading [MA]

Clark N., Perez-Trejo F. and Allen P. (1995): *Evolutionary Dynamics and Sustainable Development: A Systems Approach,* Edward Elgar, Cheltenham

Cleveland W. (1994): *The Elements of Graphic Data.* Hobart Press, Summit

Coffey W. J. and Polese M. (1987): Trade and Location of Producer Services: A Canadian Perspective, *Environment and Planning A* 19, 597-611

Cohen J. and Stewart I. (1994): *The Collapse of Chaos,* Penguin, New York

Cohen M. and Levinthal D. (1989): Innovating and Learning: The Two Faces of R&D, *Economic Journal* 99, 569-596

Cohen M. and Levinthal D. (1990): Absorptive Capacity: A New Perspective on Learning and Innovation, *Administrative Science Quarterly* 35, 128-152

Cohendet P. and Llerena P. (1997): Learning, Technical Change and Public Policy: How to Create and Exploit Diversity. In: Edquist C. (ed.) (1997) *Systems of Innovation: Technologies, Institutions, Organisations,* Pinter, London, pp. 223-242

Cooke P. (1998): Introduction: Origin of the Concept. In: Braszyk H.-J., Cooke P., Heinderich M. (eds.) *Regional Innovation Systems. The Role of Governances in a Globalized World,* UCL Press, London, pp. 2-25

Cooke P. (2000): Business Processes in Regional Innovation Systems in the European Union. In: Ács Z. (ed.) *Regional Innovation, Knowledge and Global Change,* Pinter, London, pp. 53-71

Cooke P. and Morgan K. (1993): The Network Paradigm: New Departures in Corporate and Regional Development, *Environment and Planning D* 11, 543-564

Cooke P. and Morgan K. (1998): *The Associational Economy: Firms, Regions and Innovation,* Oxford University Press, Oxford

Cooke P., Uranga M. and Etxebarria G. (1997): Regional Innovation Systems: Institutional and Organisational Dimensions, *Research Policy* 26, 475-491

Cornes R. and Sandler T. (1986): *The Theory of Externalities, Public Goods and Club Goods*, Cambridge University Press, Cambridge

Cornford J., Naylor R. and Robins K. (2000): New Media and Regional Development. In: Giunta A., Lagendijk A., Pike A. (eds.) *Restructuring Industry and Territory: The Experience of Europe's Regions*, Jessica Kingsley, London, pp. 83-108

Cornwall J. (1977): *Modern Capitalism: its Growth and Transformation*, Martin Robertson, London

Crawford V.P. (1989): Learning and Mixed-Strategy Equilibria in Evolutionary Games, *Journal of Theoretical Biology* 140, 537-550

Crevoisier O. (1999): Innovation in the City. In: Malecki E.J., Oinas P. (eds.) *Making Connections: Technological Learning and Regional Economic Change*, Ashgate, Aldershot, pp. 61-78

Crevoisier O. and Camagni R. (eds.) (2000): *Les Milieux Urbains: Innovation, Systèmes de Production et Ancrage*, EDES, Neuchâtel

Crutchfield J.P., and Hanson J.E. (1993): Turbulent Pattern Bases for Cellular Automata, *Physica D* 69, 279-301

Cumbers A. (2000): Globalisation, Local Economic Development and the Branch Plant Region: The Case of the Aberdeen Oil Complex, *Regional Studies* 34, 371-382

Cyert R.M. and March J.G. (1963): *A Behavioural Theory of the Firm*, Prentice Hall, Englewood Cliffs [NJ]

Czarnitzki D., Ebling G., Gottschalk S., Janz N. and Niggemann H. (2000): *Quellen für Innovationen: Analyse der ZEW-Innovationserhebungen 1999 im Verarbeitenden Gewerbe und im Dienstleistungssektor* 00-10, Mannheim

Daniels T.L. (1991): The Purchase of Development Rights; Preserving Agricultural Land and Open Space, *American Planning Association* 57, 421-431

Darley V. (1995): Emergent Phenomena and Complexity. In: Brooks, R.A., P. Maes (eds.) *Artificial Life IV*, Proceedings of the Fourth International Workshop on the Synthesis and Simulation of Living Systems, MIT Press, Cambridge [MA], pp. 411-416

Dasgupta P. and Stiglitz J.E. (1980): Industrial Structure and the Nature of Innovative Activity, *Economic Journal* 90, 266-293

Davelaar E. and Nijkamp P. (1990): Industrial Innovation and Spatial Systems: The Impact of Producer Services. In: Ewers H., Allesch J. (eds.) *Innovation and Regional Development*, de Gruyter, Berlin, pp. 83-122

David P. (1985): Clio and the Economics of QWERTY, *American Economic Review* 75, 332-337

David P. (1999): Towards European Innovation and Diffusion Policy for the Knowledge-Driven Economy, Experts Group on Innovation Policy, Maastricht

David P. and Foray D. (1996): Accessing and Expanding the Science and Technology Knowledge-Based Science, *Technology Industry Review* 16, 13-68

David P., Cowan R. and Foray D. (1999): The Explicit Economics of Knowledge Codification and Tacitness, EC TSER Programme

DeAngelis D. L. and Gross L.J. (eds.) (1992): *Individual-Based Models and Approaches in Ecology: Populations, Communities, and Ecosystems*, Chapman and Hall, New York

DeBresson C. (1996): *Economic Interdependence and Innovative Activity*, Edward Elgar, Cheltenham

DeBresson C. and Amesse F. (1991): Networks of Innovators: A Review and Introduction to the Issue, *Research Policy* 20, 363-379

DeLong J.B., Schleifer A., Summers L.H. and Waldmann J. (1990): Positive Feedback and Destabilizing Rational Speculation, *Journal of Finance* 45, 379-395.

Dendrinos D.S. (1992): *The Dynamics of Cities; Ecological Determinism, Dualism and Chaos,* Routledge, London, New York

References 431

Dendrinos D.S. (2000): Non-linear Dynamics, Innovation and Metropolitan Development. In: Batten D.F., Bertuglia C.S., Martellato D., Occelli S. (eds.) *Learning, Innovation and the Urban Evolution,* Kluwer, Dordrecht, Boston, pp. 75-106

Dendrinos D.S and Sonis M. (1984): Variational Principles and Conservation Conditions in Volterra's Ecology and in Urban/Relative Dynamics [extended version], Collaborative Papers Series CP-49-84, International Institute for Applied Systems Analysis, Vienna

Dendrinos D.S. and Sonis M. (1990): *Chaos and Socio-Spatial Dynamics,* Springer, Berlin, Heidelberg, New York

Deogun R., Raghavan V., Sarkar A. and Sever H. (1996): Data Mining: Trends in Research and Development. In: Lin T.Y., Cercone N. (eds.) *Rough Sets and Data Mining: Analysis for Imprecise Data,* Kluwer, Dordrecht, Boston [MA], pp. 9-45

Dillman D. (1978): *Mail and Telephone Surveys: The Total Design Method for Surveys,* Wiley, New York

Dixit A. and Stiglitz J. (1997): Monopolistic Competition and Optimum Product Diversity, *American Economic Review* 67, 297-308

Dodd S.C. (1956): Testing Message Diffusion in Harmonic Logistic Curves, *Psychometrica* 21, 191-205

Dodgson M. (1993): *Technological Collaboration in Industry,* Routledge, London, New York

Dollar D. and Wolff E. (1993): *Competitiveness, Convergence, and International Specialisation,* MIT Press, Cambridge [MA]

Dosi G. (1982): Technological Paradigms and Technological Trajectories, *Research Policy* 11, 147-162

Dosi G. and Nelson R.R. (1994): An Introduction to Evolutionary Theories in Economics, *Journal of Evolutionary Economics* 4 , 153-172

Dosi G., Freeman C., Nelson R., Silverberg G. and Soete L. (eds.) (1988): *Technical Change and Economic Theory,* Pinter, London

Dougherty J., Kohavi R. and Sahami M. (1995): Supervised and Unsupervised Discretizations of Continuous Features, *Proceedings of the 12th International Conference on Machine Learning,* Morgan Kaufmann Publishers, San Francisco. pp. 194-202

Drennan M.P. (1999): National Structural Change and Metropolitan Specialisation in the United States, *Papers in Regional Science* 78, 297-318

Dreze J.H. (1974): Axiomatic Theories of Choice, Cardinal Utility and Subjective Probability. In: Dreze J.H. (ed.) *Allocations under Uncertainty: Equilibrium and Optimality,* Macmillan, London, pp. 3-23

Dupuy C. and Gilly J.-P. (1994): *Collective Learning and Territorial Dynamics: A New Approach to the Relations Between Industrial Groups and Territories.* Institute d'Economie Régionale de Toulouse

Dupuy C. and Gilly J-P. (1995): Dynamiques Industrielles, Dynamiques Territoriales, Paper presented at the International Conference of ASRLF, Toulouse, August 30-31 to September 1, 1995

Ebeling W. (1992): Random Effects in Innovation Processes, *Journal of Scientific and Industrial Research* 51, 209-215

Ebeling W., Freund J. and Schweitzer F. (1998): *Komplexe Strukturen: Entropie und Information,* Teubner, Stuttgart

Echeverri-Carroll E. (1997): Japanese-Style Networks and Innovations in High-Technology Firms in Texas, Bureau of Business Research, Graduate School of Business, The University of Texas at Austin, Austin [TX]

Echeverri-Carroll E. L. and Brennan W. (1999): Are Innovation Networks Bounded by Proximity? In: Fischer M.M., Suarez-Villa L., Steiner M. (eds.) *Innovation, Networks and Localities,* Springer, Berlin, Heidelberg, New York, pp. 28-49

Echeverri-Carroll E.L., Hunnicutt L. and Hansen N. (1998): Do Asymetric Networks Help or Hinder Small Firms' Ability to Export? *Regional Studies* 32, 721-734

Edquist C. (1997): Systems of Innovation Approaches – Their Emergence and Characteristics. In: Edquist C. (ed.) *Systems of Innovation: Technologies, Institutions and Organisations.* Pinter, London, pp. 1-35

Edquist C. (2001): Innovation Policy – A Systemic Approach. In Archibugi D., Lundvall B.-A. (eds.) *The Globalising Learning Economy: Major Socio-Economic Trends and European Innovation Policy,* Oxford University Press, Oxford [forthcoming]

Edquist C. and Johnson B. (1997): Institutions and Organisations in Systems of Innovation. In Edquist C. (ed.) *Systems of Innovation: Technologies, Institutions and Organisations,* Pinter, London, pp. 41-63

Edquist C. and McKelvey M. (eds.) (2000): *Systems of Innovation: Growth, Competitiveness and Employment,* Edward Elgar, Cheltenham [two volumes]

Edquist C. and Riddell C. (2000): The Role of Knowledge and Innovation for Economic Growth and Employment in the IT Era. In Rubenson K., Schuetze H. (eds.) *Transition to the Knowledge Society,* Institute for European Studies, University of British Columbia, Vancouver, pp. 3-32

Edquist C., Ericsson M-L. and Sjögren H. (2000): Collaboration in Product Innovation in the East Gothia Regional System of Innovation, *Enterprise & Innovation Management Studies* 1, 37-56

Edquist C., Hommen L. and McKelvey M. (2001): *Innovation and Employment: Process versus Product Innovation,* Edward Elgar, Cheltenham [forthcoming]

Edwards K. and Gordon T. (1984): Characterization of Innovations Introduced on the U.S. Market in 1982, The Futures Group, U.S. Small Business Administration, Washington D.C.

Ellison G. (1993): Learning, Local Interaction and Coordination, *Econometrica* 61, 1047-1071

Ellison G. and Glaeser E. (1999): The Geographic Concentration of Industry: Does Natural Advantage Explain Agglomeration? *American Economic Review* 89, 311-316

Epstein J. M. and Axtell R. (1996): *Growing Artificial Societies: Social Science from the Bottom Up,* MIT Press, Cambridge [MA]

European Commission (1997): *Second European Report on S&T Indicators,* Brussels, EUR 17639

European Patent Office (EPO) (1998): Trilateral Statistical Report, Trilateral Web Site, http:www.european-patent-office.org/tsr_98/tsr_3_3.htm

European Technology Assessment Network (1998): *Technology Policy in the Context of Internationalisation of R&D and Innovation. How to Strengthen Europe's Competitive Advantage in Technology,* European Commission, Directorate-General Science, Research and Development, Brussels

Evangelista R. (1999): *Knowledge and Investment. The Sources of Innovation in Industry,* Edward Elgar, Cheltenham

Evans D.S. and Jovanovic B. (1989): An Estimated Model of Entrepreneurial Choice under Liquidity Constraints, *Journal of Political Economy* 97, 808-827

Evans D.S. and Leighton L.S. (1989): Some Empirical Aspects of Entrepreneurship, *American Economic Review* 79, 519-535

Faber M. and Proops J.L.R. (1990): *Evolution, Time, Production and the Environment,* Springer, Berlin, Heidelberg, New York

Fagan B. (1996): *The Region as Political Discourse,* Macquarie University, Sydney, Australia

Fagerberg J. (2000): Technological Progress, Structural Change and Productivity Growth: A Comparative Study, Working Paper 5/2000, Centre for Technology, Innovation and Culture, University of Oslo

Fagerberg J. and Verspagen B. (1999): Productivity, R&D Spillovers and Trade, Working Paper 3/1999, Centre for Technology, Innovation and Culture, University of Oslo

Fagerberg J., Guerrieri P. and Verspagen B. (eds.) (1999): *The Economic Challenge for Europe: Adapting to Innovation-based Growth*, Edward Elgar, Cheltenham

Famili A., Shen W., Weber R. and Simoudis E. (1997): Data Preprocessing and Intelligent Data Analysis, *Intelligent Data Analysis* 1 (1), 49-62

Farmer R.A. (1993): *The Macroeconomics of Self-Fulfilling Prophecies*, MIT Press, Cambridge [MA]

Fayyad U.M. and Irani K.B. (1993): Multi-Interval Discretization of Continuous-Valued Attributes for Classification Learning. In: Saitta L. (ed.) *Proceedings of the 13th International Conference on Machine Learning*, Morgan Kaufmann Publishers, San Mateo [CA], pp. 1022-1027

Feistel R. and Ebeling W. (1989): *Evolution of Complex Systems. Self-Organization, Entropy and Development*, Kluwer, Dordrecht, Boston

Felder J., Fier A. and Nerlinger E. (1997): Im Osten nichts Neues? Unternehmensgründungen in High-Tech Industrien. In: Harhoff D. (ed.) *Unternehmensgründungen - Empirische Analysen für die alten und neuen Bundesländer* 7, Nomos, Baden-Baden, pp. 73-110

Feldman M.P. (1994a): *The Geography of Innovation*, Kluwer, Dordrecht, Boston, pp. 73-110

Feldman M.P. (1994b): Knowledge Complementarity and Innovation, *Small Business Economics* 6, 363-372

Feldman M.P. and Audretsch D. (1999): Innovation in Cities: Science-Based Diversity, Specialisation and Localised Competition, *European Economic Review* 43, 409-429

Feser E.J. (1998): Enterprises, External Economies, and Economic Development, *Journal of Planning Literature* 12, 283-302

Fischer M.M. (1998): Spatial Analysis: Retrospect and Prospect. In: Longley P., Goodchild M.F., Maguire D.J., Rhind D.W. (eds.) *Geographical Information Systems: Principles, Technical Issues, Management Issues and Applications*, Wiley, New York, pp. 283-292

Fischer M.M. (1999): The Innovation Process and Network Activities of Manufacturing Firms. In: Fischer M.M., Suarez-Villa L., Steiner M. (eds.) *Innovation, Networks and Localities*. Springer, Berlin, Heidelberg, New York, pp. 11-27

Fischer M.M. (2001a): Innovation, Knowledge Creation and Systems of Innovation, *The Annals of Regional Science* 35 [in press]

Fischer M.M. (2001b): Spatial Analysis. In: Smelson N., Baltes P. (eds.): *International Encyclopedia of the Social & Behaviourial Sciences*, Pergamon, Amsterdam [forthcoming]

Fischer M.M. and Menschik G. (1994): *Innovationsaktivitäten in der östereichischen Industrie: Eine empirische Untersuchung des betrieblichen Innovationsverhaltens in ausgewählten Branchen und Raumtypen*, Abhandlungen zur Geographie und Regionalforschung 3, Institute for Geography, University of Vienna

Fischer, M.M. and Varga A. (2000): Technological Innovation and Interfirm Cooperation: An Exploratory Analysis using Survey Data from Manufacturing Firms in the Metropolitan Region of Vienna, *International Journal of Technology Management* [in press]

Fischer M.M. and Varga A. (2001): Geographic Knowledge Spillovers and University Research: Some Evidence from Austria, Jahresbericht 58 [forthcoming]

Fischer M.M., Fröhlich J. and Gassler H. (1994): An Exploration into the Determinants of Patent Activities: Some Empirical Evidence for Austria, *Regional Studies* 28, 1-12

Fischer M.M., Suarez-Villa L. and Steiner M. (1999): *Innovation, Networks and Localities*, Springer, Berlin, Heidelberg, New York

Fischer M.M., Haag G., Sonis M. and Weidlich W. (1990): Account of Different Views in Dynamic Choice Processes. In: Fischer M.M., Nijkamp P., Papageorgiou Y.Y. (eds.) *Spatial Choices and Processes*, Elsevier, Amsterdam, pp. 17-47

Flood R.L. and Carson E.R. (1993): *Dealing with Complexity: An Introduction to the Theory and Application of Systems Science*, Plenum Press, London [2nd edition]

Florax R., Nijkamp P. and Willis K. (2000): *Research Synthesis and Meta-Analysis*, Edward Elgar, Cheltenham

Florida R. (1995): Toward the Learning Region, *Futures* 27, 527-536

Föllmer H. (1974): Random Economies with Many Interacting Agents, *Jounal of Mathematical Economics* 1, 51-62

Forrester J.W. (1961): *Industrial Dynamics*, MIT Press, Cambridge [MA]

Freeman C. (1984): *The Economics of Industrial Innovation*, Pinter, London

Freeman C. (1987): *Technology and Economic Performance: Lessons from Japan*, Pinter, London

Freeman C. (1988): Diffusion: The Spread of New Technology of Firms, Sectors and Nations. In: Heertje, A (ed.) *Innovation Technology, and Finance. The European Investment Bank*, Basil Blackwell, Oxford, pp. 38-70

Freeman C. (1995): The 'National System of Innovation' in Historical Perspective, *Cambridge Journal of Economics* 19, 5-24

Frieberger P. and Swaine M. (1984): *Fire in the Valley: The Making of the Personal Computer*, Osborne-McGraw Hill, Berkeley [CA]

Fujita M.A., Krugman P. and Venables A.J. (1999): *The Spatial Economy*, MIT Press, Cambridge [MA]

Gabisch G. and Lorenz H.W. (1987): *Business Cycle Theory*, [Lecture Notes in Economics and Mathematical Systems 283], Springer, Berlin, Heidelberg, New York

Galam S. (1997): Rational Group Decision Making, *Physica A* 238, 66-80

Gambardella A and Malerba F. (eds.) (1999): *The Organization of Economic Innovation in Europe*, Cambridge University Press, Cambridge

Gandolfo G. (1980): *Economic Dynamics: Methods and Models*, North-Holland, Amsterdam

Gann D.M. and Salter A. (1998): Learning and Innovation Management in Project-Based Firm, Paper Presented at the 2nd International Conference on Technology Policy and Innovation, Lisbon, 3-5 August 1998

Gantmacher F.G. (1990): *The Theory of Matrices*, Chelsea Publishing, New York [2nd edition]

Gassler H. (1993): Regionale Disparitäten der betrieblichen Inventionsaktivitäten in Österreich. Eine empirische Analyse unter Verwendung von Patentdaten, *Klagenfurter Geographische Schriften* 11, 173-186

Gaudry M. and Dagenais M. (1979): The Dogit Model, *Transportation Research B* 13, 105-112

Gehrke B., Legler H., Machate-Weiß V., Schasse U., Steincke M. and Wagner F. (1997): Beitrag zur 'Berichterstattung zur technologischen Leistungsfähigkeit' im Auftrag des Bundesministeriums für Bildung, Wissenschaft, Forschung und Technologie, Materialband, Hanover

Geiger R. L. (1993): *Research and Relevant Knowledge: American Research Universities since World War II*, Oxford University Press, New York

Georgescu-Roegen N. (1971): *The Entropy Law and the Economic Process*, Harvard University Press, Cambridge [MA]

Geroski P. (1995): What Do We Know about Entry? *International Journal of Industrial Organization* 13, 421-440

Gilly J.P. and Torre A. (eds.) (2000): *Dynamiques de Proximité*, L'Harmattan, Paris

Glaeser E.L. (1994): Cities, Information, and Economic Growth, *Cityscape* 1, 9-48

Glaeser E.L. (1997): Learning in Cities, NBER Working Paper Series 6271, National Bureau of Economic Research, Cambridge [MA]

Glaeser E.L., Sacerdote B. and Scheinkman J.A. (1996): Crime and Social Interactions, *Quarterly Journal of Economics* 111, 507-549

Glaeser E.L., Kallal H., Scheinkman J. and Schleifer A. (1992): Growth of Cities, *Journal of Political Economy* 100, 1126-1152

Glass G.V. (1976): Primary, Secondary and Meta-Analysis of Research, *Educational Researcher* 5, 3-8

Goleman D. (1995): *Emotional Intelligence*, The Bantam Books, New York, Toronto

Gordon R. (1993): *Collaborative Linkages, Transnational Networks and New Structures of Innovation in Silicon Valley's High Technology Industry*, Report 1 to Datar, Paris, Silicon Valley Research Group, University of California, Santa Cruz

Gradus R. and Smulders S.A. (1993): The Trade-off Between Environmental Care and Long-term Growth: Pollution in Three Proto-Type Growth Models, *Journal of Economics* 58, 25-28

Graham H. D. and Diamond N. (1997): *The Rise of American Research Universities: Elites and Challengers in the Postwar Era*, Johns Hopkins University Press, Baltimore

Granovetter M. (1985): Economic Action and Social Structure: The Problem of Embeddedness, *American Journal of Sociology* 91 (3), 481-510

Grant R.M. (1997): The Knowledge-Based View of the Firm: Implications for Management Practice, *Long Range Planning* 30 (3), 450-454

Gregerson B. and Johnson B. (1997): Learning Economies, Innovation Systems and European Integration, *Regional Studies* 31 (5), 479-490

Griliches Z. (1958): Research Cost and Social Returns: Hybrid Corn and Related Inventions, *Journal of Political Economy* 66, 419-431

Griliches Z. (1979): Issues in Assessing the Contribution of Research and Development to Productivity Growth, *Bell Journal of Economics* 10, 92-116

Griliches Z. (1990): Patent Statistics as Economic Indicators: A Survey. *Journal of Economic Literature* 28, 1661-1707

Griliches Z. (1991): The Search for R&D Spillovers, NBER Working Paper 3768, National Bureau of Economic Research, Cambridge [MA]

Griliches Z. (1992): The Search for R&D Spillovers, *Scandinavian Journal of Economics* 94, 29-47

Grossman G.M. and Helpman E. (1991): *Innovation and Growth in the Global Economy*, MIT Press, Cambridge [MA]

Grzymala-Busse J.W. (1992): LERS. A System for Learning from Examples Based on Rough Set. In: Slowinski R. (ed.) *Intelligent Decision Support – Handbook of Applications and Advances of the Rough Set Theory*, Kluwer, Dordrecht, Boston, pp. 3-18

Gualini E. (1999): Local Development and Regional Strategies: 'New Programming' and the Influence of Transnational Discourses in the Reform of Territorial Policy in Italy, Department of Spatial Planning, University of Dortmund, Dortmund

Guerrieri P. (1999): Patterns of National Specialisation in the Global Competitive Environment. In: Archibugi D., Howells J., Michie J. (eds.) *Innovation Policy in a Global Economy*, Cambridge University Press, Cambridge, pp. 139-159

Gumilev L.N. (1994): *Ethnogenesis and Biosphere of Earth*, Gumilev's World Fund, DiDik Publ., Tanais, Moscow

Haag G. (1989): *Dynamic Decision Theory: Applications to Urban and Regional Topics*, Kluwer, Dordrecht, Boston

Hagedoorn J. (1994): *Technological Partnering in Strategic Alliances*, Paper Prepared for the Austrian Conference on R&D Co-Operation, Vienna

Hagedoorn J. (1996): Trends and Patterns in Strategic Technology Partnering Since the early Seventies, *Review of Industrial Organization* 11, 601-616

Håkansson H. (1987): *Industrial Technological Development: A Network Approach,* Croom Helm, London

Haken H. (1978): *Synergetics. An Introduction. Non-Equilibrium Phase Transitions in Physics, Chemistry and Biology,* Springer, Berlin, Heidelberg, New York [2nd edition]

Hambrecht W. R. (1984): Venture Capital and the Growth of Silicon Valley, *California Management Review* 26, 74-82

Hamel G. and Prahalad C.K. (1994): *Competing for the Future,* Harvard Business School Press, Boston [MA]

Hansen K.L.and Rush H. (1998): Hotspots in Complex Product Systems, Emerging Issues in Innovation Management, *Technovation* 18, 555-561

Hansen N. (2000): The New Economy: Implications for Peripheral Regions, Paper Presented at the Annual Meeting of the Southern Regional Science Association, Miami Beach, Florida, 13-15 April, 2000

Hansen N. and Echeverri-Carroll E. (1997): The Nature and Significance of Network Interactions for Business Performance and Exporting to Mexico: An Analysis of High Technology Firms in Texas, *The Review of Regional Studies* 27, 85-99

Hanson D. (1982): *The New Alchemists: Silicon Valley and the Microelectronics Revolution,* Little, Brown, Boston

Hardin G. (1961): Competitive Exclusion Principle, *Science* 131, 1292-1298

Harhoff D. (1995): Agglomerationen und regionale Spillovereffekte. In: Gahlen B., Hesse H., Ramser H.J. (eds.) *Standorte und Region: Neue Ansätze der Regionalökonomik,* [Wirtschaftswissenschaftliches Seminar Ottobeuren 24], Mohr, Tübingen, pp. 83-116

Harhoff D. (1997): *Innovation in German Manufacturing Enterprises: Empirical Studies on Productivity, Externalities, Corporate Finance, and Tax Policy,* Habilitationsschrift, University Mannheim

Harhoff D., Licht G., Beise M., Felder J., Nerlinger E. and Stahl H. (1996): *Innovationsaktivitäten kleiner und mittlerer Unternehmen: Ergebnisse des Mannheimer Innovationspanels,* Schriftenreihe des ZEW 8, Nomos, Baden-Baden

Harris L., Coles A.-M., Dickson K. and McLoughlin I. (1999): Building Collaboration Networks: New Product Development Across Organisational Boundaries. In: Jackson P.J. (ed.) *Virtual Working: Social and Organisational Dynamics,* Routledge, London, New York, pp. 33-45

Harrison B. (1992): Industrial Districts: Old Wine in New Bottles? *Regional Studies* 26, 469-484

Harvard Business Review (2000): Networked Incubators: Hothouses of the New Economy, September/October, pp. 74-84

Hayek F.A. (1945): The Use of Knowledge in Society, *American Economic Review* 35 (40) 519-530

Healey P., Khakee A., Motte A. and Needham B. (eds.) (1997): *Making Strategic Spatial Plans: Innovation in Europe,* UCL Press, London

Heckman J.J. (1981): Statistical Methods for Discrete Panel Data. In: Manski C.F., Mac-Fadden D. (eds.) *Structural Analysis of Discrete Data: With Econometric Applications,* MIT Press, Cambridge [MA], pp. 114-178

Hegselmann R.H. (1996): Understanding Social Dynamics: The Cellular Automata Approach. In: Troitzsch K.G., Mueller U., Gilbert G.N., Doran J.E. (eds.) *Social Science Microsimulation,* Springer, Berlin, Heidelberg, New York, pp. 282-306

Hegselmann R.H. and Flache A. (1998): Understanding Complex Social Dynamics: A Plea for Cellular Automata Based Modelling, *Journal of Artificial Societies and Social Simulation* 1, 3

Hegselmann R.H., Mueller U. and Troitzsch K.G. (eds.) (1996): *Modelling and Simulation in the Social Sciences from the Philosophy of Science Point of View*, Kluwer, Dordrecht, Boston

Helbing D. (1995): *Quantitative Sociodynamics. Stochastic Methods and Models of Social Interaction Processes*, Kluwer, Dordrecht, Boston

Held D., McGrew A., Goldblatt D. and Perraton J. (1999): *Global Flows. Politics, Economics and Culture*, Polity Press, Cambridge

Helpman E. (1992): Endogenous Macroeconomic Growth Theory, *European Economic Review* 36, 237-267

Helpman E. and Trajtenberg M. (1998): Diffusion of General Purpose Technologies. In: Helpman E. (ed.) *General Purpose Technologies and Economic Growth*, MIT Press, Cambridge [MA], pp. 85-120

Henderson J.V. (1974): The Sizes and Types of Cities, *The American Economic Review* 64, 640-656

Henderson J.V. (1985): *Economic Theory and the Cities*, Academic Press, Orlando

Henderson J.V. (1988): *Urban Development: Theory, Fact, and Illusion*, Oxford University Press, New York

Henderson J.V. (1996): Ways to Think about Urban Concentration: Neoclassical Urban Systems versus the New Economic Geography, *International Regional Science Review* 19 (1/2) 31-36

Henderson J.V. (1997): Medium Size Cities, *Regional Science and Urban Economics* 27, 583-612

Henderson J.V., Kuncoro A. and Turner M. (1995): Industrial Development in Cities, *Journal of Political Economy* 103, 1067-1090

Henry N., and Pinch S. (2000): Spatialising Knowledge: Placing the Knowledge Community of Motor Sport Valley, *Geoforum* 31, 191-208

Herzog H.W. Jr., Schlottmann, A.M. and Johnson, D.L. (1986): High-Technology Jobs and Workers' Mobility, *Journal of Regional Science* 26, 445-459

Hicks, D. and Katz, S. (1996): Systemic Bibliometric Indicators for the Knowledge-Based Economy. Paper Presented at the OECD Workshop on New S&T Indicators for a Knowledge-Based Economy, Paris, 19-21 June, 1996

Hingel A. (1992): Science, Technology and Social Economic Cohesion in the Community: A Long Term Analysis, Overall Synthesis Report, Dossier FAST: Science, Technology and Community Cohesion 1 (FOP 300), European Community, Brussels

Hippel E. von (1988): *The Sources of Innovation*, Oxford University Press, New York

Hippel E. von (1989): Cooperation Between Rivals: Informal Know-how Trading. In: Carlsson, B. (ed.): *Industrial Dynamics*, Kluwer, Dordrecht, Boston, pp. 157-175

Hippel E. von (1994): Sticky Information and the Locus of Problem Solving: Implications for Innovation, *Management Science* 40, 429-439

Hirschman A.O. (1958): *The Strategy of Economic Development*, Yale University Press, New Haven

Hobday M. (1998): Product Complexity, Innovation and Industrial Organisation, *Research Policy* 26, 689-710

Hodgson G. (1988):. *Economics and Institutions*, Polity Press, Oxford

Hodgson G. (1993): *Economics and Evolution: Putting Life Back Into Economics*, Polity Press, Oxford

Hofbauer J. and Sigmund K. (1988): *The Theory of Evolution and Dynamical Systems*, Cambridge University Press, Cambridge

Holland J.H (1988): The Global Economy as an Adaptive Process. In: Anderson P.W., Arrow K.J., Pines D. (eds.) *The Economy as an Evolving Complex System*, Addison-Wesley, Reading [MA], pp. 117-124.

Holland J.H (1998): *Emergence: From Chaos to Order*, Addison-Wesley, Reading [MA]

Holland J.H. and Miller J. (1991): Adaptive Agents in Economic Theory, *American Economic Review* 81, 365-370

Hoover E.M. (1937): *Location Theory and the Shoe and Leather Industries*, Harvard University Press, Cambridge [MA]

Hoover E.M. (1948): *The Location of Economic Activity*, McGraw Hill, New York

Horgan J. (1995): From Complexity to Perplexity, *Scientific American* 272 (6), 104-109

Horgan J. (1997): *The End of Science: Facing the Limits of Knowledge in the Twilight of the Scientific Age*, Broadway Books, New York

Hotelling H. (1929): Stability in Competition, *Economic Journal* 39, 41-57

Howells J.R. (1996): Systems of Local and Regional Innovation: Innovation Arenas, Discussion Paper, PREST, Manchester

Hudson R. (1999): The Learning Economy, the Learning Firm and the Learning Region: A Sympathetic Critique of the Limits to Learning, *European Urban and Regional Studies* 6 (1), 59-72

Hunter J.E. and Schmidt F.L. (1990): *Methods of Meta-Analysis – Correcting Error and Bias in Research Findings,* Sage, London

Huntington S.P. (1996): *The Clash of Civilizations and the Remaking of World Order*, Simon & Shuster, New York

Hutcheson P., Pearson A.W. and Ball D.F. (1996): Sources of Technical Innovation in the Network of Companies Providing Chemical Process Plant Equipment, *Research Policy* 25, 25-41

Institute of Scientific Information (1950-1995): Science Citation Index, Institute of Scientific Information, Philadelphia

Irwin D. and Klenow P. (1994): Learning by Doing Spillovers in the Semiconductor Industry, *Journal of Political Economy* 102, 1200-1227

Isard W. (1956): *Location and Space Economics,* MIT Press, Cambridge [MA]

Jacobs J. (1969): *The Economy of Cities*, Random House, New York

Jacobs J. (1984): *Cities and the Wealth of Nations*, Vintage Books, New York

Jaffe A.B. (1989): Real Effects of Academic Research, *American Economic Review* 79, 957-970

Jaffe A.B., Trajtenberg M. and Henderson R. (1993): Geographic Localization of Knowledge Spillovers as Evidenced by Patent Citations, *Quarterly Journal of Economics* 108, 577-598

Johansson B. (1997): Regional Differentials in the Development of Economy and Population. In: Sörensen, C. (1997) (ed.) Empirical Evidence of Regional Growth: The Centre-Periphery Discussion, The Expert Committee of the Danish Ministry of the Interior, Copenhagen, pp. 107-162

Johansson B., Karlsson C. and Westin L. (1994) (eds.): *Patterns of a Network Economy*, Springer, Berlin, Heidelberg, New York

Johnston R. and Lawrence P.R. (1988): Beyond Vertical Integration - The Rise of the Value-Adding Partnership, *Harvard Business Review* 66, 94-101

Jones C. (1995): R&D-Based Models of Economic Growth, *Journal of Political Economy* 103, 759-785

Judd K. (1985): On the Performance of Patents, *Econometrica* 53, 567-586

Jungmittag A., Blind K. and Grupp H. (1999): Innovation, Standardization and the Long-Term Production Function, *Zeitschrift für Wirtschafts- und Sozialwissenschaften* 119, 205-222

Kacperski K. and Hoylst J.A. (1996): Phase Transitions and Hysteresis in a Cellular Automoata-Based Model of Opinion Formation, *Journal of Statistical Physics* 84, 169-189

Kaldor N. (1970): The Case of Regional Policies, *Scottish Journal of Political Economy* 17, 337-348

Kamien M.I. and Schwartz N.L. (1982): *Market Structure and Innovation*, Cambridge University Press, Cambridge [MA]

Kandori M., Mailath G.J. and Rob R. (1993): Learning, Mutation, and Long Run Equilibria in Games, *Econometrica* 61, 29-56

Kanter R.M. (1995): *World Class*, Simian and Schuster, New York

Karandikar R, Mookherjee D., Ray D. and Vega-Redondo F. (1998): Evolving Aspirations and Cooperation, *Journal of Economic Theory* 80, 292-331

Karlsson C. (1997): Product Development, Innovation Networks, and Agglomeration Economies, *The Annals of Regional Science 31*, 235-258

Karlsson C. and Olsson O. (1998): Product Innovation in Small and Large Enterprises, *Small Business Economics* 10, 31-46

Kay J. (1993): *Foundations of Corporate Success*, Oxford University Press, Oxford

Keating M. (1998a): Is There a Regional Level of Government in Europe? In: Le Galès, P., Lequesne, C. (eds.) *Regions in Europe,* Routledge, London, New York, pp. 11-29

Keating M. (1998b): *The New Regionalism in Western Europe. Territorial Restructuring and Political Change*, Edward Elgar, Cheltenham

Keller W. (2000): Geographic Localization of International Technology Diffusion. Working Paper 7509, National Bureau of Economic Research, Cambridge [MA]

Kerber R. (1992): ChiMerge: Discretization of Numeric Attributes, *Proceedings of the 10th National Conference on Artificial Intelligence*, San Francisco, pp. 123-127

Kim E. (1998): Economic Gain and Loss from Public Infrastructure Investment, *Growth and Change* 29 (4), 445-468

Kirman A. (1993): Ants, Rationality, and Recruitment, *The Quarterly Journal of Economics* 108, 37-155

Kirzner I.M.D. (1997): Entrepreneurial Discovery and the Competitive Market Process, *The Journal of Economic Literature* 35 (1), 60-85

Klepper S. (1992): *Entry, Exit, Growth and Innovation over the Product Life Cycle*, Paper Presented at the 1992 Conference of the International Joseph A. Schumpeter Society, Kyoto

Klodt H., Maurer R. and Schimmelpfennig A. (1997): Innovations- und Beschäftigungs- potentiale im Dienstleistungssektor: Strukturpolitische Analyse und Konzeption, Schwerpunktstudie zur Strukturberichterstattung für den Bundesminister für Wirtschaft, Institut für Weltwirtschaft, Kiel

Kneller R., Bleaney M.F. and Gemmell N. (1999): Fiscal Policy and Growth: Evidence from OECD Countries, *Journal of Public Economics* 74, 171-190

Knight F.H. (1921a): *Risk, Uncertainty, and Profit*, Houghton, Mifflin, Boston

Knight F.H. (1921b): *Discovery and the Capitalist Process*, University of Chicago Press, Chicago

Kobayashi K. (1995): Knowledge Network and Market Structure: An Analytical Perspective. In: Batten D.F., Casti J., Thord R. (1995) (eds.) *Networks in Action. Communication, Economics and Human Knowledge*, Springer, Berlin, Heidelberg, New York, pp. 127-158

Kobayashi K. and Andersson Å.E. (1994): A Dynamic Input-Output Model with Endogenous Technical Change. In: Johansson B., Karlsson C., Westin L. (1994) (eds.) *Patterns of a Network Economy*, Springer, Berlin, Heidelberg, New York, pp. 243-259

Kobayashi K. and Fukuyama K. (1998): Human Contacts in Knowledge Society: An Analytical Approach. In: Beckmann M.J., Johansson B., Snickars F., Thord R. (1998) (eds.) *Knowledge and Networks in a Dynamic Economy,* Springer, Berlin, Heidelberg, New York, pp. 237-259

Kobayashi K., Sunao S. and Yoshikawa K. (1993): Spatial Equilibria of Knowledge Production with 'Meeting-Facilities'. In: Andersson Å.E., Batten D.F., Kobayashi K.,

Yoshikawa K. (1993) (eds.) *The Cosmo-Creative Society. Logistical Networks in a Dynamic Economy,* Spinger, Berlin, Heidelberg, New York, pp. 219-244

Koopmans T. (1957): *Three Essays on the State of Economic Science,* McGraw Hill, New York

Krogh G. von, Roos J. and Kleine D. (eds.) (1998): *Knowing in Firms: Understanding, Managing and Measuring Knowledge,* Sage, London

Krugman P. (1987): The Narrow Moving Band, the Dutch Disease, and the Competitive Consequences of Mrs. Thatcher, Notes on Trade in the Presence of Dynamic Scale Economies, *Journal of Development Economics* 27, 41-55

Krugman P. (1991a): *Geography and Trade,* MIT Press, Cambridge [MA]

Krugman P. (1991b): Increasing Returns and Economic Geography, *Journal of Political Economy* 99 (3), 483-499

Krugman P. (1994): Competitiveness: A Dangerous Obsession, *Foreign Affairs,* March/April, 28-38

Krugman P. (1995): *Development, Geography and Economic Theory,* MIT Press, Cambridge [MA]

Krugman P. (1996a): *The Self-Organizing Economy,* Blackwell, New York

Krugman P. (1996b): Urban Concentration: The Role of Increasing Returns and Transport Costs, *International Regional Science Review* 19 (1-2), 5-30

Krugman P. (1998): What's New about the New Economic Geography? *Oxford Review of Economic Policy* 14, 7-18

Kuznets S. (1965): *Economic Growth and Structure,* Norton, New York

Kuznets S. (1966): *Modern Economic Growth: Rate, Structure, Spread,* Yale University Press, New Haven

Laan G. van der and Tieman A.F. (1998): Evolutionary Game Theory and the Modelling of Economic Behaviour, *De Economist* 146, 59-89

Lagendijk A. (1999a): Innovative Forms of Regional Structural Policy in Europe: The Role of Dominant Concepts and Knowledge Flows. In: Fischer M.M., Suarez-Villa L., Steiner M. (eds.) *Innovation, Networks and Localities,* Springer, Berlin, Heidelberg, New York, pp. 272-299

Lagendijk A. (1999b): Regional Anchoring and Modernisation Strategies in Non-Core Regions: Evidence From the UK and Germany, *European Planning Studies* 7, 775-792

Lagendijk A. (1999c): The Emergence of Knowledge-Oriented Forms of Regional Policy in Europe, *Tidschrift voor Economische en Sociale Geografie* 90, 110-116

Lagendijk A. (2000): Regional Paths of Institutional Anchoring in the Global Economy. The Case of the North-East of England and Aragón. In: Elsner W., Groenewegen J. (eds.) *An Industrial Policy Agenda 2000 andBbeyond - New Challenges to Industrial Policy.* Kluwer, Dordrecht, Boston, pp. 775-792

Lagendijk A. and Cornford J. (2000): Regional Institutions and Knowledge - Tracking New Forms of Regional Development Policy, *Geoforum* 31, 209-218

Lagendijk A. and Rutten R.P.J.H. (2001): Associational Dilemmas in Regional Innovation Strategy Development: Regional Innovation Support Organisations and the RIS/ RITTS Programmes, Department of Spatial Planning, University of Nijmegen [unpublished manuscript, available from the authors]

Lam L. (1995): Active Walker Models for Complex Systems, *Chaos, Solitons & Fractals* 6, 267-285

Lam L. and Naroditsky V. (eds.) (1992): *Modelling Complex Phenomena,* Springer, Berlin, Heidelberg, New York

Lamming R. (1993): *Beyond Partnership: Strategies for Innovation and Lean Supply,* Prentice Hall, Englewood Cliff [NJ]

Lane D. (1992): Artificial Worlds and Economics, *Journal of Evolutionary Economics* 3, 89-107

Lane D. and Vescovini R. (1996): Decision Rules and Market Share: Aggregation in an Information Contagion Model, *Industrial and Corporate Change* 5, 127-146

Lawson C. (1999): Towards a Competence Theory of the Region, *Cambridge Journal of Economics* 23, 151-166

Lee Y., Amaral L.A.N., Canning D., Meyer M. and Stanley H.E. (1998): Universal Features in the Growth Dynamics of Complex Organisations, *Physical Review Letters* 81, 3275-3278

Legler H., Licht G. and Spielkamp A. (2000): Germany's Technological Performance, *ZEW Economic Studies* 8, Physica, Heidelberg

Levin R.C., Klevorick A.K., Nelson R.R. and Winter S.G. (1987): Appropriating the Returns from Industrial Research and Development, *Brookings Papers on Economic Activity* 1987 (3), 783-820

Levine R. and Renelt D. (1992): A Sensitivity Analysis of Cross-Country Growth Regressions, *American Economic Review* 82 (4), 942-963

Levy M., Levy H. and Solomon S. (1995): Microscopic Simulation of the Stock Market: The Effect of Microscopic Diversity, *J. Physique I (France)* 5, 1087-1107

Lewenstein M., Nowak A. and Latanè B. (1992): Statistical Mechanics of Social Impact, *Physical Review A* 45, 763-776

Lösch A. (1954): *The Economics of Location*, Yale University Press, New Haven

Lovering J. (1999): Theory Led by Policy: The Inadequacies of 'The New Regionalism', *International Journal of Urban and Regional Research* 22, 379-395

Lucas R.E. (1981): *Studies in the Business-Cycle Theory*, MIT Press, Cambridge [MA]

Lucas R.E. (1988): On the Mechanics of Economic Development, *Journal of Monetary Economics* 22 (1), 3-42

Lucas R.E.. (1993): Making a Miracle, *Econometrica* 61, 251-272

Luce R.D. (1959): *Individual Choice Behaviour*, Wiley, New York

Lundvall B.-Å. (1988): Innovation as an Interactive Process: From User-Producer Interaction to the National System of Innovation. In: Dosi G., Freeman C., Nelson R.R., Silverberg G., Soete L. (eds.) *Technical Change and Economic Theory*, Pinter, London, pp. 349-369

Lundvall B.-Å. (1992a): Introduction. In: Lundvall B.-Å. (ed.) *National Systems of Innovation. Towards a Theory of Innovation and Interactive Learning*, Pinter, London, pp. 1-19

Lundvall B.-Å. (2000): Innovation Policy in the Globalizing Learning Economy. In: Archibugi D., Lundvall B.-Å. (eds.) *The Globalizing Learning Economy*, Oxford University Press, Oxford, pp. 273-291

Lundvall B.-Å. (ed.) (1992b): *National Systems of Innovation: Towards a Theory of Innovation and Interactive Learning*, Pinter, London

Lundvall B.-Å. and Borrás S. (1997): The Globalising Learning Economy: Implications for Innovation Policy, European Commission, D.G. XII, Brussels

Lyons D. (1995): Agglomeration Economies among High Technology Firms in Advanced Production Areas: The Case of Denver/Boulder, *Regional Studies* 29, 265-278

MacDonald K.H. (1991): The Value Process Model. In: Scott Morton M.S. (ed.) *The Corporation of the 1990s*, Oxford University Press, Oxford, pp. 299-309

Machlup F. (1980): Knowledge and Knowledge Production, Princeton University Press, Princeton

MacLeod G. (1999): Reflections on the New Regionalism in Economic Development, Department of Geography, University of Durham, Durham

MacLeod G. and Goodwin M. (1999): Reconstructing an Urban and Regional Political Economy: On the State, Politics, Scale, and Explanation, *Political Geography* 18, 697-730

Maddison A. (1987): Growth and Slowdown in Advanced Capitalist Economies, *Journal of Economic Literature* 25, 649-698

Maes P. (ed.) (1991): *Designing Autonomous Agents. Theory and Practice. From Biology to Engineering and Back,* MIT Press, Cambridge [MA]

Maffin D. and Thwaites A. (1998): *Mechanical Engineering and Capital Goods Industries in the North East of England,* The Competitiveness Project, CURDS, University of Newcastle upon Tyne. http://www.ncl.ac.uk/curds/

Maffin D., Alderman N., Thwaites A., Vaughan R., Braiden P. and Hills W. (1997): Scoping Study to Explore a Framework for Business Process Research in the Context of Capital Goods, End of Award Report to EPSRC, Grant Ref. GR/K95512. CURDS, University of Newcastle upon Tyne

Maier F.H. (1998): New Product Diffusion Models in Innovation Managment – A System Dynamics Perspective, *System Dynamics Review* 14 (4), 285-308

Maillat D., Quévit M. and Senn L. (eds.) (1993): *Réseaux d'Innovation et Milieux Innovateurs: Un Pari pour le Développement Régional,* EDES, Neuchâtel

Malecki E.J. (1980): Dimensions of R&D and Location in the United States, *Research Policy* 9, 2-22

Malecki E.J., Oinas P. and Park S.O. (1999): On Technology and Development. In: Malecki E.J., Oinas P. (eds.) *Making Connections: Technological Learning and Regional Economic Change,* Ashgate, Aldershot, pp. 261-276

Malmberg A., Malmberg B. and Lundquist P. (2000): Agglomeration and Firm Performance: Economies of Scale, Localisation, and Urbanisation among Swedish Export Firms, *Environment and Planning A* 32, 305-321

Mansfield E., Rapoport J., Romeo A., Wagner S. and Beardsley G. (1977): Social and Private Rates of Return from Industrial Innovation, *Quarterly Journal of Economics* 77, 221-240

Mantegna R.N. and Stanley H.E. (2000): *An Introduction to Econophysics,* Cambridge University Press, Cambridge

Marelli E. (1981): Optimal City Size, the Productivity of Cities and Urban Production Functions, *Sistemi Urbani* 1/2, 149-163

Markusen A., Hall P. and Glasmeier A. (1986): *High Tech America: The What, How, Where, and Why of the Sunrise Industries,* Allen & Unwin, Boston

Marshall A. (1919): *Principles of Economics,* London, Macmillan, London, New York [8th edition] [1st edition 1890]

Martin F. (1997): Business Incubators and Enterprise Development Neither Tried or Tested, ICSB 42nd World Conference Proceedings, San Francisco, June 1997, pp. 255-261

Maskell P. (1999): Globalisation and Industrial Competitiveness. The Process and Consequences of Ubiquitification. In: Malecki E.J., Oinas P. (eds.) *Making Connections: Technological Learning and Regional Economic Change,* Ashgate, Aldershot, pp. 35-60

Maskell P. and Malmberg A. (1999): The Competitiveness of Firms and Regions - 'Ubiquitification' and the Importance of Localized Learning, *European Urban and Regional Studies* 6, 9-25

May R. (1973): *Stability and Complexity in Model Ecosystems,* Princeton University Press, Princeton

Maynard Smith J. and Price G.R. (1973): The Logic of Animal Conflict, *Nature* 246 (15), 15-18

Maza M. de la and D. Yuret (1995): A Futures Market Simulation with Non-Rational Participants. In: Brooks, R.A., P. Maes, *Artificial Life IV,* Proceedings of the Fourth International Workshop on the Synthesis and Simulation of Living Systems, MIT Press, Cambridge [MA], pp. 325-330.

McCallum J. (1995): National Borders Matter: Canada-U.S. Regional Trade Patterns, *American Economic Review* 85, 615-23

McKelvey B. (1982): *Organizational Systematics*, University of California Press, Berkeley

McLoughlin I.P., Alderman N., Ivory C.J., Thwaites A. and Vaughan R. (2000): Knowledge Management in Long Term Engineering Projects, Paper Presented at the Knowledge Management, Concepts and Controversies Conference, Warwick University, 10-11 February 2000

Metcalfe (1997): Technology Systems and Technology Policy in an Evolutionary Framework. In: Archibugi D., Michie J. (eds.) (1997) *Technology, Globalisation and Economic Performance*, Cambridge University Press, Cambridge, pp. 268-296

Meyer J.A. and Wilson S.W. (eds.) (1991): *From Animals to Animats*. Proceedings of the 1st International Conference on Simulation of Adaptive Behaviour, MIT Press, Cambridge [MA]

Mienko R., Stefanoski J., Tuomi K. and Vanderpoten D. (1996): Discovery-Oriented Induction of Decision Rules, Cahier du LAMSADE 141, Université de Paris Dauphine, Paris

Mills E. (1970): Urban Density Functions, *Urban Studies* 7, 5-20

Mills E. (1993): What Makes Metropolitan Areas Grow? In: Summers A., Cheshire P., Senn L. (eds.) *Urban Change in the United States and Western Europe*, The Urban Institute, Washington, pp. 193-216

Mirowski P. (1989): *More Heat than Light*, Cambridge University Press, Cambridge

Molero J. (ed.) (1995): *Technological Innovation, Multinational Corporations and the New International Competitiveness*, Hardwood, Reading [MA]

Morgan K. (1997): The Learning Region: Institutions, Innovation and Regional Renewal, *Regional Studies* 31 (5), 491-503

Mothe de la J. and Pacquet G. (1999): *Local and Regional Systems of Innovation*, Kluwer, Dordrecht, Boston

Moulaert F. and Sekia F. (1999): Innovative Region, Social Region? An Alternative View of Regional Innovation, Department of Economics, University of Lille I, Lille

Müller J.P., Wooldridge M.J. and Jennings, N.R. (eds.) (1997): *Intelligent Agents III: Agent Theories, Architectures, and Languages*, Springer, Berlin, Heidelberg, New York

Müller K.H. and Haag G. (1996): The Austrian Innovation System, Complex Modelling with NIS-Data, Final Report 5, Institute for Advanced Studies, Vienna

Murray,A.J. (1993): Restructuring Business Incubators to Support Flexible Manufacturing Organizations, Proceedings of the Third International FAIM Conference, University of Limrick, Ireland, 25-30 June 1993, pp. 336-347

Myrdal G. (1957): *Economic Theory and Under-Developed Regions*, Gerald Duckworth, London

Myrdal G. (1959): *Teoria Economica e Paesi Sottosviluppati*, Feltrinelli, Milano

Mytelka L.K. (1991): *Strategic Partnership. States, Firms and International Competition*, Pinter, London

National Academy of Sciences (1987): *Directory of Members*, National Academy Press, Washington

National Science Foundation (NSF) (2000): *Science and Engineering Indicators 2000*, U.S. Government Printing Office, Washington D.C.

Nelson R.R. (1986): Institutions Supporting Technical Advance in Industry, *American Economic Review* 76, 186-189

Nelson R.R. (ed.) (1993a): *National Innovation Systems: A Comparative Study*, Oxford University Press, Oxford

Nelson R.R. (1993b): Conclusion. In: Nelson, R.R. (ed.) *National Innovation Systems: A Comparative Study*, Oxford University Press, Oxford, p.366

Nelson R.R. (1995): Recent Evolutionary Theorizing About Economic Change, *Journal of Economic Literature* 33, 48-90

Nelson R.R. and Winter, S.G. (1977): In Search of Useful Theory of Innovation, *Research Policy* 6, 36-76

Nelson R.R. and Winter, S.G. (1982): *An Evolutionary Theory of Economic Change*, Belknap Press of Harvard University Press, Cambridge [MA]

Nerlinger E.A. (1998): *Standorte und Entwicklung junger innovativer Unternehmen, Empirische Ergebnisse für West-Deutschland*, ZEW Wirtschaftsanalysen 27, Nomos, Baden-Baden

Neumann J. von (1966): *Theory of Self-Reproducing Automata* (edited and completed by Arthur Burks), University of Illinois Press, Urbana

Nicolis G. and Prigogine I. (1977): *Self-Organization in Non-Equilibrium Systems*, Wiley, New York

Nicolis G. and Prigogine I. (1989): *Exploring Complexity*, Freeman, San Francisco

Niedersen U. and Schweitzer F. (eds.) (1993): *Ästhetik und Selbstorganisation*, Duncker & Humblot, Berlin

Nijkamp P. and Poot J. (1997): Endogenous Technological Change, Long Run Growth and Spatial Interdependence: A Survey. In: Bertuglia C., Lombardo S., Nijkamp P. (eds.) *Innovative Behaviour in Space and Time*, Springer, Berlin, Heidelberg, New York, pp. 213-238

Nijkamp P. and Poot J. (1998): Spatial Perspectives on New Theories of Economic Growth, *Annals of Regional Science* 32 (1), 7-37

Nijkamp P. and Reggiani A. (1998): *The Economics of Complex Spatial Systems*, Elsevier, Amsterdam

Niosi J. (ed.) (1999): The Internationalization of Industrial R&D, *Research Policy* 28, 107-336

Niosi J., Saviotti P., Bellon B. and Crow M. (1993): National Systems of Innovation. In Search of a Workable Concept, *Technology in Society* 15 (2), 207-227

Nonaka I. and Takeuchi M. (1995): *The Knowledge-Creating Company. How Japanese Companies Create the Dynamics of Innovation*, Oxford University Press, New York and Oxford

Normann R. and Ramirez R. (1993): From Value Chain to Value Constellation: Designing Interactive Strategy, *Harvard Business Review*, [July/August], 65-77

Norton R.D. (1999): Where Are the World's Top 100 I.T. Firms and Why? In: Fischer M.M., Suarez-Villa L., Steiner M. (eds.) *Innovation, Networks and Localities*, Springer, Berlin, Heidelberg, New York, pp. 235-258

Oakey R.P., Rothwell R. and Cooper S. (1988): *The Management of Innovation in High-Technology Small Firms: Innovation and Regional Development in Britain and the United States*, Pinter, London

Obersteiner M. and Wilk S. (1999): Determinants of Long-Term Economic Development: An Emprical Cross-Country Study Involving Rough Sets Theory and Rule Induction, Institute for Advanced Studies, Vienna

Oden M. (1997): From Assembly to Innovation - The Evolution and Current Structure of Austin's High Tech Economy, *Planning Forum* 3, 14-30

OECD (1990): *Higher Education in California*, Organisation for Economic Co-operation and Development, Paris

OECD (1997): *Internationalisation of Industrial R&D: Patterns and Trends*, Group of National Experts on Science and Technology Indicators, Organisation for Economic Co-operation and Development, Paris

OECD (1999a): *Main Science and Technology Indicators 1999*, Organisation for Economic Co-operation and Development, Paris

OECD (1999b): *The Knowledge-based Economy: A Set of Facts and Figures*, Organisation for Economic Co-operation and Development, Paris

OECD/Eurostat (1997): *Oslo Manual: Proposed Guidelines for Collecting and Interpreting Technological Innovation Data,* Organisation for Economic Co-operation and Development, Paris

Oerlemans L., Meeus M. and Boeckema F. (1998): Learning, Innovation and Proximity, Working Papers Ecis, Eindhoven

Ohlin B.G. (1933): *Interregional and International Trade*, Harvard University Press, Cambridge [MA]

Ohmae K. (1995): *The End of the Nation State*, Free Press, New York

Oinas P. (1998): The Embedded Firm? Prelude for a Revived Geography of the Enterprise, Helsinki School of Economics and Business Administration, Helsinki

Oinas P. and Malecki E.J. (1999): Spatial Innovation Systems. In: Malecki E.J., Oinas P. (eds.) *Making Connections: Technological Learning and Regional Economic Change*, Ashgate, Aldershot, pp. 7-33

Olekalns N. (1999): Demographics and Wagner's Law. Evidence from the OECD Countries, Discussion Paper, Department of Economics, University of Melbourne

Olsson O. and Frey B.S. (2000):, Entrepreneurship as Recombinant Growth, Paper Presented at the Workshop 'Institutions, Entrepreneurship and Firm Growth' in Jönköping, Sweden

Ottaviano G.I.P. and Puga D. (1997): Agglomeration in the Global Economy: A Survey of the 'New Economic Geography', Discussion Paper 356, Centre for Economic Performance, London School of Economics, London

Palander T. (1935): *Beiträge zur Standorttheorie*, Almqvist & Wiksell, Uppsala

Parisi J., Müller S.C. and Zimmermann, W. (eds.) (1998): *A Perspective Look at Non-Linear Media – From Physics to Biology and Social Sciences,* Springer, Berlin, Heidelberg, New York

Parr J. (2000): Agglomeration Economies: Some Missing Elements, Paper Presented at the 6th World Conference of the RSAI, Lugano, 16-18 May 2000

Pasinetti L.L. (1981): *Structural Change and Economic Growth*, Cambridge University Press, Cambridge

Pasinetti L.L. (1993): *Structural Economic Dynamics*, Cambridge University Press, Cambridge

Patel P. and Vega M. (1999): Patterns of Internationalisation of Corporate Technology: Location versus Home country Advantages, *Research Policy* 28, 145-157

Pavitt K. (1988): Uses and Abuses of Patent Statistics. In: Raan A.F.J. van (ed.): *Handbook of Quantitative Studies of Science and Technology*, North-Holland, Amsterdam, pp. 509-535

Pawlak Z. (1986): On Learning - A Rough Set Approach, [*Lecture Notes in Computer Science*], Springer, Berlin, Heidelberg, New, Heidelberg, New York

Pawlak Z. (1991): *Rough Sets: Theoretical Aspects of Reasoning About Data*, Kluwer, Dordrecht, Boston

Pawlak Z. (1992): Rough Sets: Introduction. In: Slowinski R. (ed.) *Intelligent Decision Support: Handbook of Applications and Advances of the Rough Sets Theory*, Kluwer, Dordrecht, Boston, pp. 1-2

Pearl R. (1925): *The Biology of Population Growth*, Knopf, New York

Penrose E. (1959): *The Theory of the Growth of the Firm*, Blackwell, Oxford

Perez C. (1983): Structural Change and the Assimilation of New Technologies in the Economic System, *Futures* 15, 357-375

Perroux F. (1955): Note sur la Notion de Pole de Croissance, *Economie Appliquée* 8, 307-320 [Translation: Livingstone I. (ed.) (1971) Economic Policy for Development. Penguin, Harmondsworth, pp. 278-289]

Perroux F. (1964): *Industrie et Creation Collective*, Volume 1, Presses Université de France, Paris:

Petzinger T. Jr. (1999): *The New Pioneers*, Simon and Schuster, New York

Pfeiffer F. (1994): Selbständige und abhängige Erwerbstätige: Arbeitsmarkt- und industrieökonomische Perspektiven, Campus, Frankfurt/Main

Picot A., Laub U.-D. and Schneider D. (1989): *Innovative Unternehmensgründungen: Eine ökonomisch-empirische Analyse,* Springer, Berlin, Heidelberg, New York

Pietrobelli C. (1998): The Socio-Economic Foundations of Competitiveness: An Econometric Analysis of Italian Industrial Districts, *Industry and Innovation* 5 (2), 139-156

Pigou A.C. (1927): *Industrial Fluctuations*, Macmillan, London

Piore M.J. and Sabel C.F. (1984): *The Second Industrial Divide: Possibilities for Prosperity*, Basic Books, New York

Pizor P.J. (1986): Making TDR Work, *American Planning Association* 52, 203-211

Polenske K., Robinson K., Hong Y-H, Lin X., Moore J. and Stedman B. (1992): *Evaluation of the South Coast Air Quality Management District's Methods for Assessing Socioeconomic Impacts of District Rules and Regulations; Volume I: Summary and Findings,* MIT Press, Cambridge [MA]

Poot J. (2000): A Synthesis of Empirical Research on the Impact of Government on Long-Run Growth, *Growth and Change* [forthcoming]

Porter M.E. (1985): *Competitive Advantage*, Free Press, New York

Porter M.E. (1990): *The Competitive Advantage of Nations*, MacMillan, London and Basingstoke

Porter M.E. (1996): Competitive Advantage, Agglomeration Economies, and Regional Policy, *International Regional Science Review* 19 (1/2), 85-94

Portugali J. (2000): *Self-Organization and the City*, Springer, Berlin, Heidelberg, New York

Powell W.W. (1990): Neither Market Nor Hierarchy: Network Forms of Organisations, *Research in Organizational Behaviour* 12, 295-336

Powell W.W. and Smith-Doerr L. (1994): Networks and Economic Life. In: Smelser N., Swedberg R. (eds.) *Handbook of Economic Sociology*, Princeton-Sage, Princeton [NJ], pp. 368-402

Prahalad C.K. and Hamel G. (1990): The Core Competence of the Corporation, *Harvard Business Review* 68, 79-91

Prigogine I. and Stengers I. (1984): *Order out of Chaos*, Fontana Paperbacks, London.

Pyka A. (1999): Der kollektive Innovationsprozeß. [*Volkswirtschaftliche Schriften* 498], Duncker & Humblot, Berlin

Rabellotti R. (1997): *External Economies and Cooperation in Industrial Districts: A Comparison of Italy and Mexico*, Macmillan, London

Radding A. (1998): *Knowledge Management: Succeeding in the Information-Based Global Economy*, Computer Technology Research Corp., Charleston [SC]

Rallet A. (1993): Choix de Proximité et Processus d'Innovation Technologique, *Revue d'Economie Régionale et Urbaine* 3, 365-386

Rapoport A. (1963): Mathematical Models of Social Interaction. In: Luce R.D., Galanger E., Bush R. (eds.) *Handbook of Mathematical Psychology* II, John Wiley, New York, pp. 495-579

Rasmussen S. and Barrett C.L. (1995): Elements of a Theory of Simulation. [=ECAL 95, *Lecture Notes in Computer Science*], Springer, Berlin, Heidelberg, New York

Ratti R., Bramanti A. and Gordon R. (eds.) (1997): *The Dynamics of Innovative Regions*, Ashgate, Aldershot

Rauch J.E. (1993): Productivity Gains from Geographic Concentration of Human Capital: Evidence from the Cities, *Journal of Urban Economics* 34, 380-400

Reynolds P.D. (1994): Autonomous Firm Dynamics and Economic Growth in the United States 1986-1990, *Regional Studies* 28, 429-442

Richardson H.W. (1979): *Regional Economics*, University of Illinois Press, Chicago

Ritaine E. (1998): The Political Capacity of Southern European Regions. In: Le Galès P., Lequesne C. (eds.) *Regions in Europe*, Routledge, London, New York, pp. 67-88

Robinson K. (1993): *A System of Economic Development Policies for Use With Air Pollution Control: The Case of Los Angeles*, Paper Presented at the October 1993 Meeting of the American Collegiate Schools of Planning, Philadelphia

Rogers E.M. (1983): *Diffusion of Innovations*, The Free Press, New York [3rd edition]

Romer P.M. (1986): Increasing Returns to Scale and Long-Run Growth, *Journal of Political Economy* 94 (5), 1002-1037

Romer P.M. (1987): Growth Based on Increasing Returns Due to Specialization, *American Economic Review* 77, 565-562

Romer P.M. (1990): Endogenous Technological Change, *Journal of Political Economy* 98, 71-102

Romer P.M. (1994): The Origins of Endogenous Growth, *Journal of Economic Perspectives* 8, 3-22

Ronning G. (1991): *Mikroökonometrie*, Springer, Berlin, Heidelberg, New York

Rosen S. (1972): Learning by Experience as Joint Production, *Quarterly Journal of Economics 86*, 366-382

Rosser J.B.Jr. (1999): On the Complexities of Complex Economic Dynamics, *Journal of Economic Perspectives* 13 (4), 169-192

Rudolph P.J. (1998): Report on the Pilot Study on Project Integration: Procurement and Supply Chain Management Issues the North East Capital Goods Industry. Report to the Northern Development Company, CURDS, University of Newcastle upon Tyne

Sakoda J.M. (1971): The Checkerboard Model of Social Interaction, *Journal of Mathematical Sociology* 1, 119-132

Sala-i-Martin X. (1997): I Just Ran Two Million Regressions, *American Economic Review* 87 (2), 178-183

Sale K. (1975): *Power Shift: The Rise of the Southern Rim and its Challenge to the Eastern Establishment*, Random House, New York

Salter W.E.G. (1960): *Productivity and Technical Change*, Cambridge University Press, Cambridge

Samuelson L. (1997): *Evolutionary Games and Equilibrium*, MIT Press, Cambridge [MA]

Samuelson P. (1954): The Transfer Problem and Transport Costs, *The Economic Journal* 64, 264-289

Sanders L., Pumain D., Mathian H., Guérin-Pace F. and Bura S. (1997): SIMPOP: A Multiagent System for the Study of Urbanism, *Environment and Planning A* 24, 287-305

Santiago L.E. and Lobo J. (1999): *Industrial Structure as a Determinant of Economic Growth: U.S. Metropolitan Economies*. Paper Presented at the Annual Meeting of the North American Regional Science Association, Montreal, November 1999

Satterthwaite M. (1992): High-Growth Industries and Uneven Distribution. In: Mills E., McDonald F. (eds.) *Sources of Metropolitan Growth*, Center for Urban Policy Research, New Brunswick, pp. 39-50

Saviotti P.P. (1988): Information, Variety and Entropy in Techno-Economic Development, *Research Policy* 17, 89-103

Saviotti P.P. (1991): The Role of Variety in Economic and Technological Development. In: Saviotti P.P., Metcalfe J.S. (eds) *Evolutionary Theories of Economic and Technological Change: Present State and Future Prospects*, Harwood, Chur, pp. 1-30

Saviotti P.P. (1994a): Variety, Economic and Technological Development. In: Shionoya Y., Perlman M. (eds.) *Technology, Industries and Institutions: Studies in Schumpeterian Perspectives*, The University of Michigan Press, Ann Arbor, pp. 27-48

Saviotti P.P. (1994b): Knowledge, Information and Organizational Structures, Paper Presented at the Conference of the 11th International Economic History Congress, Milan, 11-16 September 1994

Saviotti P.P. (1996): *Technological Evolution, Variety and the Economy*, Edward Elgar, Cheltenham

Saviotti P.P. (1998): On the Dynamics of Appropriability, of Tacit and Codified Knowledge, *Research Policy* 26, 843-56

Saviotti P.P. and Metcalfe J.S. (eds.) (1991): *Evolutionary Theories of Economic and Technological Change: Present State and Future Prospects*, Harwood, Chur

Saxenian A. (1990): Regional Networks and the Resurgence of Silicon Valley, *California Management Review* 33, 89-111

Saxenian A. (1994): *Regional Advantage: Culture and Competition in Silicon Valley and Route 128*, Harvard University Press, Cambridge [MA]

Scarbrough H. and Swan J. (1999): Knowledge Management and the Management Fashion Perspective. In: Thorpe R. (ed.) *Proceedings of the British Academy of Management, Annual Conference: Managing Diversity*, Volume II. Manchester Metropolitan University, Manchester, pp. 920-937

Schelling T.S. (1969): Models of Segregation, *American Economic Review* 59, 488-493

Schelling T.S. (1971): Dynamic Models of Segregation, *Journal of Mathematical Sociology* 1, 143-186

Schelling T.S. (1978): *Micromotives and Macrobehaviour*, Norton, New York

Scherer F.M. (1992): *International High-Technology Competition*, Harvard University Press, Cambridge [MA]

Schimansky-Geier L., Mieth M., Rosé H. and Malchow H. (1995): Structure Formation by Active Brownian Particles, *Physics Letters A* 207, 140-146

Schmookler J. (1966): *Invention and Economic Growth*, Harvard University Press, Cambridge [MA]

Schmude J. (1994): Gründungsforschung: Eine interdisziplinäre Aufgabe. In: Schmude J. (ed.) Neue Unternehmen: Interdisziplinäre Beiträge zur Gründungsforschung, *Physica*, Heidelberg, pp. 1-10

Schrader S. (1991): Informal Technology Transfer between Firms: Cooperation Trough Information Trading, *Research Policy* 20, 153-70

Schulman B.J. (1990): *From Cotton Belt to Sunbelt: Federal Policy, Economic Development and the Transformation of the South 1938-1980*, Oxford University Press, New York

Schumpeter J. (1942): Capitalism, Socialism and Democracy, George Allen and Unwin, London [3rd edition 1950, 5th edition 1976]

Schumpeter J.A. (1934): *The Theory of Economic Development*, Harvard University Press, Cambridge [MA] [original edition 1912]

Schumpeter J.A. (1939): *Business Cycles*, McGraw-Hill, New York

Schweitzer F. (1994): Natur zwischen Ästhetik und Selbstorganisationstheorie. In: Landeshauptstadt Stuttgart, Kulturamt (ed.) *Zum Naturbegriff der Gegenwart* 2, Frommann-Holzboog, Stuttgart, pp. 93-119.

Schweitzer F. (ed.) (1997a): *Self-Organization of Complex Structures: From Individual to Collective Dynamics, Part 1: Evolution of Complexity and Evolutionary Optimization*, Gordon and Breach, London

Schweitzer F. (ed.) (1997b): *Self-Organization of Complex Structures: From Individual to Collective Dynamics, Part 2: Biological and Ecological Dynamics, Socio-Economic*

Processes, Urban Structure Formation and Traffic Dynamics, Gordon and Breach, London

Schweitzer F. (1997c): Wege und Agenten: Reduktion und Konstruktion in der Selbstorganisationstheorie. In: Krug H.J., Pohlmann L. (eds.) *Evolution und Irreversibilität,* Duncker & Humblot, Berlin, pp. 113-135

Schweitzer F. (1997d): Structural and Functional Information – An Evolutionary Approach to Pragmatic Information. *World Futures, The Journal of General Evolution* 50, 533-550

Schweitzer F. (1997e): Active Brownian Particles: Artificial Agents in Physics. In: Schimansky-Geier L., Pöschel T. (eds.) *Stochastic Dynamics* [Lecture Notes in Physics 484], Springer, Berlin, Heidelberg, New York, pp. 358-371

Schweitzer F. (1998): Modelling Migration and Economic Agglomeration with Active Brownian Particles, *Advances in Complex Systems* 1, 11-37

Schweitzer F. and Holyst J. (2000): Modelling Collective Opinion Formation by Means of Active Brownian Particles, *European Physical Journal B* 15, 723-732

Schweitzer F. and Silverberg G. (eds.) (1998): *Evolution and Selbstorganisation in der Ökonomie/Evolution and Self-Organization in Economics,* Duncker & Humblot, Berlin

Schweitzer F. and Steinbrink J. (1997): Urban Cluster Growth: Analysis and Computer Simulation of Urban Aggregations. In: Schweitzer F. (ed.): *Self-Organization of Complex Structures: From Individual to Collective Dynamics,* Gordon and Breach, London, pp. 501-518

Schweitzer F. and Zimmermann J. (2000): Coordination of Decisions in a Spatial Agent Model. *Physica A* [in press]

Scott A.J. (1999): Geographic Foundations of Creativity and Innovation in the Cultural Economony, Department of Geography, UCL, Los Angeles

Segal D. (1976): Are There Returns to Scale in City Size? *Review of Economics and Statistics* 58, 239-250

Sengenberger W. and Pyke F. (1992): Industrial Districts and Local Economic Regeneration: Research and Policy Issues. In: Pyke F., Sengenberger W. (eds.) *Industrial Districts and Local Economic Regeneration,* International Institute for Labour Studies, Geneva, pp. 1-30

SFB 230 (1994): *Evolution of Natural Structures,* Proceedings of the 3rd International Symposium [Mitteilungen des SFB 230, Heft 9], Stuttgart

Shane S. and Venkataraman S. (2000): *The Promise of Entrepreneurship as a Field of Research,* Academy of Management Review [in press]

Sharp M. (2000): The Need for New Perspectives in European Commission Innovation Policy. In: Archibugi D., Lundvall B.–Å. (eds.) *The Globalizing Learning Economy,* Oxford University Press, Oxford, pp. 239-252.

Shefer D. (1973): Localization Economies in SMSA'S: A Production Function Analysis, *Journal of Regional Science* 13, 55-64

Silverberg G. (1991): Dynamic Vintage Models with Neo-Keynesian Features. In: Paye J.C. (ed.) *Technology and Productivity: The Challenge for Economic Policy. The Technology Economy Programme,* Organisation for Economic Co-operation and Development, pp. 493-507

Silverberg G. (1997): Is There Evolution after Economics? In: Schweitzer F. (ed.) *Self-Organization of Complex Structures: From Individual to Collective Dynamics,* Gordon and Breach, London, pp. 415-425

Silverberg G. and Lehnert D. (1994): Growth Fluctuations in an Evolutionary Model of Creative Destruction. In: Silverberg G., Soete L. (eds): *The Economics of Growth and Technical Change,* Edward Elgar, Cheltenham, pp. 74-108

Silverberg G. and Verspagen B. (1994): Collective Learning, Innovation and Growth in a Boundeldly Rational, Evolutionary World, *Journal of Evolutionary Economics* 4, 207-226

Simmie J. (1998): Reasons for the Development of 'Islands of Innovation': Evidence from Hertfordshire, *Urban Studies* 8, 1261-1289

Simmie J. (ed.) (2001): *Innovative Cities*, Spon, London [forthcoming]

Simmie J. and Hart D. (1999): Innovation Projects and Local Production Networks: A Case Study from Hertfordshire, *European Planning Studies* 7 (4), 445-462

Simmie J. and Sennett J. (1999): Innovative Clusters: Global or Local Linkages, *National Institute Economic Review* 4, 87-98

Simon H.A. (1947): *Administrative Behavior*, Free Press, New York

Sivitanidou R. (1999): The Location of Knowledge-Based Activities: The Case of Computer Software. In: Fischer M.M., Suarez-Villa L., Steiner, M. (eds.) *Innovation, Networks and Localities*, Springer, Berlin, Heidelberg, New York, pp. 109-156

Smelser N.J. and Almond G. (1974): *Public Higher Education in California*, University of California Press, Berkeley

Smolny W. (1999): International Sectoral Spillovers: An Empirical Analysis for German and U.S. Industries, *Journal of Macroeconomics 21*, 135-154

Soete L. (2000): The New Economy: A European Perspective. In: Archibugi D., Lundvall B.-Å. (eds.) *The Globalizing Learning Economy*, Oxford University Press, Oxford, pp. 21-44

Solomon S., Weisbuch G., de Arcangelis L., Jan N. and Stauffer D. (2000): Social Percolation Models, *Physica A* 277, 239-247

Solow R. (1956): A Contribution to the Theory of Economic Growth, *Quarterly Journal of Economics* 70, 65-94

Solow R. (1957): Technical Change in an Aggregative Model of Economic Growth, *International Economic Review* 6, 18-31

Sonis M. (1981): Diffusion of Competitive Innovations, *Modelling and Simulation* 12 (3), 1037-1041

Sonis M. (1983a): Spatio-Temporal Spread of Competitive Innovations: An Ecological Approach, *Papers of the Regional Science Association* 52, 159-174

Sonis M. (1983b): Competition and Environment - A Theory of Temporal Innovation Diffusion. In: Griffith G.A., Lea A. (eds.) *Evolving Geographical Structures*, Martinus Nijhoff, The Hague, pp. 99-129

Sonis M. (1984): Dynamic Choice of Alternatives, Innovation Diffusion and Ecological Dynamics of Volterra-Lotka Models, *Papers in Regional Science* 14, 29-43

Sonis M. (1992a): Dynamics of Continuous and Discrete Logit Models, *Socio-Spatial Dynamics* 3 (1), 35-53

Sonis M. (1992b): Innovation Diffusion, Schumpeterian Competition and Dynamic Choice: A New Synthesis. *Journal of Scientific and Industrial Reseach* 51 (3), 172-186

Sonis M. (1992c): Analytical Structure of Discrete Choice Models: Intervention of Active Environment into the Choice Process, *The Annals of Regional Science* 26, 349-360

Sonis M. (1997): Socio-Ecology, Competition of Elites and Collective Choice: Implications for Culture of Peace. In: Lasker G.E. (ed.) Advances in Sociocybernetics and Human Development, V: Culture of Peace, Human Habitat and Sustainable Living, IIASA-49-97, pp. 77-83

Sörensen C. (1997) (ed.): Empirical Evidence of Regional Growth: The Centre-Periphery Discussion, The Expert Committee of the Danish Ministry of the Interior, Copenhagen

Soros G. (1994): *The Alchemy of Finance: Reading the Mind of the Market*, Wiley, New York

Spencer H. (1892): *Essays Scientific, Political and Speculative*, Appleton, New York

Spitzer D.M. (1988): *New Business Incubators*, Proceedings of USASBE Conference 'Entrepreneurship: Bridging the Gaps Between Research and Practice', United States Association for Small Business and Entrepreneurship, Kenesaw State College, Marietta [GA], pp. 110-115

Stadtman, V. (1970): *Origin and Development of the University of California*, McGraw-Hill, New York

Stahl K. (1995): Entwicklung und Stand der regionalökonomischen Forschung. In: Gahlen B., Hesse H., Ramser H.J. (eds.) *Standorte und Region: Neue Ansätze der Regionalökonomik*, Wirtschaftswissenschaftliches Seminar Ottobeuren 24, Mohr, Tübingen, pp. 3-40

Stefanowski J. (1998): On Rough Set Based Approaches to Induction of Decision Rules. In: Skowron A., Polkowski L.(eds.) *Rough Set in Knowledge Discovering*, Volume I, Physica Verlag, Heidelberg, pp. 500-529.

Stefanowski J. and Wilk S. (1999): Minimizing Business Credit Risk by Means of Approach Integrating Decision Rules and Case Based Learning, Research Report RA-001/99

Steil F. (1999): *Determinanten regionaler Unterschiede in der Gründungsdynamik — Eine empirische Analyse für die neuen Bundesländer*, ZEW Zentrum für Europäische Wirtschaftsforschung, Wirtschaftsanalysen 34, Nomos, Baden-Baden

Steiner M. (ed.) (1998): *Clusters and Regional Specialisation*, Pion, London

Sternberg R. (1999): Innovative Linkages and Proximity: Empirical Results from Recent Surveys of Small and Medium Sized Firms in German Regions, *Regional Studies* 33, 529-540

Sternberg R., Backes-Gellner U., Moog P., Otten C. and Demirer G. (2000): GrünCol! - Gründungen in Cologne. Auswertung der Erhebungen zu 'Unternehmensgründungen und Gründungspotential im Raum Köln' im Auftrag des Amtes für Stadtentwicklungsplanung der Stadt Köln [draft, available from the authors]

Stone C. (1989): *Regime Politics. Governing Atlanta, 1946-88*, University Press of Kansas, Kansas

Storey D.J. and Tether B. (1996): Review of the Empirical Knowledge and an Assessment of Statistical Data on the Economic Importance of New Technology-Based Firms in Europe, Warwick Research Institute, Coventry

Storper M. (1997): *The Regional World. Territorial Development in a Global World*, The Guilford Press, New York, London

Stough R.R. (2000): The Greater Washington Region: A Global Gateway Region. In: Andersson Å.E., Andersson D.E. (eds.) *Gateways to the Global Economy*, Edward Elgar, Cheltenham, pp. 105-123

Suarez-Villa L. (1990): Invention, Inventive Learning, and Innovative Capacity. *Behavioral Science 35*, 290-310

Suarez-Villa L. (1993): The Dynamics of Regional Invention and Innovation: Innovative Capacity and Regional Change in the Twentieth Century, *Geographical Analysis* 25, 147-164

Suarez-Villa L. (2000a): *Invention and the Rise of Technocapitalism*, Rowman and Littlefield, Lanham [MD], New York, Oxford

Suarez-Villa L. (2000b): Southern California as a Global Gateway Region: Polycentricity and Network Segmentation as Competitive Advantages. In: Andersson Å.E., Andersson D.E. (eds.) *Gateways to the Global Economy*, Edward Elgar, Cheltenham, pp. 83-104

Suarez-Villa L. and Fischer M.M. (1995): Technology, Organization and Export-Driven R&D in Austria's Electronics Industry, *Regional Studies* 29 (1), 19-42

Suarez-Villa L. and Hasnath S.A. (1993): The Effect of Infrastructure on Invention: Innovative Capacity and the Dynamics of Public Construction Investment, *Technological Forecasting and Social Change* 44, 333-358

Suarez-Villa L. and Walrod W. (1997): Operational Strategy, R&D, and Intra-Metropolitan Clustering in a Polycentric Structure: The Advanced Electronics Industries of the Los Angeles Basin, *Urban Studies* 34, 1343-1381

Suarez-Villa, L. and Karlsson, C. (1996):, The Development of Sweden's R&D-Intensive Electronics Industries: Exports, Outsourcing, and Territorial Distribution, *Environment and Planning A 28*, 783-817

Susmaga R. (1997): Analyzing Discretization of Continuous Attributes Given a Monotonic Discriminations Function, *Intelligent Data Analysis* 1 (3), 118-128

Sveikauskas L. (1975): The Productivity of City Size, *Quarterly Journal of Economics* 89, 393-413

Sveiskauskas L., Gowdy J. and Funk M. (1988): Urban Productivity: City Size or Industry Size, *Journal of Regional Science* 28 (2), 185-202

Swyngedouw E. (1997): Neither Global Not Local: 'Glocalization' and the Politics of Scale. In: Cox K. (ed.) *Spaces of Globalization. Reasserting the Power of the Local*, Guildford, New York, pp. 137-166

Tan K.C., Kannan V.R. and Handfield R.B. (1998): Supply Chain Management: Supplier Performance and Firm Performance, *International Journal of Purchasing and Materials Management*, 34 (3), pp. 2-9

Tassey G. (1991): The Functions of Technology Infrastructure in a Competitive Economy, *Research Policy* 20, 329-343

Teece D.J. (1981) The Market for Know-How and the Efficient International Transfer of Technology, *Annals of the American Association of Political and Social Sciences* 458, 81-86

Teece D.J. (1986): Profiting from Technological Innovation, *Research Policy* 15, 285-305

Teece D.J. (1998): Capturing Value from Knowledge Assets: The New Economy, Markets for Know-How, and Intangible Assets, *California Management Review* 40, 55-79

Thom R. (1975): *Structural Stability and Morphogenesis*, Benjamin, Reading

Tichy G. (1998): Clusters: Less Dispensable and More Risk Than Ever. In: Steiner M. (ed.) *Clusters and Regional Specialisation*, Pion, London, pp. 226-237

Tolbert C.M. and Sizer M. (1990): U.S. Commuting Zones and Labour Market Areas: A 1990 Update, Rural Economy Division, Economic Research Service, U.S. Department of Agriculture, Staff Paper No. AGES-9614, 1996

Towe J.B. and Wright D.J. (1995): Research Published by Australian Economics and Econometrics Departments 1988-1993, *Economic Record* 71, 8-17

Trippi R. (1995): *Chaos and Non-Linear Dynamics in the Financial Markets*, La Jolla, Irwin

Troitzsch K.G., Mueller U., Gilbert G.N. and Doran J.E. (eds.) (1996): *Social Science Microsimulation*, Springer, Berlin, Heidelberg, New York

Tushman M.L. and Anderson P. (1986): Technological Discontinuities and Organizational Environments, *Administrative Science Quarterly* 31, 439-465

Tyson L.D. (1992): Who's Bashing Whom? Trade Conflict in High-Technology Industries, Institute for International Economics, Washington D.C.

U.S. Bureau of the Census (1950-1995): Statistical Abstract of the United States. U.S. Government Printing Office, Washington

U.S. Bureau of the Census (1975): *Historical Statistics of the United States, Colonial Times to 1970*, U.S. Government Printing Office, Washington

U.S. Bureau of the Census (1981): *Construction Reports: Value of New Construction Put in Place in the United States, 1964 to 1980*. U.S. Government Printing Office, Washington

Udayagiri N.D. and Schuler D.A. (1999) Cross-Product Spillovers in the Semiconductor Industry: Implications for Strategic Trade Policy, *The International Trade Journal* 13, 249-271

UNESCO (1996): *Statistical Yearbook*, Geneva

Vallacher R. and Nowak A. (eds.) (1994): *Dynamical Systems in Social Psychology*, Academic Press, New York

Varga A. (1998): *University Research and Regional Innovation: A Spatial Econometric Analysis of Academic Technology Transfers*, Kluwer, Dordrecht, Boston

Varga A. (1999): Time-Space Patterns of U.S. Innovation: Stability or Change? In: Fischer M.M., Suarez-Villa L., Steiner M. (eds.) *Innovation, Networks and Localities*, Springer, Berlin, Heidelberg, New York, pp. 215-234

Varga A. (2000): Local Academic Knowledge Spillovers and the Concentration of Economic Activity, *Journal of Regional Science* 40, 289-309

Varga A. (2001): Universities and Regional Economic Development: Does Agglomeration Matter? In: Johansson B., Karlsson C., Stough R. (eds) *Theories of Endogenous Regional Development: Lessons for Regional Policies*, Springer, Berlin, Heidelberg, New York [forthcoming]

Veit H. and Richter G. (2000): The FTA Design Paradigm for Distributed Systems, *Future Generation Computer Systems* 16, 727-740

Venkataraman S. (1997): The Distinctive Domain of Entrepreneurship Research, in Advances in Entrepreneurship, *Firm Emergence and Growth* 3, 119-138

Verhulst P. (1838): Notice Sur la Loi que la Population Suit Dans son Accroissement, *Correspondence Mathématique et Physique* 10, 113-121

Verspagen B., Moergastel T. and Slabbers M. (1994): *MERIT Concordance Table*: IPC-ISIC (rev.2). MERIT Research Memorandum 2-94-004, Maastricht

Vivarelli M. and Pianta M. (eds.) (2000): *The Employment Impact of Innovation*, Routledge, London, New York

Volterra V. (1927): The Calculus of Variations and the Logistic Curve. In: Scudo F.M., Ziegler J.R. (eds.) *The Golden Age of Theoretical Ecology: 1923 – 1940.* [Lecture Notes in Biomathematics 22], Springer, Berlin, Heidelberg, New York

Ward M. (1999): *Virtual Organisms*, Macmillan, London

Weber A. (1929): *Theory of the Location of Industries*, Chicago University Press, Chicago

Weibull J.W. (1995): *Evolutionary Game Theory*, MIT Press, Cambridge [MA]

Weidlich W. (1972): The Use of Statistical Models in Sociology, *Collective Phenomena* 1, 51-59

Weidlich W. (1991): Physics and Social Science – the Approach of Synergetics, *Physics Reports* 204, 1-163

Weidlich W. (2000): *Sociodynamics. A Systematic Approach to Mathematical Modelling in the Social Sciences*, Harwood, London

Weidlich W. and Haag G. (1983): *Concepts and Models of a Quantitative Sociology: The Dynamics of Interacting Populations*, Springer, Berlin, Heidelberg, New York

Wheeler J.O. and Mitchelson R.L. (1991): The Information Empire, *American Demographics 13*, 40-42

Wheeler J.O., Aoyama Y. and Warf B. (2000) (eds.): *Cities in the Telecommunications Age. The Fracturing of Geographies*, Routledge, London, New York

Wiener N. (1961*): Cybernetics or Control and Communication in the Animal and the Macine*, MIT Press, Cambridge [MA] [2nd edition]

Wiig H. and Wood M. (1997): What Comprises a Regional Innovation System? Theoretical Base and Indicators. In: Simmie J. (ed.) *Innovation Networks and Learning Regions?* Jessica Kingsley, London, pp. 66-98

Wilkins M. (1974): *The Maturing of Multinational Enterprise: American Business Abroad from 1914 to 1970*, Harvard University Press, Cambridge [MA]

Williamson, O.E. (1975): *Markets and Hierarchies: Analysis and Antitrust Implications*, The Free Press, New York

Winter S.G. (1984): Schumpeterian Competition in Alternative Technological Regimes, *Journal of Economic Behaviour and Organizations* 5, 287-320

Wirtschaftskammer Österreich (1992): *Forschung und Entwicklung in Österreich*, Vienna

Witt U. (1995): Moralität versus Rationalität - Über die Rolle von Innovation und Imitation in einem alten Dilemma. In: Wagner A., Lorenz H.W. (eds.): *Studien zur Evolutorischen Ökonomik III*, Duncker & Humblot, Berlin, pp. 11-33

Woeckener B. (1993): Innovation, Externalities and the State. A Synergetic Approach, *Journal of Evolutionary Economics* 3 (3), 225–248

Woeckener B. (1995): *Hotelling-Modelle der Konkurrenz und Diffusion von Netzeffektegütern, deterministische und stochastische Ansätze zur Erklärung der Ausbreitung neuer Kommunikations- und Gebrauchgüter-Systeme*, Francke, Tübingen, Basel

World Bank (2000): *World Development Indicators*, Query Database, Washington D.C.

World Intellectual Property Organization (WIPO) (1998): *Industrial Property Statistics*, Geneva

Wright J.B. (1993): Conservation Easements. An Analysis of Donated Development Rights, *American Planning Association* 59, 487-493

Wright, V. (1998): Intergovernmental Relations and Regional Government in Europe: A Sceptical View. In: Le Galès, P., Lequesne, C. (eds.) *Regions in Europe*, Routledge, London, New York, pp. 39-49

Yeung H.W-C. (1994): Critical Reviews of Geographical Perspectives on Business Organisations and the Organisation of Production: Towards a Network Approach, *Progress in Human Geography* 18 (4), 460-490

Yin R.K. (1994): *Case Study Research*, Sage, London

Young A. (1991): Learning by Doing and the Dynamic Effects of International Trade, *Quarterly Journal of Economics 106*, 369-406

Young, H.P. (1993): *The Evolution of Conventions*, Econometrica 61, 57-84

Zhang W.B. (1990): *Economic Dynamics: Growth and Development*, Springer, Berlin, Heidelberg, New York

Zhang W.B. (1999): *Capital and Knowledge: Dynamics of Economic Structures with Non-Constant Returns*, Springer, Berlin, Heidelberg, New York

Zucker L.G, Darby M.R. and Armstrong J. (1998): Geographically Localized Knowledge: Spillovers or Markets? *Economic Inquiry 36*, 65-86

Zucker L.G, Darby M.R. and Brewer M.B. (1998): Intellectual Human Capital and the Birth of U.S. Biotechnology Enterprises, *American Economic Review 88*, 290-306

Zysman J. (1996): The Myth of the 'Global' Economy: Enduring National Foundations and Emerging Regional Realities, *New Political Economy 1*, 157-184

List of Figures

List of Tables

Subject Index

agent-based computational economics, 309
  design, 279-295
  simulation, 297-298, 299-302, 306-309. 311-314
agglomeration, 150, 156, 158-159
  economies, 150-156, 181, 182
    impact on innovation capacity, 182, 204, 275, 404
    for firms, 81, 84, 91, 92, 113-114, 125
    for high-tech firms, 113, 146-162
    spatial, 183
artificial economics, 311-313, 314
  intelligence tool, 374
Austria, analysis of spatial data (knowledge spillovers), 125-129
  knowledge creation in, 124, 130-142
  regional clustering, 132-136
  R&D in manufacturing industry, 137-141

benchmarking, 55
biological organisms, properties of, 27-28, 275
biology, as a metaphor, 27-28, 29, 50, 265, 275, 318, 320

catastrophe theory, 4
catch-up strategies, 34-37, 45
cellular automata, 14, 302-305
  and socio-dynamics, 303
  applied to neighbourhood behaviour, 304-305
chaos, in economics, 356
theory, 4
city, as a milieu, 187, 191-197, 203
club goods, 96
cluster analysis, 190-194, 195, 196-197
clusters, of innovation behaviour in metropolitan cities, 191-195

  weakening of competitive advantage, 225
clustering, 96, 111
  promotion of, 210
  strategies (criticism of), 225
co-evolution, 314
collective learning (see *learning*)
communicating agents, model of, 283-286
competence-based development, 83
  approach, 86, 94, 100
competitiveness, 42, 47, 82, 89, 94, 98
  regional, 90, 99, 100, (loss of), 225
complex, behaviour, 13, 317
  systems, 13, 275-276, 276-277, 295
complexity, 317, 345, 369-371
  modelling of, 13, 26, 310, 311
  of production, 81
  theory, 4, 23-29
concentration of economic activities, 101, 113
co-operation strategies (modelling of), 269-273
core-periphery models, 146, 147, 148, 159
  new, 149-151
  traditional, 148, 151
corporate knowledge, 2
creative destruction, 42, 47
cumulative growth, 160

decentralisation of political power, 88
decision-making strategies, 269-273, 291-292
  spatial co-ordination of, 296
decision support systems, 2
demand pull, 15
derivatives markets, 354, 364-365, 366-368
development, sustainable, 47

Author Index

List of Contributors

Zoltán J. Ács
Department of Economics and Finance
Robert G. Merrick School of Business
University of Baltimore
1420 N-Charles Street
Baltimore, MD 21201
USA

Neil Alderman
Centre for Urban and Regional
Development Studies
University of Newcastle
Newcastle upon Tyne NE1 7RU
United Kingdom

Daniele Archibugi
Italian National Research Council
Via Cesare De Lollis, 12
I-00185 Rome
Italy

David Batten
Temaplan Group
Applied Systems Analysis for Industry and
Government
P.O. Box 3026
Dendy Brighton Victoria 3186
Australia

Roberta Capello
Department of Economics
University of Molise
and
Department of Economics and Production
Politecnico di Milano
Piazza Leonardo da Vinci 32
I-20133 Milano
Italy

Dimitrios S. Dendrinos
School of Architecture and Urban Design
The University of Kansas
302 Marvin Hall
Lawrence, Kansas 66045 – 2250
USA

Elsie Echeverri-Carroll
The Red McCombs School of Business
The University of Texas at Austin
Austin, Texas 78713
USA

Charles Edquist
Department of Technology and Social
Change
Linköping University
S-58183 Linköping
Sweden

Dirk Engel
Center of European Economic Research
Department of Industrial Economics and
International Management
P.O. Box 103443
D – 68034 Mannheim
Germany

Andreas Fier
Center of European Economic Research
Department of Industrial Economics and
International Management
P.O. Box 103443
D – 68034 Mannheim
Germany

Manfred M. Fischer
Department of Economic Geography &
Geoinformatics
Vienna University of Economics and
Business Administration
Roßauer Lände 23/1
A – 1090 Vienna
Austria

Josef Fröhlich
Austrian Research Centre Seibersdorf
Business Division Systems Research
Technology-Economy-Environment
A – 2444 Seibersdorf
Austria

Helmut Gassler
Austrian Research Centre Seibersdorf
Business Division Systems Research
Technology-Economy-Environment
A – 2444 Seibersdorf
Austria

Günter Haag
Steinbeis Transfer Centre
Applied Systems Analysis
Rotwiesenstr. 22
D–70599 Stuttgart
Germany

Charlie Karlsson
Jönköping International Business School
Jönköping University
P.O. Box 1026
S–55111 Jönköping
Sweden

Arnoud Lagendijk
Faculty of Policy Studies
University of Nijmegen
PO Box 9108
NL–6500 HK Nijmegen
The Netherlands

Philipp Liedl
Steinbeis Transfer Centre
Applied Systems Analysis
Rotwiesenstr. 22
D–70599 Stuttgart
Germany

Agostino Manduchi
Jönköping International Business School
Jönköping University
P.O. Box 1026
S–551 11 Jönköping
Sweden

Peter Nijkamp
Department of Spatial Economics
Free University Amsterdam
De Boelelaan 1105
NL – 1081 HV Amsterdam
The Netherlands

Jacques Poot
School of Economics and Finance
Victoria University of Wellington
Murphy Building, Kelburn Parade
Wellington
New Zealand

Pier Paolo Saviotti
Institut National de la Recherche
Agronomique
Départment d'Economie et Sociologie
Rurales
Université Pierre Mendès
BP 47
F-38040 Grenoble, Cédex 9
France

Frank Schweitzer
Real World Computing Partnership -
Theoretical Foundation GMD Laboratory,
Schloss Birlinghoven
D-53754 Sankt Augustin
Germany

477

Michael Sonis
Department of Geography
Bar – Ilan University
ISR – 52900 Ramat Gan
Israel

Luis Suarez-Villa
School of Social Ecology
University of California
202 Social Ecology I Bldg.
Irvine, California 92697 – 7075
USA

Dr. Attila Varga
Department of Economic Geography &
Geoinformatics
Vienna University of Economics and
Business Administration
Roßauer Lände 23/1
A – 1090 Vienna
Austria

Gabriella Vindigni
Dept. Economia e Terri Torio
Facolta di Economia
Universita di Catania
Corso Italia 55
I-95129 Catania, Italia

Jörg Zimmermann
Real World Computing Partnership -
Theoretical Foundation GMD Laboratory,
Schloss Birlinghoven
D-53754 Sankt Augustin
Germany

A. Reggiani (Ed.)

Spatial Economic Science

New Frontiers in Theory and Methodology

This volume aims to provide an overview of new frontiers in theoretical/methodological studies and research applications concerning the space-economy.

It is a focussed selection of ideas and reflections put forward by scientists exploring new insights and channels of research, where the quantitative synthesis of spatial systems is the integrative framework. The conclusion drawn from the book is that the fast-changing socio-economic structures and political landscapes are pushing spatial economic science in various „evolutionary" directions. From this perspective, the valuable heritage of the discipline, built up over fifty years, constitutes the solid methodological basis from which to proceed.

2000. XI, 457 pp. 94 figs., 26 tabs. (Advances in Spatial Science) Hardcover * **DM 169**; £ 58.50; FF 637; Lit. 186.640; sFr 146 ISBN 3-540-67493-4

P.W.J. Batey, P. Friedrich, (Eds.)

Regional Competition

Many parts of the world are currently experiencing the outcome of processes of economic integration, globalization and transformation. Technological advances in telecommunications and in transport facilities have opened up new possibilities for contracts and exchanges among regions. External effects among regions have increased in importance. As a result, competition among regions has intensified. Except some pioneering work by regional scientists and scholars of public finance and economics, the phenomenon of regional competition has yet to attract the attention it warrants, despite its importance for policy-making. The present volume is intended to remedy this neglect by providing high-level contributions to the three main topics of the book, the theory of regional competition, methods of analysis of regional competition and policies of regional competition.

2000. VIII, 290 pp. 32 figs., 29 tabs. (Advances in Spatial Science) Hardcover * **DM 149**; £ 51.50; FF 562; Lit. 164.550; sFr 129 ISBN 3-540-67548-5

D.G. Janelle, D.C. Hodge (Eds.)

Information, Place, and Cyberspace

Issues in Accessibility

This book explores how new communication and information technologies combine with transportation to modify human spatial and temporal relationships in everyday life. It targets the need to differentiate accessibility levels among a broad range of social groupings, the need to study disparities in electronic accessibility, and the need to investigate new measures and means of representing the geography of opportunity in the information age. It explores how models based on physical notions of distance and connectivity are insufficient for understanding the new structures and behaviors that characterize current regional realities, with examples drawn from Europe, New Zealand, and North America. While tradional notions of accessibility and spatial interaction remain important, information technologies are dramatically modifying and expanding the scope of these core geographical concepts.

2000. XII, 381 pp. 77 figs., 27 tabs. (Advances in Spatial Science) Hardcover * **DM 149**; £ 51.50; FF 562; Lit. 164.550; sFr 129 ISBN 3-540-67492-6

M.C. Keilbach

Spatial Knowledge Spillovers and the Dynamics of Agglomeration and Regional Growth

When considering the dynamics of regional growth rates, one usually observes growth convergence on spatial aggregates but non-convergence or even divergence within smaller regions of different type. This book suggests various approaches to investigate this puzzle. A formal model, merging approaches from growth theory and new economic geography, shows that spatial knowledge spillovers might be the driving force behind this behavior. To analyze an arbitrary number of regions, the model is implemented on a locally recursive simulation tool - cellular automata. Convergence regressions from different runs of the automaton confirm previous findings. Finally, the existence of spatial knowledge spillovers is tested. Regressions give strong evidence for spatial knowledge spillovers. All the relevant literature and spatial econometric methods are surveyed.

2000. X, 192 pp. 43 figs., 21 tabs. (Contributions to Economics) Softcover * **DM 85**; £ 29.50; FF 321; Lit. 93.880; sFr 75 ISBN 3-7908-1321-4

Please order from
Springer · Customer Service
Haberstr. 7 · 69126 Heidelberg, Germany
Tel: +49 (0) 6221 - 345 - 217/8 · Fax: +49 (0) 6221 - 345 - 229
e-mail: orders@springer.de
or through your bookseller

Springer

* Recommended retail prices. Prices and other details are subject to change without notice.
  In EU countries the local VAT is effective. d&p · BA 41969/3

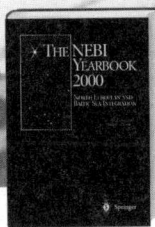

The Annals of
Regional Science

An International Journal of Urban,
Regional and Environmental Research
and Policy

Official Journal of the Western Regional
Science Association

Editors-in-Chief:
B. Johansson, Jönköping International
Business School, Sweden;
T.J. Kim, University of Illinois, Champaign, IL,
USA;
R.R. Stough, George Mason University,
Fairfax, VA, USA

The Annals of Regional Science is a quarterly journal in the interdisciplinary field of regional and urban studies. Its purpose is to promote high quality scholarship on the important theoretical and empirical issues in regional science. The Annals publishes papers which make a new or substantial contribution to the body of knowledge in which the spatial dimension plays a fundamental role, such as regional economics, resource management, location theory, urban and regional planning, transportation, and communication, human geography, population distribution, and environmental quality. The Annals particularly seeks thoughtful, carefully-written articles which are addressed to the broad audience of regional scientists, not just to the author's peers in a subspecialty. Commissioned articles in the journal are intended to focus on what general insights may be gleaned or what general lessons may be learnt from an important line of research. The journal publishes original research, commissioned articles, survey papers, book reviews, and one special issue per annum.

Subscription information 2001:
Volume 35 (4 issues) DM 640,00
ISSN 0570-1864 (print) Title No. 168
ISSN 1432-0592 (electronic edition)

Please order from
Springer · Customer Service
Haberstr. 7
69126 Heidelberg, Germany
Tel: +49 (0) 6221 - 345 - 239
Fax: +49 (0) 6221 - 345 - 229
e-mail: subscriptions@springer.de
or through your bookseller

Plus carriage charges. Price subject to change without notice.
In EU countries the local VAT is effective. d&p · 7738.MNTZ/SF

Springer

The Journal of Geographical Systems, a journal dedicated to geographical information, analysis, theory, and decision, aims to encourage and promote high-quality scholarship on important theoretical and practical issues in regional science, geography, the environmental sciences, and planning. One of the distinctive features of the journal is its concern for the interface between mathematical modelling, the geographical information sciences, and regional issues. An important goal of the journal is to encourage interdisciplinary communication and research, especially when spatial analysis, spatial theory and spatial decision systems are the themes. In particular, the journal seeks to promote interaction between the theorists and users of the geographical information sciences and practitioners in the fields of regional science, geography, and planning.

Journal of Geographical Systems

Geographical Information, Analysis, Theory, and Decision

Editors:
M.M. Fischer, University of Economics and Business Administration, Vienna, Austria
A. Getis, San Diego State University, San Diego, CA, USA

Editorial Board:
L. Anselin, R.G.V. Baker, R.S. Bivand, B. Boots, P.A. Burrough, A.U. Frank, M.F. Goodchild, G. Haag, R.P. Haining, K.E. Haynes, W. Kuhn, Y. Leung, P. Longley, B. Macmillan, P. Nijkamp, A. Okabe, S. Openshaw, Y.Y. Papageorgiou, D. Pumain, A. Reggiani, G. Rushton, F. Snickars, V. Tikunov, D. Unwin, G.G. Wilkinson

Subscription information 2001:
Volume 3, 4 issues
DM 348,–
ISSN 1435-5930 (print) Title No. 10109
ISSN 1435-5949 (electronic edition)

Please order from
Springer · Customer Service
Haberstr. 7
69126 Heidelberg, Germany
Tel: +49 (0) 6221 - 345 - 239
Fax: +49 (0) 6221 - 345 - 229
e-mail: subscriptions@springer.de
or through your bookseller

Plus carriage charges. Price subject to change without notice.
In EU countries the local VAT is effective. d&p · BA 41969/1

Springer